ASSESSMENT AND CONTROL OF NONPOINT SOURCE POLLUTION OF AQUATIC ECOSYSTEMS

A Practical Approach

MAN AND THE BIOSPHERE SERIES

Series Editor J.N.R. Jeffers

VOLUME 23

ASSESSMENT AND CONTROL OF NONPOINT SOURCE POLLUTION OF AQUATIC ECOSYSTEMS

A Practical Approach

Edited by

Jeffrey A. Thornton
International Environmental Management Services, Wisconsin, USA

Walter Rast
United Nations Environment Programme, Nairobi, Kenya

Marjorie M. Holland
University of Mississippi, USA

Geza Jolankai
VITUKI – Research Centre for Water Resources, Budapest, Hungary

Sven-Olof Ryding
Federation of Swedish Industries, Stockholm, Sweden

PUBLISHED BY

PARIS

AND

The Parthenon Publishing Group

International Publishers in Science, Technology & Education

Published in 1999 by the United Nations Educational, Scientific and Cultural Organization
7 Place de Fontenoy, 75700 Paris, France – UNESCO ISBN 92-3-103407-3
and
Published in the USA by
The Parthenon Publishing Group Inc.
One Blue Hill Plaza
PO Box 1564, Pearl River,
New York 10965, USA–ISBN 1-85070-384-1
and
Published in the UK and Europe by
The Parthenon Publishing Group Limited
Casterton Hall, Carnforth,
Lancs LA6 2LA, UK–ISBN 1-85070-384-1

British Library Cataloguing in Publication Data
Assessment and control of nonpoint source pollution of aquatic ecosystems : a practical approach. – (Man and the biosphere ; v. 23)
1. Nonpoint source pollution
I. Thornton, Jeffrey A.
363.73'394
ISBN 1-85070-384-1

Library of Congress Cataloging-in-Publication Data
Assessment and control of nonpoint source pollution of aquatic ecosystems : a practical approach / edited by Jeffrey A. Thornton . . . [et al.].
    p.   cm. – (Man and the biosphere series ; v. 23)
Includes bibliographical references and index.
ISBN 1-85070-384-1
1. Nonpoint source pollution.   2. Water–Pollution.   1. Thornton Jeffrey A.
II. Series
TD424.8.A87   1998
628.1'68–DC21   97-9771
                                                                      CIP

Typeset by H&H Graphics, Blackburn, UK
Printed and bound by Butler & Tanner Ltd., Frome and London, UK

CONTENTS

PREFACE

UNESCO's Man and the Biosphere Programme

Improving scientific understanding of natural and social processes relating to human interactions with the environment; providing information useful to decision-making on resource use; promoting the conservation of genetic diversity as an integral part of land management; enjoining the efforts of scientists, policy-makers, and local people in problem-solving ventures; mobilizing resources for field activities; strengthening of regional cooperative frameworks – these are some of the generic characteristics of UNESCO's Man and the Biosphere (MAB) Programme.

The MAB Programme was launched in the early 1970s. It is a nationally based, international programme of research, training, demonstration, and information diffusion. The overall aim is to contribute to efforts for providing the scientific basis and trained personnel needed to deal with problems of rational utilization and conservation of resources and resource systems and problems of human settlements. MAB emphasizes research for solving problems. It thus involves research by interdisciplinary teams on the interactions among ecological and social systems, field training, and application of a systems approach to understanding the relationships among the natural and human components of development and environmental management.

MAB is a decentralized programme with field projects and training activities in all regions of the world. These are carried out by scientists and technicians from universities, academies of sciences, national research laboratories, and other research and development institutions under the auspices of more than one hundred MAB National Committees. Activities are undertaken in co-operation with a range of international governmental and non-governmental organizations.

Man and the Biosphere Book Series

The Man and the Biosphere Book Series was launched to communicate some of the results generated by the MAB Programme and is aimed primarily at upper level university students, scientists, and resource managers, who are not necessarily specialists in ecology. The books are not normally suitable for undergraduate text books but rather provide additional resource material in the form of case studies based on primary data collection and written by the researchers involved; global and regional syntheses of comparative research conducted in several sites or countries; and state-of-the-art assessments of knowledge or methodological approaches based on scientific meetings, commissioned reports, or panels of experts.

Assessment and control of nonpoint source pollution of aquatic ecosystems

Over the last forty or so years, the effects of pollution have been a major mobilizer of public concern with environmental issues. An example from the late 1950s was the demonstration that unexplained sickness in human communities in Minamata, Japan, was linked to an organo-mercury compound in wastes discharged by local chemical factories into the sea, polluting fish and shellfish consumed by local inhabitants. The first diseases whose cause was recognized as industrial pollution of seawater, 'Minamata disease' aroused worldwide concern and did much to stimulate the development of the environmental protection movement. At about the same time, fungicidal seed dressings of methyl mercury were associated with depletion of seed-eating song-bird populations in North America and Europe and played an important part in the *Silent Spring* of Rachel Carson.

In terms of pollution control, initial emphasis tended to be given to point sources of pollution, such as the discharges from chimneys and pipes and effluents from wastewater treatment plants or industrial factories. Such sources can be readily identified and the volume of water and waterborne materials relatively easy to quantify. In recent decades however, it has become increasingly clear that a large part of environmental contamination originates not from specific point sources, but rather enters soils and waters by soaking into or moving over the slope of the land surface or being transported in the atmosphere.

The present volume is designed to review present and significant knowledge and experience regarding the assessment and control of nonpoint source pollution, and bring this knowledge together in a form that is useful for both scientific and management purposes. The underlying goals of the book are to be practical, to be applicable in both developed and developing countries, to be understandable both to scientific/technical individuals and policy makers/ managers, and be useful in as many environmental settings as possible. As such, it is hoped that the book will be of interest to many people concerned with the problems of nonpoint source pollution, including those involved in shaping and making management decisions regarding pollution control and prevention.

The focus of the book is primarily with nonpoint source pollution from agricultural, urban construction, and forested areas, including the health of aquatic ecosystems and human use of water resources. Pollutants considered include eroded sediments, nutrients, heavy metals, organic and inorganic chemicals, chlorides and micro-organisms. Attention is given to transport mechanisms as well as to the various factors and processes controlling nonpoint source pollution of aquatic ecosystems. The book also provides information for policy-makers in chapters dealing with management frameworks, available source control options, and selection of management strategies.

The roots of the book lie in work within the MAB Programme on the effects of human activities on freshwater ecosystems (MAB Project Area 5), and more practically in efforts to take stock of selected water quality problems around the world. As such, the elaboration of the book has been undertaken in close collaboration with UNESCO's International Hydrological Programme (IHP), with several of the contributors to the book providing direct links with various IHP projects and with the broader international community's concern with water related issues. Several members of the team of contributing authors and editors were earlier responsible for the very first volume in the Man and the Biosphere series, which dealt with The Control of Eutrophication of Lakes and Reservoirs. A number of features of that first volume find their reflection in the structure of the present book, including chapters on the development and implementation of pollution control programmes, the use of models, and the availability of measures and techniques for addressing pollution.

In publishing this volume in the Man and the Biosphere Series, UNESCO would like to thank all the contributors and more especially the five-person core editorial team – Jeffrey Thornton, Walter Rast, Marjorie Holland, Geza Jolánkai and Sven-Olof Ryding – for seeing the volume through to print. The cover montage is by Ivette Fabbri, based on an original engraving by José Chan and an aerial photograph by Yann Arthus Bertrand/Earth from Above/UNESCO.

MAN AND THE BIOSPHERE SERIES

1. The Control of Eutrophication of Lakes and Reservoirs, S.-O. Ryding; W.Rast (eds.), 1989.

2. An Amazonian Rain Forest. The Structure and Function of a Nutrient Stressed Ecosystem and the Impact of Slash-and-Burn Agriculture. C.F. Jordan, 1989.

3. Exploiting the Tropical Rain Forest: An Account of Pulpwood Logging in Papua New Guinea. D. Lamb, 1990.

4. The Ecology and Management of Aquatic-Terrestrial Ecotones. R.J. Naiman; H. Décamps (eds.), 1990.

5. Sustainable Development and Environmental Management of Small Islands. W. Beller; P. d'Ayala; P. Hein (eds.), 1990.

6. Rain Forest Regeneration and Management. A. Gómez-Pompa; T.C. Whitmore; M. Hadley (eds.), 1991.

7. Reproductive Ecology of Tropical Forest Plants. K. Bawa; M. Hadley (eds.), 1990.

8. Biohistory: The Interplay between Human Society and the Biosphere – Past and Present. S. Boyden, 1992.

9. Sustainable Investment and Resource Use: Equity, Environmental Integrity and Economic Efficiency. M.D. Young, 1992.

10. Shifting Agriculture and Sustainable Development: An Interdisciplinary Study from North-Eastern India. P.S. Ramakrishnan, 1992.

11. Decision Support Systems for the Management of Grazing Lands: Emerging Issues. J.W. Stuth; B.G. Lyons (eds.), 1993.

12. The World's Savannas: Economic Driving Forces, Ecological Constraints and Policy Options for Sustainable Land Use. M.D. Young; O.T. Solbrig (eds.), 1993.

13. Tropical Forests, People and Food: Biocultural Interactions and Applications to Development. C.M. Hladik; A. Hladik; O.F. Linares; H. Pagezy; A. Semple; M. Hadley (eds.), 1993.

14. Mountain Research in Europe: An Overview of MAB Research from the Pyrenees to Siberia. M.F. Price, 1995.

15. Brazilian Perspectives on Sustainable Development of the Amazon Region. M. Clüsener-Godt; I. Sachs (eds.), 1995.

16. The Ecology of the Chernobyl Catastrophe: Scientific Outlines of an International Programme of Collaborative Research. V.K. Savchenko, 1995.

17. Ecology of Tropical Forest Tree Seedlings. M.D. Swaine (ed.), 1996

18. Biodiversity of Land–Inland Water Ecotones. J. Lachavanne; R. Juge (eds.), 1997.

19. Population and Environment in Arid Regions. J. Clarke; D. Noin (eds.), 1998

20. Forest Biodiversity Research, Monitoring and Modeling. F. Dallmeier; J.A. Comiskey (eds.), 1998.

21. Forest Biodiversity in North, Central and South America, and the Caribbean. F. Dallmeier; J.A. Comiskey (eds.), 1998.

22. Commons in a Cold Climate – Coastal Fisheries and Reindeer Pastoralism in North Norway: The Co-management Approach. S. Jentoft (ed.), 1998.

CHAPTER 1

INTRODUCTION

W. Rast

'We do not inherit the land from our ancestors
. . . we borrow it from our children'

– old American Indian proverb

BACKGROUND

In many countries, degradation of the natural environment is an issue of high priority. In fact, never before in the history of humanity has society as a whole shown as much interest in its environment as it does today. The awareness that humans and their physical environment may be exposed to threats that are invisible and subtle, yet potentially devastating in their impacts, has generated powerful emotional responses in many cases. In specific instances, it also has generated intense commitment to attempt to correct such problems.

Nevertheless, despite massive efforts made in many countries (primarily the developed nations) to reduce the discharges of contaminants to the air, land and waters, environmental problems ironically seem to have increased in number and intensity in recent decades. Today's environmental problems also have changed in character from previous years – from easily-visible, localized problems to less-easily recognized regional and global problems, and from traditional substances to more exotic, and often toxic, substances. Furthermore, developing countries typically exhibit a greater spectrum of environmental problems than the developed countries. Unfortunately, developing countries usually also face more severe economic and technological constraints than the developed countries. One consequence of this reality is that centralized, co-ordinated pollution control strategies are not feasible for many developing countries, despite their urgent need.

Another unfortunate reality is that every newly-recognized environmental degradation problem seems to raise a battery of new questions. How large an area is affected? What caused it? What can be done about it and how much will it cost? Will a given control measure be effective over the long term? A realistic answer to these questions increasingly is being linked to the phenomenon of nonpoint source pollution. An initial impetus was the lack of positive responses to control measures directed at point sources of pollutants (i.e. water quality did not improve significantly following implementation of point source control measures, suggesting other possible pollutant sources). Further, it was found that pollutants can travel long distances from their sources, being transported by winds and waters. Indeed, at the same time that it was being noticed that degraded, polluted rivers and lakes were being discovered at locations far from logical pollution sources (e.g. industrial plants, urban settlements), it was becoming harder and harder to identify a single factor or source responsible for the pollution. This reality underlies and highlights the complex nature of nonpoint source pollution.

There are many factors underlying the widespread and increasingly serious environmental problem of nonpoint source pollution of aquatic ecosystems. These factors include (1) lack of basic knowledge and understanding of the extent of nonpoint source pollution in developed and developing countries, (2) lack of knowledge and experience regarding the multidisciplinary and multi-institutional approach necessary to solve nonpoint source pollution, and (3) the fact that nonpoint source pollution closely links the activities of individual humans to such variables as soil types, climate, drainage basin topography, types and quantities of material usage in the basin, land use patterns and land management practices, and drainage basin demography. In fact, individuals have a more direct role in the generation of nonpoint source contaminants than in industrial or municipal waste discharges. Indeed, individual actions may seem insignificant; however, cumulatively they can have substantial negative impacts on the quality of the environment. Thus, strictly speaking, everyone has a potential role to play in addressing nonpoint source contamination problems – they are not problems just to be addressed by politicians, administrators or managers.

THE CONCEPT OF NONPOINT SOURCE POLLUTION

For the purposes of this book, the term 'pollution' refers to the contamination of the atmosphere, soil, and/or water by noxious substances. The sources of contaminating materials found in natural waters (lakes, rivers, estuaries) comprise two basic categories: (1) natural, and (2) anthropogenic (human-induced). As previously noted, pollution control efforts in past years were directed primarily toward the more obvious smokestack or pipeline sources. The rationale for this approach was that such sources were readily identifiable, and the volume of water and waterborne materials was easier to quantify. Obvious examples of these types of contaminants are effluents from municipal wastewater treatment plants or industrial factories being discharged to a stream or river.

In recent years, however, it has become clear that a large portion of the contaminant load to surface and ground waters originates from the land around these waters (e.g. farm fields, city streets), rather than from specific point sources. This is because nonpoint source pollution is intimately linked to the hydrologic cycle, and particularly drainage from storm events. Rainfall that reaches the land surface (as well as water in its other forms; e.g. snowmelt, hail) has a limited number of subsequent pathways. It either can soak into, or move over the top of, the land surface. The force of moving stormwater (runoff) over the land surface provides the energy for the movement of contaminants to receiving waters (e.g. rivers, lakes, estuaries). The only other mode of movement of nonpoint source contaminants would be winds; however, this transport mode would result in significantly smaller nonpoint source pollutant loads to waterbodies. Indeed, aquatic ecosystems are the ultimate sinks for virtually all natural and anthropogenic materials from the land surface, as well as those transported in the atmosphere. Thus, recognizing the direct relationship between precipitation and subsequent stormwater drainage is fundamental to understanding nonpoint source pollution of aquatic ecosystems.

Also, nonpoint source contaminants are not necessarily significantly different from point source contaminants. In fact, nonpoint sources can generate both traditional and toxic contaminants. In pristine areas, far from anthropogenic influences, aquatic ecosystems receive inputs of natural nonpoint source contaminants (e.g. sediments). However, because rainfall and snowmelt events are the primary driving forces, the movement of nonpoint source contaminants is more erratic and less frequent, and the duration is shorter, than the typically more continuous point source discharges.

In spite of uncertainties about many of its complicated facets, there is little disagreement that nonpoint source pollution is a major cause of water quality degradation on a global scale. For example, in the first comprehensive study directed primarily toward nonpoint source pollution, the International Joint Commission (1969) reported that 'the current conditions in the lakes [Lake Erie, Lake Ontario and the International Section of the St Lawrence River] could not be related entirely to pollutant loadings from readily-identifiable point sources'. To illustrate this point, it was found that approximately 30 and 40 percent of the total phosphorus loads to Lakes Erie and Ontario, respectively, were from nonpoint sources in the drainage basins. In a subsequent study, the Commission's International Reference Group on Pollution of the Great Lakes (PLUARG, 1978) reported the Great Lakes were being polluted from 'land drainage [nonpoint] sources by phosphorus, sediments, some industrial organic compounds, some previously-used pesticides and, potentially, some heavy metals'.

Such findings dictate that agencies and individuals concerned with water quality degradation must refine their traditional notions about the nature and control of water pollution processes. As noted above, a fundamental link exists between the hydrologic cycle and the generation and transport of materials from the land surface to receiving waters (also see Chapter 3), even in drainage basins unsettled by humans. The quality of such nonpoint-source runoff is related to natural phenomena, such as the weathering of minerals and natural erosion processes. In fact, the quality of runoff from undisturbed drainage basins (especially forested basins), typically represents the 'background' or baseline condition. However, it is becoming more and more difficult to find areas which have not been affected in some manner by anthropogenic contamination, particularly airborne contaminants.

CHARACTERISTICS OF NONPOINT SOURCE POLLUTION

Most previous experience regarding nonpoint-source contamination has focused on plant nutrients (phosphorus, nitrogen) and sediments. Therefore, this book reflects this bias to some degree. However, it also attempts to address other nonpoint source contaminants, including heavy metals and synthetic organic chemicals; these latter substances may be even more important, especially in regard to their potential effects on human health. Even so, many of these non-traditional pollutants occur in concert with their more traditional counterparts. This is particularly the case in urban and agricultural areas, where pesticides and herbicides, nutrients, heavy metals and macropollutants can be components of nonpoint source runoff. A comparison of important features of point and nonpoint-source pollution is provided below in Table 1.1.

It is noted that the magnitude of nonpoint-source contaminant loads following storm events depends on such factors as (1) drainage basin physiography,

Table 1.1 Comparison of point and nonpoint sources of pollution (modified from PLUARG, 1978; Novotny and Chesters, 1981; Vignon, 1985)

Point sources	*Nonpoint sources*
Fairly steady volume and quality	Highly dynamic; occurs at random intervals closely related to hydrologic cycle
Variability of values typically less than one order of magnitude	Variability of values can range across several orders of magnitude, one order of magnitude and between regions
Most severe water quality impacts typically occur during low-flow summer period	Most severe water quality impacts occur during or after storm events
Enters receiving waters at identifiable points, usually *via* pipelines or channel sources	Entry point to receiving waters usually cannot be identified; typically arises from extensive land areas
Can be quantified with traditional hydraulic techniques	Difficult to quantify with traditional techniques
Primary water quality parameters of concern are BOD, dissolved oxygen, nutrients, suspended solids, and sometimes heavy metals and synthetic organics	Primary water quality parameters are sediments, nutrients, heavy metals, synthetic organics, pH (acidity), and dissolved oxygen
Control programmes typically applied by governmental agencies	Control programmes involve individuals not normally considered in pollution control programmes (e.g. farmers, urban homeowners)

(2) type and chemistry of soil, (3) type and extent of vegetative cover, (4) density of drainage channels, (5) types and quantities of materials applied to the land surface, (6) duration of the dry period preceding the rainfall event, and (7) volume, intensity and quality of rainfall. These, and other factors, are discussed further in Chapters 4 and 5.

TYPES OF NONPOINT SOURCE POLLUTANTS

The primary nonpoint source contaminants can be grouped into several general categories, as follows:

(1)   Sediments (erosion);
(2)   Plant nutrients (phosphorus, nitrogen);
(3)   Biodegradable organic matter (BOD, COD);
(4)   Heavy metals;
(5)   Synthetic organic chemicals;
(6)   Dissolved solids (salinity);
(7)   Microorganisms (bacteria) and metabolic products;
(8)   Acidifying compounds; and
(9)   Macropollutants (large debris).

Sediments are the most widespread nonpoint source contaminant, constituting the greatest volume (by weight) of materials transported in land drainage to receiving waters. Sediment runoff results from any activity which disturbs the land surface or leaves soils bare, including streambank erosion, livestock activities, and construction sites. Pollution problems associated with sediment inputs to aquatic systems include: (1) decreased water transparency (increased turbidity), (2) contaminants which can move with sediments (e.g. synthetic organic chemicals), (3) destruction of aquatic habitat, (4) decreased aesthetic quality of waterbodies, (5) decreased water storage capacity in lakes and reservoirs, and (6) decreased water depth in harbours and navigable waterways.

Nutrient enrichment (eutrophication) ranks as probably the most pervasive water quality problem on a global scale (Thornton and Rast, 1997; Ryding and Rast, 1989). Eutrophication was the impetus for the previously cited study by the International Joint Commission of the Great Lakes, providing early evidence of the significant role of nonpoint source contaminants in degrading the quality of natural waters. Phosphorus and nitrogen are the nutrients of most concern, because of their primary role as potential growth limiting factors for phytoplankton (algae) and aquatic plants in waterbodies. Significant nonpoint nutrient sources include agricultural and urban runoff, intensive livestock activities, and atmospheric deposition (precipitation and dry fallout).

The decomposition of natural organic materials by microorganisms requires oxygen. In biologically-productive waterbodies, this process can result in the available dissolved oxygen being completely consumed (particularly in the bottom waters). In turn, this oxygen depletion can affect other oxygen-demanding biochemical processes. The primary oxygen-consuming processes in natural waters include (1) biodegradation of organic materials by microorganisms, (2) aquatic plant respiration, and (3) oxidation of ammonia and organic nitrogen to nitrate (nitrification). Nonpoint sources of oxygen-consuming materials include natural organic matter (detritus), runoff of manure fertilizer from agricultural fields, animal feedlot runoff, urban drainage, etc. Major impacts of oxygen depletion in aquatic ecosystems include the death of aquatic organisms, and the remobilization of nutrients and heavy metals from bottom sediments (so-called 'internal loading').

Novotny and Chesters (1981) and Novotny and Olem (1994) identified 17 heavy metals they believed constituted significant nonpoint source contaminants. This list included antimony, bismuth, cadmium, cobalt, copper, nickel and zinc. In addition, bacteria in the bottom sediments of waterbodies can transform lead, mercury, palladium, platinum, selenium, silver, tellurium, tin and zinc to their methylated chemical forms, which are toxic to aquatic and terrestrial organisms (so-called 'biomethylation'). Novotny and Chesters (1981) also suggested that antimony, cadmium, copper, mercury, nickel, lead, silver, tin and zinc enter the aquatic environment at greater rates than expected from natural geological processes. Nonpoint sources of these heavy metals include runoff from agricultural, urban and industrial sites, and atmospheric deposition.

The nonpoint source synthetic organic chemicals of primary concern include pesticides and industrial chemicals. Organic chemicals can concentrate in the tissues of aquatic organisms in food chains, a phenomenon known as 'biomagnification'. They also can be carcinogenic to humans. In fact, the negative impacts of synthetic organic chemicals on surface and ground waters has only become evident in recent decades. Primary nonpoint sources of synthetic organic chemicals (especially pesticides and herbicides) include runoff from agricultural, urban and industrial sites, and atmospheric deposition. In fact, some synthetic organic chemicals entering natural waters arise almost exclusively from nonpoint sources, examples being DDT and PCBs.

As used in this book, the term 'salinity' refers to the quantity of dissolved solids (salts) in natural waters. The origin of the minerals constituting salinity is primarily the weathering of rocks on the land surface. As water moves over the land surface, it dissolves certain minerals (salts), which subsequently are carried in the surface runoff to receiving waters. Evaporation of water results in an increased concentration of dissolved materials in the remaining water; sometimes, the salinity can increase to very high concentrations. The specific dissolved minerals in a waterbody constitute its chemical 'type'. The major dissolved constituents of salinity include calcium, magnesium, sodium, potassium, sulfate, chloride, fluoride, nitrate, bicarbonate, carbonate, and silica. Excessive salinity can interfere with many water uses, including human and livestock drinking water supply, irrigation, and industrial uses. Anthropogenic sources of salinity include irrigation drainage waters, excessive fertilization of agricultural lands, and runoff from urban areas.

The term 'macropollutants' refers primarily to large pieces of litter or debris (trash). It includes a wide range of materials, from discarded litter on farms and in cities, to materials thrown onto a street from an automobile or bus, etc. This type of nonpoint source contaminant includes the litter that lines a street or road along the curbside. Such materials usually are not toxic to humans or livestock, in the same sense as toxic or hazardous chemicals. The exceptions would be empty containers whose previous contents were toxic substances (e.g. biocides, CFC-based refrigerants, etc.) or where the materials themselves are ingested or become entangled and cause subsequent mortality (e.g. plastics which suffocate). Macropollutants typically are washed into rivers and lakes in storm runoff. In urban areas, their transport in storm runoff is enhanced by the presence of large areas of impervious land surface. Such materials also may be deliberately dumped into rivers or lakes by individuals not wishing to transport them to (or in the absence of) appropriate landfill sites. Such litter in waterbodies can be very unsightly, constituting a source of considerable aesthetic degradation of water resources. Macropollutants are found in virtually all human settlements, and are especially significant in agricultural and urban areas.

SOURCES OF NONPOINT POLLUTANTS

One of the most perplexing problems in regard to controlling nonpoint source pollution is the identification of the potential nonpoint sources in a drainage basin (see Chapters 5 and 6). The very essence of nonpoint source contamination is as diffuse as their introduction of contaminants to aquatic ecosystems. Nevertheless, it has become reasonably clear that the majority of nonpoint source contaminants originate in agricultural and urban areas. These two broad categories can be further sub-divided into additional categories, some being significant sources of specific contaminants.

Major agricultural nonpoint source contaminants include sediments, nutrients, and synthetic organic chemicals (pesticides, herbicides). These materials arise from agricultural activities relating to the disturbance of the land surface (e.g. tillage, clear-cutting, etc.), fertilizer usage and pesticide application. Some specific agricultural sources can produce especially large quantities of pollutants per unit area of land, examples being livestock feedlots and application of municipal sewage wastes and sludges (as fertilizers) on the land surface. Although not agricultural activities in the strict sense, some investigators group several non-urban land uses within the broader agricultural label, including rangelands, pastures, managed forests (silviculture), etc. These latter land-use activities usually produce a smaller quantity of pollutants per unit area than do agricultural (cropping) activities, at least in developed countries. Nevertheless, in developing countries, clear-cutting of tropical forests, and overgrazing of rangelands by livestock, cause major nonpoint-source contamination problems.

The largest quantities of many nonpoint source contaminants per unit area of land surface originate in urban regions, especially those regions having an industrial, manufacturing and/or mining economic base. Major urban contaminants include nutrients, heavy metals, and synthetic organic chemicals. The process of covering a large land surface area with impervious surfaces (streets, parking lots, roofs), inherent in the urbanization process, can significantly disrupt the normal storm drainage patterns in a drainage basin (see Chapter 4). The presence of impermeable surfaces inhibits rainfall from penetrating or percolating into the soils. Thus, the flow of urban stormwater over the land surface to receiving waters is accelerated. A consequence of this accelerated drainage is that materials washed off urban surfaces are also transported rapidly over the land surface to receiving waters, where they can cause significant water quality degradation.

NONPOINT SOURCE POLLUTION IN DEVELOPED *VERSUS* DEVELOPING NATIONS

The above discussions apply both to developed and developing countries. However, it is a reality that most nonpoint source pollution studies, as well as practical experiences with nonpoint source control programmes, have occurred in the developed nations. This is due in part to previous experiences in developed countries in dealing with degradation of aquatic ecosystems (see Novotny, 1995). The traditional approach has been to look for specific pipeline discharges or smokestack emissions of the materials causing problems. As a result, most pollution-control efforts in the past were directed toward the conventional pollutants (e.g. BOD, microorganisms, suspended solids, nutrients, etc.). After isolating and identifying these pollutant sources, agencies and municipalities designed effluent treatment programmes, diversion projects or other point-source remedial programmes to reduce or eliminate pollutant inputs to aquatic ecosystems. This approach was the basis of most pollution control efforts in the more industrialized, developed nations. Where it was economically and technically feasible to do so, this general approach also has been used in developing nations.

However, as noted above, control of nonpoint source pollution does not lend itself readily to this 'pipeline' approach. Nonpoint source pollution of aquatic ecosystems is more diffuse and subtle in character. Many nonpoint source pollutants also may be more toxic than the traditional sewage-related pollutants. Recognition of nonpoint sources as causes of serious environmental degradation was further enhanced by the fact that many aquatic ecosystems continued to exhibit serious degradation, even after initiation of point source control programmes. The efforts necessary to identify and study nonpoint source pollution tend to be more technically complex, time consuming and expensive than point source control efforts. As a result, identification of nonpoint source pollution problems, and initiation of necessary control programmes, often is hindered in developing countries. An unfortunate reality is that scarce resources of many developing countries must be put to the often more immediate problems of human health, nutrition and shelter. As a result, nonpoint source pollution problems may occupy a subordinate role to these concerns, in spite of the reality that the need for nonpoint source control programmes may be greater for developing countries than for developed countries. Unfortunately, nonpoint source pollution can affect the very basis on which the survival and well-being of the developing nations depend, and particularly the loss of productive agricultural lands.

It is true that many features of nonpoint source pollution are similar in developed and developing countries. For example, one can find considerable litter (trash) in the streets of cities in developed and developing countries. One also can see the effects of bare land and associated erosion in virtually all nations. Thus, some nonpoint source pollution problems are similar in character in both settings. However, it also is true that the ability to assess and/or control these types of problems can differ considerably between developed and developing nations.

The most compelling causative factor for nonpoint source pollution probably is poverty. Lack of financial resources dictates that local inhabitants use the least expensive land-use and land-management techniques. Unfortunately, this often is the most environmentally degrading approach. For example, the slash-and-burn farming techniques practiced in some developing countries is dictated primarily by the local economic situation. Lacking the more sophisticated equipment and resources of developed nations, local farmers must use available techniques in their farming activities. As noted above, however, this results in the environmental degradation characteristic of this farming technique. In contrast, the relatively greater financial and technical resources of developed nations allows them the 'option' of attempting to adequately address nonpoint pollution. This is not to say the nonpoint source problem will be adequately addressed even in this latter case, but rather that the prospect of its adequate treatment is likely to be greater in a more affluent setting. This means that greater efforts must be made to develop nonpoint source pollution control methods applicable to the typically less-affluent economic settings of developing nations.

This latter possibility clearly seems reasonable. For example, the nonpoint source study of the North American Great Lakes Basin, conducted by the Pollution From Land Use Activities Reference Group (PLUARG, 1978) of the US–Canada International Joint Commission has demonstrated that some nonpoint pollutant control methods can be inexpensive. They can even be profitable in some instances. For example, farmers can test soils for fertilizer needs (rather than indiscriminately applying fertilizers), or plough manure fertilizers into the soil (rather than simply applying them to the land surface where they are subject to storm-mediated runoff). These techniques are simply 'good stewardship' of the land, and typically reduce nonpoint source pollution of receiving waters.

This approach also is in keeping with the spirit of the United Nations

Conference on the Environment and Development (UNCED; UN, 1993). Clearly, the developing nations must be viewed as more than 'dumping grounds' for the toxic wastes of developed nations. Although debated among the 'have' and 'have-not' nations, it also seems clear that the ability of developing nations to adequately address nonpoint source pollution will be limited by the relatively poor economic situation existing in these nations. Some efforts on the part of the developed nations to develop nonpoint source control measures applicable in the economically-depressed nations of the world will help enhance the ability of these latter countries to more effectively address serious land, water and air pollution problems. In the long run, this approach should prove less expensive than simply ignoring the problem, and will work to the environmental betterment of the entire Earth.

CHAPTER 2

PURPOSE OF THE BOOK AND HOW IT CAN BEST BE USED

M.M. Holland

PURPOSE OF THE BOOK

This book is designed to review present and significant past knowledge and experience regarding the assessment and control of nonpoint source pollution, and bring this knowledge together in a form that is useful for both scientific and management purposes. The book is not meant to be an extensive technical treatise on the topic of nonpoint source pollution, but rather it is intended to provide sufficient information to allow interested persons to acquire a broad, general knowledge of the assessment, causes and control of nonpoint source pollution. In this way, both decision-makers and technical staff can diagnose and formulate site-specific solutions to nonpoint source pollution problems. Through reference to the published literature cited in this book, persons requiring more in-depth knowledge of particular approaches or techniques can develop a detailed understanding of specific topics.

GOALS OF THE BOOK

The goals of this book are:
– To be practical,
– To be applicable in both developed and developing countries,
– To be understandable both to scientific/technical individuals and policy makers/managers,
– To be useful in as many environmental settings as possible.

To accomplish these goals, the book will:
(1) Review recent and significant past knowledge on the nature of nonpoint source pollution of freshwater systems;
(2) Identify significant nonpoint sources in a drainage basin, and help quantify pollutant contributions from different types of land-use and land-management practices in the basin;
(3) Present quantitative guidelines for monitoring nonpoint source pollution and its impacts on aquatic ecosystems;
(4) Identify scientific, engineering, socioeconomic and political elements to be considered in designing nonpoint pollution control programmes;

(5) Present an overall framework for developing effective nonpoint source pollution control programmes; and

(6) Highlight information gaps and research needs in regard to nonpoint source pollution.

SCOPE OF THE BOOK

The book deals primarily with nonpoint source pollution from agricultural, urban, construction, and forested areas, including the health of aquatic ecosystems and human use of water resources. It includes atmospheric deposition as a mode of transport for nonpoint and point source pollutants.

The book considers such pollutants as eroded sediments, nutrients, heavy metals, organic and inorganic chemicals, chlorides and microorganisms. It discusses the various factors and processes controlling nonpoint source pollution of aquatic ecosystems. The book also discusses the important scientific/ engineering and socioeconomic/political factors to consider in developing nonpoint source control programmes, and provides a general framework for developing and implementing such programmes. Examples provided and references included in this book highlight past and present successes and failures in the treatment of nonpoint source pollution problems, and provide guidance for decision-makers and technical staff in designing site-specific solutions to local nonpoint source pollution problems.

Special attention is directed toward providing information for policy-makers in chapters dealing with management frameworks, available nonpoint source control options, and selection of nonpoint source management strategies.

CHARACTERISTICS OF NONPOINT SOURCE POLLUTION

Most experience to date in regard to nonpoint source pollution involves plant nutrients (phosphorus and nitrogen) and sediments (IJC, 1969; PLUARG, 1978; Novotny and Chesters, 1981; Vignon, 1985), reflecting the historical linkage between nonpoint source pollution control and eutrophication management in the northern hemisphere. This book necessarily reflects this bias. However, this book also addresses other nonpoint source pollutants, including heavy metals and synthetic organic chemicals (Novotny and Olem, 1994).

Past scientific studies have provided ample evidence that, technically, effective control of nonpoint source pollution of lakes and reservoirs is possible (Ryding and Rast, 1989). Whether or not effective control also is economically feasible, however, depends on the dominant factors in specific situations (see Chapters 3 and 10). During the last several decades, a number of studies focusing on the nonpoint source pollution problem, and the relationship between nutrient load and responses in waterbodies of different character, have been undertaken. Many types of nonpoint source pollution relationships and mathematical models, both simple and complex, have been developed and presented in the scientific literature. Because nonpoint source pollution continues to be a problem in many countries, there is an urgent need to make practical use of such studies, both for the accurate assessment and the effective long-term control of lake and reservoir nonpoint source pollution.

Many scientific and technical nonpoint source pollution studies are referenced in this book as sources of relevant nonpoint source pollution information. However, it is emphasized that primary attention in this document is given to methodologies which have been shown, over time, to be useful and practical in the assessment and control of nonpoint source pollution. Accordingly, the main purpose of this book is to bring together current knowledge and experience on this complex topic in a form which is both usable and practical for the control of nonpoint source pollution. Thus, some recent or controversial studies may receive only minor attention in this document, in favour of those emphasizing items and topics that have received adequate technical review and have been shown to be useful over the longer term.

This book focuses primarily on lakes, both natural and impounded, and reservoirs. It does not directly address other water systems (e.g. flowing waters or estuarine systems). These latter systems are no less important than lakes and reservoirs. However, even though the same basic principles would apply to a large degree, the available knowledge for dealing effectively with flowing and estuarine waters on a practical level is less extensive than that pertaining to lakes and reservoirs.

It is emphasized that this document focuses primarily on the average conditions in the whole waterbody, rather than on seasonal dynamics or specific areas in a given waterbody. This focus may not be adequate in some cases; for example, in situations where the specific timing and/or location of algal blooms constitute the primary concern. The present state of knowledge regarding the factors that control the seasonal and temporal dynamics of nonpoint source pollution processes is still rudimentary. In most cases, accurate knowledge of the annual, average conditions in a lake or reservoir will be sufficient for the purpose of nonpoint source pollution management. Consequently, the practical nature of this book dictates the focus on average, lake-wide conditions.

INTENDED AUDIENCE FOR THE BOOK

It is the intention that virtually anyone concerned with the problems of nonpoint source pollution will find this book to be useful. It should be useful to individuals in local soil and water conservation authorities, private consultants, state agencies, etc., concerned with technical aspects of nonpoint source pollution. It also should be useful to individuals responsible for making management decisions regarding nonpoint source pollution, including chief elected officials, legislators, managers, and senior administrators. For these latter individuals, the book concentrates on items important or relevant to the decision-making process. The book also should be useful to individuals with a general interest in the topic of nonpoint source pollution.

The book should be especially useful to individuals in local water authorities, private companies, state agencies, etc., working on the control of nonpoint source pollution of lakes and reservoirs (e.g. in development of monitoring programmes, collection and evaluation of data, etc.). It discusses available techniques and strategies for attempting to control nonpoint source pollution, and contains practical guidance for assessing the potential outcome of various nonpoint source pollution control alternatives.

An important audience for this book is individuals with responsibility for making management decisions regarding the control of nonpoint source pollution. This group of individuals includes *chief elected officials, legislators, managers, senior administrators, company presidents*, etc. For them, the book concentrates on those items important or unique to the decision-making process, including guidelines for decision making related to nonpoint source pollution control, some indications of the costs associated with different control strategies, and a bibliography of documented case-studies.

Following the chapter dealing with the primary concerns of the decision-maker (Chapter 3), the book is roughly divided into two broad categories. The first major part (Chapters 4 to 6) is relatively descriptive in nature, and provides details regarding the processes and factors affecting nonpoint source pollution. This portion of the document should be useful to *scientists, engineers and educators*.

The second major part of the book (Chapters 7 to 11) deals with the practical aspects for measurement, modelling, control or management of nonpoint source pollution. This latter portion of the book is intended to be especially useful for the work of project managers and technicians, as well as of scientists and engineers.

FOCUS OF THE BOOK

A schematic guide to the contents of this document is presented in Figure 2.1. The main topics addressed in each chapter are listed in a logical sequence for nonpoint source pollution management considerations. The specific topics and related materials are discussed in the referenced chapters. *Why cultural nonpoint source pollution is a problem* is discussed in *Chapters 1 and 5*. These chapters (especially Chapter 5) describe the causes and symptoms of nonpoint source pollution. They also discuss the differences and similarities of the nonpoint source pollution process in various regions of the world. Information on water quality, as related to intended water uses, is also provided.

Topics that are of special importance to the decision-maker are dealt with in *Chapter 3*. The role of the decision-maker is discussed in terms of identifying nonpoint source pollution management goals, determining the extent of available information, identifying different management options and cost–benefit considerations, assessing the adequacy of the existing legislative/regulatory framework, and selecting specific nonpoint source pollution control strategies.

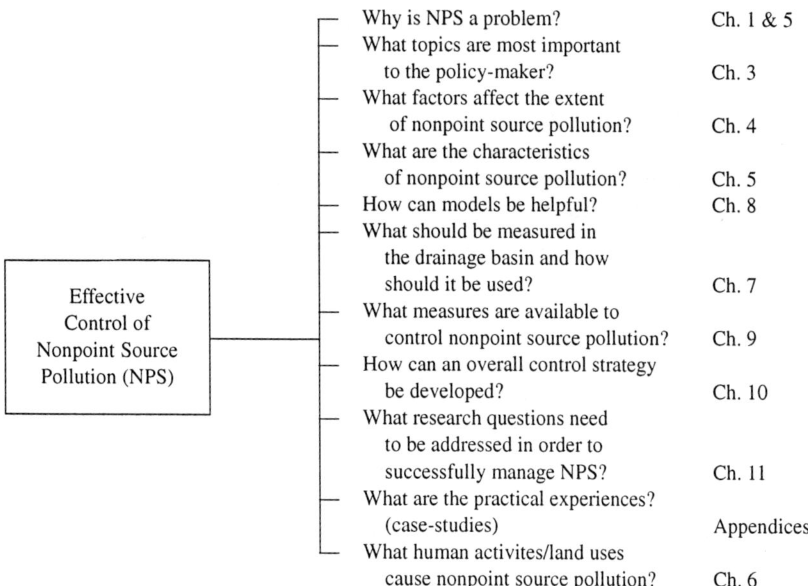

Figure 2.1   Flow chart approach to using this book

Factors that can affect the extent of nonpoint source pollution are presented in *Chapters 5 and 6*. These factors are related both to the catchment area and to the waterbody. In the first case, both natural- and human-produced factors are considered. In the latter case, various in-lake characteristics are identified. These factors include physical (e.g. morphology of lake basin, light intensity), chemical (e.g. in-lake nutrient sources) and biological (e.g. plankton and macrophyte production) components.

Guidelines for assessing nutrient inputs from different point and nonpoint sources in a drainage basin (i.e. what should be measured in the watershed and how it should be used) are presented in *Chapter 7*. This chapter discusses many relevant point and nonpoint nutrient sources, and their potential impacts on the nonpoint source pollution process. This chapter also presents a simple approach to determining the total, component nutrient sources (point and nonpoint) in the drainage basin, as well as the general reliability of nutrient load estimates.

The use of mathematical models in assessment and management of nonpoint source pollution problems is considered in *Chapter 8*. This chapter presents an overview of relevant nonpoint source pollution models (both watershed and waterbody models). It also presents guidelines for using such models to estimate pollutant loads. The book stresses the use of simple, practical relationships and models.

Available methods for the treatment of nonpoint source pollution are identified and discussed in *Chapter 9*. The methods discussed include the control of the nutrient runoff, external nutrient supply, as well as different management-control methods. This chapter also provides a basis for developing sound land-use practices which focus on the management of nutrient inputs to lakes and reservoirs. Different water quality problems treatable by various in-lake nutrient control measures are also identified.

The development of a nonpoint source pollution control strategy is discussed in *Chapter 10*. This chapter focuses on general considerations, and presents an overall approach for the selection of a specific nonpoint source pollution control programme. It considers both the control of external point and nonpoint nutrient sources, and the possible use of in-lake control measures. It also discusses special limnological conditions and situations where such measures should, and should not, be applied. Chapter 10 outlines the selection of nonpoint source pollution control programmes in some detail. The Rhine River is included as a detailed example.

Chapter 11 presents *research needs for the future*. Information on the past experiences of different control measures is very useful, prior to making specific decisions on the development and implementation of nonpoint source pollution control measures. *Appendix 1* provides a list of documented case-studies regarding various nonpoint source pollution control programmes around the world.

This book represents a multinational approach to the practical application of our present knowledge of the nonpoint source pollution process, and generally accepted nutrient load lake and reservoir response relationships for nonpoint source pollution management. It is recognized that other publications on the subject of nonpoint source pollution control do exist. Examples include the reports of Novotny and Chesters (1981), Novotny and Olem (1994), IJC (1969), USEPA (1987), PLUARG (1978), and Vignon (1985). However, the existing publications do not appear to cover the full range of experiences documented thus far. In addition, these latter publications generally focus only on certain types of waterbodies, or on specific geographic areas in the world. In contrast to the above, this book attempts to present a broader view of the practical control of nonpoint source pollution around the world, and in most of the lake and reservoir situations likely to be encountered.

CHAPTER 3

DEVELOPMENT AND IMPLEMENTATION OF A NONPOINT SOURCE POLLUTION CONTROL PROGRAMME

J. Chandler, M.M. Holland, D.L. Forster, D.D. Southgate,
J.A. Thornton and A. Weinberg

A MANAGEMENT FRAMEWORK FOR POLICY-MAKERS

Although countries are different, to varying degrees, all are subject to many global pressures. These include increasing populations, rapid urbanization, and economic competition that is increasingly global. As these pressures exert themselves on individual countries, new stresses are put on the ecosystem and ultimately on the quality of life. These stresses often lead to increased nonpoint pollution. For example:

- as rapid industrialization occurs, local air pollution problems are often addressed through the use of taller stacks, leading to deposition of pollutants in new, often widely dispersed areas;
- as soil erodes due to over-farming, more and newer pesticides and fertilizers are used to maintain yields, thus increasing polluted runoff into surface streams;
- as urbanization accelerates, unnecessary soil erosion occurs due to poorly planned and executed construction projects.

As point sources of water pollution have been controlled over the last decade, the remaining problem of nonpoint sources of water pollution has emerged more clearly. Nonpoint source pollution is a leading cause of surface water quality problems and is also a source of groundwater contamination in many areas of the world. Further progress in improving the world's water quality will require accelerated implementation of nonpoint source management programmes to control diffuse sources of pollutants in addition to ongoing point source control efforts.

There are many sources of nonpoint pollutants and many different ways to address them. There are, however, some general strategies that have proven repeatedly successful in developing nonpoint source management programmes (Figure 3.1). This chapter presents a series of steps that may be used by policy-makers and managers in developing nonpoint source management programmes; these may be summarized in the following management framework for policy-makers/managers:

17

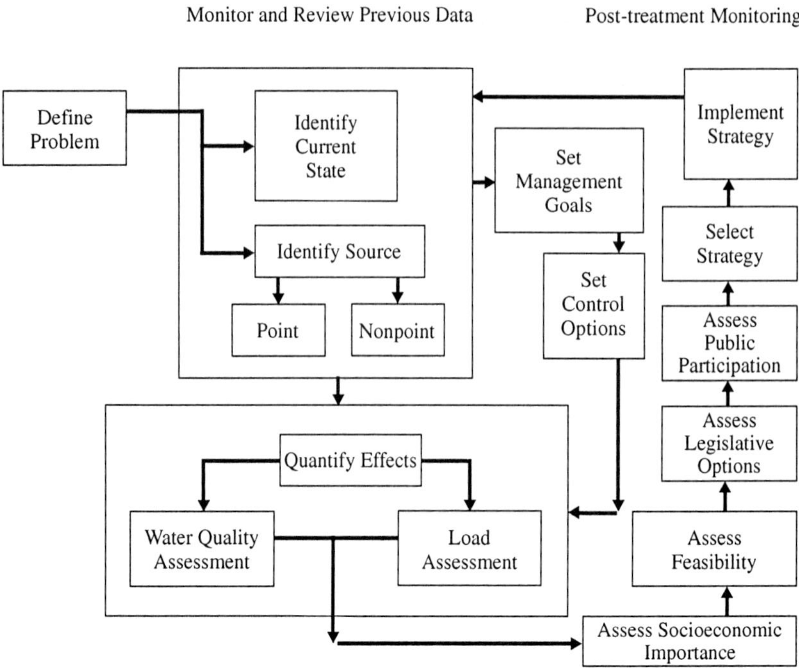

Figure 3.1 Sequence of decisions to be made in the development and implementation of nonpoint source pollution control programmes

(1) Relate nonpoint pollution problems to water use interference (e.g. lack of sufficient quantity of good quality water):
  a. define the problem,
  b. develop an overview of situations in which nonpoint pollution could cause such interference.
(2) Identify specific sources/causes of nonpoint pollution problems:
  a. who has viewed situation as a problem?
  b. what evidence is there for identification of the problem as nonpoint pollution?
(3) Review previous data; check similar case-studies:
  a. what studies have already been conducted on this problem?
  b. are other case-studies available addressing similar problems?
(4) Develop management goals:
  a. disseminate relevant information to appropriate people,
  b. identify the constituency affected.
(5) Identify control options available to address problem:
  a. examine structural/mechanical options,

> b. consider economic options,
> c. assess the technical ramifications of outcomes.

(6) Assess socioeconomic, political and other non-technical concerns relevant to the various control options:
> a. is the option consistent with the country's cultural heritage?
> b. does the nonpoint pollution problem cross political boundaries or is it within one political unit?

(7) Assess adequacy of the existing legislative/regulatory framework:
> a. what (if any) laws apply to the situation?
> b. what (if any) regulations are in place?
> c. disseminate appropriate information to relevant agencies and chief elected officials within local government.

(8) Assess the possibilities for public participation:
> a. what does the public already know/understand?
> b. is there a citizens advisory committee already in place?
> c. can this programme build on existing education programmes, and/or compliment existing environmental programmes?

(9) Select a specific control strategy for a given problem and implement a pollution control programme:
> a. what happens if nothing is done?
> b. what are the technical options?
> c. what are the relative costs?

(10) Adjust the control programme in light of post-treatment monitoring:
> a. what follow-up studies are necessary?
> b. what steps are necessary to prevent nonpoint pollution problems from recurring?
> c. revise/update control programmes as necessary; establish mechanisms for the ongoing maintenance/monitoring/review of the selected programme,
> d. establish feedback mechanisms to make sure good suggestions do not disappear or die!; see that they are incorporated into the programme,
> e. conduct a post-implementation environmental audit (economic accountability).

Each of these steps is briefly discussed below. While the specific steps may need to be adjusted to fit a particular political system or cultural setting, this process can provide a structured approach to developing an effective nonpoint source pollution abatement programme with sufficient grass-roots support from the affected constituency to ensure its long term utility and survival. This process also appeals to technical staff because of its scientific grounding, to educators because of its educational focus, to citizens because of their inclusion in the design and implementation of the controls, and to officials and agencies because of its cost-effective approach and sound legal basis.

IDENTIFY WATERS IMPACTED BY NONPOINT SOURCES OF POLLUTION

Before a country, state or local community (hereinafter referred to as a 'community') can begin to solve its nonpoint source pollution problems, it needs to identify which waters are impacted by nonpoint sources, or which waters are failing to support their desired uses (see e.g. Busch and Sly, 1992; Hess, 1993; Davis and Simon, 1995). Then a determination of why there is this failure to support their desired uses must be made; e.g. the problems must be assessed as naturally occurring, point source dominated or nonpoint source related. If the problem is nonpoint-related, the types of nonpoint sources which are causing the impact should be identified; e.g. agricultural or urban nonpoint sources. Once a community has defined the extent of its nonpoint source problems, it can then prioritize its nonpoint source management needs. In setting such priorities, citizens' groups, such as watershed associations and river basin councils, can be extremely helpful.

In many cases, judgments about whether surface waters are meeting their uses will be based on best professional judgment. In some cases, this best professional judgment will be supplemented by fishery surveys, historical records, citizen complaints, or chemical and biological monitoring data.

Nonpoint source impacts on groundwater may also be defined in terms of impacts on beneficial uses (Boulding, 1995). When groundwater is used for drinking, it must be of the highest quality. Groundwater is also used extensively for irrigation and industrial purposes which likewise have specific quality requirements, especially in terms of its salinity. Assessing nonpoint source impacts on groundwater may be conceptually difficult as, in general, communities tend to have chemical monitoring data for major public water supplies but little or no data for smaller public supplies or private wells. Also, while water supplies may be routinely or periodically monitored for conventional pollutants, monitoring is generally not done for organic chemicals, such as pesticides, which readily accumulate in groundwaters.

A good first step in beginning to identify nonpoint source-impacted waters, is for communities to identify surface water and groundwater resources which are considered to have a high potential to be impacted by nonpoint sources. For example, areas with intensive agriculture and/or urban development, steep slopes, and erosive soil types would generally be expected to have a greater potential for nonpoint source pollution problems. Conversely, forested areas with low slopes and non-erosive soil types would generally be expected to have a lower potential for nonpoint source pollution problems.

IDENTIFY TYPES OF WATER QUALITY IMPACTS CAUSED BY NONPOINT SOURCES

As mentioned above, communities will next need to identify the major types of nonpoint sources which are affecting their nonpoint source-impacted waters. It is best to identify the specific types of nonpoint sources affecting each nonpoint source-impacted water. If this is not possible, it is desirable to at least obtain a good, general idea of what types of nonpoint sources are most likely to be causing a majority of the nonpoint source water quality problems (see Chapter 6).

Sediment is a prime example of a nonpoint source pollutant impacting water resources throughout the world (Lal and Stewart, 1994). Excessive sedimentation may be caused by a variety of sources including forestry activities, urban development, agriculture, and mining. Sediment deposited in waters as a result of these activities may produce a variety of water quality and other impacts including:

- decreased light transmission which obscures food, habitat, refugia and nesting sites;
- effects on the respiration and digestion of aquatic species (e.g. gill abrasion);
- decreased survival rates of eggs and, therefore, decreases in the size of fish populations;
- changes in species composition;
- increased drinking water costs; and
- interference with navigation and other water uses (US EPA, 1984).

If a community determines that sedimentation is a major problem in a particular water resource, it will need to determine the source(s) of the sediment. A good first step in defining the source(s) is to inventory the land-use and management activities in the watershed. This inventory may be done by using a combination of aerial photographs, soil surveys and field interviews. Often several activities in a watershed such as deforestation and agricultural activities will be causing the sedimentation.

Another type of nonpoint source pollution is the leaching of nitrates into groundwater. Nitrate pollution has been found to result from the excessive application of fertilizer, both fertilizer produced as such and manure as a by-product of animal husbandry activities (see Chapters 5 and 6). Fertilizers are increasingly used in agricultural production; world fertilizer use skyrocketed 9.5 times between 1950 and 1986 (Brown *et al.*, 1987). Fertilizer use frequently leads to excessive loadings of nitrogen as well as phosphorus, which may be leached by rainfall into groundwater or carried by runoff to surface waters. Increasing concentrations of nitrates in drinking water, which pose a health threat to infants, have been reported in Hungary, Denmark, France, The Netherlands, United Kingdom, parts of Africa, and Germany (Brown *et al.*, 1987). At the levels of fertilizer applied by European farmers, as much as 30 to 45 kg of nitrogen may be lost per hectare – more fertilizer than is applied to cropland in many developing countries (Brown *et al.*, 1987). Increasing concentrations of nitrates in drinking water have also been reported in the United States (USGS, 1985). Determining whether groundwater is impacted by nitrates will require groundwater monitoring. Communities with particularly vulnerable aquifers and with intensive agricultural activities relying on fertilizer use may be advised to investigate this type of nonpoint source impact.

The final example of a type of nonpoint pollution is air pollution. Pollutants such as nitrogen oxides, sulfur oxides, trace metals, pesticides, phosphorus, and other organic compounds may be transported long distances *via* the atmosphere and deposited in areas far removed from their source. In the case of this type of nonpoint source, management solutions may need to be applied far from the water resource of concern.

The most well-known example of atmosphere-borne (gaseous) substances affecting water resources is the problem of acid rain. Acid rain causes substantial declines in the acid neutralizing capacity (ANC) of some water resources, possibly leading to a decline in pH (acidification) and, in the worst cases, biological damage. Acid rain has already caused widespread acidification of many water resources in the northeastern United States, Canada, Norway, Sweden and the United Kingdom (Schindler, 1988). These airborne acids can be traced to three primary sources: sulfur dioxide (SO_2), nitrogen oxides (NO and NO_2, collectively referred to as NO_x), and ammonia (NH_3). The main sources of sulfur dioxide are coal- and oil-burning power plants; nitrogen oxides come largely from the exhausts of cars and trucks; and atmospheric ammonia is almost entirely a by-product of animal waste on farms. Sulfur dioxide is the biggest problem, accounting for 60 percent of the acidification in Europe; nitrogen oxides account for 21 percent, and ammonia 19 percent. But the situation varies widely from country to country (IIASA, 1990). The total economic value of damages incurred as a result of the loss of unprocessed wood, its basic industrial value and non-wood benefits caused by sulfurous damage to European forests has been estimated at \$30.4 billion each year for the next 100 years (although the true cost may be considerably higher given the conservative nature of the assumptions made). Even so, \$30 billion is about two and a half times as much as European governments have agreed to spend annually to control air pollution – and damage to forests is just a fraction of the total damage inflicted by acidic pollutants (IIASA, 1990).

More recently, greenhouse gases have become of greater concern. Examples of land-use effects on emissions of greenhouse gases are plentiful. Tropical deforestation has been estimated to contribute a substantial net CO_2 flux to the atmosphere and may also be a significant cause of the increasing concentration of atmospheric NO_x (Luizao *et al.*, 1989). Increasing rice paddy extent and production may represent a substantial source of the global increase in CH_4, as do the growing populations of livestock. These concerns are also relevant to global warming and sea level rise, phenomena which may significantly alter the hydrological cycle driving the transport of, and dispersing, nonpoint source contaminants.

These are only a few examples of the types of sources and impacts of nonpoint pollutants, although definitions of what constitutes a nonpoint source may vary from country to country. Chapter 5 presents a more detailed discussion of the major types and sources of nonpoint source contaminants.

IDENTIFY PRIORITY MANAGEMENT NEEDS AND GOALS

Once a community has defined the extent, type and general causes of its nonpoint source pollution problems, it will then need to prioritize its nonpoint source management options. There are a number of possible ways to approach such a task (see Laenen and Dunnette, 1996). These are briefly set out below:

(1) Focus available resources on a limited number of nonpoint source-impacted waters which are particularly important water resources for the community. Experience in many demonstration projects has shown that concentrating pollution control efforts on a particular water-resource problem can produce results in a reasonably short period such as 5 to 10 years (Maas *et al.*, 1987). (In contrast, implementing national or sub-national nonpoint source controls may result in no visible water quality improvements and may not sustain public support for additional nonpoint source control efforts.) A resource-specific focus can overcome these limitations by creating local 'ownership'.

(2) Focus management efforts on the protection of waters with the highest total value. As noted above, an important first step in nonpoint source pollution control is an assessment of the degree to which waters are degraded or threatened by nonpoint source pollution. Surface water quality degradation is often defined in terms of the impact of pollution sources on a water's 'beneficial use' such as fishing, swimming and the propagation of aquatic life, but can include water supply, hydropower generation and other economic uses. In terms of this approach, communities focus their efforts on waters that are not supporting their beneficial uses, and/or waters that are presently meeting their uses but which are likely to be impaired by nonpoint pollution in the future. For example, a stream might have historically supported a coldwater trout fishery, but recent changes in farming practices may have destroyed that fishery. Such a stream may still have the potential to support a trout fishery and might reach this potential with better management of the agricultural nonpoint sources. Similarly, a community may wish to impose nonpoint source pollution controls in the watersheds of water supply reservoirs rather than pay increased treatment costs as water quality deteriorates in response to increased urbanization over time.

(3) Focus management efforts on a particular type of nonpoint source pollution problem such as overgrazing or excessive fertilizer use. Such an approach would not necessarily be limited to a few watersheds but might, for example, entail regional educational, incentive or regulatory programmes. Concentrating on control of a particular nonpoint source pollutant over a large area may be a useful variant of this approach.

(4) Focus management efforts on implementing cost-effective management actions that will meet the management goal at lowest cost. In contrast to focussing on a specific water resource or pollutant, as above, this approach

targets specific management measures such as grassing waterways, providing technical assistance to encourage contour farming, and supporting anti-littering/recycling campaigns on a national or sub-national basis that will yield a measure of benefit at minimal cost. As implied, this approach generally builds on existing programmes and encourages grass-roots action on the part of local communities.

(5) Focus management efforts on individual action. Some economists would analyze nonpoint pollutants as externalities, or costs which should be borne by society as a whole (given the fact that the whole of society benefits from an unpolluted environment), rather than as internal costs of production. While there is likely to be some portion of the cost of nonpoint source controls that should be borne by society at large – such as control of pollutants in road runoff – a significant degree of control could be achieved by encouraging/requiring householders and landowners to adopt 'good housekeeping practices' that will minimize contaminant washoff from their properties/operations. Recent amendments to the US Clean Water Act implement just such an approach to stormwater management.

(6) Focus management efforts on transboundary pollution in accordance with international conventions, agreements, regulations, etc. Numerous regional and/or river basin conventions have been agreed between nations, which, together with the land-based sources of pollution provisions of the United Nations Convention on the Law of the Sea (UN, 1983), require signatories to manage water quality in transboundary waters. Examples include the Rhine River Basin Convention, the Aral Sea Agreement, the Lake Chad Basin Convention, the Rio de la Plata Convention, and the Mekong River Basin Agreement which impose legally binding duties on member states. Addressing related nonpoint source pollution in terms of such agreements will fulfill in part these legal obligations.

(7) Focus on the investment value of water resource restoration efforts. Good water quality improves the quality of life for the entire community. In the case of nonpoint source pollution, a modicum of prevention (or good housekeeping on the part of individuals and corporations) can save a substantial public investment in clean-up costs and infrastructure. This reduces everyone's costs (= taxes) over the longer term and often results in more efficient corporate operations (see the non-structural approaches discussed in Chapter 9).

In any given community, a variety and/or combination of these and other approaches to nonpoint source pollution control can be employed to meet community goals. A key determinant of a successful nonpoint source control programme in every case, however, is the ability to select the correct approach in the context of community norms. Preparing a written plan is a good way to organize a management programme. Plans are best prepared for specific drainage basins. If resident associations exist, it may be especially productive to work with them in developing a basin plan.

REVIEW EXISTING DATA AND DETERMINE ADDITIONAL DATA NEEDS

As implied above, the full scope of the problem needs to be evaluated before an effective nonpoint source management programme can be developed. Although some control actions can be initiated in the absence of such knowledge, subsequent refinement of management alternatives often requires more knowledge than initially available. Therefore, a review of previous monitoring data and relevant case histories should be an integral part of the development of a nonpoint source control programme. Potential sources of such information include local drinking water and wastewater agencies, universities and research centers, and scientific or engineering literature dealing with the waterbody of concern or with similar waterbodies (Rast and Holland, 1988). Also, listen to community complaints as these may give a good indication of community concerns. By incorporating these concerns into a water quality management plan and action programme, agencies can not only appear to be responsive to the public but also garner much public support for, and participation in, subsequent management actions.

Regardless of the control programme ultimately selected, up-to-date information regarding the characteristics of the waterbody of concern, as well as the surrounding drainage basin, is usually needed in the development of the programme. This information would normally include such items as the depth, volume and water-flushing rate of the lake, the concentrations of nutrients and algae in the lake, the occurrences of nuisance growths of algae and other aquatic plants, the occurrence of oxygen-depleted bottom waters in the lake and related fish kills, the annual nutrient loadings to the lake, and the population and land-use characteristics of the drainage basin. If existing data are not sufficient to provide the necessary information for assessment or management purposes, an initial drainage basin or in-lake monitoring programme should be initiated (Rast and Holland, 1988; Ryding and Rast, 1989).

Several reasons for examining or collecting such monitoring data in regard to establishing a nonpoint source pollution management programme are:

(1) To establish past and present baseline conditions, in order to confirm a problem and/or provide a reference against which progress can be assessed;
(2) To identify significant information gaps; and
(3) To develop a cost-effective monitoring programme for the long term (Rast *et al.*, 1989).

Ideally, a monitoring programme would accommodate the maximum number of these needs. Ryding and Rast (1989) provide a detailed discussion of relevant data needs, as well as methods for estimating sampling costs, for the assessment of lake and reservoir eutrophication. These methods can also be applied to nonpoint source monitoring programmes.

DEVELOP APPROPRIATE CONTROL OPTIONS/ALTERNATIVES

Once a community has identified some of its highest priority nonpoint source management needs, it will then need to develop strategies to control the identified problems (see Chapter 9). Whether the community decides to focus on one or a limited number of waterbodies or watersheds, or on a particular nonpoint source problem such as excessive fertilizer use, it will need to develop plans and programmes to guide the clean-up activities.

Plans may be used to summarize the water quality problems, the control approaches to be used, and who will do what. Table 3.1 lists several of the key items that need to be addressed in a drainage basin plan. The plan will need to define the institutional responsibilities for the project; in other words, the plan should clearly state who is going to do what. Experience has shown that it is important to include all potential agencies and organizations which might be able to provide assistance during the planning phase of the project in order to get the project off to a good start (Maas *et al.*, 1987). A community will need to ensure that the legislative mandates of the agencies involved are adequate for

Table 3.1 Idealized key steps in drainage basin planning (adapted from Maas *et al.*, 1987)

Define institutional responsibilities for the project.
Refine the description of the water resource and its impairment.
Identify the drainage basin, its area and land uses.
Develop a drainage basin inventory.
Establish water quality goals and objectives and determine pollutant reductions needed to achieve water quality goals.
Rank individual nonpoint sources or source areas within the drainage basin and establish a priority system for installation of best management practices.
Determine appropriate nonpoint source control options.
Evaluate methods for obtaining agency and citizen participation.
Determine the types of assistance needed to implement nonpoint source pollution controls and establish the roles and responsibilities of the participants, public and private (this determination should include consideration of cost-sharing and financing schemes, and include input from local organizations; i.e., the non-technical aspects of the programme).
Establish an information and education programme.
Establish an implementation schedule with annual goals and milestones.
Carry out a nonpoint source implementation and monitoring programme.
Maintain records on the progress of the watershed project.
Revise the implementation programme based on the results of the annual evaluations and citizen inputs.

---

carrying out a drainage basin project as well as evaluating the capabilities and resources of the various agencies and organizations. For example, people with expertise in soils and water resource management will be needed to prepare and implement the more technical aspects of the plan. While many agencies and individuals are likely to be participants in the project, it is recommended that one agency be designated to coordinate the project. A project coordinator should also be designated (Maas *et al.*, 1987). Ideally, water quality and land management should be important objectives of the designated agency. Also, ideally, the project co-ordinator should be involved from the beginning of the project and be responsible for preparing and implementing the plan.

Once a project co-ordinator has been selected, a good first step in preparing a drainage basin plan is to refine the description of the water resource, its desired use(s) and its degree of impairment. This may involve compiling existing water quality data or perhaps doing some new drainage basin or in-lake water quality monitoring. Obtaining a better understanding of the water quality problem will facilitate a more accurate selection of control measures. For example, if sedimentation is the main water quality problem then it would be appropriate to install erosion control measures.

A good next step would be to prepare a drainage basin inventory. This inventory should identify all the potential pollutant sources in the drainage basin including both point and nonpoint sources. Based on this inventory, and considering the biological availability and/or impact of the specific pollutants generated by these sources, major pollutant sources can be identified. The drainage basin inventory should be tailored to identify the causes of the observed use impairment (Maas *et al.*, 1987). For example, if agricultural nutrients are a problem in a lake, then it would be important to collect information on such sources as fertilizer usage and manure management in the drainage basin and evaluate their impacts on the drainage system.

Once the water resource impairment has been identified and the watershed inventory completed, water quality goals should be established for the project. To the extent possible, these goals should be quantitative and measurable as such goals provide important benchmarks for evaluating the progress and success of the project. An important related task is to determine the amount of pollutant reduction needed to achieve water quality goals. In general, the larger the pollutant reduction needed, the larger the critical area or greater the number of sources which must be treated. Within the critical area, the largest sources of biologically-active substances – the priority pollutants, and, generally, the sources closest to impacted waters – should be treated first in an environmentally- and cost-effective manner (Maas *et al.*, 1987; Ryding, 1992).

The next step is to determine the type of nonpoint source controls needed in the watershed project area. Nonpoint source controls are often referred to as 'best management practices' (BMPs). BMPs are defined by the United States Environmental Protection Agency (1985) as 'the methods, measures or practices to prevent or reduce water pollution, including structural and nonstructural controls and operation and maintenance procedures'. Ideally, BMPs should be applied as a system of practices rather than a single practice. Also, BMPs should be selected on the basis of site-specific conditions that reflect local technical, political, social, and economic feasibility (US EPA, 1985). For example, if agricultural activities are the major nonpoint source in a particular watershed then a variety of agricultural BMPs should be considered. Common agricultural BMPs include conservation tillage, terraces, diversions, nutrient and pesticide management, contour farming, cover cropping, animal waste management, and grassed waterways. One or a combination of these BMPs might be needed on any particular farm to reduce nonpoint source pollution.

ASSESS SOCIOECONOMIC, CULTURAL, AND OTHER NON-TECHNICAL CONCERNS

Consideration of socioeconomic, political and cultural issues is critical to the development of an effective strategy to reduce nonpoint pollution. Nonpoint pollution does not occur in a vacuum. It occurs within the context of the increasingly complex social and economic relationships that have developed within each country and throughout the world. While the specific technical causes of nonpoint pollution are often quite similar, solutions that work in one location may not work in another. All too frequently, the reasons for the failure of such technology transfers lie not so much with technical application of the methodologies (or in the palpable differences between ecoregions – although these do introduce real constraints to technology transfers; Thornton and Rast, 1993; Thornton *et al.*, 1992) but in the degree of local acceptance of the methods, their cost and the ability of countries to implement and maintain the practices (Holland, 1996).

Nonpoint source problems in many different countries stem from the perception (and/or necessity) of the short-term economic benefit accrued from a specific type of economic and land-use activity. This perception of benefit is encouraged by the many political and economic priorities inherent in economic growth and development that are viewed as more urgent than the nonpoint pollution that may result from such uncontrolled growth; for example, selling public land results in an immediate capital gain, a longer term tax income, job creation and reduced maintenance costs – however, the sale also creates additional demands for public services, the need for additional (often public) infrastructure, reduced amounts of public open space, and enhanced nonpoint source pollution. Notwithstanding, the perception of benefit is further reinforced by the fact that the creator of the nonpoint pollution is often not the one who is adversely affected by it (or paying to remediate the cosequences of it; it is considered an externality). Exacerbating this situation is the fact that most nonpoint pollution results from the actions of many individuals – while each specific action may cause little damage, the cumulative effect is often quite severe (Chandler, 1988).

Economic factors

Nonpoint source pollution is a consequence of economic activity, most importantly in terms of area covered, crop and livestock production. Recognizing this, a community would seldom opt for the total elimination of that form of environmental degradation, and would do so only if the community has an alternative source of supply or is provided with sufficient money to adopt some other type of activity. The more typical problem is likely to be the reduction of nonpoint source pollution without eliminating gainful economic activity, thereby approaching a socially optimal, yet still positive, level. In order to develop policies that accomplish this objective, it is essential to understand why individuals choose to 'produce' levels of nonpoint source pollution that are, from a social standpoint, excessive. An interdisciplinary approach is fundamental to this process.

Among the reasons why economic performance, in general, can be suboptimal is that individual economic agents often fail either to pay, or even recognize, the full social costs of their activities or to capture the full social benefits of pollution controls (Meade, 1973). In particular, when at least some part of the social costs of natural resource degradation (or the social benefits of resource conservation) are 'external' or shared more widely than the individual or individual activity, then natural resource management will be suboptimal, from an overall social standpoint. To wit, research has shown that many, if not most, North American farmers and ranchers could make sizable reductions in nonpoint source pollution without sacrificing much, if any, income. In one Ohio watershed, for example, Forster and Becker (1979) found that sediment yield could be reduced by one-half by simply switching from conventional tillage involving the frequent use of a moldboard plough to conservation or reduced tillage. Farm income would probably rise as the cost of crop production fell (primarily because conservation tillage does not greatly affect yields but is less labour intensive than conventional tillage, although new or modified ploughing equipment may be needed). This notwithstanding, however, US soil conservation policy has stressed subsidization (or, cost-sharing – within the limitations of regulations and taxes – along with technical assistance), which has been popularly conceived of, and run, as more of an income-transfer scheme than a resource conservation initiative creating a significant disincentive for modifying traditional ploughing practices. (Indeed, a significant amount of controversy over the efficacy of this programme can be traced to disagreements over the specific natural resource problems, if any, that the payments to farmers are meant to address.) Even when the choice between accepting lower yields and taking compensatory measures is wholly or partially averted, and adverse downstream impacts contained, Ribaudo (1986) suggests that the choice between the allocation of limited funds for subsidizing soil conservation and the mitigation of nonpoint source pollution problems is influenced greatly by the relative importance attached to agricultural yield impacts and nonpoint sources of pollution. Prices for inputs and outputs also affect land use, and government policies that affect prices may have significant implications for nonpoint source pollution. For example, the decision by the US Department of Agriculture to support domestic farm prices and to implement a cropland diversion programme can have a significant positive impact on pollution abatement, although, for years, the converse was true, with subsidies being provided for increased production. This is still the case in many countries faced with food shortages.

This tendency toward sub-optimal economic performance is further aggravated by the difficulties involved in accurately quantifying nonpoint source impacts using traditional economic theory. For example, one might suppose that the value of lost crop production due to soil erosion is significant. However, an economic analysis by the American Agricultural Economics Association (AAEA) would suggest that soil erosion does not seriously threaten agricultural earnings, at least in the United States. In contrast, other analyses that apply the full cost of resource depletion to the traditional balance sheet calculations of the AAEA total several times the amount reported by that association. The higher cost is certainly valid in other [African] countries (Lal, 1985). On the other hand, the value of non-traditional costs, such as the value of nutrients carried away with eroded soil (Larson *et al.*, 1983), can easily be overestimated. (The recent decision by the US General Accounting Office (GAO) to phase in the cost of resource depletion as a component of the US gross domestic product (GDP) during the 1995 and 1996 fiscal years could obviate some of these concerns by providing a standardized methodology for making such determinations.) Notwithstanding, the traditional (AAEA) value, adjusted for the cost of resource depletion, would result in an estimate of the annual cost associated with soil erosion in the United States of only about $5/hectare/year when averaged over the total cultivated area of 150 to 200 million hectares – of course, where erosive crops are planted on erodible soils in the United States, the on-farm costs of soil erosion are much higher (see below).

Aside from underestimating the actual on-farm costs of soil erosion, such averaged information on the cost of soil erosion in the United States also contributes a misleading perspective on the severity of global erosion–productivity issues. Wolman (1985) points out that only a relatively small share of the land used for crop production in the United States is subject to high rates of soil loss. Other countries are considerably more dependent on land that is steeply sloped, with relatively thin topsoil layers and/or more exposed to erosive winds and rains. Buringh and van Heemst (1977), for example, found that 6 percent of North America's grain is grown on 'intermediate' quality land (note that dividing the total cost of soil erosion by this significantly lower area results in an estimated cost of about $83/hectare/year, more than an order of magnitude higher than the averaged value!). By contrast, 40 percent of South America's production and fully two-thirds of Africa's production is harvested from the same category of land resource. These observations suggest that soil erosion might be a much more serious problem for farmers living outside countries like the United States, where the linkage between soil erosion and the economic productivity of agriculture has been studied most carefully.

To overcome these difficulties, some economists would try to find the 'optimal level' of pollution reduction; i.e. to reduce the level of pollution until the marginal cost of pollution control equals the marginal benefits derived from reduced pollution. Economic incentives, in the form of tax credits or subsidized investment loans, can be helpful in modifying the point at which this optimal level occurs and in encouraging more rapid implementation of nonpoint pollution control measures. Nevertheless, at some point, the marginal cost to the farmer of reducing nonpoint source pollution grows quite high.

Socio-political factors

In fashioning a strategy to control, or otherwise reduce, nonpoint pollution, governmental and other leaders should strive to ensure that the resulting programme is in tune with the unique economic, social, political and cultural characteristics of the country or community involved. Because the preponderance of nonpoint pollution results from the actions of large numbers of individuals, experience throughout the world has demonstrated that it is important for these strategies to include the development of an alliance between government and the public.

In the control of nonpoint source pollution, regulatory measures alone are generally of limited utility. Unlike point source controls, there are simply too many opportunities for individuals to disregard regulatory requirements when they conflict with the individual's perceived self interest (which are determined by the cost of not following the rules and the risk of being caught breaking the rules). Less direct approaches appear more promising. For example, thoughtfully prepared educational and technical assistance programmes are necessary components of a successful strategy. These programmes should be targeted at particular segments of the population in order to address the most significant sources of nonpoint pollution. They are most successful when more than one social goal is advanced. For example, education and demonstration projects can illustrate the link between biological rather than chemical pest control methods and the long-term productivity of agricultural lands. Such an approach complements the interests of the farmers (as well as of the nation) in maintaining a stable agricultural sector while at the same time reducing the amount of pollutants that reach surface waters. Similarly, educational programmes dealing with nonpoint source pollution can complement related public information efforts such as recycling and litter prevention campaigns (affecting both terrestrial and aquatic biomes). World Environment Day, Arbor Day, and related events provide convenient pegs on which to hang citizen participation in nonpoint source abatement campaigns.

Nevertheless, direct governmental action may, at times, be required. This is particularly true when new projects or methods are being introduced. For example, to encourage integrated nutrient and pest management, restrictions may be placed on pesticides either by means of a tax, a licensing requirement, or limitations on the amount of the product that can be purchased. This approach is in contrast to the current practice in some countries where pesticide use is subsidized or even carried out by government. (While agricultural pesticide use is generally discouraged in this book in favour of integrated nutrient and pest management, it is recognized that public health uses of such chemicals may be a necessary trade off in many, largely tropical countries where waterborne diseases such as malaria, trypanosomiasis and onchocerciasis can create severe public health problems.) Likewise, licenses for mining or forestry operations can be of limited term with renewal contingent upon implementation of nonpoint pollution control measures.

Other cultural factors affecting nonpoint source pollution

In addition to those methods and approaches which directly influence the control of nonpoint source pollution, other, indirect factors can have important effects on its extent. Public policies typically influence these 'other factors', and, in the end, may have more of an impact on nonpoint source pollution prevention than explicit pollution controls (Henderson and Barrows, 1984). One of the most important of these factors is the demand for downstream use of water. Without such downstream demand, the economic incentive to reduce pollution is nonexistent. While this fact seems obvious, the downstream demand for water is often lost in nonpoint source pollution debates. Recognizing the importance of such demand implies that the 'proper' level of pollution is driven as much by downstream user demands for water with low pollutant concentrations as it is by factors within the upstream community. In practice, the definition of 'good' and 'bad' water quality is determined by legislation, although, in principle, its actual attributes may change from one watershed to the next both in a relative, use-based sense as well as in an absolute, scientific sense. Settlement patterns, and policies promoting economic and population growth in certain areas, have important implications for water quality demand. This is especially true in developing countries where economic development areas are typically hewn out of virgin bush; a fact that provides opportunities to install nonpoint source abatement practices before land uses become established in the watershed.

The economic activities that produce nonpoint source pollutants are shaped, to some extent, by society's changing values. A number of contemporary social movements and groups – including 'green' political parties – have recognized the basic principles of ecological science, and thus stress that nature is to be tended and nurtured rather than dominated; technology must be considered in relationship to the natural environment, and the effect of production activities on nature must be considered in assessing their worth. These values affect nonpoint source pollution through government policies, industrialists' attitudes toward the means of production (where alternatives exist) and farmers' attitudes toward soil conservation, land preservation, and nutrient and pest management. In many nations, these values are codified as environmental law (*cf.* Schlickman *et al.*, 1994), or, as set out above, are modified through the use of economic policies that either encourage or discourage certain types of activities (e.g. US farm subsidies, South African development zones, etc.). All of these reflect societal values concerning the environment.

Another influence on nonpoint source pollution is consumer preferences. For example, consumers in the US and other developed countries currently have high enough incomes to effectively assert their strong preferences for red meats. These preferences cause grazing on vast areas of rangeland that otherwise would have little agricultural use. These same preferences result in higher prices for feed grains, which further encourage the use of marginal land for their production, leading to soil erosion and nonpoint source pollution.

Of course, technology also influences the extent of nonpoint source pollution. It was technological advances that spurred the European industrial revolution that led to sweeping social changes, massive urbanization and the demand for both consumer goods and more efficient means of agricultural production. In the latter case, technological advances on and off the farm demonstrably affect nonpoint source pollution. Over the past several decades, technological changes in agriculture in developed countries have led to more intensive cropping systems requiring larger machines, more fertilization, and more use of a wider variety of chemicals. Off farm technology similarly affects nonpoint source pollution; for example, land used in agriculture is affected by transportation systems, irrigation systems and water development projects, and drainage and flood control projects, all of which change the character of the natural drainage patterns and encourage the presence and greater washoff of contaminants.

The role of non-technical factors in strategy formulation

It is clear that no single approach to nonpoint source pollution control will work in all places, even within a single country. However, experience from around the world suggests that the most successful strategies are those that are consistent with the social, political and economic goals of the community, and build upon the strengths and concerns of the society involved. Unlike the mainly technical approaches to the control of point source pollution familiar to engineers and technicians throughout the world, voluntary societal – and individual – commitment and action are extremely important in the control and management of nonpoint pollution; an investment of time and effort by governmental and other leaders in the development of a grassroots consensus regarding nonpoint pollution control goals and strategies, while perhaps not a prerequisite, is at least an essential consideration for a successful nonpoint source management programme.

ASSESS THE ADEQUACY OF LEGISLATION/REGULATION

More often than not, the failure of an individual to consider all of the consequences of their economic activity in controlling (or failing to control) nonpoint source contaminant washoff is explained in terms of limitations on his/her property rights (or 'land tenure'). A tenant farmer, for example, might perceive that s/he will cease occupying a parcel past the terminal date of his lease. Accordingly, s/he will neglect the benefits to be derived from soil conservation after that date; indeed, s/he could conceivably consider any investment in conservation activities as 'lost' income and be actively disinclined to implement actions whose effects/benefits will accrue well beyond his/her tenancy. This sort of temporal limitation in land tenure reduces the marginal internal social benefits of soil conservation and, hence, widens the gap between socially optimal and privately optimal benefits. In other words, if you own it, you will pay more attention than if you rent it.

A more serious 'legal' cause of land degradation and nonpoint source pollution arises in developing countries where individuals jeopardize (or gain) property rights by making certain changes in land use. For example, agricultural use rights are the norm along the typical developing country agricultural frontier. That is, settlers win informal, or even formal, tenure of a parcel by clearing it for crop or livestock production. Alternatively, an individual considering reforestation in such a setting, for example, recognizes that this land use carries the risk that an agricultural colonist will usurp his/her property right. In the context of this example, this latter user treats the disutility associated with the risk of losing his/her property right as an internal cost of soil conservation (Forster and Southgate, 1988). Of course, any increase in marginal cost of production (i.e. the additional cost of reforestation *versus* cropping or livestock production; the internal cost associated with the risk of loss of property described above) causes a reduction in the privately optimal level of benefit through erosion control. In a more general analysis, Southgate and Pearce (1987) have shown that, where agricultural use rights prevail, the disincentive promoting deforestation is dominant and inadequate conservation measures are applied.

The tenure of farmers and size of farming operations also influence land use and nonpoint source pollution in other ways. Farmers having short-term tenure arrangements not only have less incentive to prevent erosion than farmers whose future income depends largely on soil conservation, but farmers operating on vast acreages, employing large amounts of labour and capital – and carrying large debt loads – may be more likely to approach soil conservation from a purely economic perspective. On the other hand, debt-free farmers, using labour and capital supplied by family members, may be more willing to introduce conservation and environmental elements into their decision-making process, albeit still tempered by economic considerations. In this regard, the managerial talents of farmers become important. Increased education, improved entrepreneurial abilities, and improved information systems make farmers far more flexible in using resources, less resistant to change, and more willing to consider nonpoint source pollution abatement, given the proper incentives and statutory support.

Notwithstanding these specific considerations, however serious, most nations in the world continue to employ more direct means of regulating land use and development controls to achieve water quality (and economic development) goals. In parallel with their experiences in managing point source pollution, many countries establish standards or other measurements of quality which are applied to point sources, areas of land or waterbodies (see Chapters 5 and 9). In some cases, these controls are established through economic levies or allowances designed to promote certain activities (such as agricultural or industrial development). While developing countries may have little experience with the use of such non-technical policy instruments (e.g. regulations, differential taxation, and selective subsidies) to encourage the control of nonpoint sources of pollution, and given that the limitations on their effectiveness (discussed above) would undoubtedly transfer with them to developing countries, such policy instruments should not be disregarded. In fact, one area of policy not typically considered in the more affluent parts of the world deserves serious attention from policy-makers in many developing countries: overhauling land tenure regimes. As indicated above, limitations on property rights are common in developing countries, and are especially so in the environmentally fragile hinterlands of many of these nations. Furthermore, circumscribed private land tenure inhibits resource conservation because the legal arrangement encourages land users not only to ignore the downstream costs of contamination but also to ignore aspects of productivity impacts as well. It stands to reason, then, that part of the solution to land degradation and nonpoint source pollution problems in developing countries is to strengthen individual 'ownership' in the land they use. (Depending on a variety of political and cultural factors – see above – 'ownership' in this context can range from actual title to other, less tangible forms of participation in land use decision-making, co-operatives being an example of one such form of ownership.)

By adopting such complementary but non-regulatory approaches, governments can facilitate public awareness and voluntary compliance with nonpoint source pollution control programmes. For example, in the United States, much of the public soil conservation and nonpoint pollution control effort is based on a government-led research–education approach. Land grant universities, agricultural experiment stations, government and university extension services, and the US Department of Agriculture (principally, through its Natural Resources, formerly Soil, Conservation Service) work together to develop new soil conserving management practices, and to inform farmers about the effects of soil erosion on lost productivity and the means to avoid such losses. The Natural Resources Conservation Service, for example, assists individual farmers in developing soil conservation plans for their farmsteads, while other federal government farm programmes aimed at enhancing farm incomes and managing farm production underscore this effort by requiring these plans before making payments to participating farmers. On a global scale, the debt-for-nature swaps negotiated by The World Bank typify this type of incentive.

PUBLIC PARTICIPATION

Encouraging and enlisting public participation in the control of nonpoint source pollution problems should be an important component in the identification of such problems and the search for solutions. As noted above, the diffuse nature of nonpoint source pollution, and the individual contributions to nonpoint pollution problems, requires as comprehensive an approach as possible. While public involvement may not always be feasible in every instance, it is important that, where there is such involvement, a readily identifiable forum for obtaining this viewpoint be identified. One example of such a forum is the citizens' advisory committee (Holland, 1996). This type of committee is generally comprised of civic leaders and other individuals who can provide (additional) insights into the extent of a given nonpoint pollution problem, and its social and political consequences. As noted earlier, the policy-maker often must balance the interests of the advocates of long-term benefits against those wishing more expedient solutions (Rast *et al.*, 1989).

Interested citizens can be an asset in the development of effective nonpoint pollution management programmes. Where water quality data are scarce, say at the beginning of a control programme, narrative descriptions of prior conditions provided by elder citizens and leaders can serve as an initial reference point on which a control programme can be designed and its potential effectiveness assessed. In the broader sense, such interactive communication can have at least two beneficial effects:

(1)   Knowledge gained through a lifetime of observations of a waterbody can be documented for use in developing management programmes;
(2)   Persons encouraged to participate in the development of a programme are more likely to become advocates for the programme (Rast *et al.*, 1989).

Newspaper archives and other media reports, historical society and museum archives (including those of the colonial powers in the case of many developing countries), and diarized historical accounts also form useful repositories for supplemental historical data and information.

Further, as the costs of specialized manpower rise and/or where such manpower is spread thinly, citizen monitoring can be an effective means of acquiring quantitative scientific information. In the United States, citizen monitors collect data on various water quality variables ranging from Secchi disc transparency measurements to nutrient chemistry data to biological monitoring information. Elsewhere, such data gathering is integrated with environmental education curricula in schools and technical colleges. In the latter case, these programmes have the added advantage of promoting the training of skilled technical staff. In some developing countries, in particular, where financial constraints may limit the use of large structural solutions to nonpoint source pollution problems, governments may wish to make maximum use of community-based programmes to engender individual grassroots action to manage nonpoint source pollution; in such cases, a community education or communications specialist can be a valuable asset both to disseminate information and act as a focal point for citizen involvement (Rast *et al.*, 1989). Effective public participation requires that government officials be accessible, honest in their presentation of information, and responsive to the views expressed to them. Nothing can be more damaging to public confidence in a new government initiative than a feeling by the public that the government did not listen to those participating in the process.

SELECT AND IMPLEMENT A SPECIFIC CONTROL STRATEGY

For those locally identified watersheds where water quality concern has been documented, a comprehensive and integrated water resource management plan should be developed at the drainage basin or sub-regional level. This plan should identify the types of pollutants, their sources, and magnitudes of their nonpoint source loadings. The impact of each source on the receiving water body should be evaluated and priorities established with the most severe sources receiving the more immediate attention. Based on desired use, achievable quality, community demands and other considerations (such as the degree of water resource impairment, the primary pollutants of concern, transport considerations, etc.; Maas *et al.*, 1987), appropriate measures to address the sources of perceived and potential problems can be developed. Unfortunately, these measures are unlikely to be as technically clear-cut as the construction of a wastewater treatment plant and are likely to be less easily quantified – in restoring an eroding streambank, for example, numerous techniques can be used with varying levels of 'natural' materials, heavy equipment, and contract labour, ranging from the replanting of streambank vegetation to the reconstruction of the stream bed and placement of rip-rap. Costs will vary accordingly.

Once the major nonpoint sources and the control options (i.e. BMPs) are identified, a strategy needs to be developed to obtain landowner participation and acceptance. A variety of means including education, technical and financial assistance as well as regulation may be used to encourage participation in the watershed project. In the case of voluntary, agricultural programmes, for example, the practice chosen for the project must integrate with the farmer's production considerations (Maas *et al.*, 1987). Also, financial assistance may be needed to assure participation in the programme. Installation of nonpoint source BMPs can be expensive, so early consideration of cost-sharing and project financing is important in the development of a comprehensive watershed plan. Similarly, a project funded by the United States Agency for International

Development in the Dominican Republic found that the availability of government support services also appears to be an important factor in determining the success of a nonpoint source management programme (Hansen *et al.*, 1987). They found that credit is necessary in the short term to facilitate the purchase of equipment and materials to implement land conservation activities and practices. They also found that participation in educational programmes is important so that farmers (and others) develop technical expertise and, in some cases, have access to the technologies required to introduce specific practices.

After all of the above steps are completed, the watershed project may reasonably proceed. During the implementation phase, landowners or users will be encouraged by a variety of means to install or use the necessary nonpoint source BMPs to improve water quality in the watershed area. Some of these means include taxes, subsidies, regulation, and tradable pollution permits (i.e. bubble concept) (see Chapters 9 and 10). Watershed projects may take from several to 10 years to complete depending on the size of the watershed and the control needs. Ideally, a monitoring programme should be instituted at the beginning of the project to assist in evaluating the project (including identifying any 'mid-course corrections' that might be needed in the programme to better attain the project goals; see e.g. Kleiss, 1996). The project co-ordinator should maintain records to track the progress of the watershed project.

REVIEW THE PROGRAMME

In evaluating the results of a nonpoint source control programme, progress against the overall programme objectives should be assessed annually. Water quality monitoring results, and best management programme implementation and maintenance, in specific projects should be evaluated as part of the programme review process. Results of the evaluation should be included in annual reports and disseminated to interested individuals and the mass media. Where appropriate, the results of such evaluations should be used to improve project targeting efforts and other programme components.

In order to obtain sufficient information for a judicious evaluation of nonpoint source control measures, on-going studies of the chemical and biological conditions of the waterbody of concern and its tributaries are recommended. Even after nonpoint source control programmes have been planned, initiated and completed, post-treatment studies should be continued for several years. This should be done to assess conditions in the waterbody before and after the application of nonpoint source control measures, and to ascertain whether or not the forecast conditions obtained from model calculations have actually been achieved. Only then can one be certain of the degree to which corrective actions have achieved the desired result and determine that the investment decision was financially responsible (Rast *et al.*, 1989).

The period of time necessary for carrying out such post-treatment measurements is dependent on each individual case: the longer a lake is expected to take to recover, the longer the period of time over which such measurements will have to be taken. Even after a prompt recovery of a given lake, it may still be necessary to continue monitoring studies for several years in order to be sure that the situation has been correctly assessed. This is necessary especially in those cases in which large annual differences in water quality occurred before nonpoint source control measures were initiated. Examples are lakes and reservoirs with relatively short water retention times, and those located in regions characterized by rapid, dramatic shifts in weather conditions.

Should it occur that, in spite of very careful planning and use of all available knowledge, the results obtained fall short of those expected, post-treatment measurements can still be used to improve the model predictions in question. This will also decrease the uncertainty of model predictions in future planning programmes. In this regard, post-treatment monitoring and evaluation also provides valuable information to others concerned with similar nonpoint source management problems. Such information helps guide future water quality protection and management efforts (e.g. building the information and experience base for improved lake and reservoir management; Rast *et al.*, 1989).

Although following these steps in developing and implementing a watershed plan and nonpoint source pollution control project does not guarantee success, this process does permit an objective assessment of nonpoint source-related water quality problems, and the formulation of a community-based action plan. In the management of nonpoint source pollution, these elements have proven to be a good place to start.

CHAPTER 4

THE HYDROLOGIC CYCLE AND FACTORS AFFECTING THE GENERATION, TRANSPORT AND TRANSFORMATION OF NONPOINT SOURCE POLLUTANTS

G. Jolánkai and W. Rast

INTRODUCTION

Land and water are two of the most essential human resources, being prerequisites of human life. Land forms the basis of food production, while water is needed for sustaining life, as well as for all kinds of economic activities (agricultural and industrial production, power generation, irrigation, etc.). Indeed, land and water constitute the Earth's surface. The atmosphere also has a fundamental role to play, as a means of transporting natural and human-induced materials. These components interact with each other by means of (1) the hydrologic cycle and its human-induced modification, and (2) relationships between land and water uses. Even considering only the generation, transport, and transformation of nonpoint source pollutants, hydrologic processes continue to have a fundamental role. For example, nonpoint source pollutants can be released over, or into, the (1) land surface, (2) atmosphere, and (3) rivers, lakes and groundwaters. The fate and further movement of these pollutants, in any of these three basic environmental media, is fundamentally associated with the hydrologic cycle.

THE HYDROLOGIC CYCLE

The hydrologic cycle can be viewed as a series of connected pathways for the movement of water in its various forms through the physical (air, land and water) environment. Being a 'cycle', it has no specific starting or ending point. Rather, the component of concern in the hydrologic cycle usually reflects the specific interests of the observer. The major components of the hydrologic cycle that define the movement of water through the environment include (1) precipitation, (2) interception, (3) evaporation, (4) infiltration, (5) evapotranspiration, (6) surface runoff, (7) percolation, (8) sub-surface runoff, and (9) storage. These various components (Figure 4.1) are discussed further in the following sections. A description of the relative quantities of water flowing through the components of the hydrologic cycle is provided in Figure 4.2.

The total amount of fresh water on Earth is estimated to be approximately 40.7×10^{12} m^3 (33 trillion acre-feet). Of this portion, only a relatively minuscule

41

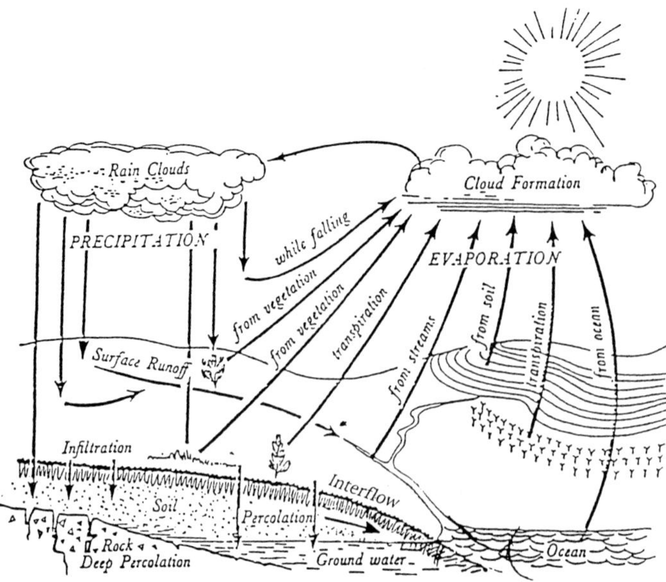

Figure 4.1 The hydrologic cycle – a descriptive representation (after Ackermann *et al.*, 1973)

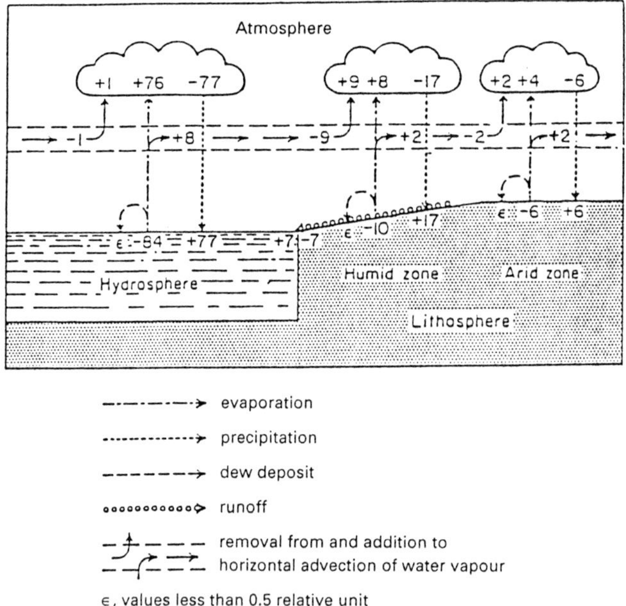

Figure 4.2 The hydrologic cycle – a quantitative representation. 100 relative units = 85.7 g/cm²/year or 857 mm, the global annual mean precipitation

amount actually exists as a fresh (non-salty) liquid in rivers and lakes, and in the Earth's crust. Considering all fresh waters, the relative proportions are as follows: polar ice and glaciers – 75 percent; groundwater – 25 percent; rivers and lakes – 0.6 percent. Even these estimates only provide an indication of the relative global water distribution at any given time. For example, the atmosphere is estimated to contain only about 0.035 percent of the Earth's fresh water. As noted above, however, the atmosphere represents the Earth's fresh water source and sink, and huge quantities of water pass through the atmosphere annually *via* the mechanisms of evaporation and precipitation.

More detailed illustrations of this process are shown in Figure 4.3. Water evaporating from the ocean and from land surface sources (ponds, rivers, lakes, estuaries and plants) condenses into clouds that can move inland. The clouds release the water vapour as precipitation (rainfall, snowfall, hail) onto the land and water surfaces. Precipitation either returns to the atmosphere as water vapour

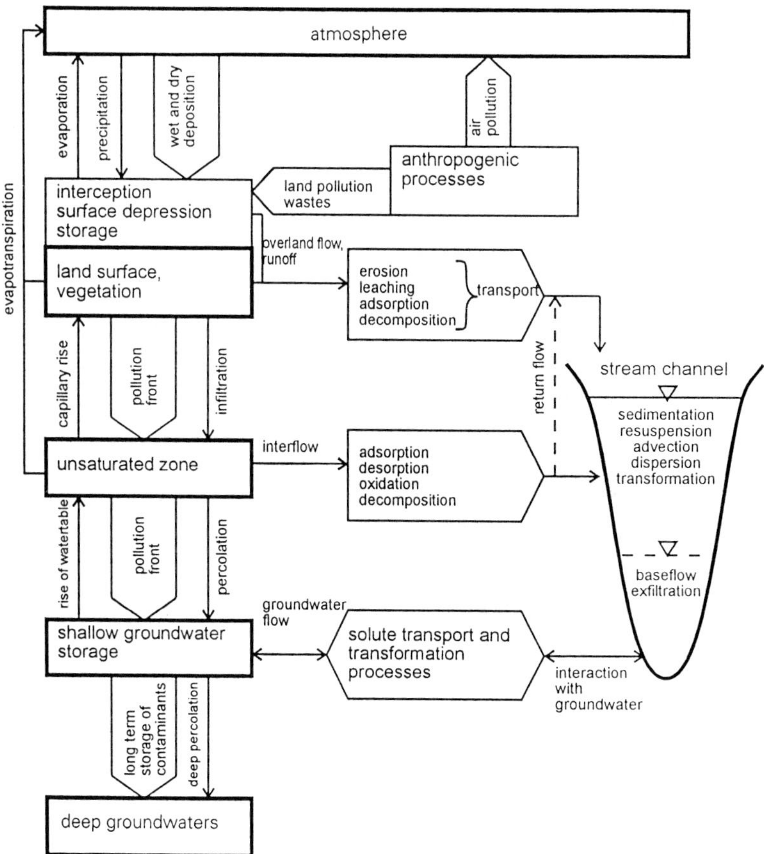

Figure 4.3 Conceptual model of hydrological processes with the indication of nonpoint source pollution

(evaporation), flows over the land surface (surface runoff), or soaks into the soil (infiltration) where it may transpire to the atmosphere, percolate to deeper layers, or flow toward surface water channels (groundwater flow). Portions of the precipitation may be stored, for longer or shorter periods of time, in the various storage volumes along the flow-transport routes indicated in Figure 4.3 (surface depression storage, interception storage, groundwater storage, etc.).

Runoff moves rapidly along the soil surface in many small rivulets flowing to stream channels, which subsequently drain to rivers, lakes and, ultimately, the ocean. Drainage water from the unsaturated zone moves slowly through the soil (percolation) eventually entering a groundwater aquifer, causing the water table to rise. The deepest depressions on the land surface above the aquifer (stream channels) will contain water. Shallow saturated zones discharge water to stream channels, while deeper saturated zones (far below the land surface) receive infiltration seepage from shallow ground and surface waters. In very arid regions, most groundwater is in deeper aquifers, and stream flows are depleted by evaporation and water percolation into the water table.

The interrelationship between the above briefly-described components of the hydrologic cycle can be mathematically expressed, either in a time-varying or steady-state manner, for a given land area. For example, the simplest steady-state formulation is the water-balance equation:

$$\mathrm{Pxx} = \mathrm{P} - \mathrm{ET} - \mathrm{S} - \mathrm{R} = \Delta V_1 + \Delta V_2 \tag{4.1}$$

where: P = precipitation;
            ET = evapotranspiration;
             S = subsurface runoff;
            R = surface runoff;
        ΔV_1 = changes in surface storage (= depression storage + interception storage + storage in form of snow and ice) during the time period t; and
        ΔV_2 = changes in sub-surface storage (= soil moisture storage + groundwater storage + deep groundwater storage) during time period t

If the water balance is made for a specific land unit (areal unit), then all the terms in Equation 4.1 can be expressed in terms of length (e.g. water-column mm/year). The terms R (surface runoff) and S (sub-surface runoff) directly affect the fate of contaminants present in the water. They provide the means of pollutant transport, as well as the media for their chemical and biological transformations. Nevertheless, the other processes of the hydrologic cycle are also important. For example, rainfall (precipitation) provides the energy for displacing soil particles and any attached contaminants. Infiltration and penetration (percolation) provide water for leaching natural chemicals and human-induced contaminants from the soil. Evapotranspiration (the major component of the water balance) determines the quantity of water available for surface and sub-surface runoff, etc.

Prior to describing each of these above-identified hydrologic processes, and their effects on nonpoint source pollution, some general considerations on the types of nonpoint sources, and the major transport and transformation mechanisms, are discussed in the following section.

HYDROLOGICALLY-AFFECTED TRANSPORT AND TRANSFORMATION OF NONPOINT POLLUTANTS

Nonpoint pollutants on the land surface

Pollutants generated on, or applied to, the land surface move and/or undergo transformation reactions when water is present (the only exception is wind transport). The water normally must flow over, and below, the land surface. Thus, water usually is the main carrier or transporting medium of nonpoint pollutants. In addition, the variety of complex chemical, biological and biochemical processes (pollutant transformation processes) which take place in the soil–plant–water (the so-called 'biogeochemical') system is controlled largely by the availability of water. Consequently, the monitoring, quantification and modelling of nonpoint source pollutant transport and transformation reactions are based largely on processes inherent in the hydrologic cycle.

The generation and transport of nonpoint source pollutants is not necessarily a result only of the introduction (application, spreading, etc.) of human-generated wastes or polluting substances. In fact, such substances usually also are natural constituents of soil and rock. As with human-generated pollutants, the dissolution or movement of natural substances is fundamentally associated with the presence and movement of water. This can result in undesirable concentrations or accumulations of non-human-induced pollutants in, or over, land formations and/or in surface and ground waters. Examples of processes subsequently amplified by human activities include erosion and the nutrient enrichment (cultural eutrophication) of lakes and reservoirs.

The primary sources of polluting substances may be unidentifiable and inseparable. In estimating the total pollutant load in the outflow section of a watershed, for example, it often is not possible to identify and quantify the exact proportion of the natural phosphorus or nitrogen content of the soil, compared to human inputs (e.g. fertilizers). However, estimation of the magnitude of anthropogenic alterations of flow (runoff) and transport routes, along with those of human-induced erosion, may enable planners to design their nonpoint source pollution control strategies without knowing the exact contribution of the natural and human-mediated pollutant sources. Thus, nonpoint source pollutant control strategies can be closely associated with (1) the regulation of water runoff processes, and (2) the processes inherent in the soil–plant–water system associated with the runoff process as described further in Chapter 9.

Atmospheric nonpoint pollutants

Pollutants discharged into the atmosphere by human-induced (point or nonpoint) sources will eventually find their way to the land and water surfaces, via dry and/or wet deposition. From the point of view of the land and water surfaces as pollutant recipients, atmospheric deposition can be considered a nonpoint source pollution transport medium. Wet deposition is closely associated with precipitation (rainfall, hail, snowfall) and, at least in humid regions, is the dominant component of the total atmospheric pollutant load.

Nonpoint pollutants within a waterbody

Pollutants being transported to a river or lake will take part in various hydrologic processes until they arrive at their final destination, usually the ocean or the inactive spaces of groundwater storage. Transport processes that dominate the fate of nonpoint source pollutants once they reach waterbodies include:

(1)   Advective transport which is the movement of suspended or dissolved substances with the flow of water (determined by the velocity of the water flow);

(2)   Molecular and turbulent diffusion which results in the effect commonly called 'mixing'. Molecular diffusion is the 'spreading' of a pollutant in water as a result of the thermally-dependent motion of the pollutant molecules (so-called Brownian motion), while turbulent diffusion is the 'spreading' of a pollutant due to the random, pulsating fluctuation of flow velocity components around an observed mean value. The joint effect of molecular and turbulent diffusion is termed dispersion. Dispersion is usually dominated by turbulent diffusion in flowing waters, while molecular diffusion usually dominates in standing waters and during calm, windless periods.

Advective and dispersive transport actually are general hydraulic processes. Thus, they occur in any flowing water, including land runoff and groundwater flow.

QUANTIFICATION OF HYDROLOGIC PROCESSES AFFECTING NONPOINT SOURCE POLLUTION

Sufficient long-term records of hydrologic elements (rainfall, evaporation, channel flow, etc.), and their associated hydro-meteorological parameters (temperature, wind speed, vapour content of air, etc.) exist in many countries and allow the reliable quantification of water balances. In many cases, one also can reliably predict the runoff in drainage channels, tributaries, etc. However, additional data are needed to characterize some important aspects of nonpoint source pollution. While hydrology is an ancient science, it, alone, cannot provide sufficient information about some processes of major interest to permit the analysis and quantification of nonpoint source pollution. This lack of information is due largely to the spatial scale of the involved processes and measurements.

Processes influencing nonpoint source pollution occur in, and between, watersheds in a highly-variable, spatial manner. They are a function of the various land uses in a watershed; for example, the generation and/or transport of nonpoint pollutants from a ploughed field usually is very different from that from a meadow or rural settlement. Pollutant export rates also vary as a function of the specific crop(s) grown, the quantities of fertilizers and other chemicals applied, and the cultivation method used. Nevertheless, the pollutant generation still is determined by, or related to, hydrologic processes (quantity and intensity of rainfall, interception, infiltration, evapotranspiration) and the resultant runoff. The amount of nonpoint source pollutants (sediments, nutrients, organic compounds) moved from the land surface to a receiving waterbody will depend on such factors as the type of usage to which the land is being subjected, the frequency and duration of precipitation, and antecedent moisture level. For example, streams normally carry the greatest material loads at high flows, which occur for only a short period of time each year. Furthermore, in many parts of the world, summer storms have sufficient duration and intensity to cause flooding and subsequently permit streams to move even larger quantities of sediments and other nonpoint source materials.

A fundamental problem in assessing the generation of nonpoint source pollutants is that the relevant hydrological and hydrometeorological parameters usually are measured at stations located in a drainage basin on a larger spatial scale than the specific pollutant source areas. For example, in Hungary, the density of existing rain gauges is found to be approximately one gauge/100 km^2. The measurement of some basic hydrologic parameters (rainfall, runoff) is relatively easy at this scale. However, other elements of the water balance are difficult to measure at this scale. For example, measurements of evapotranspiration are highly uncertain because many factors, including ecological parameters, influence the process. Evapotranspiration measurements also are technically difficult to make because they require measurement of the complete water balance of a vegetated soil column (e.g. lysimeter studies). This problem is somewhat alleviated by the fact that all hydrologic and related pollutant transport parameters usually can be reliably measured in a plot size or small watershed-size study. Therefore, both hydrologic, and material transport and transformation, processes can be described and predicted on this scale with any of several widely-used nonpoint source models (see Chapter 8).

A shortcoming of the plot-sized scale is that the results usually are only valid for the study site. They are not necessarily valid for a larger area, or even the specific location to which practical, management-oriented control measures are to be applied within a given watershed or national boundary. This means that the values measured in smaller-scale studies must be extrapolated to, and assumed valid in, larger areas. It is possible to perform a series of plot-scale experiments on all the important land-use forms in a larger area, as well as for differing soil types and topographic conditions, but this translates into a large number of field experiments (e.g. 10 to 15 plot-scale studies, sub-watershed studies), even for only a medium-sized drainage basin (a few 100 km^2). Time and funding constraints virtually always exclude such intensive research/ monitoring efforts. Thus, it usually is not feasible to acquire detailed, site-specific descriptions of hydrologic and material transport processes. (Unfortunately, even assuming that all hydrologic sub-processes could be properly described and modelled with a single experimental expression, and still remain valid for the whole watershed, such lumped-parameter hydrologic and related nonpoint source modelling efforts may not yield the necessary information for planning effective nonpoint source control strategies.) The almost antagonistic relationship between the needed data and available data greatly hinders the use of detailed models (see Chapter 7). Difficulties in quantifying and assessing nonpoint source pollution processes remain the principal reason why these processes are often ignored in favour of point sources (and other readily identifiable and quantifiable processes).

A brief description of each hydrologic element involved in the nonpoint source runoff process, as well as methods available for estimating and measuring their associated parameters, now follows. Special emphasis is given to the influence of ecological factors, since many of the hydrologic processes are strongly dependent on them.

Precipitation

The term precipitation includes rainfall, snowfall, hail and fog. Of these components, surface water and groundwater flows generally are governed by rainfall and snowfall.

Rainfall

In regard to rainfall, important factors governing the amount and rate of water runoff, as well as of sediment and sediment-associated pollutant transport, include the amount, distribution, intensity, and energy load of the rainfall.

1. *Intensity* – Surface runoff simply represents the difference between the rates of rainfall and infiltration into the soil. Yet, rain falling over a short period of time results in more runoff than the same quantity falling over a longer period of time. In general, tropical rains fall at higher intensities than rains in the temperate zones. In fact, some tropical rains can attain a short-time (5 to 10 min) intensity of 150 to 200 mm/h (Lal and Stewart, 1994). High rainfall intensity usually is characterized by (1) a relatively large raindrop size and (2) a greater number of raindrops falling per unit area per unit of time. In contrast, rainfall intensities generally are low in the temperate climatic zone of Europe. Nevertheless, even in this zone,

isolated, intense storms can produce up to 90 percent of the annual runoff and sediment transport. In fact, when compared to Europe, North America and temperate Australia experience even a greater number of high intensity rains. For example, short-time intensities exceeding 100 mm/h have been observed in Ohio (USA).

2. *Energy load* – Raindrops hitting the land surface during an intense storm event can structurally degrade the soil, thereby decreasing the soil's infiltration capacity over time. The kinetic energy of impacting raindrops is a function of their drop size and terminal velocity. The upper limit of drop size in a natural rainstorm is about 6 mm. The median volume drop diameter (termed D_{50}) refers to a drop diameter at which 50 percent of the volume of the rain falls in the shape of smaller-diameter drops and 50 percent as larger-diameter drops. Intense rains of short duration have a relatively high proportion of big drops, with D_{50} values ranging from 2 to 4 mm. The median drop size is affected by (1) type of rain, (2) amount of rain, (3) duration of rain, (4) rainfall intensity, and (5) wind velocity accompanying the rain. In general, the larger the quantity of rain per storm event, and the larger the intensity ranges, the bigger the median raindrop size. The D_{50} is related to the rainfall intensity, as follows:

$$D = aI^b \tag{4.2}$$

where: I = rainfall intensity; and
         a,b = constants

The drop size also affects the terminal velocity of rainfall in that the terminal velocity increases with increasing drop size. Thus, the kinetic energy (the product of the mass of rain falling per unit time and the square of the terminal velocity) generated by rainfall also increases with increasing drop size. Kinetic energy (E) can be estimated from rainfall intensity using one of the following equations:

$$E = a + b \log_{10} I \tag{4.3}$$
$$E = c(b - a\,I^{-1}) \tag{4.4}$$
$$E = bI - a \tag{4.5}$$

where: a,b,c = empirical constants

The most commonly-used equation relating kinetic energy to rainfall intensity is:

$$E = 118.9 + 87.3 \log_{10} I \tag{4.6}$$

where: E = total energy (10 Joules/ha/mm of rain); and
          I = rainfall intensity (mm/h)

In addition, the amount and rate of surface runoff is also related to rainfall momentum. Momentum (the product of mass of rain falling and terminal velocity) is affected by drop size distribution and terminal velocity. Both kinetic energy and momentum can be drastically altered by the wind velocity accompanying rainfall.

The erosion potential of a rainstorm can be greatly altered by overland flow. The presence of shallow overland flow increases the soil detaching and splash capacities of a raindrop. Soil detachability and sediment load generally increase with water layer depth, up to a threshold approximately equal to the rainfall diameter, and decrease thereafter (Palmer, 1963). Thus, the energy available for soil detachment consists of (1) rainfall, and (2) runoff (Onstad and Foster, 1975; Monke *et al.*, 1977). The combined energy term is:

$$E' = 0.5\, P_{st} + 15\, Qq_p^{1/3} \qquad (4.7)$$

where: E' = combined energy term;

$\quad\quad P_{st}$ = storm rainfall factor (expressed in EI units in the USLE; Wischmeier and Smith, 1978);

$\quad\quad Q$ = storm runoff volume; and

$\quad\quad q_p$ = storm peak runoff rate

Although overland flow depth influences sediment detachment, sediment transport or deposition is a function of the water flow velocity. The kinetic energy of overland flow depends on velocity. The velocity is related to hydraulic radius and to slope gradient, according to Manning's equation:

$$V = 1.486\, [(R^{2/3}S^{1/2})/n] \qquad (4.8)$$

where: V = velocity (m/s);

$\quad\quad R$ = hydraulic radius (m);

$\quad\quad S$ = slope gradient (m/m); and

$\quad\quad n$ = roughness coefficient

When the flow becomes turbulent at high velocities, its detachment and transport capacities increase as a power function of the velocity.

Hail

Hail storms can contribute significantly to erosion potential, degradation of soil structure, and increased surface runoff (especially in northern latitudes and tropical highlands). Because of its larger size and higher terminal velocity, hail can substantially damage the soil structure and infilterability. The diameter of most hailstones is less than 5 mm, although some may exceed 20 mm. The average hailstorm energy ranges from 0.06 to 2.0 Newton-cm/cm^2.

Snow

Snowmelt causes considerable runoff in the 'snow belt' of the northern latitudes. For example, in the US Pacific Northwest, it is estimated that up to 90 percent of the erosion on steeply-rolling wheatland is caused by runoff associated with surface thaws and snowmelt (Wischmeier and Smith, 1978). Peak sediment loads in the early spring generally are caused by the snowmelt. In addition to these runoff effects, the thermal stresses that accompany snowfall can alter the surface structure of soils by repeated freezing and thawing, increasing soil wetness, of the soil moisture profile. Runoff and sediment transport associated with snowmelt, thaw, or light rain on frozen soil can be approximately 1.5-times greater than that of normal precipitation received during the winter months.

Precipitation as an atmospheric nonpoint source

Air pollution, *via* dry and wet deposition from the atmosphere, provides a primary component of nonpoint source water pollution. Contaminants transported by the atmosphere are carried by surface run-off, and deposited directly onto water surfaces. Precipitation is the transport mechanism of 'wet' deposition. This is the source of the largest part of the total atmospheric inputs for many constituents. To illustrate this effect as documented in the scientific literature, the concentration ranges of selected substances in rainfall are summarized in Table 4.1 (Jolánkai, 1983). A further comparison of the importance of atmospheric sources of cadmium, one of the more dangerous pollutants, is given in Table 4.2 which illustrates the primary nonpoint source inputs to the Rhine River drainage basin (Jolánkai, 1990).

Table 4.1 Concentrations and loading rates of selected chemical parameters in rainfall (after Jolánkai, 1983)

Parameter	Concentration (mg/l)	Areal loading rate (kg/ha/year)	Remarks
NH_4-N	0.6–2.7	0.7–15.0	
NO_3-N	0.1–1.2	0.9–5.0	
Total-N	1.0–7.0	1.2–28	
PO_4-P	0.0–0.3	0.0–2.0	
Total-P	0.0–0.5	—	
Total suspended solids	1.3–11[x]	100[x]	[x]only these data
COD	9.0–16[x]	124[x]	
SO_4	0.0–28	2.0–114	
Ca	0.2–20.0	1.0–52	
Cl	0.2–7.0	1.5–29	
K	0.1–6.0	1.4–12	
Na	0.1–2.5	0.4–35	
Mg	—	1.1–17	

Table 4.2 Ranges of cadmium inputs, from various primary nonpoint sources, to the Rhine River drainage basin, Germany (after Jolánkai, 1990)

Nonpoint source of cadmium	Input load (g/ha/year)
Atmospheric deposition	0.5–1.76
Sewage sludge application	0.05–1.57
Inorganic fertilizer application	1.1–4.0

Interception

There are several possible fates that might befall rain water reaching the land surface. A portion of the rainfall can be intercepted and/or retained by the vegetative canopy (trees, plants, crops) before reaching the ground. Interception loss represents the part of the precipitation retained by the aerial portion (leaves, stems, twigs) of this vegetation. It is either (1) absorbed by the vegetation, or (2) returned to the atmosphere *via* evaporation. The portion of the rainfall that reaches the ground by falling through the intervegetative spaces, or by dripping from the vegetation, to the ground is termed 'throughfall', while the portion of the intercepted water that reaches the ground by running down the stems is called 'stemflow'. Therefore, the net precipitation that actually reaches the ground (and contributes to surface runoff and other processes) is the sum of throughfall and stemflow.

Interception can vary widely with (1) species composition, (2) age and density of vegetation, (3) season of the year, and (4) distribution and intensity of rainfall. Thus, it is highly influenced by the successional stage of an ecosystem. As the successional stage of the vegetation in a given region approaches its climax community, the available growing space will be more completely utilized because of the greater variety of species. Thus, the vegetation will intercept more precipitation than that of a preclimax community. Coniferous trees in full leaf will intercept more rainfall than deciduous trees. Grasses and herbs also can intercept large quantities of rainfall. In fact, their leaf-area-to-ground-area ratio can approach that of forest vegetation. The characteristics of a storm can also significantly influence interception. For example, the interception of low-intensity rainfall (expressed as a percentage of total precipitation) is higher than for heavy rainstorms. The annual temporal distribution of the rainfall is also very important. If the annual precipitation consists of small storms, separated by periods of clear weather, the interception may be significantly increased, compared to fewer and longer-lasting rainfalls. This is due to the fact that interception essentially occurs during the first part of a rain storm, before the foliage becomes saturated with water.

Interception losses can be very significant. A rough estimate of annual interception losses in temperate climates ranges from 20 to 60 percent of the total precipitation. Examples of species- and crop-dependent interception losses are provided in Table 4.3. Unfortunately, because of the many complex ecologic, climatologic and hydrologic factors that contribute to interception losses, it is very difficult to accurately estimate such losses without conducting direct measurements under a variety of ambient conditions. Nevertheless, the available literature does contain some suggested equations for estimating interception losses, an example of which is the modified Horton equation (Meriam, 1960):

$$I = S_v (1 - \exp - P_o/S_v) + REt \qquad (4.9)$$

where: S_v = storage capacity of the vegetation over the area of concern;

        P_o = total precipitation over the area;

Table 4.3 Approximate values of seasonal net interception (modified from Ven Te Chow, 1964)

| | *Rain %* | | |
Description	*with full leaves*	*without leaves*	*Snow*
Northern hardwood	15	7	10
Aspen–birch	10	4	7
Spruce–spruce-fir	32		35
White pine	26		25
Hemlock	23		25
Red pine	29		30

| | *Development of vegetation %* | |
Crop	*high*	*low*
Alfalfa	35.8	21.9
Corn	15.5	3.4
Soybeans	14.6	9.1
Oats	6.9	3.1

R = ratio of the surface area of vegetation to the projected area;
E = evaporation rate from the vegetation surfaces; and
t = time

A significant problem with these types of interception equations, however, is that they include factors and parameters whose values are as difficult to determine as that of the interception loss itself. Examples in Equation 4.9 include the terms R, S and E.

In regard to the effects of interception on nonpoint source pollution processes, the following observations can be made:

(1)   To a considerable degree, interception affects the water available for land runoff and infiltration, thereby influencing the runoff of detached pollutants of all kinds;

(2)   Interception also captures the substances present in rainwater (i.e. atmospheric deposition), and, at the same time, can create a negative feedback effect. This is because air pollution can degrade vegetation which, in turn, can decrease interception, resulting in increased runoff and ultimately increased pollutant and sediment transport (erosion);

(3)   Interception also reduces the kinetic energy of through-falling rainwater, thereby decreasing detachment of sediment and pollutant particles.

To summarize, interception has both positive and negative effects on the generation and transport of nonpoint pollutants. Quantification of the resultant nonpoint source impacts is difficult, especially in larger drainage basins. The effect of interception, however, will be indirectly involved in calculations relating sediment and pollutant export rates to runoff, which is the usual method of quantifying nonpoint pollutant loads as emphasized in Chapter 8.

Evapotranspiration

On a global scale, about 70 percent of the annual precipitation onto the Earth's land surface is returned to the atmosphere *via* evaporation and transpiration (Raudkivi, 1979). As a practical matter, it is impossible to differentiate between evaporation and transpiration under field conditions. As a result, the two parameters usually are lumped together as evapotranspiration. Thus, the term evapotranspiration describes the combined effect of evaporation from wet surfaces and transpiration by plants. It represents that part of the precipitation returned to the atmosphere (in contrast to that portion going into runoff). Evapotranspiration is a dominant element of the hydrologic cycle, as well as in the watershed processes of primary concern in regard to the generation and transport of nonpoint source pollutants.

Evaporation rates depend on the availability of water and energy. They also depend on other factors, such as atmospheric vapour pressure, humidity, wind, vegetation, etc. The energy needed to evaporate water is obtained from solar radiation, which varies globally with latitude and season. The actual transport of moisture away from an evaporating surface, such as the surface of a lake, depends mostly on wind and its turbulence. Evaporation from the land surface depends largely on the soil–plant–water system, in part because the leaves, twigs and stems of plants (1) provide the evaporating surfaces, (2) shade the soil from solar radiation, and (3) usually define the effectiveness of wind in carrying away moisture from evaporating surfaces.

Transpiration represents the loss of water from a plant. It differs from evaporation in that it is (1) subject to the effects of the structural and functional features particular to a specific plant, and (2) strongly influenced by light. The rate at which moisture can be taken up by a plant depends on the root system, and the characteristics and moisture content of the soil. Yet, only a small fraction of the water needed by a plant is retained in the plant tissues. Most of it escapes from the leaves through the stomata and is transpired into the atmosphere (stomatal transpiration). At night, the stomata close up and very little moisture leaves the plant. Water can also evaporate from the moist membranes of plants through the cuticle (cuticular transpiration). When transpiration is occurring at a low rate, excess water also can be forced out of a plant through special organs called hydathodes (guttation).

Regardless of the specific mechanism, transpiration is a very important ecological process because it (1) transports nutrients and water to the upper part of a plant, and (2) cools the leaves. However, excessive transpiration can damage plants by desiccating the protoplasm below its critical minimum water content.

Because the transpiration rate depends on the evaporative power of the air, it is also determined in part by air temperature, wind, humidity saturation deficit, and the amount of light. The latter component partly controls the opening of the stomata and the availability of water in the leaf tissue, which, in turn, depends on soil-moisture availability.

As may be assumed, the amount of water involved in transpiration is the most important component of evapotranspiration during the growing season, while evaporation is responsible for most of the moisture loss during the dormant season. It is noted, however, that a distinction can be drawn between potential evapotranspiration and that which actually occurs. Potential evapotranspiration is the rate at which water is lost from the soil–plant–water system **assuming water is abundant and does not limit either evaporation or transpiration**. There are many methods for estimating potential and actual evapotranspiration. However, none have general applicability. Furthermore, experimental measurements (using lysimeters) are cumbersome, and their results are applicable only to the plant species and environmental conditions extant during the period of measurement. An example of an experimental expression for estimating evapotranspiration rates is that of Penman (1948):

$$U = (AH + 0.27\,E)/(A + 0.27) \tag{4.10}$$

where: U = daily evapotranspiration (mm);

   E = daily evaporation (mm) = $0.35\,(e_a - e_d)\,(1 + 0.0069\,w_2)$;

   H = daily heat budget (mm of water) =
     $R\,(1 - r)\,(0.18 + 0.558\,S) - B(0.56 - 0.092\,e_d^{0.5})\,(0.1 + 0.90\,S)$;

   A = slope of the saturated vapour pressure curve of air;

   e_a = saturation vapour pressure at mean air temperature (mmHg);

   e_d = actual vapour pressure at mean dew point (mmHg);

   w_2 = mean wind velocity at 2 m above ground (km/day);

   S = estimated ratio of actual duration of sunshine to maximum possible duration of sunshine;

   r = estimated percentage of reflecting surface; and

   R = mean monthly extraterrestrial radiation (mm of water evaporated/day)

In spite of the fact that evapotranspiration can be experimentally determined or estimated from the parameters and factors given in Equation 4.10, the Penman equation contains no expressions that account for the plant species or other ecological processes that actually affect (if not dominate) the daily evapotranspiration.

Another set of expressions, developed for conditions in Hungary, does take the effects of vegetation into account:

$$ET_{ref} = 0.9\,(E - e)^{0.7}\,(1 + at)^{4.6}\ (mm/day) \tag{4.11}$$

$$ET_{act} = aET_{ref} \qquad (4.12)$$
$$a = (g + B)/(1 + B) \qquad (4.13)$$
$$g = (w - w_o)/(n - w_o) \qquad (4.14)$$

where: ET_{ref} = reference evapotranspiration;
ET_{act} = actual evapotranspiration;
E = saturation vapour pressure at temperature t (^oC);
e = daily average value of the actual vapour pressure;
a = 1/273;
B = vegetation constant (given in tabulated form);
w, w_o = incidental and minimum soil moisture content of the upper soil layer (percent by volume); and
n = porosity of soil

Nevertheless, the expected accuracy of the calculated value of ET_{act} is rather questionable because of the many variables involved, and the site-specific character of the derived formulae. Consequently, the usual method of determining evapotranspiration is to calculate its value as the difference between the measured data for the other terms (e.g. P, S and R) in the water balance (Equation 4.1).

Because evapotranspiration basically defines the fraction of the precipitation available for runoff, it eventually contributes to the definition of pollutant and sediment export rates as well. The more water that is evaporated and transpired, the less there is available for inducing runoff export. This, together with the interception role of the vegetation cover, suggests a fundamental role for flora in governing and controlling hydrologic processes. Obviously, this role also fundamentally influences associated nonpoint pollutant processes. If one also considers that vegetation forms a major barrier to surface and sub-surface fluxes of nutrients and other water contaminants (by physical detention and biochemical uptake processes), the role of vegetative cover becomes even more important. In fact, many nonpoint source pollutant control strategies are based on utilizing this property of the ecosystem (e.g. no-tillage cultivation methods; see Chapter 9).

Infiltration

The portion of precipitation that does not evaporate becomes runoff or infiltrates into the soil. Infiltrated water can enter the atmosphere again, *via* the transpiration process. Infiltration is a complex process affected by many factors, especially soil texture and structure, and vegetative cover. Thus, the infiltration capacity of the same soil can be significantly different in different ecosystems. Infiltration initially exhibits a high rate, decreases rapidly, then more slowly, until it approaches a constant rate (termed the 'infiltration capacity', f_c) in a period of a few hours. This sequence of events occurs because infiltration initially occurs at a high rate when the soil is dry; the water first fills up the spaces between soil particles. As soil colloids swell, the pore spaces become full, restricting the rate of entry of additional water to decline to a lower, but uniform, level.

Infiltration generally occurs at its maximum rate under the undisturbed canopy and on the floor of forests. The major factor influencing infiltration in this environment is the thickness of the forest floor litter which provides protection for the soil. In general, therefore, the greater the forest cover of a watershed, the higher the infiltration rate. As long as protective forest floor cover is maintained, silviculture activities (thinning, cutting, reforesting, etc.) generally have little effect on infiltration rate. However, tractor-logged silviculture techniques can greatly disturb this protective cover and reduce the permeability of forest soils, thereby decreasing infiltration and increasing runoff.

The root systems of plants (specifically, the soil cavities formed after the decay of the roots) strongly influence the soil infiltration capacity. For example, fibrous-rooted vegetation usually is much more effective in increasing infiltration than tap-rooted annual weeds. Large populations of burrowing animals, as well as insects and their larvae, can also significantly increase infiltration over considerable areas, primarily because their burrowing activities provide passages for water movement into the soil. Another major factor influencing infiltration is the degree to which the surface soil is compressed. For example, cultivated fields which have been compressed by farm vehicles can have much lower infiltration than nearby woodlands. Plant succession can also significantly affect infiltration capacities. For example, a succession from old pasture to Virginia pine stands resulted in the doubling of the infiltration capacity in one study.

Infiltration rates can be determined experimentally with instruments called infiltrometers. There are two basic types: (1) the flooding type, and (2) the rainfall simulator type. Other methods applicable to smaller, homogeneous watersheds are based on the analysis of rainfall and runoff records (hydrograph analysis). These latter methods include the detention-flow relationship method, and the time-condensation method (Ven Te Chow, 1964). However, both the instrumental and analytical techniques provide only estimates of infiltration. Therefore, the results of such techniques should be used with caution when extrapolating infiltration rates to larger, nonhomogeneous watersheds. The elaboration of standard infiltration rate curves for various soil–land use complexes may be useful in attempting to overcome this type of problem.

The complexity of the infiltration process excludes exact solutions. Of the semi-empirical methods for calculating infiltration, the Horton equation is probably is the most widely-used method (Horton, 1945):

$$f = f_c + (f_0 - f_c)\, e^{-kt} \tag{4.15}$$

where: f = infiltration rate;

f_0 = initial infiltration rate;

f_c = infiltration capacity;

k = infiltration rate coefficient; and

t = time

In an elaboration of this relationship, Holtan's (1971) equation allows one to take the vegetation into account in calculating infiltration. This equation relates infiltration to the depletion of the water storage capacity of the soil:

$$f = a F_p + f_c \tag{4.16}$$

where: $F_p = S - F$;

S = water storage capacity of the soil at the beginning of the rainfall (mm of water) = difference in water content at saturation and at permanent wilting point); and

F = accumulated infiltration

Estimates of the value of the vegetative parameter 'a' are presented in Table 4.4.

It is also noted that infiltration is the phase of the hydrologic cycle which determines the rate at which contaminants present on the soil surface or in the topsoil (e.g. fertilizers, manures, sewage sludges, atmospheric deposits, etc.) are leached into the ground, and travel toward the groundwater table, or move with the interflow. On the other hand, precipitation that does reach the land surface, and does not infiltrate into the ground, is available as surface runoff. Thus, infiltration, by influencing the quantity of surface runoff (as well as sub-surface runoff), plays a dominant role in subsequent water-transport processes.

The infiltration rate is also a deterministic factor. It defines the time available for physical, chemical and biochemical processes that govern the composition of both leachate water and surface runoff water (e.g. adsorption–desorption processes, plant uptake, etc.). Thus, the period of time during which water is in contact with natural and human-introduced substances present in the upper soil layer is defined by the infiltration rate.

Runoff

Surface runoff

The water available for surface runoff is the portion of the precipitation which (1) is not lost *via* evapotranspiration, and (2) does not penetrate the surface soil layer to infiltrate into groundwater. Like other components of the hydrologic

Table 4.4 Estimates of the vegetative factor 'a' of the Holtan equation (modified from Raudkivi, 1979)

Land use or cover	*Condition and rating of soil surface*			
Fallow	after row crop	0.1	after sod	0.3
Row crops	poor	0.1	good	0.2
Small grains	„	0.2	„	0.3
Hay (legumes)	„	0.2	„	0.4
Pasture (bunch grass)	„	0.2	„	0.4
Hay (sod)	„	0.4	„	0.6
Temporary pasture (sod)	„	0.4	„	0.6
Permanent pasture (sod)	„	0.8	„	1.0
Wood and forests	„	0.8	„	1.0

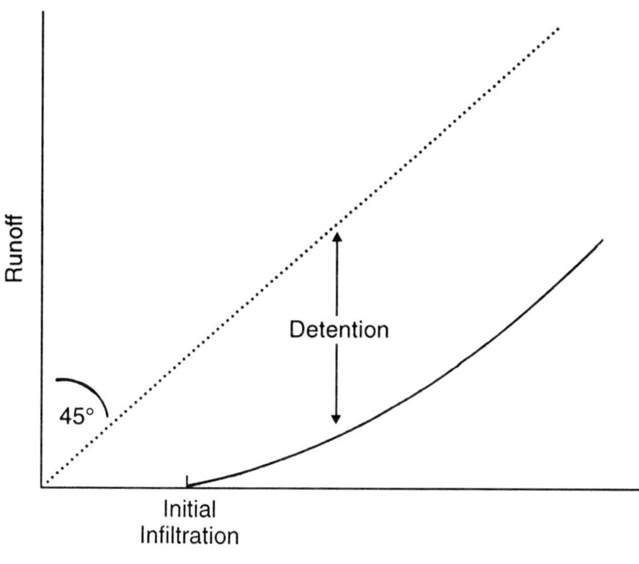

Figure 4.4 Runoff response to a precipitation event

cycle, surface runoff involves complex processes of storage and flow in the interface between soil and water (i.e. the soil–plant–water–air system).

Water traveling down a slope, over the land surface, is termed 'overland flow', or 'sheet flow', until it moves onto a flat portion of the land surface. However, completely flat surfaces seldom exist in nature. Rather, the land surface consists of a micro-pattern of small depressions (recesses and hollows) which result in a considerable water storage effect, termed 'depression storage'. Thus, overland flow (or land runoff) actually only occurs when the amount of precipitation available on the land surface exceeds (1) the infiltration capacity, and (2) the depression storage capacity.

Runoff typically follows the response to precipitation illustrated in Figure 4.4. In the case of an unprotected, perfectly-impervious surface, runoff would follow the 45-degree dashed line in Figure 4.4; all of the precipitation would run off the catchment surface. In reality, however, runoff often exhibits the pattern shown by the solid curve. The early portions of precipitation are captured by depression storage, and initial infiltration. As the rainfall event proceeds, precipitation begins to run off. With subsequent precipitation, the slope of the runoff function increases as the water saturates the soil and fills the depression storage. The simplest expression of this relationship between precipitation and runoff is the runoff coefficient (the ratio of direct runoff, RD, to precipitation, P):

$$a = (RD)/P \text{ when } 0 < a < 1 \qquad (4.17)$$

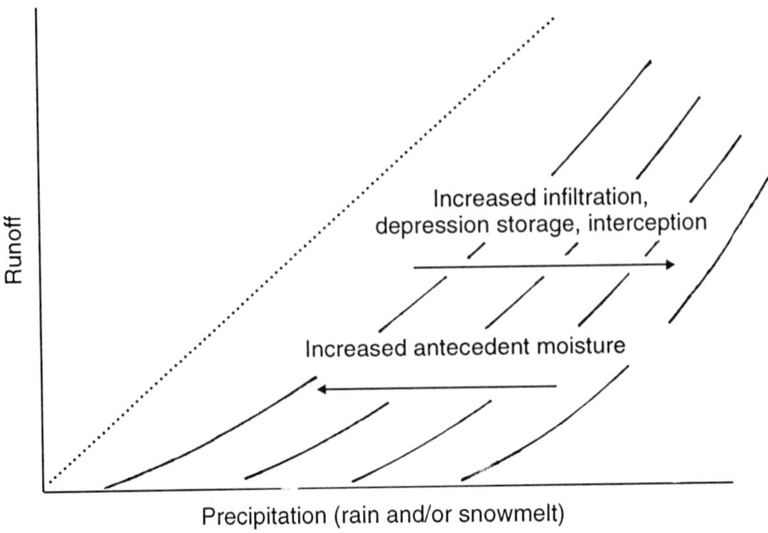

Figure 4.5   Effects of infiltration and antecedent moisture on runoff

RD is calculated from the area below the runoff curve in a hydrograph measured in the receiving water stream; the area corresponding to the base flow is subtracted from the runoff curve. The value of *a* depends on many factors. Some of the most important are rainfall and snowmelt intensity and duration, antecedent soil moisture, the slope of the terrain, and the vegetative cover of the land surface. Figure 4.5 illustrates the effects of land usage, antecedent soil moisture, and the time of precipitation on the value of the runoff coefficient.

The quantities of runoff and infiltration resulting from a precipitation event (rain, snowmelt, irrigation) are also strongly influenced by the wetness (antecedent moisture) of the catchment just prior to the event. If the catchment is very dry, most of the water input may be detained by interception, depression storage and dry soil. Thus, little percolation or runoff will occur. In contrast, for a wet catchment, there is little room for additional moisture in any of these detention mechanisms. Therefore, most of the precipitation water can run off and/or infiltrate.

Antecedent moisture depends on a catchment's recent history of precipitation and evapotranspiration. It is similar to a reservoir volume in its mediating effect; that is, if a large storm occurs when the reservoir is full, a flood can result. Alternatively, if the reservoir is empty, the stormwater is stored for later use. A major effect of antecedent moisture is that the very largest precipitation events do not necessarily produce the most nonpoint source pollution. In fact, a sequence of moderate precipitation events often can result in the generation and transport of more pollution than a single large event.

Table 4.5 Guideline values of the peak runoff coefficient a_p (after Dyck, 1980)

| | *Type of soil* | | | | | |
| | Clay | | Loam | | Sand | |
Topography	*fallow*	*forest*	*fallow*	*forest*	*fallow*	*forest*
Flat land	0.5	0.4	0.4	0.3	0.2	0.1
Hilly land	0.6	0.5	0.5	0.4	0.3	0.2
Mountainous land	0.7	0.6	0.6	0.5	0.4	0.3

Another important measure of water runoff is its peak discharge in the stream channel. The ratio of peak discharge (HQ) to the intensity of precipitation (PI), both expressed in units of depth per time (e.g. mm/h), is termed the 'peak-runoff coefficient' (a_p):

$$a_p = (HQ)/PI \tag{4.18}$$

Guideline values of both a and a_p can be found in tabulated form in hydrologic handbooks, allowing estimation of the expected runoff quantity or rate. Examples of some peak runoff coefficients are given in Table 4.5.

Precipitation which does not runoff is detained in the catchment. This detention represents the difference between precipitation (maximum potential runoff) and actual runoff (Figure 4.6). Several factors influence catchment detention and, hence, the shape of the precipitation–runoff function. For example, increased antecedent moisture shifts the function to the left, producing greater quantities of runoff. Conversely, increases in catchment infiltration, depression storage, and interception reduce runoff. Control of runoff involves management of all three factors. Another method of expressing the effects of various factors

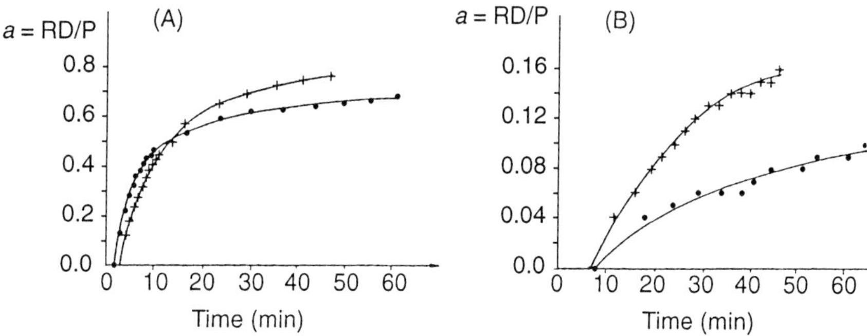

Figure 4.6 Experimental relationship between the runoff coefficient and time of precipitation for (A) ploughed land and (B) forest lots; '+' corresponds to an antecedent precipitation index (API) of 45 mm and '•' corresponds to an API of 30 mm (after Dyck, 1980)

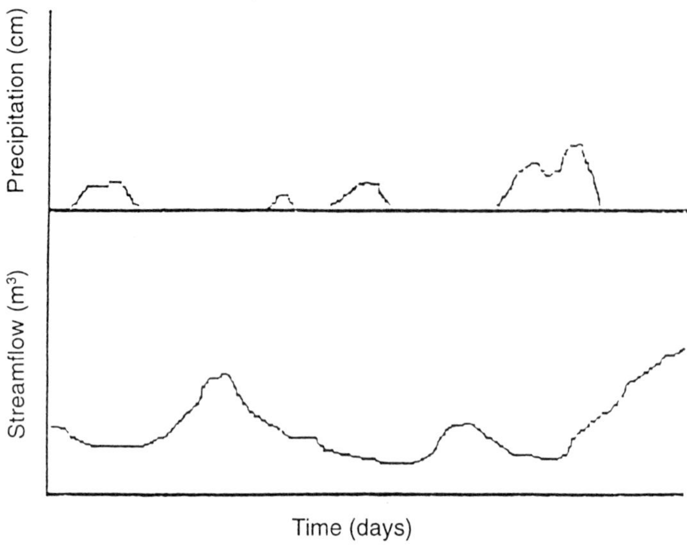

Figure 4.7   Streamflow response to precipitation

(e.g. rainfall intensity, antecedent soil moisture, etc.) on the magnitude of runoff is presented in Figure 4.7. For example, the higher the rainfall intensity, and the higher the antecedent (initial) soil moisture, the larger the sum of surface and sub-surface runoff.

Sub-surface water flow also can play an important role in transferring contaminants to both surface and sub-surface waters. In the generation of sub-surface flows, infiltration and subsequent water percolation play the roles.

Percolation

Percolation is the vertical drainage of soil water by gravity. Soil consists of solid particles with pore spaces of various sizes between them. Infiltrating water is held in pores by (1) capillary forces or (2) negative pore pressures inversely related to pore size. Water from the largest soil pores drains readily as pore pressures are weak. The stronger capillary forces of smaller pores resist gravity and, hence, these pores drain slowly, if at all. However, much of this water is available to plants and can be depleted by evapotranspiration. Under these conditions, soil moisture will remain only in the very smallest pores.

Although the physics of soil water retention and movement are complex, the processes can be approximated by the conceptual model illustrated in Figure 4.8, which relates soil moisture to the three approximate measures of (1) saturation, (2) field capacity, and (3) wilting point. At saturation, all soil pores are filled with water. Percolation will occur as long as the soil is above the field capacity. At the wilting point, soil water is bound so tightly that it cannot be

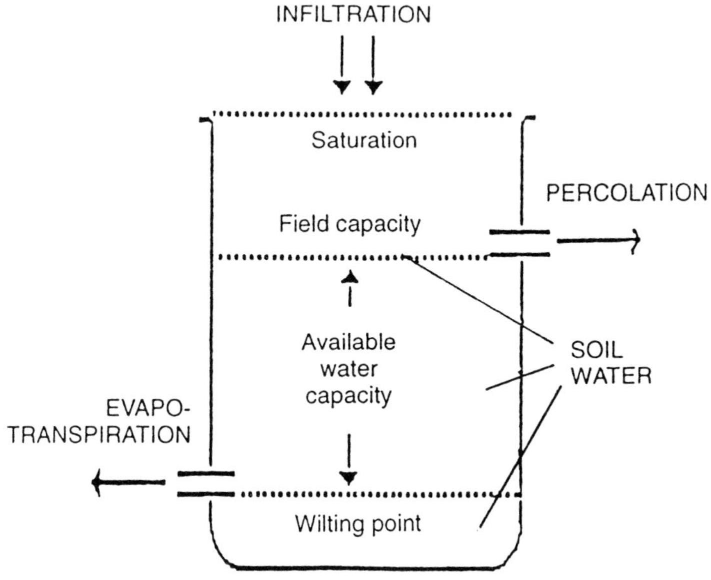

Figure 4.8   Model of soil moisture processes

removed by evapotranspiration. A fourth parameter, the available water capacity (the difference between field capacity and wilting point), measures the maximum water available for plant growth. These soil moisture parameters are primarily influenced by soil texture (Table 4.6). However, they are also affected by tillage and organic matter content. The water content of soil is a direct measure of the catchment antecedent moisture. (Following the earlier analogy with reservoirs, saturation can be considered the reservoir capacity, the field capacity the reservoir

Table 4.6   Soil moisture parameters (cm/cm) as a function of texture

Texture	*Saturation*	*Field capacity*	*Wilting point*	*Available water capacity*
Sand	0.44	0.09	0.03	0.06
Loamy sand	0.44	0.13	0.06	0.07
Sandy loam	0.45	0.21	0.10	0.11
Loam	0.46	0.27	0.12	0.15
Silt loam	0.50	0.33	0.13	0.20
Sandy clay loam	0.40	0.26	0.15	0.11
Clay loam	0.46	0.32	0.20	0.12
Silty clay loam	0.47	0.37	0.21	0.16
Sandy clay	0.43	0.34	0.24	0.10
Silty clay	0.48	0.39	0.25	0.14
Clay	0.48	0.40	0.27	0.13

volume at the spillway level, and the wilting point the volume at a water supply intake.)

Percolation reflects seasonal infiltration and evapotranspiration. During wet and cool seasons, percolation is frequent, in response to elevated moisture levels produced by low evapotranspiration and/or high infiltration. Percolation is generally much less significant during the rest of the year, primarily because infiltration must replenish the moisture depleted by evapotranspiration.

The ultimate disposition of percolation water depends on the nature of the soil zone saturated with water. In humid areas with hilly topography or subsoil strata of low permeability, the shallow saturated zone may flow toward surface waters, converting a potential groundwater pollution problem (if the percolating water is contaminated with salts or other pollutants) to a surface water problem. Comparable problems are created when pipe or tile drains are installed (primarily) in agricultural areas to lower the groundwater table. In these situations, water drainage to surface channels is fairly rapid. In other situations, polluted percolation waters eventually reach the deep saturated zones, where they remain until withdrawn *via* water supply or other wells. Thus, both surface and sub-surface flows can be considered the final response of a catchment to all natural hydrologic and anthropogenic inputs, both in terms of its quantity and its solid and dissolved constituents. For example, substantial quantities of percolation often are produced in irrigated areas, where control of salinity requires sufficient water applications to leach accumulated salts from the soils. Descriptions of other sources of nonpoint contaminants, and the ways and means by which these substances get into runoff waters are discussed elsewhere (see Chapters 5 and 6), while some of the hydrologically-related factors that affect these processes are discussed below.

NATURAL FACTORS AFFECTING THE GENERATION, TRANSPORT AND TRANSFORMATION OF NONPOINT SOURCE POLLUTANTS

Geology

Under natural conditions (undisturbed by humans), the chemical composition and quantity of runoff water is determined, to a large extent, by the geology of the drainage basin. The outer crust of the Earth, where it is exposed above the level of the oceans, exerts a direct deterministic influence on the composition of natural terrestrial waters. Table 4.7 shows the average composition of igneous and sedimentary rocks. Because the distribution of these rocks varies widely, the chemical composition of both surface runoff and groundwaters is highly variable. As a result, global average water quality values, although widely quoted, have little real significance, except, perhaps, as a basis for comparison. Some of these global average values of the chemical composition of river waters are given in Table 4.8.

Table 4.7 Average composition of igneous and sedimentary rocks

Element	Igneous rocks	Sedimentary rocks		
		Resistates (sandstone)	Hydrolyzates (shale)	Precipitates (carbonates)
Si	285,000	359,000	260,000	34
Al	79,500	32,100	80,100	8,970
Fe	42,200	18,600	38,800	8,190
Ca	36,200	22,400	22,500	272,000
Na	28,100	3,870	4,850	393
K	25,700	13,200	24,900	2,390
Mg	17,600	8,100	16,400	45,300
Ti	4,830	1,950	4,440	377
P	1,100	539	733	281
Mn	937	392	575	842
F	715	220	560	112
Ba	595	193	250	30
S	410	945	1,850	4,550
Sr	368	28	290	617
C	320	13,800	15,300	113,500
Cl	305	15	170	305
Cr	198	120	423	7.1
Rb	166	197	243	46
Zr	160	204	142	18
V	149	20	101	13
Ce	130	55	45	11
Cu	97	15	45	4.4
Ni	94	2.6	29	13
Zn	80	16	130	16
Nd	56	24	18	8.0
La	48	19	28	9.4
N	46		600	
Y	41	16	20	15
Li	32	15	46	5.2
Co	23	.33	8.1	.12
Nb	20	.096	20	.44
Ga	18	5.9	23	2.7
Pr	17	7.0	5.5	1.3
Pb	16	14	80	16
Sm	16	6.6	5.0	1.1
Sc	15	.73	10	.68
Th	11	3.9	13	.20
Gd	9.9	4.4	4.1	.77
Dy	9.8	3.1	4.2	.53
B	7.5	90	194	16
Yb	4.8	1.6	1.6	.20
Cs	4.3	2.2	6.2	.77
Hf	3.9	3.0	3.1	.23
Be	3.6	.26	2.1	.18
Er	3.6	.88	1.8	.45

Table 4.7 Continued

Element	Igneous rocks	Sedimentary rocks		
		Resistates (sandstone)	Hydrolyzates (shale)	Precipitates (carbonates)
U	2.8	1.0	4.5	2.2
Sn	2.5	.12	4.1	.17
Ho	2.4	1.1	.82	.18
Br	2.4	1.0	4.3	6.6
Eu	2.3	.94	1.1	.19
Ta	2.0	.10	3.5	.10
Tb	1.8	.74	.54	.14
As	1.8	1.0	9.0	1.8
W	1.4	1.6	1.9	.56
Ge	1.4	.88	1.3	.036
Mo	1.2	.50	4.2	.75
Lu	1.1	.30	.28	.11
Tl	1.1	1.5	1.6	.065
Tm	.94	.30	.29	.075
Sb	.51	.014	.81	.20
I	.45	4.4	3.8	1.6
Hg	.33	.057	.27	.046
Cd	.19	.020	.18	.048
In	.19	.13	.22	.068
Ag	.15	.12	.27	.19
Se	.050	.52	.60	.32
Au	.0036	.0046	.0034	.0018

Geology also defines the nature of the soils in a watershed which, in turn, have a pronounced effect on the surface vegetation, also a function of hydrologic and climatic factors. All these factors can affect the erosion and transport of sediment (and sediment-associated pollutants).

Physiography

The physical features of the landscape (e.g. slope of the terrain, shape and geometric dimension of the drainage basin) also substantially influence the magnitude and dynamics of land runoff. Thus, they also influence nonpoint source pollution. Generally, the steeper the slopes, and the shorter the pathways to the nearest receiving water stream, the higher the potential of runoff waters to transport dissolved and sediment-associated pollutants from a primary source area to a downstream area. Further, the more dense the drainage channel network, the higher the rate of pollutant 'delivery' (i.e. the higher the fraction of initially displaced contaminating substances that will reach a downstream point). Thus, the delivery ratio depends on the total length of the drainage pathways. This

Table 4.8 Average composition of world river waters

Constituent	1		2		3		4	
	July 16, 1963		*Oct. 1, 1964 – Sept. 30, 1995*					
	mg/l	meq/l	mg/l	meq/l	mg/l	meq/l	mg/l	meq/l
Silica (SiO$_2$)	7.0		7.9		13		10.4	
Aluminium (Al)	.07							
Iron (Fe)	.06		.02					
Calcium (Ca)	4.3	.215	38	1.896	15	.749	13.4	.669
Magnesium (Mg)	1.1	.091	10	.823	4.1	.337	3.35	.276
Sodium (Na)	1.8	.078	20	.870	6.3	.274	5.15	.224
Potassium (K)			2.9	.074	2.3	.059	1.3	.033
Bicarbonate (HCO$_3$)	19	.311	113	1.852	58	.951	52	.852
Sulfate (SO$_4$)	3.0	.062	51	1.062	11	.239	8.25	.172
Chloride (Cl)	1.9	.054	24	.677	7.8	.220	5.75	.162
Fluoride (F)	.2	.011	.3	.016				
Nitrate (NO$_3$)	.1	.002	2.4	.039	1	.017		
Dissolved solids	28		232		89		73.2	
Hardness as CaCO$_3$	15		138		54		47	
Noncarbonate	0		45		7		5	
Specific conductance (micromhos at 25°C)	40		371					
pH	6.5		7.4					
Colour			10					
Dissolved oxygen	5.8							
Temperature (°C)	28.4							

1. Amazon at Obidos, Brazil. Discharge, 216,000 m³/s (7,640,000 cfs) (high stage)
2. Mississippi at Luling Ferry, LA, USA (17 miles west of New Orleans). Time-weighted mean of daily samples
3, 4. Mean composition of river water of the world (estimated). Dissolved solids computed as sum of solute concentrations, with HCO$_3$ converted to equivalent amount of CO$_3$

means it also depends on the size of the drainage basin; the larger the watershed area, the smaller the delivery ratio (Figure 4.9).

The slope of the terrain affects the velocities of surface runoff and groundwater flows (interflows). Surface runoff flow velocities can be calculated using Manning's equation (Equation 4.8). In contrast, sub-surface flow velocities are defined by the gradient (difference) of water pressures according to Darcy's Law. The water pressure gradient, in turn, is influenced by the slope of the terrain and of the underground formations. Thus, the slope of the terrain determines the energy of the water flows and, hence, their flow velocities. In turn, these determine the rates at which dissolved and suspended substances are transported by water. The sediment transport rate is related to the slope by many experimental and semi-experimental formulae, of which the Wischmeier and Smith (1978) Universal Soil Loss Equation is the most widely used (see Chapter 8). Other formulae of interest are those of Moeyerson and DePloey (1976) and Bryan (1976):

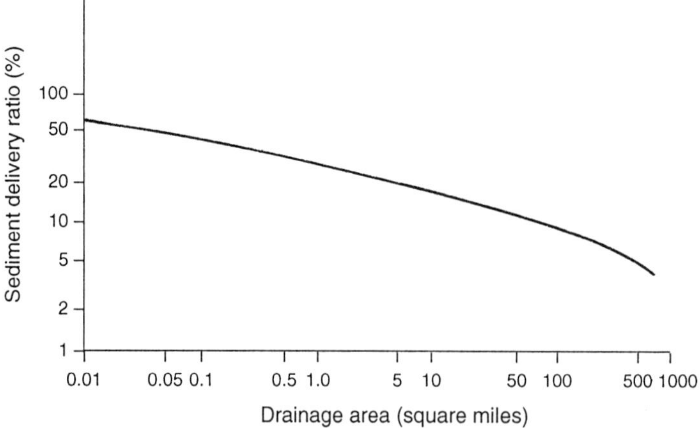

Figure 4.9 A generalized relationship between drainage area and sediment delivery ratio based on the findings of several US workers (after Novotny and Chesters, 1989)

$$A_R = a\,(E)^b\,(S)^c \qquad\qquad (4.19)$$
where: A_R = sediment transport;
   E = kinetic energy of rain; and
   S = slope gradient

Morgan (1978) also reported an empirical expression for relating sediment transport to the kinetic energy of rainfall and slope gradient:

$$A_R = E^{0.84} + S^{2.3} - 4.5 \qquad\qquad (4.20)$$

In addition to these topographic factors, the generation of nonpoint source pollutants in surface and sub-surface flows also depends on the physical, chemical and biological properties of the soil. For example, the susceptibility of soil to detachment and transport by rainfall, snowmelt and/or overland flow is related to (1) soil particle size distribution, (2) soil organic matter content, (3) degree and strength of soil aggregates, (4) surface area and clay mineral content, and (5) water retention characteristics. The most easily-detached soil particles are those in the size range of 238 to 1,041 µm. These soil aggregates are bound together mainly by a colloidal complex consisting of clay and soil organic matter. Thus, sediment detachment or availability is positively related to rainfall intensity and kinetic energy, and negatively related to clay content (Bubenzer and Jones, 1971):

$$A = 7.50\,(I)^{0.40}(E)^{1.14}(Cl)^{-0.52} \qquad\qquad (4.21)$$
where: A = sediment detachment;
   I = rainfall intensity;
   E = rainfall kinetic energy; and
   Cl = clay content (percent)

In addition to the size distribution of the primary particles, sediment detachment and transport also depends on the resistance of soil aggregates to dispersion and entrainment in overland flow. The resistance is a function of (1) the forces of cohesion and adhesion binding the particles together, (2) the rate of wetting, and (3) the effects of entrapped air, antecedent soil moisture potential and heat of wetting.

Another hydrologically-related physiographic factor influencing the temporal impacts of nonpoint source pollution is delayed catchment response. Simply stated, nonpoint source pollution caused by a specific rainfall or snowfall event may not become apparent until days, months or even years after the event. Such delayed reactions to precipitation events are most often associated with snowmelt, groundwater and sediment transport. Snowmelt often is the major source of annual runoff in colder climates (Canada, northern Europe, northern Asia, northern United States). Precipitation in the form of snow produces no runoff until the temperature is sufficiently warm to melt the snow. Thus, the runoff response to snow may occur many days or weeks after the snowfall event. In fact, the melting of an accumulated snowpack frequently will provide much more water than the largest rainfalls of warmer seasons. Moreover, since antecedent moisture conditions usually are high during snowmelt periods, a very large portion of the melt will occur as runoff and/or percolation (although this latter may be restricted if the ground beneath the snowpack is still frozen).

In humid areas, groundwater discharge to streams can account for over 80 percent of the total streamflow. Because water moves more slowly in the soil than on the surface, streamflows in such regions also can exhibit delayed responses to rainfalls or snowmelt (Figure 4.10). The time lag between the precipitation event and its subsequent streamflow response complicates the assessment of nonpoint source pollution. For example, contaminants from urban source areas close to stream channels can be washed into a stream before the stream flows have increased sufficiently in response to the precipitation to provide significant dilution of the contaminants – leading to the oft-cited 'first flush' phenomenon – while the slower rate of groundwater movement can impose a delay of several years before dissolved chemicals leached by percolation events reach water supply wells.

In contrast, sediment transport often exhibits a lagged reaction to precipitation and subsequent erosion. Intensive rain storms are major initiating mechanisms for soil erosion because of their ability to pulverize and dislodge soil particles. However, when antecedent moisture is low, the runoff arising from the storm may be insufficient to transport the eroded soil from the site. Later storms subsequently may provide the necessary runoff to move the soil (and associated pollutants) to a stream channel. The time lag between erosion and sediment transport, in such cases, can be many months. For example, in the eastern United States, most soil erosion is due to late spring and summer rains falling on unprotected soils. However, the actual transport of the sediment to streams is often delayed until the occurrence of higher runoff events during the following late winter and early spring.

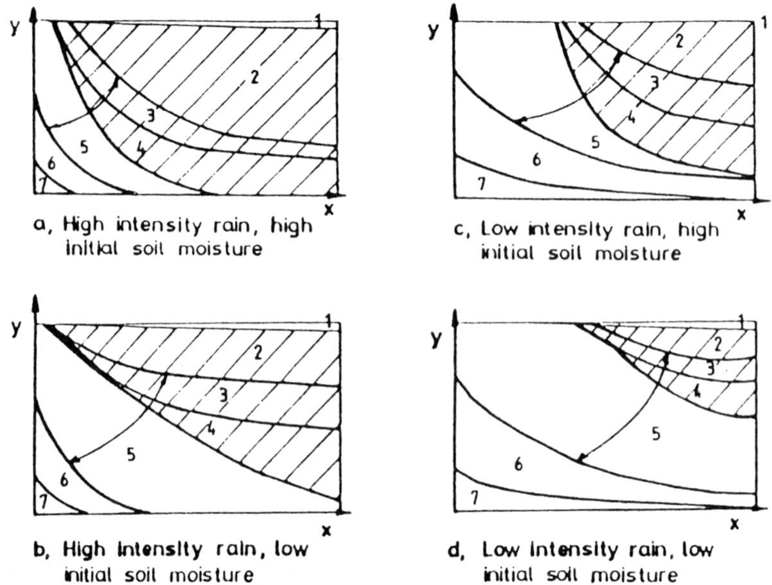

Figure 4.10 Effects of rainfall intensity and initial soil moisture on the components of runoff generation; the x-axis is the time after the beginning of the rainfall and the y-axis is the percentage of the rainfall onto a unit drainage basin area during the unit period of time, where 100% is comprised of the sum of (1) precipitation onto water surfaces, (2) surface runoff, (3) interflow, (4) groundwater penetration or recharge, (5) soil moisture storage, (6) depression storage, and (7) interception (after Dyck, 1980)

HUMAN-INDUCED FACTORS AFFECTING THE HYDROLOGIC CYCLE, AND THE RESULTANT TRANSPORT AND TRANSFORMATION OF NONPOINT SOURCE POLLUTANTS

Human modifications of natural hydrologic processes are the direct causes of much nonpoint source pollution of aquatic ecosystems. Major human-influenced alterations of hydrologic elements include the following:

(1)   Precipitation patterns being changed on a global scale by emissions of carbon dioxide and other pollutants, which, when discharged to the atmosphere, alter elements of the hydrologic cycle (i.e. climate change). Indirect effects, such as alteration of interception and evapotranspiration by changes to vegetation (e.g. acid rain), also can occur;

(2)   Other human activities and interventions which change nonpoint source runoff patterns including:

(a)   Engineered/technical interventions in stream channels (e.g. river channel alterations, damming streams, etc.), and in the drainage basin (e.g. sub-soil drainage, reservoir construction, channel regulation);

(b)   Modification of soil coverage (vegetation) by agricultural activities,

70

including soil-structure modifications caused by cultivation machinery and chemical dosage;

(c) Creation of relatively large areas of impermeable surface areas (e.g. asphalt and concrete pavements, buildings);

(d) Forestation which can deplete groundwater levels *via* increased evapotranspiration; and deforestation which is one of the main causes of increased erosion and sediment transport;

(e) Artificially-changed soil moisture and runoff conditions caused by irrigation and drainage;

(f) Changes in drainage pathways, due to construction of transportation corridors, irrigation and navigation canals, hydropower diversion tunnels, flood levees, etc.;

(3) Substantial changes in ecosystem structure resulting from the many direct and indirect human influences (e.g. desertification) that locally, or even globally, alter elements of the hydrologic cycle.

Human impacts, which are mediated primarily through hydrological processes, and which contribute to or substantially alter the generation, transport and transformation of nonpoint pollutants are discussed in Chapter 6.

CHAPTER 5

TYPES OF AQUATIC POLLUTANTS, IMPACTS ON WATER QUALITY AND DETERMINATION OF CRITICAL LEVELS

S.-O. Ryding and J.A. Thornton

INTRODUCTION

In Chapter 1, pollution was defined as the contamination of soil, water, or the atmosphere by the discharge of noxious substances. Thus, a pollutant can be defined as something which corrupts or contaminates, especially harmful chemicals or waste materials discharged into receiving waters or the atmosphere (Guralnik, 1970). This means that virtually any unwanted material can become a pollutant if its presence is deleterious to a component of the natural environment, aquatic ecosystems in this case.

This chapter identifies and discusses major classes of nonpoint source pollutants which can affect aquatic ecosystems, including (1) sediments, (2) nutrients (phosphorus, nitrogen), (3) biodegradables and organic matter (oxygen-consuming substances), (4) heavy metals and radioisotopes, (5) organic chemicals, (6) salts, (7) acids/bases, (8) microorganisms (bacteria), and (9) macro-pollutants (litter, organic debris). Appendix 5.1 provides a summary of those contaminants considered by the United States government to be priority pollutants in terms of potential human impact through drinking water supplies; this table gives some idea of the range and complexity of the possible contaminants to be considered in terms of nonpoint source runoff (van der Leeden *et al.*, 1990). Further details of specific nonpoint pollutant sources are given in Chapter 6.

As a preface, it is noted that pollutants usually are transferred to receiving waters (rivers, lakes, estuaries) as a function of the quantity and intensity of precipitation (see Chapter 4). Thus, the quantity of pollutant transferred to receiving waters can be limited either by (1) the availability of the material being transported, or (2) the availability of the medium in which the material is being transported. Limitation resulting from depletion of the material being transported is referred to as 'supply limitation'; limitation resulting from depletion of the transporting medium is referred to as 'transport limitation'. Transport limitation is more prevalent in regions of arid and semi-arid climates with highly seasonal rainfall, while supply limitation is typical of regions characterized by higher, more constant rainfall (Grobler, 1985; Gregory and Walling, 1985).

73

SEDIMENTS

Erosion

Sediments (as suspended particles in the water column) can be classified either as (1) inorganic or (2) organic (primarily decomposed vegetation and organic matter; see also biodegradables, below). The accumulation of sediment and organic matter in waterbodies over a geologic time scale represents a natural process, relating to the ontogeny of lakes through swamps and bogs to terrestrial ecosystems (Wetzel, 1975). However, this natural process can be greatly accelerated by human processes which lead to the erosion of terrestrial soils. In fact, in terms of both severity and frequency, soil erosion (primarily inorganic sediments) constitutes the most significant form of nonpoint source pollution affecting waterbodies on a global scale (also see Chapter 4). In addition to acting as a transport mechanism for nutrients and other sediment-associated pollutants, the influx of inorganic sediments to lakes and reservoirs leads to the related problem of increased turbidity in the water column, as well as to the siltation of lake basins.

The clearing of forests and rural areas, as well as urbanization and construction activities, can enhance the erosion of soil and its subsequent transport to receiving waters (Gregory and Walling, 1985). The transport of sediments is commonly affected by precipitation, being a function of rainfall intensity and volume, and the level of organic matter present in the soil structure (Lal and Stewart, 1994) (see Chapter 4). There is a direct relationship between rainfall intensity and the quantity of soil transported to receiving waters (FAO, 1965); the Universal Soil Loss Equation (USLE) provides a general form for this relationship, while Fournier's equation provides a practical means of estimating sediment discharges in a variety of catchments (see Chapters 4, 7 and 8). The USLE has been widely used by the US Department of Agriculture (USDA), Natural Resources Conservation Service (NRCS), to manage agricultural land use in the United States, and has been successfully applied in other temperate zone countries (Gregory and Walling, 1985). Versions of this model have been computerized (US EPA, 1987) and many are freely available from United States government agencies under the public domain laws of that country. However, extreme caution should be used when applying these relationships in other climatic regions (*cf.* Schmidt and Schulze, 1987).

Fournier's relationship contains some of the same elements as the USLE but has been validated using a global data base and contains a seasonality factor which makes the model more widely applicable than the USLE in unmodified form. Table 5.1 gives some indication of the magnitudes of erosional processes experienced in various parts of the world (see also Figure 5.1). These, and other land-use related processes, are discussed in detail in the following chapters.

The relationship between erosion and the organic content of soil is inverse; soils with little organic content tend to erode more readily than organic-rich soils. Thus, any mechanism which adds organic matter to the soil (e.g. mulching)

Table 5.1 Rates of soil loss (t/km^2/year) from lands under varying intensities of usage (after van der Leeden *et al.*, 1990; Lal and Stewart, 1994; Thornton, 1982)

Land use	Location	Climate zone	Sediment yield
Natural forest	Nigeria	Humid tropics	0+
	Ivory Coast	Humid tropics	2–45
	United States	Temperate	9–4633
Grassland agriculture	United States	Temperate	92
	West Africa	Humid tropics	300–9000
	United States	Temperate	1853
Surface mining	United States	Temperate	926–18 532
Urban construction	Zimbabwe	Semi-arid tropics	1–25
	United States	Temperate	18 532

represents a positive means of attempting to control soil erosion. However, once movement of sediment particles has been initiated, their behaviour is then primarily controlled by their settling velocities relative to the hydrodynamic regime (Krumbein and Sloss, 1963; Gregory and Walling, 1985). This relationship is summarized in terms of Hjulstrom's diagram (Figure 5.2).

Further, wind erosion represents another common method of transporting soils from the land surface to receiving waters. This latter problem is especially noteworthy in areas of unconsolidated and disturbed terrestrial sediments. For example, in a recent news release, the USDA (1991) reported that wind erosion in the Great Plains of the United States damages over 33 000 km^2 of cropland (90 percent of the total area) and rangeland (about 10 percent of the total area) annually. Much of this latter damage was ascribed to loss of vegetative ground cover due to drought.

Other consequences of soil erosion include (1) altered patterns of deposition and scour, (2) habitat destruction, and (3) altered flow patterns in streams (Davies, 1979). For example, because of enhanced sediment deposition in reservoir basins, the outflow waters from an impoundment can have a higher silt-carrying capacity than inflow waters to the impoundment. The result is that the impoundment outflow encourages downstream erosion, which can significantly affect riverine deltas and floodplains. This altered pattern of sediment scour and deposition can be catastrophic to the flora and fauna (including humans) in a drainage basin, which had previously depended on seasonal patterns of flooding and sediment deposition.

Large quantities of suspended particles (inorganic turbidity) can markedly reduce light penetration into the water column. In addition to decreasing the visual clarity (transparency) of the water column, this can decrease the euphotic zone of a lake, the water layer in which algae and aquatic plants typically grow. Theoretically, this can reduce the biological productivity of the waterbody. However, evidence from Mexico and southern Africa suggests the productivity profile in the waterbody may be merely compressed (Davalos *et al.*, 1989;

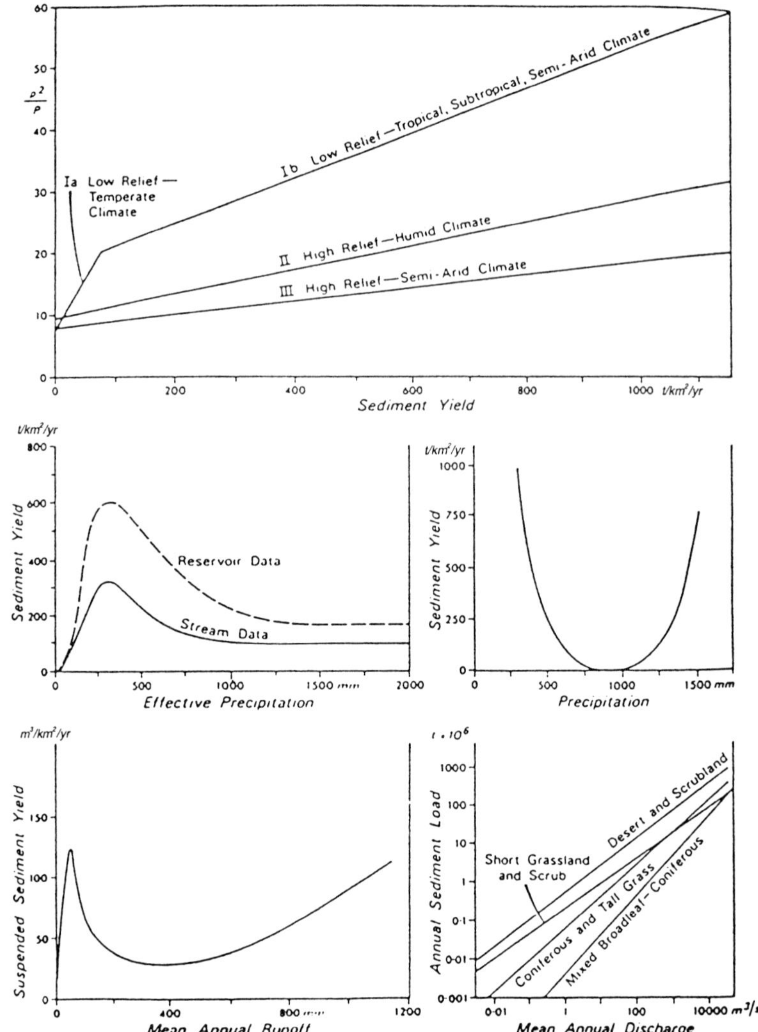

Figure 5.1 Fournier's analysis of sediment yields in various climatic zones (after Gregory and Walling, 1985)

Stegmann, 1982). Thus, the specific role of turbidity may lie more in the area of determining where in the water column productivity maxima occur, and, to a greater or lesser extent, in determining the algal class that may be selected under these conditions. Thornton and Rast (1993) have provided some evidence on a global scale that blue-green algae may be better able to compete in those tropical/sub-tropical systems where high turbidities occur; this finding is consistent with the work of Robarts and Zohary (1985), Robarts *et al.* (1985), Reynolds (1975) and others working on specific lakes within these regions.

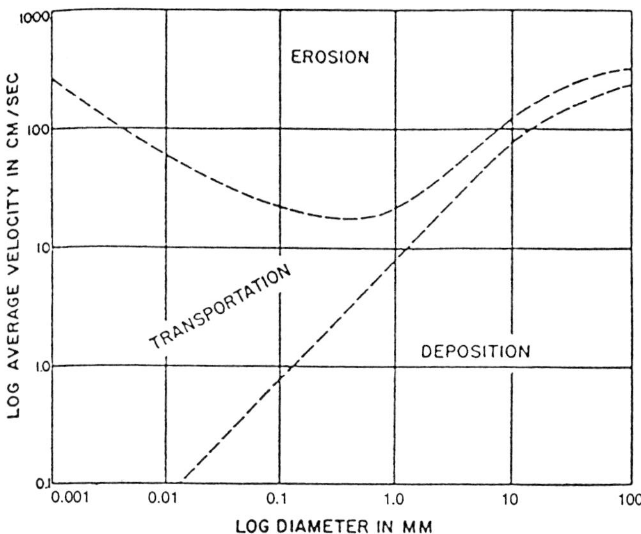

Figure 5.2 Hjulstrom's diagram of the relationship between sediment erosion, transport and deposition (after Krumbein and Sloss, 1963)

Suspended particles in the water column can also act as substrates for nutrients, metals and biocides. The effect can be either to reduce or enhance the biological availability of these materials to phytoplankton and other biological organisms in the waterbody. It has also been shown that co-flocculation of suspended particles and bacterial/phytoplankton cells can be an important mechanism for reducing the concentrations of both materials in aquatic ecosystems (Melack, 1985; Pedros-Alio and Brock, 1983; Britton and Marshall, 1980). Increased organic turbidity can also affect larger animals by reducing water-column visibility and by smothering benthic fauna and fish (Bruton, 1985).

In regard to human impacts, soil erosion has potentially severe repercussions for terrestrial agricultural production, particularly in developing countries. It also can lead to (1) reduced active lives for water-supply impoundments, (2) difficulties in water treatment, and (3) altered stream flows. The latter impact can result from sediment deposition which blocks culverts, etc., thereby increasing flood hazards. The meandering of streams as a result of altered deposition and scour patterns also can lead to property losses and losses of water uses. High concentrations of suspended inorganic particulates also can reduce the visually aesthetic quality of waterbodies used for recreational purposes (*cf.* Thornton *et al.*, 1989), and adversely affect the efficiency of spray irrigation systems, etc. Suspended solid concentrations of less than 10 mg/l are required, in temperate waters, for most beneficial uses of water; recreational use (swimming), waterfowl use, and coolant uses can tolerate slightly higher concentrations (50 mg/l; van der Leeden *et al.*, 1990). In semi-arid regions, a Secchi disc transparency of 1 m is considered acceptable for public recreational use of a waterbody and its classification as non-eutrophic (Thornton and McMillan, 1989; Thornton and Rast, 1989).

In-lake resuspension of bottom sediments

A special problem with shallow lakes and reservoirs is that in-lake turbidity (water transparency) can be aggravated by the resuspension of bottom sediments, independent of external sediment inputs. This resuspension can occur as winds stir up the water column (Krumbein and Sloss, 1963), and also as fish and other organisms stir up bottom sediments in shallow regions of a waterbody ('bioturbation'; Wetzel, 1975). Benthivorous (bottom feeding) fishes such as the common carp, *Cyprinus carpio*, commonly contribute to this process in natural and man-made waterbodies. Because nutrients ('internal loading') and other materials can be sorbed or otherwise associated with the bottom sediments, such resuspension can result in the addition of these materials to the water column at a higher rate than would be possible due to diffusive processes alone (Thornton, 1979a; Grobler and Davies, 1981). The internal loading process is described in more detail in Chapter 6.

NUTRIENTS

Nutrients, primarily nitrogen and phosphorus, can stimulate the growth of phytoplankton, periphyton and aquatic plants in waterbodies, often to levels (e.g. algal blooms) which interfere with many water uses (especially water supply and recreational uses). In natural systems (i.e. waterbodies relatively isolated from, and undisturbed by, human activities), nutrients are commonly derived from the weathering and leaching of nutrients from rocks and soils (see Chapter 4). However, nutrient inputs to aquatic ecosystems can be greatly accelerated by human activities, resulting in nutrient enrichment and its accompanying interferences with many water uses. This constitutes 'cultural' or 'anthropogenic' eutrophication, which is in contrast to the natural, ontogenic eutrophication or aging process observed in waterbodies (Wetzel, 1975; Ryding and Rast, 1989).

In the recent past, point sources (e.g. municipal wastewater treatment plant effluents) often constituted the major nutrient inputs to waterbodies. As a result, eutrophication control measures, including phosphorus removal from effluents, often were directed to these sources (*cf.* Alabaster, 1980; Fuggle and Rabie, 1983; Dejoux, 1988; Goldfarb, 1988; Tarlock, 1990; Schlickman *et al.*, 1994). However, it has become apparent that the point source controls established in many countries were not universally successful in eliminating cultural eutrophication of lakes and reservoirs (*cf.* Davis, 1989; Thornton and Day, 1989). Rather, a significant component of the nutrient budget of a waterbody has been found to arise from diffuse, or nonpoint, sources within a drainage basin.

Examples of the latter include the excessive application of natural and artificial fertilizers to agricultural lands, motor vehicle exhaust emissions and urbanization, and industrial processes. It has been found that both water and wind serve as transport mechanisms for nutrients, with aerial deposition accounting for ten percent and greater of the annual nutrient input to many waterbodies (Ryding and Rast, 1989). Table 5.2 summarizes some typical nutrient exports from varying types of land usages.

The consequences of cultural eutrophication have been described in detail by Ryding and Rast (1989), and include (1) the formation of algal scums, (2) excessive growths of algae and aquatic plants, (3) undesirable changes in the species composition of plant and animals species in and around a waterbody, (4) tastes and odours in drinking waters drawn from such waterbodies, (5) losses of recreational water uses, (6) increased treatment requirements prior to water use, and (7) deoxygenation of the bottom waters of lakes and reservoirs. Although enhanced biological growth is beneficial in some cases (especially for aquaculture), excessive productivity usually is detrimental to the productive, multi-purpose utilization of waterbodies. In addition, indirect consequences can arise from nutrient enrichment, including the spread of disease vectors and/or production of carcinogens. An example of the latter is the occurrence of trihalomethanes (THMs), produced by the chlorination of municipal wastewaters and their organic components, in waterbodies suffering from cultural eutrophication (van Steenderen *et al.*, 1987, 1988).

Eutrophication also can result in the accumulation of organic-rich sediments in waterbodies. In the same manner as siltation, the accumulation of decomposing

Table 5.2 Land use and nutrient export summarized from the literature; exports given in mg/m²/year (Thornton, 1986; van der Leeden *et al.*, 1990)

Land use	Location	P-export	N-export
Natural	Southern Africa	14	31
Forest	United States	20	250
Rangeland	South Africa	3–26	2–333
Pasture	United States	3–9	1450
Cropland	United States	5–320	11–1345
Rural (combined)	Southern Africa	1–38	–
	United States	28	1310
Residential	Southern Africa	14–60	85–400
	United States	123–605	717–1000
Industrial	Zimbabwe	280	584
Urban (combined)	Southern Africa	16–162	–
	United States	101	2140
Mixed	Gambia	7	116
Atmosphere	United States	4–6	560–1100

organic matter in lake and reservoir basins can reduce the volume of these waterbodies. In fact, the admixture of inorganic sediments and organic matter can form an ooze carpet on the bottom of a lake basin. The subsequent generation of hydrogen sulfide, methane and other gases under such conditions can begin the process of bog formation, as part of the natural ontogeny of lakes (Wetzel, 1975). Although a rare occurrence, it also can lead to catastrophic events, an example being the 'eruption' of a cloud of noxious gas as recently witnessed at Lake Nias in Cameroon, West Africa (Jorgensen *et al.*, 1996). The concomitant process of deoxygenation of the bottom waters can result in fish kills, as well as the in-lake selection of undesirable organisms (disease vectors), such as mosquito larvae, which can withstand low oxygen conditions. Organically rich sediments also encourage the rooting of nuisance aquatic plants, particularly exotic invaders which have the potential to decimate native species in some habitats (*cf.* Macdonald *et al.*, 1986; Dejoux, 1988).

Thermal stratification and subsequent hypolimnetic deoxygenation can affect the chemical structure of a lake or reservoir. Elements such as iron, manganese and phosphorus can become soluble under anaerobic conditions in the hypolimnion, thereby being regenerated and transported back into the water column. One consequence is that when in-lake mixing does occur at the cessation of thermal stratification, regenerated phosphorus can again become available to algae and aquatic plants, thereby accentuating eutrophication problems. In severe cases of hypereutrophy (= hypertrophy), these mixing events can result in a 'boom and bust' cycle of oxygen super-saturation and water column deoxygenation, and/or in the cataclysmic rise and fall of algal dominance alternating with macrophyte dominance (Barica and Mur, 1980). In other cases, the quantities of phosphorus supplied to a waterbody in this manner can be sufficient to supply the nutrient requirements of algae for the remainder of the growing season (*cf.* Thornton, 1986).

Setting standards for nutrient concentrations in waterbodies has traditionally followed the engineering approach of the wastewater sciences (Thornton and Boddington, 1989). The initial approaches adopted within the Organization for Economic Cooperation and Development (OECD) eutrophication study were of this type (OECD, 1982). It was soon recognized, however, that a more probabilistic approach was needed to overcome the inherent variability of the lake ecosystems studied. From this approach arose the open-ended classification system shown in Figure 5.3. While this system is far more flexible and realistic than its closed-ended predecessor, it is still temperate zone dominated (*cf.* Williams, 1988; Thornton *et al.*, 1991). To overcome this bias, similar schemes have been developed for the semi-arid (Thornton and Rast, 1989) and humid (Salas and Martino, 1990) zones (Figures 5.4 and 5.5). Typically, these latter schemes recognize the importance of watershed geography in the setting of in-lake water quality standards, the eutrophic break points of which range from greater than 0.02 mg/lP and from greater than 0.20 mg/1N (Ryding and Rast, 1989; see also Figures 5.3 to 5.5).

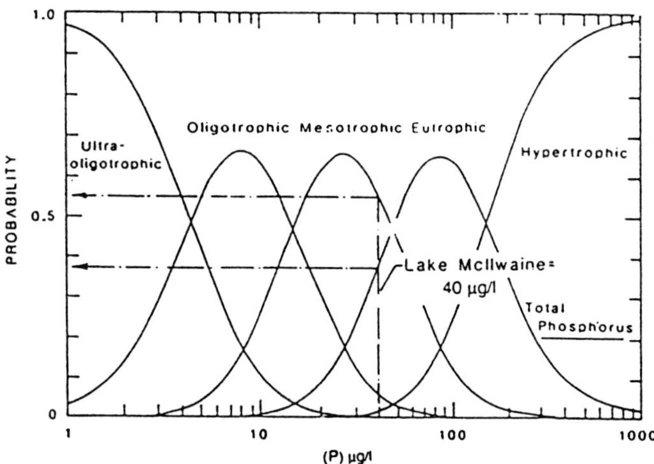

Figure 5.3 Probability-based open-ended phosphorus-trophic classification scheme for primarily temperate lakes (after OECD, 1982)

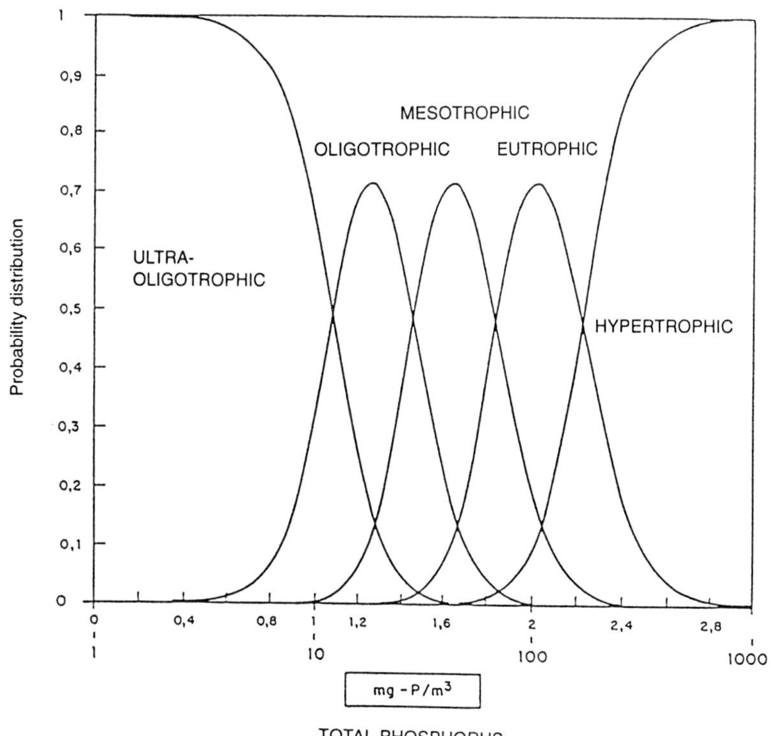

Figure 5.4 Probability-based open-ended phosphorus-trophic classification scheme for warm water lakes (after Salas and Martino, 1990)

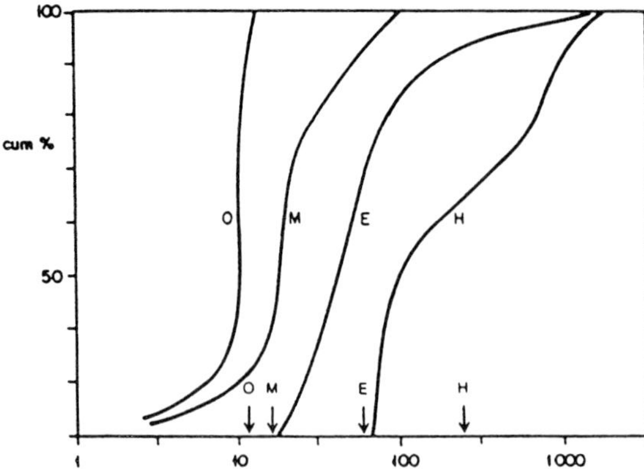

Figure 5.5 Cumulative probability-based phosphorus-trophic classification scheme for semi-arid zone lakes (redrawn from Thornton and Rast, 1993)

BIODEGRADABLES AND ORGANIC MATTER

Biodegradable and non-biological oxygen-consuming substances comprise a class of primarily organic pollutants that can be introduced to waterbodies through both point and nonpoint sources. Typically, these substances are measured as chemical and biochemical oxygen demands using standard techniques of water and wastewater analysis (*cf.* APHA, 1985). The former (COD) refers to the amount of dissolved oxygen taken up in the oxidation of the chemical constituents of a sample during oxidation with an oxidizing agent such as permanganate or dichromate in acid solution, while the latter (BOD) refers to the amount of oxygen taken up through microbiologically-mediated (bacterial) decomposition processes (Stumm and Morgan, 1970; see Chapter 7). Stumm and Morgan (1970) point out that a better measure of the organic content of a sample would be in terms of total organic carbon, as neither of the two tests mentioned above result in the complete oxidation of all organic matter in the sample. Nevertheless, the former methods are regularly used in both wastewater engineering and effluent discharge standards (*cf.* Alabaster, 1980; van der Leeden *et al.*, 1990) and form a convenient way to collectively measure oxygen-consuming substances. In reality, however, these substances comprise a wide variety of naturally-occurring and man-made substances ranging from simple compounds to complex molecules.

Some common sources of these substances include various refining and distilling processes (producing a 5 d, 20°C BOD of between 300 and 20 000 mg/l, dairy and meat packing processes (300 to 2 000 mg/l BOD), laundering processes (300 to 1000 mg/l BOD), and textile, tannery and paper producing processes (50 to 25 000 mg/l BOD)(van der Leeden *et al.*, 1990). Of these, pulp and paper processes using various sulfite-based production methodologies are perhaps amongst the worst producers of oxygen-consuming substances commonly discharged to waterways. Domestic wastewater averages 200 mg/l BOD, and 500 mg/l COD, in the United States (van der Leeden *et al.*, 1990), while leachates from solid waste disposal sites average about 15 000 mg/l BOD and between 80 and 22 500 mg/l COD. Stormwaters can also contain significant levels of BOD and COD, ranging upwards from 500 mg/l and 3500 mg/l, respectively (Dejoux, 1988; van der Leeden *et al.*, 1990).

Oxygen-consuming substances are relatively easily controlled by aerating the effluents prior to discharge into natural water courses, and a variety of technologies exist to do this. Trickling biofilters are the most common means of reducing BOD/COD in domestic wastewaters, although activated sludge processes are becoming more common in many countries (van der Leeden *et al.*, 1990), while recent advances in stormwater detention/retention basin technologies have proved useful in reducing stormwater BOD/COD (Novotny, 1988; Novotny and Chesters, 1981). Further details on these techniques appear in later chapters; however, most discharge standards for effluents require < 100 mg/l oxygen to be consumed after release of the effluents into the environment.

HEAVY METALS AND RADIONUCLIDES

Heavy metals

Heavy metals include elements such as copper, cobalt and cadmium, which occur naturally in waterbodies as the result of leaching and weathering of rocks and soils. These elements are typically multi-valent cations which are potentially toxic to biota when present in greater than trace quantities. Trace quantities of most heavy metals are required, however, for the growth and reproduction of aquatic flora and fauna, as well as for humans (Wetzel, 1975). Contamination of waterbodies with larger than trace concentrations of heavy metals is enhanced by anthropogenic activities, examples being urbanization and its concomitant increases in motor vehicle exhaust emissions, and mining activities. In addition, the use of algicides such as copper sulfate, and flocculent aids such as aluminium sulfate (alum), can also aggravate heavy metal contamination of waterbodies. Nevertheless, it is important to note that some lakes are naturally rich in heavy metals, especially the saline lakes of north Africa (Dejoux, 1988).

The primary negative impacts of heavy metals on waterbodies are often detected principally at two levels of aquatic food webs. On the one hand, heavy metals generally are not soluble in well-oxygenated waters, being removed from the water column fairly rapidly by silt deposition. Because of their high positive valences, these metals are usually extremely reactive with negatively-charged particles (e.g. clays). The result is that flocculated contaminants tend to accumulate in sediments at the bottom of a lake, where they can affect the growth of rooted plants and benthic (bottom-dwelling) fauna. This is especially true in urban areas where there is typically a ten-fold increase in concentrations between the aqueous and solid phases (van der Leeden *et al.*, 1990). On the other hand, the major impact of heavy metal contamination is at the tertiary consumer level of the aquatic food web. This is because heavy metals can undergo accumulation (biomagnification) in aquatic food webs (de Wet *et al.*, 1990; Bezuidenhout *et al.*, 1990). For example, aquatic birds, as well as humans, are especially susceptible to heavy-metal toxicity resulting from the consumption of bottom-feeding fish. Heavy metal toxicity in humans is linked to a number of diseases, including Parkinson's disease (suspected aluminium toxicity) and Minamata disease (mercury toxicity) (van der Leeden *et al.*, 1990).

Many countries legislate against metal contamination of surface and ground waters. Both Zimbabwe (Rowe, 1982) and South Africa (DWA, 1986), developing countries having base metal extraction as a major primary industry, have promulgated effluent discharge standards for total heavy metals as well as maximum discharge concentrations for specific contaminants such as arsenic, copper, lead and zinc. In terms of this legislation, discharges are limited to between 0.05 mg/l and 2.0 mg/l depending on the country and element. Average background levels of these elements range from 2.5 to 56 mg/l in Zimbabwean stream sediments (Hatherly and Viewing, 1982). In contrast, concentrations of heavy metals in natural river and lake waters contain up to three orders of magnitude less in both the developed and developing worlds; the higher values being recorded in heavily impacted rivers draining industrial areas of Europe and the United States (Dejoux, 1988; van der Leeden *et al.*, 1990) (Table 5.3). Higher levels yet have been recorded in groundwaters (van der Leeden *et al.*, 1990). By way of comparison, Whyte and Burton (1980) have reported a range of drinking water standards based on guidelines published by a number of agencies and nations, including those in the developing world.

Radionuclides

Nuclear energy is still not a common form of power production in most countries of the world. Nevertheless, contamination of the aquatic environment with radionuclides does occur, both as the result of natural processes and of atmospheric fall-out from nuclear bomb testing conducted since the late 1940s and more recent nuclear accidents (e.g. Chernobyl; Ryding, 1992). Like the closely-related heavy metals, radioactive elements such as strontium-90 and

Table 5.3 Observed ranges in heavy metal concentrations in various aquatic environments; concentrations in mg/m^3 (Whyte and Burton, 1980; Dejoux, 1988; van der Leeden *et al.*, 1990)

Environment	[Pb]	[Cd]	[Cr]	[Cu]
Surface waters	0–235	0.04–6.4	1.7–24.4	0.91–53.2
Groundwaters	–	–	–	–
Potable waters	40–100	0.5–100	10–50	< 0.01–10

iodine-131 have been measured in lake sediments. The environmental consequences of high levels of radiation arising from geologic and artificial sources can also be substantial in some cases, especially in countries such as Namibia and Australia where such radionuclides are mined and refined. Radionuclides also can be bioaccumulated.

Excessive levels of radionuclides in the environment have public health implications, and may cause cancers and birth defects in some cases. They also can render a water supply unfit for human consumption, and possibly produce genetic mutations. Furthermore, recent evidence of high levels of radon-222 gas (a precursor of plutonium) in groundwater in portions of North America and Europe have reinforced the potential human-health hazards inherent in naturally-occurring radionuclide contamination (van der Leeden *et al.*, 1990). Like heavy metals, many countries have adopted a legislative approach to regulate against radionuclide contamination. The United States has included alpha and beta particle activity, radium-226 and radium-228, radon and uranium in its safe drinking water standards, requiring maximum contaminant levels (MCLs) to be below 4 mrem in the case of alpha and beta activity and less than 5 pCi/l (van der Leeden *et al.*, 1990). Radon-222 concentrations in natural groundwaters in the United States range from < 100 pCi/l to > 4500 pCi/l, while surface water supplies can contain up to 10 000 pCi/l (van der Leeden *et al.*, 1990). In respect to other nuclides, Dejoux (1988) cites recent data from South Africa which suggest that emissions from the Pelindaba nuclear research facility amount to less than 6.2 pCi/l ^{40}K annually. Potassium-40 accounts for about 80 percent of the radioactivity generated from this station. Potassium-40 concentrations in the downstream Hartbeespoort Dam drop to under 4 pCi/l (Dejoux, 1988).

SYNTHETIC ORGANIC CHEMICALS

Organic chemicals include a range of naturally-occurring and man-made chemicals. Of primary concern are synthetic organic chemicals, many of which have been shown to be harmful to humans and other biological organisms when their concentrations exceed trace levels. Prominent examples include polychlorinated biphenyls (PCBs), trihalomethanes (THMs), organochlorinated (e.g. DDD, DDE and DDT), organophosphate pesticides (e.g. malathion and parathion), and herbicides (e.g. 2,4-D). The more commonplace soaps, oils and detergents are also included in this category although not as toxic. Some of these organic compounds, typically the organochlorinated compounds, are concentrated in lake and reservoir bottom sediments, while others, such as the organophosphates, are concentrated in the aqueous phase. Many of these compounds can be bioaccumulated through aquatic food webs. Greichus (1982), Symoens *et al.* (1981) and Dejoux (1988) give various ranges for toxic compound concentrations in African waters which highlight these points, ranging from trace amounts in the aqueous phase to upwards of 13 mg/g total organochlorines in piscivorous birds. Groundwaters can also be adversely affected by organic compounds leached from surfacial sediments. Typical groundwater concentrations of these pesticides in the United States range from 0.1 to 700 mg/m^3 depending on locality and compound measured. In addition to being carcinogens, genetic damage and other human health hazards from such synthetic organic chemicals have been reported (van der Leeden *et al.*, 1990).

Sources of synthetic organic chemicals include human activities such as agricultural application of pesticides and herbicides, industrial processes, and production of potable waters. It also is noted that the use of organic chemicals for pest control is widespread in many developing countries (Dejoux, 1988; Symoens *et al.*, 1981). In these latter cases, there exists a conflict between the need to control pest organisms for reasons of public health and/or increased agricultural production on the one hand, and the potential environmental and long-term public health damage on the other hand. Unfortunately, at present, there is no readily-apparent, and widely-applicable, solution to this problem. Even when banned, these chemicals tend to continue to be distributed by atmospheric export (see Chapter 7). While many countries mandate zero discharge levels for biocides into the environment (Rowe, 1982; DWA, 1986), it is clearly not realistic to expect complete conformance given the acute need to control agricultural and human health pests in many parts of the world. Nevertheless, research into alternative (usually biological) pest control methods, and integrated nutrient and pest management control methods (to minimize chemical applications), is continuing on many fronts (Dejoux, 1988).

Industrial processes also produce an abundance of synthetic substances ranging from heavy metals used in plating and manufacturing processes (see above), to polyaromatic hydrocarbons (PAHs; also known as polycyclical hydrocarbons or PCHs) used in lubrication and internal combustion engines, to polychlorinated biphenyls (PCBs) used in the electrical manufacturing and plastics industries, to phenols and other solvents, to industrial by-products such as dioxins and furans. All of these substances can be discharged into natural waters or deposited through aerial deposition (Korth, 1990). Few of these substances occur naturally and virtually all are toxic, some being carcinogenic and bioaccumulative (e.g. dioxins; van der Leeden *et al.*, 1990). Few countries permit the discharge of more than trace amounts of these chemicals, yet even remote waterbodies would appear to have detectable levels of these contaminants present (Dejoux, 1988). PCB concentrations, for example, ranged from 0.001 µg/g in water to over 3 µg/g in fishes in results reported from a number of African lakes (Greichus, 1982; Dejoux, 1988). Similar values are reported from Europe (Petts, 1989) and the United States (van der Leeden *et al.*, 1990). Many of these contaminants are only marginally affected by even advanced wastewater treatment methods, requiring chemical stripping and activated carbon filtration for effective removal (van der Leeden *et al.*, 1990).

A further source of these contaminants can be categorized as household chemicals – many of these are common cleaning materials, oils and garden chemicals stored and used around homesteads throughout the world. This frequently overlooked source of aquatic contaminants has begun to be quantified through neighborhood 'clean sweep' programmes conducted in the United States. The object of these exercises is to collect the partially used containers of paints, oils, herbicides, fertilizers, insecticides, detergents and cleaning fluids that accumulate in homesteads – recent 'clean sweeps' conducted in the mid-western states (communities of < 100 000 persons) of the US have netted up to 100 tonnes of such materials that would have typically met one of three ends: either the substances would be used up for their intended purposes, disposed of in landfill sites (where the contaminants can leach into ground and surface waters), or dumped into sanitary or storm sewers where they can make their way to natural water courses. Avoiding improper disposal of these substances is a major educational component of the 'clean sweep' programme. For example, recent estimates made by the US National Solid Waste Management Association suggest that just under 20 percent of all lubricating products purchased in that country become nonpoint contaminants through improper disposal. This is in addition to those contaminants that reach waterways as urban and highway runoff carried by stormwaters.

SALTS

Salination, sometimes called salinization or mineralization, is the process whereby the concentration of total dissolved solids (TDS) is increased. Similar to natural eutrophication, salination is a natural process in most lakes, occurring as inorganic anions and cations are washed into a waterbody over time. An example is the endorheic saline lakes of East Africa (Symoens *et al.*, 1981; Livingstone and Melack, 1984). Salination is primarily associated with an increase in the concentration of chloride salts (e.g. sodium chloride, potassium chloride), although carbonate salts also have been shown to dominate in some aquatic systems, especially in Africa (Symoens *et al.*, 1981). Where natural salination occurs, the underlying geochemistry of the rocks and soils present in the drainage and lake basin usually determines whether an aquatic system is dominated by chloride or carbonate salts. In contrast, artificial salination usually is chloride-salt based (*cf.* Symoens *et al.*, 1981; Hammer, 1996).

Similar to cultural eutrophication, the salinization process can be greatly accelerated by human intervention. This is especially the case in regions exhibiting arid and semi-arid climates, which are characterized by low rainfall and a preponderance of well-leached, surfacial soils. In these latter regions, irrigation-based agriculture is a necessity. However, the process of irrigation can change the soil infiltration rates, bringing soluble salts deep in the soil into contact with the 'wetting front' of [groundwater] flows (Williams and Noble, 1984). Runoff generated under such conditions is more saline than that resulting from natural conditions.

Saline runoff also can be generated in urban areas, especially where frigid climates require the use of large quantities of road de-icing salts (van der Leeden *et al.*, 1990). In the United States, de-icer applications add 2.5 million tons of (primarily) calcium chloride to road surfaces annually (van der Leeden *et al.*, 1990). Furthermore, municipal wastewaters and leachates are usually relatively high in salt content, compared to most surface waters. Average concentrations of chlorides in wastewaters and leachates in the United States range from 50 mg/l Cl in wastewaters to between 200 and 750 mg/l Cl in leachates (van der Leeden *et al.*, 1990). Although the principle mode of salt transport to receiving waterbodies is runoff and irrigation waters, salt also can enter aquatic systems *via* atmospheric deposition and point source discharges (*cf.* Symoens *et al.*, 1981; Fuggle and Rabie, 1983; DWA, 1986; van der Leeden *et al.*, 1990). Discharge of chlorides in effluents may or may not be regulated in terms of water pollution control legislation, but most countries generally regulate total dissolved solids (TDS) concentrations in effluents and potable supplies (*cf.* Whyte and Burton, 1980). TDS concentrations in natural waters in southern Africa ranged from 100 mg/l to over 2000 mg/l in highly mineralized lakes in the eastern Cape Province of South Africa (DWA, 1986). Naturally saline lakes have even higher salinities (Thornton, 1986; Hammer, 1996).

The principal impact of salination on aquatic ecosystems relates to the physiological effects of high-salinity waters on the osmo-regulatory systems of plant and animals cells. Cotton tends to be the most salt-resistant crop typically planted in warmer climates, tolerating up to 3000 mg/l TDS, while most fruits and row crops succumb to salt pollution at TDS concentrations of below 750 mg/l (DWA, 1986). Water tends to flow from solutions of low salt concentration to high salt concentration. Thus, by creating a steeper salt gradient between cells and the surrounding waters, the cells are in effect dried out, or else undergo rupture ('lysis'), as the water moves from areas of low salt concentration to high concentration, often resulting in the death of aquatic organisms. In fact, few freshwater plants or animals can readily adapt to large fluctuations in salinity, the primary exceptions being estuarine species or species which require freshwater conditions at one point in their life cycle, and marine conditions at another part of their life cycle (e.g. salmonids).

Impacts of salination on human water uses include impairment of irrigation and potable water supplies. Less obvious impacts are subtle changes to the biological community structure of the biota of lakes and reservoirs. For example, Williams and Noble (1984), Hammer (1996), Davies and Walmsley (1985), and Macdonald and Crawford (1988) reported changes to the flora and macrobenthos of Australian and South African rivers as a result of increased salinity. However, Williams and Noble (1984) also noted that few reports of the environmental impacts of salination have been published; they believe this to be due to the difficulties involved in distinguishing salinity effects from other types of aquatic pollution.

It is also noted that one positive feature of salination is the possibility of improved water clarity as a result of enhanced flocculation of particulates in more saline waters. This has been reported for the Vaal River, South Africa (Grobler *et al.*, 1987).

MICROORGANISMS AND METABOLIC PRODUCTS

Introduction

This class of nonpoint source pollutants consists of a variety of waterborne microorganisms (especially pathogens) and chemicals produced as a result of their metabolism. For this report, these components have been categorized as (1) bacteria and viruses, (2) metabolic toxins, (3) parasites, and (4) macro-biological pollutants. As a general observation, pathogenic microorganisms cover the full range of public health concerns as they relate to water. It is also noted that many pathogenic problems are more widely found in developing countries of the world, where sanitation and water treatment are often not as advanced as found in the developed nations. While these organisms and related chemical substances are not usually considered as nonpoint contaminants, we mention them here for completeness; in many cases, the spread of parasitic microorganisms and the production of metabolic by-products is related to cultural eutrophication arising from nonpoint source contamination of waters.

Bacteria and viruses

In regard to bacteria and viruses in natural waters, the concerns for waterbodies located in temperate climatic zones are different from those of waterbodies located in sub-tropical/tropical zones. Diseases such as cholera and typhoid are more common in the latter than in the former (Symoens *et al.*, 1981). Although this fact is related in part to immunization and disease eradication programmes in developing countries, it is also related to the state of sanitation and the economic status of the countries. In the developed countries, sufficient resources and technology usually exist to conduct monitoring and disease-eradication programmes. As a result, most health authorities in developed countries usually conduct some degree of bacterial monitoring of water resources on a continuing basis. In contrast, there may not be sufficient resources in developing nations to undertake the necessary work, or even to maintain wastewater sewers and sewerage works in good operating conditions (*cf.* Thornton *et al.*, 1991). Spillage of waters containing human excrement wastes is common under these conditions. Therefore, human pathogens can readily be spread *via* contaminated drinking water. Furthermore, developing countries typically have large populations, relative to more limited employment opportunities. This usually means there are large numbers of unemployed, homeless persons who regularly make use of public streets for defecation (*cf.* van der Leeden *et al.*, 1990). In addition, livestock and wildlife also contribute to the typically high bacterial (faecal coliform) content in natural waters.

Compared to bacterial concerns, monitoring of viruses is much more difficult to conduct. Therefore, viral monitoring is not as commonly undertaken, even in developed countries (Grabow *et al.*, 1984). However, recent outbreaks of viral-induced illnesses require that health authorities place greater emphasis on monitoring these latter pathogens. As one example of this need, the rotaviruses and hepatitis viruses are reaching epidemic proportions in many developed countries (Geldreich, 1996; Hattingh and Nupen, 1976).

It is important to note that there are many types of bacteria present in natural waters. NIWR (1985), for example, reports that bacterial populations in Hartbeespoort Dam, South Africa, ranged from 10^6 to 10^7/ml. The bacteria were mainly coccoid heterotrophs that played a major role in the in-lake carbon cycle (although they were less important to the in-lake phosphorus cycle). Few of these bacteria were suspected human pathogens (e.g. coliform bacteria) based on the *Escherichia coli:Streptococcus faecalis* ratio, despite the high population numbers. Typically, bacteria and viruses are present in human faeces at a concentration of between 10^6 and 10^8 organisms/g, and have a survival rate of days to no more than 3 months (van der Leeden *et al.*, 1990). Few viruses can reproduce in the environment, but most bacteria can. Infective doses (ID_{50}) range from 10^4 to 10^9 organisms (van der Leeden *et al.*, 1990). Fortunately, wastewater treatment, beyond the primary level, is relatively effective in removing most bacterial, and some viral, contaminants (van der Leeden *et al.*, 1990). Most countries have statutory controls on the discharge of bacteria from point sources which range from zero to about 10 coliforms/ml (Dejoux, 1988; DWA, 1986; Rowe, 1982).

Bacterial and viral pathogens, in and of themselves, may have a limited direct impact on lake and reservoir water quality as many such organisms tend to be host-specific. However, the presence of bacteria and viruses can contaminate fish and shellfish populations in coastal waters causing severe disruption for food supplies and/or local industries (van der Leeden *et al.*, 1990). Furthermore, where extensive pre-treatment is not possible, these organisms can reduce the availability of potable water supplies to local communities (van der Leeden *et al.*, 1990). While such contamination rarely affects the local population to the same extent as visitors (due to the build up of a natural immunity through prolonged exposure), this can be devastating to centres having a tourism- or fishery-based economy (Grabow, 1979). Thus, the potential water quality impacts of bacteria and viruses as water pollutants probably lies more in the realm of impaired human usage than in direct damage to the lake or reservoir ecosystem. This is in contrast to the water quality impacts of many other nonpoint source pollutants where impacts were expressed both in terms of environmental damage and human impacts.

Toxins

The presence of carcinogenic compounds, such as trihalomethanes (THMs) and chloro-/bromoforms in potable water, which can result from over-chlorination of algal-laden raw water from a lake or reservoir, was previously mentioned. However, in addition to this type of water-quality contaminant, there is also a host of other primarily organic compounds occurring in natural waters that can be considered toxins. A prominent example of such compounds is the organic toxins produced by variants of the blue-green alga *Microcystis aeruginosa*, as well as by other cyanophytes (Carmichael, 1981). These toxins are thought to occur throughout the world. However, the tropical and sub-tropical regions of the world remain primary areas of concern (van Steenderen *et al.*, 1987). These blue-green algal toxins can cause cattle and wildlife deaths in these regions, with obvious economic consequences. In addition, they have been implicated as causal agents in gastro-intestinal infections in humans (Zilberg, 1966; Falconer *et al.*, 1983), and subcutaneous, photo-sensitive growths (R.E. Lee, pers. comm.). Further epidemiological evidence is needed to more clearly define the full extent of the public health threat of such bio-toxins (Toerien *et al.*, 1976).

Convincing evidence that the occurrence of *Microcystis* toxins is linked to eutrophication has been provided from studies on southern African lakes (van Steenderen *et al.*, 1987, 1988). Thus, it appears that a major water-quality impact of nutrient enrichment of lakes and reservoirs can include the formation of potentially carcinogenic compounds harmful to humans. At present, various agencies have promulgated drinking water standards which require that the THM/TOHp (total organohalogen potential) content of potable waters not exceed approximately 0.10 mg/l; taste and odour thresholds for phenolic compounds have also been promulgated at between 0.01 and 60 mg/l in natural waters, or between 0.01 and 20 µg/l for chlorinated waters (van Steenderen *et al.*, 1987; van der Leeden *et al.*, 1990). Measured values in the Vaal River, South Africa, an urban impacted perennial waterway, averaged 10 µg/l for phenolics and 169 µg/l for volatile halogenated hydrocarbons (van Steenderen *et al.*, 1987).

Parasites

The same eutrophic conditions that lead to the production of toxic metabolic by-products can also encourage the spread of waterborne diseases and parasites. There are a number of water-related parasitic diseases that can affect humans and animals. These include schistosomiasis (bilharzia), onchocerciasis, trypanosomiasis and malaria. Most of these diseases occur predominantly in the sub-tropical and tropical regions of the world, with typically warm climates. As in the case of bacterial and viral diseases, they are related in part to such phenomena as overpopulation, and improper sanitation and water treatment (van der Leeden *et al.*, 1990). Man-made developments, such as hydroelectric schemes, have been known to aggravate the transmission of these parasitic diseases by altering streamflow patterns, as well as the surrounding demography of the region (Hunter *et al.*, 1983). However, it is also clear that parasitic diseases are not confined to developing countries. For example, recent evidence suggests that parasitic organisms such as giardiasis are becoming rampant even in regions where sanitation and water supplies are maintained at a high level (Geldreich, 1996; van der Leeden *et al.*, 1990). These latter organisms are usually resistant to normal treatment methods; hence, they are difficult to eradicate.

As with bacteria and viruses, parasites can diminish the usability of water resources. Parasitic diseases are usually related to the need for standing waters and an intermediate host (often with very specific substrate requirements). For example, effective spread of the parasitic disease schistosomiasis requires (1) standing water, (2) a specific snail host, and (3) the presence of rooted macrophytes as a habitat for the snail host. Unfortunately, these characteristics are found in areas with shallow waters where infected persons congregate to draw drinking water, bathe and/or defecate/urinate, thus completing the life cycle of the parasitic organism. Given that some of these parasites are easily controlled using such simple, low-technology techniques as coarse filtration of drinking waters, the failure to break this cycle of infection/re-infection illustrates the prime importance of good public education as the single most important control measure for preventing the spread of these diseases (Thornton *et al.*, 1991). This is in contrast to the tendency among international agencies to consider economic development as the prime means of controlling the spread of parasitic diseases (in other words, creating the correct economic climate to support high technology treatment plants, etc.) (Thornton *et al.*, 1991). Building a hydroelectric project on a tropical river, for example, would create the necessary developmental state to support water and wastewater treatment plants. It also creates the low flow conditions necessary for the growth of rooted macrophytes, and draws potentially infected persons to the site to take advantage of the employment opportunities created by the development. Because these persons are likely to be semi- or unskilled, in the developing world situation described, they will, in all probability, be unable to afford to use the potable water and sanitary facilities developed as an adjunct the dam-building project. Hence, they will probably become yet another cog in the spread of waterborne parasites among the host population. In this manner, a seemingly innocuous, even beneficial, action like building a dam on a river actually can have (potentially) severe social and environmental impacts, primarily as a result of nonpoint source pollution.

MACROBIOTA AND NON-NATIVE SPECIES

Dam construction can also have macrobiological effects. Davies (1979; Petitjean and Davies, 1988), reviewing the impacts of stream regulation in southern Africa, has quantified several species transfers from the Orange River system into other waterways. Principally, these unintentional transfers of biota between water systems as the result of transfer schemes have involved fishes. While inter-system transfers of fish species still occur with regularity (i.e. stocking programmes that have introduced and artificially maintained such species as the various trout and bass genotypes in waters throughout the world; *cf.* Coke, 1988), many historical introductions have involved plants (*cf.* Macdonald *et al.*, 1986). Nuisance species such as the water hyacinth (*Eichhornia crassipes*) native to South America and the milfoil (*Myriophyllum aquaticum* and *Myriophyllum spicatum*) native to Europe are now widespread in Africa and North America where their populations have reached epidemic proportions. Much effort continues to be devoted to the control of these macrophytes (e.g. Olem and Flock, 1990), much of it being based on chemical controls using the herbicide 2,4-D.

Recently unintentional introductions of European species to the North American Great Lakes have gained prominence. The river ruffe (*Gymnocephalus cerna*), spiny water flea (*Bythotrephes cedertoemi*) and zebra mussel (*Dreissena polymorpha*) have created new threats to the inland fishery and coastal industries (IJC, 1990). While some of these species may have been spread as the result of natural processes (e.g. by natural migrations of fishes and waterfowl or through natural range expansions), most have been the result of human interventions (e.g. through stocking, ballast discharges and inter-basin transfers). To combat the human-induced spread of these organisms, chemical controls have been mooted (IJC, 1990), some involving the introduction of chlorine, for example, and some as simple as exchanging freshwater ballast for seawater prior to vessels entering the Great Lakes. Elsewhere, introductions of exotic species have also reached nuisance proportions; for example, the introduction of the Nile perch (*Lates niloticus*) into the African Great Lakes has virtually destroyed the native fishery that pre-dated the introduction of the perch (Ogutu-Ohwayo, 1992), while the presence of stocked trout in southern African streams has decimated the indigenous species in some cases (Coke, 1988). Similarly, the introduction of eels (*Anguilla anguilla*) in Lake Balaton has had the same effect on the fishery of this European lake (Biro, 1977).

While macrobiotic introductions are seldom considered to be nonpoint contaminants, such anthropogenic actions do have the effect of being deleterious to natural systems and, as such, meet the definition given in Chapter 1. And while many countries attempt to regulate this type of species introduction by requiring permits for the transportation and stocking of exotics, recent developments in the world's great lakes, for example, suggest that other measures may be required. Many of these involve the use of biocides that, in themselves, create new hazards for the aquatic environment.

ACIDIFICATION

Power generation using fossil fuels is common in many countries of the world. Unfortunately, stack emissions from power plants which burn coal with a high content of sulfur inject high concentrations of sulfur dioxide into the atmosphere. Atmospheric oxidation processes can then convert the sulfur dioxide to sulfuric acid. In a similar manner, automobile emissions introduce nitrogen compounds into the atmosphere (see Chapter 6). Other activities which contribute to the formation of sulfuric and nitric acid in the atmosphere include (1) industrial processes, (2) slash-and-burn agriculture, and (3) mining-related activities.

The sulfuric and nitric acids formed in the atmosphere subsequently can be transported down to the land surface in precipitation and snowfall. The resultant acidic waters (low pH, typically less than 5.0; van der Leeden *et al.*, 1990) enter natural water courses either as rainfall runoff or as flood waters from snowmelts. If the geologic make up of the drainage basin, and/or the quantities and types of catchment soils, are not sufficient to buffer (commonly referred to as 'acid neutralizing capacity', or ANC, measured as total alkalinity) the acidic drainage waters, the inputs of acidic waters may be sufficient to acidify the receiving waters (lakes, reservoirs) (*cf.* Smol *et al.*, 1986; Bosman and Kempster, 1985). Obvious examples of this phenomenon include, but are not limited to, many lakes in the Scandinavian countries, where lakes with pH values as low as 4.0 to 4.5 have been measured (Smol *et al.*, 1986). Less obvious, but equally serious consequences of acidic precipitation and runoff are the corroding effects of the atmospheric acids on buildings and infrastructure, the leaching of metals from soils, and the destruction of soil structure leading to enhanced erosivity which indirectly contribute to other water quality problems discussed above (e.g. erosion, heavy metal contamination, etc.; Ryding, 1992).

In addition to atmospheric sources, mining activities can contribute to lake acidification, primarily in the form of wastewaters and leachates from mine dumps (Thompson, 1980; Dejoux, 1988). Recently, ammonia emissions from cattle manure and leachates from decomposing terrestrial vegetation and leaf litter as the result of clear-cutting of forests have been implicated in the production of acidic pH values in the terrestrial environment. pH values from municipal landfill leachates and wastewaters in the United States average between 5.2 and 7.3 in the former, and 8.0 in the latter (van der Leeden *et al.*, 1990).

The consequences of acidification of natural waters include (1) fish mortality, and (2) enhanced release of heavy metals mainly from watershed and lake bottom sediments and subsequent toxicity to various lake biota. Ironically, acid-polluted waterbodies typically have very clear (transparent) waters; under extremely acidified conditions, a lake can be devoid of biological organisms. In order to mitigate these long-range impacts, many countries have adopted air quality (Davidson and Delogu, 1989; Fuggle and Rabie, 1983; Dejoux, 1988; Whyte and Burton, 1980) and effluent discharge (DWA, 1986; Rowe, 1982; Dejoux, 1988; Davidson and Delogu, 1989) standards, although the latter tend to set maximum pH values rather than minimum values or ranges.

MACRO-POLLUTANTS, LITTER AND GARBAGE

This class of nonpoint source pollutant consists primarily of the ubiquitous litter and organic refuse that can enter receiving waters from stormwater runoff and/or illegal dumping. While the erosion of soils and the wash-off of other natural and man-made (dissolved or suspended) materials has been previously discussed, this section will introduce those larger, visible items that many people, the world over, commonly view as constituting 'pollution', usually to the exclusion of most of the other types of pollutants already discussed (Thornton *et al.*, 1989). Few areas of the world are free from some type of litter or debris. When present in large quantities, litter is not only unsightly, but can obstruct stormwater drainage systems, leading to localized flooding and related water damage to human developments, and even create small-scale habitats (e.g. tyres) suited to the breeding of various species acting as host to a number of common human parasites (e.g. mosquitoes and snails which can facilitate the spread of malaria and schistosomiasis). Litter that is not biodegradable (largely plastics) forms a major component of urban and industrial wastes (including the agricultural and fishing industry) (van der Leeden *et al.*, 1990). Introduction of this type of waste into receiving waters is obviously a function of the frequency of refuse collection and street sweeping. In developing countries, this often is very infrequent (Hayuma, 1983).

Inorganic litter is typically made up of (1) durable goods (such as appliances, tyres and furnishings), (2) non-durable goods (such as newspapers and the like), and (3) containers and packaging materials. The occurrence of these materials in waterways is usually due to deliberate dumping which can be aggravated by a lack of refuse removal or by lack of provision for small-scale disposal (e.g. many countries have waste management contractors serving larger producers of waste materials and local authorities that serve household producers; small businesses do not fall into either category, even though they are numerically dominant as industrial producers, and hence have little alternative but to either store the waste on their premises or dispose of it illegally). In-stream disposal of these items can introduce numerous contaminants to water bodies, including compounds such as refrigerants and volatile organics and hydrocarbons, which affect both human and environmental health. In the United States, inorganic debris accounts for the majority of the macro-pollutant load, comprising 71 percent of the total municipal waste stream of 160 million tons/year (van der Leeden *et al.*, 1990), much of which (over 35 percent) could be potentially recycled to recover its metal, glass, plastic and cellulose content.

Organic debris can be subdivided into (1) unwanted human litter (e.g. cabbage leaves, carrot tops and vegetable matter often left over in marketplaces at the end of the business day), and (2) naturally-occurring organic debris (e.g. leaf litter). Both categories of organic debris can enter watercourses, as the result of urbanization and the associated increase in impervious surface. In the United States, these organic components of the waste stream account for 29 percent of the annual municipal waste load (van der Leeden *et al.*, 1990); virtually none of which is recovered or composted. While it is noted that leaf litter can be beneficial to in-stream fauna, such as Caddis flies which prey on undesirable insect species (Hynes,1972), the input of excessive quantities of such debris into receiving waterbodies, as a result of stormwater drainage, also can destroy the populations of benthic-dwelling organisms by increasing the biological oxygen demand (BOD) associated with bacterial decomposition of the materials (see above). Organic debris also generates humates which affect both pH and waterbody nutrient status. Wetzel (1975) observes that humic waters typically have a pH of about 4.5 which benefits bacterial and other decomposition processes, but which can place stresses on aquatic fauna and flora. Putting garbage into streams can produce the same type of impact. In addition, large quantities of inorganic litter can cause blockages in drainage channels, leading to flooding.

Water quality impacts of macro-pollutants generally centre around aesthetic problems. However, as noted above, they can also include reduced oxygen concentrations (due to high BODs), deleterious changes in benthic communities, modification of stream-flow regimes, and the enhanced spread of human disease vectors. Nonbiodegradable litter also has the potential to destroy higher trophic level organisms. For example, plastic packaging is widely implicated in some cases of strangulation and hobbling of wildlife and domestic animals. Even when properly disposed of, leachates from landfills and disposal sites often contain many of the nutrients, toxics, and oxygen-consuming substances implicated in water quality deterioration in the United States (among other countries where this has been documented; van der Leeden *et al.*, 1990). Hence careful selection of such sites is indicated, as is provision of the means to treat or otherwise control the inevitable leachates that will be generated from such sites (this is especially true in terms of the location of the disposal site relative to groundwater recharge areas). Many communities are beginning to find that composting, recycling, alternative packaging, and energy recovery programmes are becoming increasingly cost effective as the global resource-base declines and the world economy shrinks. Even developing countries, like Brazil, are finding that waste recovery carries with it substantial and sustainable economic and employment-related benefits that have repercussions throughout the country (for example, composting of biodegradables can result in an effective, natural fertilizer that not only reduces the country's dependency on imported artificial fertilizers but also helps stabilize organic-poor soils having a high erosion potential; care must be exercised, however, when dealing with waste streams containing industrial effluents to avoid heavy metal contamination: see Chapter 9).

ATMOSPHERIC POLLUTANTS

A final consideration in assessing the typology of pollutants from nonpoint sources is the notion that the atmosphere is a pollutant 'source'. In fact, with the possible exception of some specific chemical reactions related to the hydrolysis of acidic emissions from industrial sources mentioned above, the atmosphere does not constitute a 'source' of any pollutants - rather, it should be viewed instead as a transport mechanism for pollutants released to the atmosphere. Nevertheless, we include it here for the simple reason that the atmosphere is often the most overlooked and neglected conduit for pollutants entering aquatic systems.

One of the first studies that identified the substantial role of the atmosphere in the transport of nonpoint-source pollutants to waterways was that of the Pollution From Land Use Activities Reference Group of the International Joint Commission on nonpoint-source pollution in the North American Great Lakes Basin (PLUARG, 1978). Among the results of this multi-year study was the observation that the atmosphere provided the transport mechanism for a number of nonpoint-source pollutants, including nutrients (phosphorus, nitrogen), heavy metals, some organic chemicals and silt/particulates, to the Laurentian Great Lakes. The role of the atmospheric transport of pollutants was particularly significant for the lakes with sparsely-settled drainage basins. Indeed, one of the unexpected results of the PLUARG (1978) study was that the majority of the annual phosphorus load to Lake Superior (the most pristine of the North American Great Lakes) entered the lake *via* atmospheric deposition (rainfall, dry fallout). Because of the greater inputs from other human-induced sources (e.g. municipal wastewater treatment plants, urban and agricultural runoff), the atmosphere was assumed to be a relatively insignificant source of these materials to the other Great Lakes. Subsequent studies conducted in other regions of the globe have clearly demonstrated that the atmosphere can be a major transport mechanism for many different types of contaminants entering aquatic systems.

Yet the atmosphere is more than a simple conduit for contaminants to surface waters. These contaminants can also pollute the atmosphere itself and create a range of other 'side-effects' in addition to the impacts on the aquatic environment. Severe air pollution, for example, can cause public health problems, as was observed in 1930 in the Meuse Valley, Belgium; in 1940 in Donova, Pennsylvania; and, in 1952 in a well-known episode in London called 'The Great Smog'. The public health consequences of these events were manifested through increased cardiovascular and pulmonary failure-related deaths. These episodes were characterized by high concentrations of sulfur dioxide and particulates from coal combustion. Air pollution problems were also associated with photochemical smog – which was fist observed in the 1940s in the Los Angeles, California area. The contaminant in these events was airborne oxidizing agents, such as nitrogen oxides, ozone and peroxyacetyl nitrates (PANs) which arose from internal combustion engine emissions, principally from cars. The health effects included eye irritations, elevated susceptibilities to infection, pulmonary disease, impaired pulmonary function, and nose and throat irritations. This type of problem has been increasingly reported lately from other cities and areas as well, especially those areas subject to atmospheric temperature inversions.

Many of the world's big cities have air pollution problems, which can chronically affect human health (see Figure 5.6). So-called smog alerts have become common in many cities with heavy traffic. For example, in Mexico City in 1986, smog levels exceeded WHO standards on 312 days out of 366.

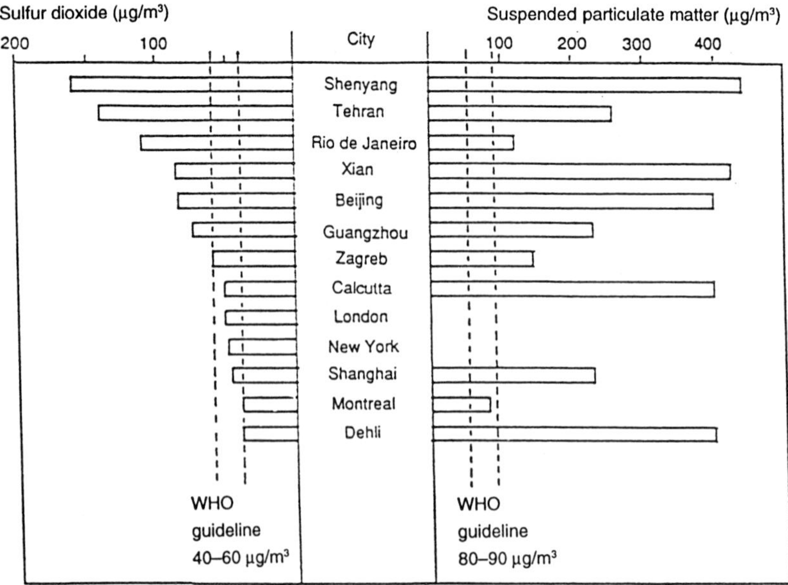

Figure 5.6 Pollution of sulfur dioxide and suspended particulate matter in selected cities, 1980–84 (after Atlas of the Environment, 1990)

The following January, the air quality was so bad that school children were given a month off; the air pollution situation was so bad as to be compared to smoking two cigarettes a day. In the United States, it has been reported that 150 million people breathe unhealthy air. In the Mediterranean, six times as many people die in Athens on heavily polluted days as during clean air periods. In the former Soviet Union, air pollution levels in more than a hundred cities are often ten-times above the accepted levels and air pollution affects more than 50 million people, while, in Eastern Europe, residents in a majority of major cities – such as Krakow (Poland), Leipzig (Germany), Bratislava (Slovenia) and Miskolc (Hungary) – are exposed to such high levels of sulfur and nitrogen oxides, heavy metals and soot, ash and dust that their health is endangered:

(1) Three million Poles in Upper Silesia live in up to four-times the maximum permitted level of dust fallout;

(2) Residents of the Katowice province in Silesia have a 15 percent higher incidence of circulatory illnesses, a 47 percent higher rate of respiratory ailments and 30 percent more cancers than other Poles; and

(3) Between 35 to 40 percent of the population of Hungary live in severe air pollution, with every 17th death and 24th disability being due to bad air quality.

Heavily-populated cities in developing countries are even more polluted. Cities as diverse as Beijing, Jakarta, Bangkok, Manila, Tehran and Lagos are often locked inside gigantic 'bags' of pollution:

(1)   City dwellers in major cities in China are four to six-times more likely to die of lung cancer than people living in the countryside – the world's most seriously-polluted city (according to currently available data) is Benxi near the North Korean border, which often vanishes from satellite photographs under a pall of pollution;

(2)   Air pollution levels in Kuala Lumpur, mainly due to motor vehicle exhausts, are two- to three-times as bad as in cities of the United States; and

(3)   In Cubatao near Sao Paulo, Brazil, locally known as 'the Valley of Death', the levels of industrial particulates in the atmosphere are often twice the level that the WHO judges to be lethal.

In urban areas, motor vehicle traffic is generally the largest single source of air pollution. Vehicle exhaust pipes are situated close to the breathing zone of people. Under these conditions, the concentrations of pollutants to which pedestrians are exposed may become extremely high. The obvious negative effects of motor traffic are the nuisances caused by noise, soot and odour. Asthmatics claim to be more susceptible to the traffic exhausts than healthy persons. A large number of studies have shown an association between symptoms and diseases affecting the respiratory tract and the level of air pollution. In the case of motor vehicle exhausts, nitrogen dioxide is considered the most important component with limit values recommended by the WHO often being exceeded in streets with heavy traffic flows. In developing countries, the continued use of wood and/or fossil fuels for heating and cooking purposes also contributes to the discharge of nitrogen dioxide and particulate aerosols, exacerbating the occurrence of pulmonary diseases. Carbon monoxide is another product of (incomplete) combustion. After inhalation, carbon monoxide binds much more strongly than oxygen to haemoglobin in the blood, thereby leading to the impairment of the oxygen supply to body tissues. The cardiovascular and central nervous systems are particularly sensitive to carbon monoxide-induced stresses. Here, too, the limit values recommended by the WHO may often be exceeded in narrow streets with heavy traffic volumes or in heavily populated urban areas dependent on wood or fossil fuels.

Lead is used as an additive in gasoline in many countries and, in cars not equipped with catalytic converters, is emitted in the exhaust. The critical health effect of lead is damage to the nervous system. Children are more sensitive than adults. The concentrations of lead in the blood of urban residents have declined in many countries after reductions in the lead content of gasoline. However, the largest intake of lead normally comes from food, reflecting the biomagnification of the metal in foodstuffs produced in the vicinity of urban centres. For example, the highest blood lead levels yet measured in the world have been found in

Mexico City, clearly indicating an additional health risk for the population of that city.

Motor vehicles emit a large amount of organic material, due to incomplete combustion. Among them are cyclic and polycyclic organic compounds, a group containing several known or suspected carcinogenic substances (e.g. benzene, ethene and propene). For many carcinogens, there is reason to believe that there are no thresholds below which effects are tolerable; even at low concentrations, there may still be a significant, if small, health risk. Traffic is in most cases the dominating source of carcinogens in the ambient air in cities, and 5 to 10 percent of lung cancer incidences have been estimated to be due to air pollution. Short-term mutagenicity studies, using bacterial test systems, have shown that diesel-powered vehicles emit more mutagenic substances than do gasoline-powered vehicles, and that cars equipped with catalytic converters generally have the lowest emissions.

In summary, then, elevated levels of contaminants in the atmosphere affect human health both directly and indirectly (e.g. through contaminated food). These effects can either be short or long term. Considering the direct effects on human health, the influences of air pollutants range from mere inconvenience and discomfort to increased frequencies of respiratory diseases and death, especially after serious pollution episodes. There is still much to learn about the specific effects of various airborne pollutants; however, it is clear that carbon monoxide lowers the rate of blood oxygen uptake, causing injury to the heart and the central nervous system, and can contribute to fetal distress. Sulfur dioxide and soot can irritate the respiratory system, causing the swelling of the mucous membranes and the formation of mucus. Nitrogen oxides also affect the respiratory organs and can increase susceptibility to respiratory infections. Lead, a common additive in motor fuels, can injure blood-creating organs and the nervous system, especially impairing brain functions. And all of these compounds can enter aquatic ecosystems where their effects are further magnified through acidification and heavy metal contamination (see above).

STANDARDS, GUIDELINES, CRITERIA AND CRITICAL LEVELS OF CONTAMINANTS

Numerous compilations of criteria, guidelines and standards exist. While an extended discussion of the research methodologies used to derive these limits is beyond the scope of this publication, a brief review of the terminology may be useful for the manager considering some form of regulatory control of nonpoint source contaminants. Each of the terms mentioned above has a definite legal meaning which has direct bearing on the degree of enforcement needed and protection provided by the numerical value assigned to any given contaminant concentration. In addition, the terms imply a specific philosophical approach to water quality management, which is set forth below.

Standards

Standards are legally prescribed limits which shall not be exceeded without penalty. Some standards, such a faecal coliform standards promulgated in many countries, do permit some degree of exceedance to occur; to wit, faecal coliform standards are often written in the following terms:

> 'the membrane faecal coliform count may not exceed 200 per 100 ml as a geometric mean based on not less than 5 samples per month, nor exceed 400 per 100 ml in more than 10 percent of all samples during any month' (SEWRPC, 1993).

Standards are legally enforceable requirements, usually promulgated by statute or other rule-making authority, set by governments to protect public health, safety and welfare. The most common types of standards are drinking water standards and effluent discharge standards.

In the former case, the standards proscribe the inclusion of substances in potable water that may be deleterious to the public health. The US EPA, for example, has primary responsibility for establishing and enforcing safe drinking water standards in the United States. The standards are promulgated under the authority of the Safe Drinking Water Act of 1974. Samples taken and analyzed to show conformance with these standards must be collected as set out in the law (or in supplementary guidance) and analyzed using standard methods of analysis by a certified laboratory. The need for a standard methodology stems from the legal desire for fairness or equity between all parties affected by the law, and has resulted in the publication of handbooks of standard methodologies – the most well known being the APHA *Standard Methods . . .* handbook (see Chapter 7). Laboratory certification is usually done by the state or other nominated organization, typically the bureau of standards or similar body. Inter-laboratory comparisons, run using standard or 'known' solutions, are often conducted as part of the certification process. Certification may be awarded for different levels of precision; for example, in the United States, it is common to certify laboratories conducting phosphorus analyses as certified for high level (e.g. wastewater samples) or low level (e.g. environmental samples) analyses. Standards are usually applied at a specific point in the process (e.g. drinking water standards are usually applied at the tap or point of delivery) and when exceeded usually require some form of response, such as the issuance of health advisories, warnings or other notification (e.g. notification of the public health authority). Drinking water standards in the United States relate to the public health (stated as maximum contaminant levels or MCLs) and aesthetic qualities of the finished product (secondary maximum contaminant levels or SMCLs) – only the MCLs are legally enforceable as the SMCLs, while affecting the appearance or taste of the water, would not result in adverse health effects over the short term.

Wastewater standards share many of the same characteristics as drinking water standards, being legally enforceable at the point of discharge from the treatment plant or factory premises, requiring specified methods of sample collection and analysis, and reporting of exceptions. Such standards typically specify a maximum concentration level that must not be exceeded in the discharge – this type of standard is one of the oldest as well as one of the most common on the statute books of many countries. The range in parameters specifically controlled ranges from BOD_5, total suspended and/or dissolved solids, and toxic substances to more broadly-based standards covering total and/or dissolved phosphorus, nitrogen, and other elements and substances (e.g. soaps, oils and greases, radioactive substances, metals, and synthetic organic chemicals). The shortcoming of such standards is that they do not take cognizance of the potential impact on a receiving waterbody in any quantitative way; an almost unlimited load of a given substance can be discharged provided it meets the concentration standard. To overcome this limitation, many countries are presently experimenting with waste load allocation standards – also known as total maximum daily load (TMDL) allocation standards – which specify a waterbody-related, maximum load of a contaminant to be released. This can be specified either in terms of a total load per discharger, or in terms of a performance standard-related water quality to be met in the receiving waterbody. The former option introduces a number of free market mechanisms into waste discharge regulation, including waste load banking, where unused waste load allocation units are stored up to provide for future expansion of operations, or waste load trading, where waste load allocation units are sold by one operator who has adequate wastewater controls to another operator who perhaps cannot meet the level of waste reduction specified (see Novotny, 1988, and Ryding, 1992 for further discussion of these alternatives).

While these latter mechanisms are generally viewed as experimental, similar approaches have been adopted in the regulation of land use in the United States. Analogous to waste load allocation banking and trading – also known as transfers – are the use of impact fees (moneys paid to an authority by a developer to mitigate, usually off-site, the impact of a development on the environment, similar to the existing requirements in many countries that housing estate developers deed over a certain percentage of the land as public open space for common use within a development), the spread of planned unit developments (PUDs) (where greater residential densities are permitted on part of a site to allow the greater portion of the site to be retained in an undisturbed state), and performance zoning techniques that leave a large measure of discretion to the developer as to how s/he will meet the requirements of the law (this promotes individuality and initiative in the planning of estates but does specify a minimum level of acceptability). Indeed, the State of Oregon (USA) has set water quality standards for receiving waters that guide their allocation of wastewater discharge permits within each of their major river basins, which the Republic of South Africa is adopting for a waste load allocation-based standards programme for the heavily-industrialized Vaal River basin. Many states in the United States – Florida, Maryland and New Jersey, amongst them – have adopted this type of standard for stormwater runoff; initially these standards have been applied to the volume of runoff (e.g. 'no net increase in runoff from the site after development'), but, with the promulgation of stormwater quality standards, are being increasingly applied to water quality as well. As many water authorities are adopting a more holistic view of the resource they manage, the move toward performance standards and watershed-based waste load allocations will probably gain momentum, and such standards will replace the older, concentration-based standards mentioned above and presented in the appendices to this chapter.

Guidelines

Guidelines, recommended levels, and acceptable or admissible concentrations are all terms applied to the regulatory assessment of water quality. Although similar in many respects to the standards discussed above, guidelines, while legally established in many instances, are generally not enforceable under the law and perform an advisory function only. The better known guidelines include the World Health Organization (WHO) drinking water guidelines used by many countries as the basis for their national legislation. While these limits cannot be imposed on any government, they do provide for the fullest possible protection of the public health, safety and welfare. The limits suggested can be equivalent to or exceed those specified in terms of standards, but provide a more flexible option for regulatory authorities to impose on water suppliers or wastewater dischargers. Further, instead of specifying a maximum level that shall not be exceeded, guidelines suggest a minimum level that should be attained, putting the onus on the operator to produce a final product that is well within the limits set by the guidelines and encouraging producers of potable water or effluents to better the specified level. In short, guidelines can be a positive means of expressing the same desire so often viewed negatively in the case of standards; whereas an operator has little incentive to do more than meet a standard, the incentive exists to keep well within a guideline.

Criteria and critical levels of contaminants

Many criteria are based on bioassay experiments whose results are given in terms of both a lethal concentration at which level 50 percent of the test organisms succumb (the so-called LC_{50} level) and time of exposure. Unlike the standards which have force of law and apply at specific points in a process, and guidelines which are advisory, criteria and critical limits are much more specific both in terms of geographical extent and the species covered, and are use-oriented. The levels specified are related to an intended use – preservation of the aquatic ecosystem being the intended use in cases where specific organisms are being protected – and apply to discrete waterbodies such as lakes, rivers, bays, estuaries or harbors – examples of the use of criteria include inter-governmental agreements on the Potomac River Estuary–Chesapeake Bay system and Puget

Sound system in the United States. Appendices 5.2 and 5.3 give a selection of critical levels as they have been determined for some of the common micro-pollutants previously described (van der Leeden *et al.*, 1990). Most agencies will find it convenient and economical to make use of these existing tabulations as their compilation is both expensive and time-consuming. Agencies such as the US Environmental Protection Agency, the World Health Organization of the United Nations, the Canadian Council of Resource and Environment Ministers, and the European Economic Community regularly publish such compilations and issue criteria for both effluent discharges and domestic water supplies (*cf.* van der Leeden *et al.*, 1990) and various other users. Nevertheless, agencies wishing to adopt locally relevant criteria, guidelines or standards are referred to Chapter 7.

CHAPTER SUMMARY

This chapter has been written to provide some insight into what constitutes nonpoint source pollution by identifying specific contaminants and providing some indication of the potential problems or impacts that might result from their presence (i.e. the manner in which the contaminant affects waterbody structure and function). The approach adopted has been to outline nine broad categories of contaminants, ranging from the microscopic and elemental to the macroscopic, which commonly impact natural waters throughout the world, and to give some indication of the quantities observed in various locales, typically in the developed and developing worlds. No comment has been offered on the levels at which these contaminants become nuisances, other than through the closing discussion of standards, guidelines and other criteria used to establish critical levels, as these are usually site and use specific. However, further discussion of these critical levels is contained in Chapter 6 relative to specific human activities.

Appendix 5.1   Priority pollutants identified by the US EPA for assessment and control (after van der Leeden *et al.*, 1990)

Inorganics
   Aluminium
   Antimony
   Arsenic[†]
   Asbestos
   Barium[†]
   Beryllium
   Cadmium[†]
   Chromium[†]
   Copper
   Cyanide
   Fluoride[†]
   Lead[†]
   Mercury[†]
   Molybdenum
   Nickel
   Nitrate[†]
   Selenium[†]
   Silver[†]
   Sodium
   Sulfate
   Thallium
   Vanadium
   Zinc
Microbiology and turbidity
   Giardia lamblia
   Legionella
   Standard plate count
   Total coliforms[†]
   Turbidity[†]
   Viruses
Organics
   Acrylamide
   Adipates
   Alachlor
   Aldicarb
   Atrazine
   Carbofuran
   Chlordane
   2,4-D[†]
   Dalapon
   Dibromochloropropane
   Dibromomethane
   1,2-Dichloropropane
   Dinoseb

Organics, continued
   Dioxin
   Diquat
   Endothall
   Endrin[†]
   Epichlorohydrin
   Ethylene dibromide
   Glyphosate
   Hexachlorocyclopentadiene
   Lindane[†]
   Methoxychlor[†]
   Pentachlorophenol
   Phthalates
   Pichloram
   Polychlorinated biphenyls
   Polycyclic aromatic hydrocarbons
   Simazine
   2,4,5-TP[†]
   Toluene
   Toxaphene[†]
   1,1,2-Trichloroethane
   Vydate
   Xylene
Radionuclides
   Beta particle and photon activity[†]
   Gross alpha particle activity[†]
   Radium-226 and radium-228[†]
   Radon
   Uranium
Volatile organic chemicals
   Benzene[†]
   Carbon tetrachloride[†]
   Chlorobenzene
   cis-1,2-Dichloroethylene
   Dichlorobenzene[†]
   1,2-Dichloroethane[†]
   1,1-Dichloroethylene[†]
   Methylene chloride
   Tetrachloroethylene
   trans-1,2-Dichloroethylene
   Trichlorobenzene
   1,1,1-Trichloroethane[†]
   Trichloroethylene[†]
   Vinyl chloride

Seven substitutions are permitted. †Already regulated

Appendix 5.2 Drinking water standards and guidelines adopted by the US EPA, WHO, EEC and Canada (after van der Leeden *et al.*, 1990)

Substance	US maximum contaminant level* mg/l	Canadian maximum acceptable limit[†] mg/l	EEC maximum admissable concentration[†] mg/l	WHO guideline value mg/l
Inorganics				
Arsenic	0.05	0.05	0.05	0.05
Barium	1.0	1.0	0.1	NS
Cadmium	0.01	0.005	0.005	0.005
Chromium	0.05	0.05	0.05	0.05
Fluoride	4.0	1.5	NS	1.5
Lead	0.05	0.05	0.05	0.05
Mercury	0.002	0.001	0.001	0.001
Nitrate	10.0	10.0	50	10.0
Selenium	0.01	0.01	0.01	0.01
Silver	0.05	0.05	0.01	NS
Microbials				
Coliforrns – *organisms/100 ml*	< 1	10	0	0
Turbidity – *ntu*	1–5	5.0	0–4	< 1
Organics				
2,4-D	0.1	0.1	NS	0.001
Endrin	0.0002	0.0002	NS	NS
Lindane	0.0004	0.004	NS	NS
Methoxychlor	0.1	0.1	NS	0.001
Pesticides (total)	NS	0.1	0.005	NS
Toxaphene	0.005	0.005	NS	NS
2,4,5-TP silvex	0.01	0.01	NS	NS
Trihalomethanes	0.10	0.35	0.001	0.03 ($CHCl_3$ only)
Radionuclides				
Beta particle and photon activity	4 mrem	NS	NS	1.0 Bq/l[§]
Gross alpha particle activity	15 pCi/l	NS	NS	0.1 Bq/l[§]
Radium-226 + radium-228	5 pCi/l	1 Bq/l[§]	NS	NS
Volatile organic chemicals				
Benzene	0.005	NS	NS	0.01
Carbon tetrachloride	0.005	NS	NS	0.003
1,1-Dichloroethylene	0.007	NS	NS	0.003
1,2-Dichloroethane	0.005	NS	NS	0.01
para-Dichlorobenzene	0.075	NS	NS	NS
1,1,1-Trichloroethane	0.2	NS	NS	NS
Trichloroethylene	0.005	NS	NS	0.03
Vinyl chloride	0.002	NS	NS	NS

*Enforceable; [†]nonenforceable; NS = no standard; [§]Becquerels per litre

Appendix 5.2 Continued

| | US secondary maximum contaminant level* | Canadian maximum acceptable limit* | EEC | | WHO guideline value |
			Guide level*	Maximum admissible concentration	
Substance					
Chloride	250 mg/l	250 mg/l	25 mg/l	NS	250 mg/l
Colour	15 cu	15 cu	1 mg Pt-Co/l	20 mg Pt-Co/l	15 cu
Copper	1 mg/l	1.0 mg/l	100 µg at treatment plant; 3000 µg after 12 hours in piping	NS	1.0 mg/l
Corrosivity	noncorrosive				
Fluoride	2 mg/l	1.5 mg/l		Water should not be aggressive Varies according to average temperature in the area	1.5 mg/l
Foaming agents	0.5 mg/l	NS	NS	NS	
Iron	0.3 mg/l	0.3 mg/l	50 µg/l	300 µg/l	0.3 mg/l
Manganese	0.05 mg/l	0.05 mg/l	20 µg/l	50 µg/l	0.1 mg/l
Odour	3 TON		0 dilution number	2 dilution number at 54°F (12°C)	
pH	6.5–8.5	6.5–8.5	6.5–8.5	NS	6.5–8.5
Sulfate	250 mg/l	500 mg/l	25 mg/l	NS	400 mg/l
Total dissolved solids	500 mg/l	500 mg/l	NS	NS	1000 mg/l
Zinc	5 mg/l	5 mg/l	100 µg at treatment plant; 5000 µg after 12 hours in piping	NS	5.0 mg/l

*Nonenforceable; NS = no standard; EEC = European Economic Community; WHO = World Health Organization

Appendix 5.3   Water quality criteria for aquatic life (after van der Leeden *et al.*, 1990)

Pollutant	*Criteria*	*Ref.[b]*
Acenaphthene	Acute toxicity occurs as low as 1700 µg/l in freshwater species and 970 µg/l in salt-water species. Freshwater algae are affected by 520 µg/l, saltwater algae at 500 µg/l. Chronic toxicity occurs in saltwater species as low as 710 µg/l.	1
Acrolein	Acute toxicity occurs as low as 68 µg/l in freshwater species and 55 µg/l in saltwater species. Chronic toxicity occurs in freshwater species as low as 21 µg/l.	1
Acrylonitrile	Acute toxicity occurs as low as 7550 µg/l in freshwater species. Mortality occurred in freshwater fish exposed for 30 days at 2600 µg/l.	1
Aluminium	For protection of saltwater species an application factor of 0.01 is recommended to be applied to the 96-hour LC50 for sensitive organisms. Concentrations exceeding 1500 µg/l constitute a hazard in the marine environment, and levels less than 200 µg/l present minimal risk of deleterious effects.	2
Ammonia (un-ionized)	For marine species, an application factor of 0.1 is recommended. Concentrations equal to or exceeding 400 µg/l consitute a hazard to marine biota. Levels below 10 µg/l present minimal risk of deleterious effects. (Insufficient data for 1984 criterion.)	2
Antimony	Acute toxicity occurs a low as 9000 µg/l in freshwater species and is toxic to freshwater algae at 610 µg/l. Chronic toxicity occurs in freshwater species as low as 1600 µg/l.	1
	For protection of saltwater species an application factor of 0.01 is recommended to be applied to the 96-hour LC50 for sensitive organisms. Concentrations exceeding 0.2 µg/l constitute a hazard in the marine environment.	2
Arsenic[c] (trivalent)	For freshwater aquatic life in each 30 consecutive days the average concentration of arsenic shall not exceed 72 µg/l, the maximum concentration shall not exceed 140 µg/l, and the concentration may be between 72 µg/l and 140 µg/l for up to 96 hours. For saltwater aquatic life in each 30 consecutive days the average concentration of arsenic shall not exceed 63 µg/l, the maximum concentration shall not exceed 120 µg/l, and the concentration may be between 63 µg/l and 170 µg/l for up to 96 hours.	3
Barium	For protection of saltwater species an application factor of 0.05 is recommended to be applied to the 96-hour LC50 for sensitive organisms. Concentrations equal to or exceeding 1000 µg/l constitute a hazard in the marine environment, and levels less than 500 µg/l present minimal risk of deleterious effects.	2
Benzene	Acute toxicity occurs as low as 5300 µg/l in freshwater species and 5100 µg/l in saltwater species. Adverse effects occur in saltwater fish exposed for 168 days as low as 700 µg/l.	1
Benzidine	Acute toxicity occurs as low as 2500 µg/l in freshwater species.	1
Beryllium	Acute toxicity occurs as low as 130 µg/l in freshwater species. Chronic toxicity occurs in freshwater species as low as 5.3 µg/l. Hardness has a substantial effect on acute toxicity. For protection of saltwater species an application factor of 0.01 is recommended to be applied to the 96-hour LC50 for sensitive organisms. Concentrations equal to or exceeding 1500 µg/l constitute a hazard in the marine environment, and levels less than 100 µg/l present minimal risk of deleterious effects.	1

Appendix 5.3 Continued

Pollutant	Criteria	Ref.[b]
Boron	For protection of saltwater species an application factor of 0.1 is recommended to be applied to the 96-hour LC50 for sensitive organisms. Concentrations equal to or exceeding 5000 µg/l constitute a hazard in the marine environment, and levels less than 5000 µg/l present minimal risk of deleterious effects.	2
Bromate	It is recommended that ionic bromine in the form of bromate be maintained below 100 000 µg/l in the marine environment.	2
Bromine	It is recommended that free (molecular) bromine in the marine environment not exceed 100 µg/l.	(free)
Cadmium	For freshwater aquatic life, the concentration of active cadmium shall not exceed a level equal to 1.16 (ln hardness mg/l) – 3.841 due to acute and chronic toxicities being nearly the same. For saltwater aquatic life in each 30 consecutive days the average concentration of cadmium shall not exceed 12 µg/l, the maximum concentration shall not exceed 38 µg/l, and the concentration may be between 12 µg/l and 38 µg/l for up to 96 hours.	3
Chlorine	For freshwater aquatic life in each 30 consecutive days the average concentration of chlorine shall not exceed 8.3 µg/l, the maximum concentration shall not exceed 14 µg/l, and the concentration may be between 8.3 µg/l and 14 µg/l for up to 96 hours. For saltwater aquatic life in each 30 consecutive days the average concentration of chlorine shall not exceed 7.4 µg/l, the maximum concentration shall not exceed 13 µg/l, and the concentration may be between 7.4 µg/l and 13 µg/l for up to 96 hours.	3
Chromium[c] (hexavalent)	For freshwater aquatic life in each 30 consecutive days the average concentration of chromium shall not exceed 7.2 µg/l, the maximum concentration shall not exceed 11 µg/l, and the concentration may be between 7.2 µg/l and 1100 µg/l for up to 96 hours. For saltwater aquatic life in each 30 consecutive days the average concentration of chromium shall not exceed 5.4 µg/l, the maximum concentration shall not exceed 1200 µg/l, and the concentration may be between 5.4 µg/l and 1200 µg/l for up to 96 hours.	3
Chromium[c] (trivalent)	For freshwater aquatic life in each 30 consecutive days the average concentration of chromium shall not exceed 0.819 (ln hardness mg/l) to 537, the maximum concentration shall not exceed 0.819 (ln hardness mg/l) + 3.568. No saltwater criterion were derived, but levels of 10 300 µg/l are lethal to the eastern oyster.	3
Copper[c]	For freshwater aquatic life in each 30 consecutive days the average concentration of copper shall not exceed 0.905 (ln hardness) – 1.705, the maximum concentration shall not exceed 0.905 (ln hardness mg/l) + 3.568, and the concentration may be between the average and the maximum for up to 96 hours. For saltwater aquatic life in each 30 consecutive days the average concentration of copper shall not exceed 2 µg/l, the maximum concentration shall not exceed 3.2 µg/l, and the concentration may be between 2 µg/l and 3.2 µg/l for up to 96 hours.	3
Cyanides (sum of HCN and CN)	For freshwater aquatic life in each 30 consecutive days the average concentration of cyanides shall not exceed 4.2 µg/l, the maximum concentration shall not exceed 22 µg/l, and the concentration may be between 4.2 µg/l and 22 µg/l for up to 96 hours. For saltwater aquatic life in each 30 consecutive days the average concentration of cyanides shall not exceed 0.57 µg/l, the maximum concentration shall not exceed 1 µg/l, and the concentration may be between 0.57 µg/l and 1 µg/l for up to 96 hours.	3
2,-4-Dinitrotoluene	Acute toxicity occurs as low as 330 µg/l in freshwater species and 590 µg/l in saltwater species. Chronic toxicity occurs in freshwater species as low as 230 µg/l. A decrease in saltwater algal cell numbers occurs as low as 370 µg/l.	1

Appendix 5.3 Continued

Pollutant	Criteria	Ref.[b]
1,2-Diphenyl-hydrazine	Acute toxicity occurs as low as 270 µg/l in freshwater species.	1
Ethylbenzene	Acute toxicity occurs as low as 32 000 µg/l in freshwater species and 430 µg/l in saltwater species.	1
Fluoranthene	Acute toxicity occurs as low as 3980 µg/l in freshwater species and 40 µg/l in saltwater species. Chronic toxicity occurs in saltwater species as low as 16 µg/l.	1
Fluorides	For protection of saltwater species an application factor of 0.1 is recommended to be applied to the marine 96-hour LC50. Concentrations equal to or exceeding 1500 µg/l constitute a hazard to the marine environment, and levels less than 500 µg/l present minimal risk of deleterious effects.	2
Iron	Concentrations equal to or exceeding 300 µg/l constitute a hazard to the marine environment, and levels less than 50 µg/l present minimal risk of deleterious effects.	2
Isophorone	Acute toxicity occurs as low as 117 000 µg/l in freshwater species and 12 900 µg/l in saltwater species.	1
Lead[c]	For freshwater aquatic life in each 30 consecutive days the average concentration of lead shall not exceed 1.34 (ln hardness mg/l) –5.245, the maximum concentration shall not exceed 1.34 (ln hardness mg/l) – 2.014, and the concentration may be between the average and the maximum for up to 96 hours. For saltwater aquatic life in each 30 consecutive days the average concentration of lead shall not exceed 8.6 µg/l, the maximum concentration shall not exceed 220 µg/l, and the concentration may be between 8.6 µg/l and 220 µg/l for up to 96 hours.	3
Manganese	For protection of saltwater species an application factor of 0.02 is recommended to be applied to the marine 96-hour LC50. Concentrations equal to or exceeding 100 µg/l constitute a hazard to the marine environment, and levels less than 20 µg/l present minimal risk.	2
Mercury[c]	For freshwater aquatic life in each 30 consecutive days the average concentration of mercury shall not exceed 0.2 µg/l, the maximum concentration shall not exceed 1.1 µg/l, and the concentration may be between 0.2 µg/l and 1.1 µg/l for up to 96 hours. For saltwater aquatic life in each 30 consecutive days the average concentration of mercury shall not exceed 0.1 µg/l, the maximum concentration shall not exceed 1.9 µg/l, and the concentration may be between 0.1 µg/l and 1.9 µg/l for up to 96 hours.	
Molybdenum	It is recommended that the concentration in seawater should not exceed 0.05 of the 96-hour LC50 at any time for the most sensitive species and that the 24-hour average not exceed 0.02 of the 96-hour LC50.	1
Naphthalene	Acute toxicity occurs as low as 2300 µg/l in freshwater species and 2350 µg/l in saltwater species. Chronic toxicity in freshwater species occurs as low as 620 µg/l.	1
Nickel	For freshwater aquatic life, total recoverable nickel should not exceed 1100 µg/l at any time assuming a hardness of 50 mg/l as $CaCO_3$. For saltwater species the concentration should not exceed 140 µg/l at any time. The 24-hour average freshwater criterion is 56 µg/l for a hardness of 50 mg/l. The 24-hour saltwater criterion is 7.1 µg/l.	1
Nitrobenzene	Acute toxicity occurs as low as 27 000 µg/l in freshwater species and 6680 µg/l in saltwater species.	1
Nitrophenols	Acute toxicity occurs as low a 230 µg/l in freshwater species and 4850 µg/l in saltwater species. Toxicity to freshwater algae occurs as low as 150 µg/l.	1

Appendix 5.3 Continued

Pollutant	Criteria	Ref.[b]
Nitrosamines	Acute toxicity occurs as low as 5850 µg/l in freshwater species and 3 300 000 µg/l in saltwater species.	1
Phenol	Acute toxicity occurs as low as 10 200 µg/l in freshwater species and 5800 µg/l in saltwater species. Chronic toxicity occurs in freshwater species as low as 2560 µg/l.	1
2,4-Dimethyl phenol	Acute toxicity occurs as low as 2120 µg/l in freshwater species.	1
Phenolics (phenolic compounds)	For freshwater species, an application factor of 0.05 is recommended to be applied to the 96-hour LC50 for important sensitive species. No concentration greater than 100 µg/l is recommended at any time or place.	2
Phthalate esters	Acute toxicity occurs as low as 940 µg/l in freshwater species and 2944 µg/l in saltwater species. Chronic toxicity occurs in freshwater species as low as 3 µg/l. Toxicity to one species of saltwater algae occurs as low as 3.4 µg/l.	1
Phosphorus (elemental)	For protection of saltwater species an application factor of 0.01 is recommended to be applied to the marine 96-hour LC50. Concentrations equal to or exceeding 1 µg/l constitute a hazard to the marine environment.	2
Polychlorinated biphenyls	Acute toxicity probably will only occur at concentrations above 2.0 µg/l for freshwater species and above 10 µg/l for saltwater species. The 24-hour average freshwater criterion is 0.014 µg/l. The 24-hour saltwater criterion is 0.030 µg/l.	1
Polynuclear aromatic hydrocarbons	Acute toxicity occurs as low as 300 µg/l for saltwater species.	1
Selenium (as inorganic selenite, Se^{2+})	For freshwater aquatic life, total recoverable inorganic selenite should not exceed 260 µg/l at any time. For saltwater species, the concentration should not exceed 410 µg/l at any time.	1
Selenium (as inorganic selenite, Se^{2+})	The 24-hour average freshwater criterion is 35 µg/l. The 24-hour saltwater criterion is 54 µg/l.	1
Selenium (as inorganic selenite, Se^{4+})	Acute toxicity occurs as low as 760 µg/l in freshwater species.	1
Silver	For freshwater aquatic life, total recoverable silver should not exceed 1.2 µg/l at any time, assuming a hardness of 50 mg/l as $CaCO_3$. For saltwater species the concentration should not exceed 2.3 µg/l at any time. Chronic toxicity in freshwater species occurs a low as 0.12 µg/l.	1
Sulfide	For protection of saltwater species an application factor of 0.1 is recommended to be applied to the marine 96-hour LC50. Concentrations equal to or exceeding 10 µg/l constitute a hazard to the marine environment, and levels les than 5 µg/l present minimal risk of deleterious effects with pH maintained within a range of 6.5 to 8.5.	2
Hydrogen sulfide (undissociated)	For freshwater species, a level assumed to be safe for all aquatic organisms including fish is 2 µg/l. It is recommended that the concentration of total sulfides not exceed 2 µg/l at any time or place.	2
Thallium	Acute toxicity occurs as low as 1400 µg/l in freshwater species and as low as 2130 µg/l in saltwater species. Chronic toxicity occurs as low as 40 µg/l in freshwater species, and one freshwater fish is affected after 2600 hours as low as 20 µg/l.	1
	For salt species, because of a chronic effect of long-term exposure, tests should be conducted for at least 20 days to determine harmful, sublethal concentrations. The concentration in seawater should not exceed 0.05 of this concentration. Concentrations equal to or exceeding 100 µg/l constitute a hazard to the marine environment, and levels less than 50 µg/l present minimal risk of deleterious effects.	2

Appendix 5.3 Continued

Pollutant	Criteria	Ref.[b]
Toluene	Acute toxicity occurs as low as 17 500 µg/l in freshwater species and as low as 6300 µg/l in saltwater species. Chronic toxicity occurs in saltwater species as low as 5000 µg/l.	1
Uranium	For protection of saltwater species an application factor of 0.01 is recommended to be applied to the marine 96-hour LC50. Concentrations equal to or exceeding 500 µg/l constitute a hazard to the marine environment, and levels less than 100 µg/l present minimal risk of deleterious effects.	2
Vanadium	It is recommended that the concentration of seawater not exceed 0.05 of the 96-hour LC50 for the most sensitive species.	2
Zinc	For freshwater aquatic life, total recoverable zinc should not exceed 180 µg/l at any time. assuming a hardness of 50 mg/l as $CaCO_3$. For saltwater species the concentration should not exceed 170 µg/l at any time. The 24-hour average criterion for freshwater is 47 µg/l for a hardness of 50 mg/l. The 24-hour saltwater criterion is 58 µg/l.	1

*In addition to the pollutants listed in the Appendix, certain pesticides and numerous halogenated organics are addressed by the US EPA 1980 Water Quality Criteria for protection of aquatic life and/or levels at which toxicity occurs are specified.

Pesticides	**Halogenated organics**
Aldrin/dieldrin	Carbon tetrachloride
Chlordane	Chlorinated benzenes
DDT	Chlorinated ethanes
Endosulfan	Chloroalkyl ethers
Endrin	Chlorinated naphthalene
Heptachlor	Chlorinated phenols
Toxaphene	Chloroform
	2-Chlorophenol
	Dichlorobenzenes
	Dichlorobenzidine
	Dichloroethylenes
	2,4-Dichlorophenol
	Dichloropropanes/propenes
	Haloethanes
	Halomethane
	Hexachlorobutadiene
	Hexachlorocyclohexane
	Hexachlorocyclopentadiene
	Pentachlorophenol
	Tetrachloroethylene
	Trichloroethylene
	Vinyl chloride

[b] References:
1. US EPA, Water Quality Criteria, *Federal Register,* November 28, 1980 (with updates).
2. NAS/NAE, Water Quality Criteria 1972. Prepared for the US Environmental Protection Agency by the National Academy of Sciences, National Academy of Engineering, National Academy of Sciences, Washington, DC, EPA-R3-73-933.
3. US EPA (1984) Water Quality Criteria; Request for Comments, *Federal Register*, Volume 49, No. 26, 4551–4554, February 7, 1984.

[c] For arsenic, chromium, copper, lead, and mercury, the chemical is defined as the dissolved fraction that passes through a 0.45 µm membrane filter.

ln = natural logarithm

CHAPTER 6

HUMAN ACTIVITIES IN THE DRAINAGE BASIN AS SOURCES OF NONPOINT POLLUTANTS

W.B. Clapham Jr, L. Granat, M.M. Holland, H.M. Keller,
L. Lijklema, S. Löfgren, W. Rast, S.-O. Ryding
J.A. Thornton and L. Vermes

INTRODUCTION

Basin development and nonpoint source pollution

All forms of land use potentially can affect the water quality of stormwater runoff from the land surface to receiving waters. In an undeveloped area, naturally-occurring physical, chemical and biological processes interact to recycle most waterborne materials in stormwater runoff. As a drainage basin becomes more developed, however, these processes which ameliorate the potential environmental effects of natural pollutants are disrupted. Humans contribute to this disruption by adding additional polluting materials to the land surface, including fertilizers (nutrients), pesticides and synthetic organic chemicals, animal wastes, and heavy metals. As stormwater runoff washes these materials off the land surface, the nonpoint source pollutant loads carried to receiving waters by stormwater runoff can increase significantly. It is the purpose of this chapter to identify the types of nonpoint pollutants from different land uses, and to provide an overview of the quantities of the pollutants likely to be generated and/or transported to receiving waters.

Unit area loads

In any discussion of nonpoint source pollutants, it is useful to consider the concept of nutrient export coefficients or unit area loads (UALs), as a means of providing a quantitative description of nonpoint pollutant loads. It also provides a means of comparing the pollutant loads from different types of land uses.

The unit area load of a pollutant is a quantitative expression of the mass of the pollutant generated from a given area of land during a specified time interval. Accordingly, it typically is expressed in units such as kg/ha/y or $g/m^2/y$. The concept of the unit area load is based on the observation that, under average hydrologic conditions, a watershed will generate a relatively constant quantity of a given pollutant over the annual cycle. This qualification obviously means that, if the watershed experiences 'wet' or 'dry' years, the actual UAL can vary

113

significantly from that calculated on the basis of an 'average' hydrologic input. For the same reason, one typically would not calculate a UAL value based on a single storm event. Rast and Lee (1978), PLUARG (1978), Novotny and Chesters (1981), Ryding and Rast (1989) and Novotny and Olem (1994) discuss these, and other environmental and anthropogenic, limitations on the estimation of UALs in greater detail.

The development of a UAL value involves utilization of the information available on the land uses within a watershed, and the pollutant coefficients applicable to those land uses. Because the UAL is a quantitative expression of the mass of a pollutant generated from a given land surface area over a given time interval, it is typically calculated as the product of water drainage and pollutant concentrations from specific land uses (calculation of total pollutant loads, and the resultant UALs, is discussed in further detail in Chapters 7 and 8). Further, the UAL value of a pollutant from watersheds impacted by human activities typically will be greater than that of a pollutant from watersheds that remain relatively undisturbed.

Land use

Simply stated, the term 'land use' refers to the purpose for which a given area of land is being used. Some individuals also use the term to indicate land use or management practices being carried out on a given parcel of land. These terms obviously are complementary. Further, in many cases, a given parcel of land may be used for multiple purposes, which may be catalogued to varying levels of precision depending on the scale of the inventory. For example, a typical farming operation might include a barnyard or animal enclosure, a homestead, row-cropped garden, and animal grazing area. At a small scale (e.g. 1 cm = 10 km), this land-use may be catalogued as agricultural; at a larger scale (e.g. 1 cm = 0.1 km), the same land-use might appear as agricultural, residential, and pasture. One typically uses the term land use to describe the prevailing activity within a given land area for most applications using UALs.

The concept of land use is fundamental to assessing nonpoint source pollution, primarily because it denotes the activity on land (or use of the land) that generates a pollutant(s) of concern. In its simplest designation, land can be categorized as (1) forest, (2) agriculture, or (3) urban. However, there are many delineations and sub-delineations of land usage. In fact, planning agencies and similar institutions have developed many dozens of specific land use categories, most of which are simply further, more-defined subdivisions of the three major categories identified above. Many countries have an established land use coding system.

An example of a reasonable land use classification system was developed by PLUARG (1978). PLUARG began with a basic designation of two major categories, specifically (1) rural and (2) urban. Rural lands were further subdivided into (1) general agriculture – a usage broadly encompassing all non-

Table 6.1 Factors affecting the generation of nonpoint pollutants (modified from Novotny and Chesters, 1981)

[Degree of correlation with land use]		
Strong	*[Weak]*	*No correlation*
Population density	Condition of street surface	Meteorology
Atmospheric deposition	Degree of impervious	Soil characteristics
Degree of impervious area	area directly connected	and composition
correlated with population	to drainage channel	Permeability
density	Pollutant delivery ratio	Slope of land
Vegetative cover	Surface storage	Geographic factors
Street litter accumulation rate	Soil organic and nutrient	
Traffic density	content	
Curb height and density		
Street-cleaning methods		
Pollutant conveyance systems		

urban lands, (2) cropland, (3) pasture, (4) land used for receiving municipal sewage fertilizers, (5) idle land – rural land not used for active agricultural purposes, including wetlands and perennial grasslands, and (6) woodland – primarily forests and brushland. Urban lands were subdivided into (1) general urban – a usage broadly encompassing all non-rural lands, (2) residential – land used for housing, (3) commercial, (4) industrial – land used for manufacturing purposes, including mining activities, (5) developing urban land – primarily construction sites, and (6) transportation corridors – primarily land used for road and rail transport. Techniques for identifying specific types of land uses are discussed below.

The utility of delineating specific land uses, as discussed below, is that different types of land uses generate different types and quantities of pollutants. However, factors other than land use can also significantly affect the production of nonpoint pollutants. To cite one such example, based on an international study of the North American Great Lakes Basin, Novotny and Chesters (1981) have summarized the importance of a number of such factors (Table 6.1). Another example, involving nutrient export from major land uses, is provided in Table 6.2. This latter example is based primarily on the climatic conditions found in the northern climatic zone of the United States.

Other considerations regarding land use and nonpoint source pollution

Before moving on to the discussion of different types of land-use activities, and the types and quantities of pollutants produced on these land areas, several other factors regarding nonpoint source pollution merit mention; namely, the calculation of baseline conditions and consideration of the atmosphere as a transport mechanism and source of nonpoint source contaminants.

Table 6.2 General characteristics of nutrient generation from specific land uses, primarily in the north central region of North America (modified from Reckhow *et al.*, 1980)

---

(1) **FOREST**

Forest watersheds with sandy soils overlying granitic igneous watersheds produce half the phosphorus export of forest watersheds with loam soils overlying sedimentary formations

Young (< 5 years) forests produce greater water and sediment-phosphorus runoff than old forests

Areas with warm climates and high rainfall produce greater water and higher phosphorus export than other climatic regions

Deforestation or timber harvest activities generally cause greater nutrient export than undisturbed forested watersheds

Forest fires cause a temporary increase in phosphorus export; however, the levels usually return to pre-fire conditions relatively quickly

(2) **AGRICULTURE**

(A) **Cropland**

Phosphorus export *via* runoff from sandy or gravel soils generally is small

Phosphorus export in runoff from clay soils and organic soils generally is large

The time of fertilizer application is critical; for manure-type fertilizers applied to frozen soils, the phosphorus export following snowmelt and high rainfall periods is large; incorporating applied fertilizers into the soil results in reduced nutrient export

Fertilizer application above recommended levels for existing soil conditions produced increased nutrient export

Conventional tillage methods (e.g. fallow land during non-growth season, harvest removal of crop residues) results in large nutrient export

Conservation tillage practices reduce nutrient export; however, they can result in increased herbicide usage and export

Row crops produce much larger nutrient exports than non-row crops

(B) **Pasture and rangeland**

Limiting livestock grazing time on a given parcel of land reduces nutrient export, in contrast to continuous grazing on the same parcel

Fertilized pastures often exhibit increased nutrient export

The greater the animal density, the greater the potential for increased nutrient export

(C) **Feedlot and manure storage**

The greater the extent of impervious surface, the greater the nutrient export

The greater the animal density, the greater the nutrient export

The greater the roof area:feedlot area ratio, the smaller the nutrient runoff

The use of a detention basin decreases nutrient export

(3) **URBAN**

The greater the extent of impervious surface, generally the greater the nutrient export

Increases in housing density, fertilizer applications, and pet density result in increased nutrient export

Decreases in grass and vegetative cover result in increased nutrient export

Commercial, business, and industrial areas often have considerable vehicular/ pedestrian traffic, resulting in greater nutrient export than residential areas

---

Table 6.2 Continued

(4) ATMOSPHERE

In agricultural areas, increased nutrient export can result from ammonia volatilization from feedlots and fertilizers, and from wind erosion of fertilized soils

Increased nutrient loads from agricultural areas often coincide with fertilization and tilling periods

Increased nutrient export can result from boiler and furnace operations, and from automotive and aviation emissions, in urban areas

Forested watersheds as a baseline condition

Implicit in the following discussions of different types of nonpoint source pollutants is the notion that, as a drainage basin becomes more settled by humans, the generation and export of pollutants increases. Thus, in order to judge the extent of the impact of human activities on the pollutant load from a settled or developed land area, it is necessary to have a baseline or reference value. However, the reality is that there is virtually no location on the Earth that has not been impacted to some degree (even if small) by human activities. As a minimum, even isolated regions can receive inputs of anthropogenic materials, *via* their long-range transport in the atmosphere (see below).

Nevertheless, undisturbed (recognizing the limitations of this term) forested watersheds commonly are used to provide a baseline value thought to represent an 'undisturbed' watershed. This is because undisturbed forests typically exhibit low hydrologic activity which results in a low export of waterborne pollutants. This low hydrologic activity is due to a number of factors, including high water storage in the forest vegetation, the nature of the terrain, and organic content of the soils. Consequently, the export of sediment (and sediment-associated pollutants) is much reduced when compared to other types of land uses. As an example, sediment loads typically are less than 100 kg/ha/y from forested lands in the North American Great Lakes Basin. Disturbed basins in the same region have a significantly higher export of sediment. Studies of several well-studied forested regions throughout the temperate zone provide substantive evidence for this low pollutant export. Thus, forested watersheds are presumed, within the limitations expressed above, to represent baseline conditions in assessing pollutant loadings from human-impacted drainage basins. Elsewhere, watersheds primarily under natural vegetation have been used to similar effect.

The atmosphere as a source of nonpoint pollutants

Another consideration in assessing the generation and export of pollutants from nonpoint sources is the notion of the atmosphere as a pollutant 'source'. In fact, with the possible exception of some specific chemical reactions related to the hydrolysis of acidic emissions from industrial sources, the atmosphere does not constitute a 'source' of any pollutants. Rather, as noted in Chapter 5, it should be viewed instead as a transport mechanism for pollutants released to the atmosphere.

One of the first examples of the substantial role of atmospheric transport of nonpoint source pollutants was the international study of the Pollution From Land Use Activities Reference Group of the International Joint Commission on nonpoint source pollution in the North American Great Lakes Basin (PLUARG, 1978). Indeed, one result of the PLUARG (1978) study was the finding that the majority of the annual phosphorus load to Lake Superior (the most pristine of the North American Great Lakes) entered the lake *via* atmospheric deposition (rainfall, dry fallout). Elsewhere, Balon and Coche (1974) have noted a similar role for the atmosphere relative to nitrate-nitrogen inputs to Lake Kariba (Zimbabwe/Zambia). Its relative significance, in terms of pollutant impacts, depends primarily on the magnitude of the pollutant load from other pollutant sources in a drainage basin. Nevertheless, until demonstrated to be a minor factor, the atmosphere should be considered in any assessment of pollutant loads from nonpoint (and point) sources.

The relative contribution of pollutants from point and nonpoint sources

One other factor to consider in assessing pollutant loads from nonpoint sources is the pollutant load from point sources. This is of particular importance when considering alternative pollutant control measures. In some cases, point source pollutant loads should receive the initial concern. A prime example arises in urban areas; the relative phosphorus load from a municipal wastewater treatment plant typically surpasses that from urban runoff. In such a case, the larger quantities of biologically-available phosphorus in the wastewater effluent should receive the major initial attention. Consideration should be given to nonpoint sources only after the major point sources have been adequately addressed. This conclusion is based on both engineering and economic considerations. Simply stated, humans have had considerable experience in removing phosphorus from domestic wastewater, to the extent that removal can be readily achieved at relatively lower cost than that of nonpoint source controls. In other cases, however, consideration should be given to control of nonpoint sources of a given pollutant. Examples of the latter would include oil and grease from urban areas, smokestack emissions from industrial plants, and sediment runoff from construction sites. The primary control programme must be based on the specific conditions of a given situation.

Examples of the relative inputs of different types of pollutants from various point and nonpoint sources are tabulated in Table 6.3. It lists the ranges in concentrations of several pollutants observed in rural and urban runoff, and compares these values to those measured in municipal wastewaters.

Table 6.3 Pollutant characteristics from different point and nonpoint sources

Parameter[1]	*Urban stormwater*	*Rural stormwater*	*Untreated wastewater*	*Secondary-treated wastewater*
BOD_5	1–400	1–200	100–350	10–30
Suspended solids	2–7340	3–726	100–350	10–30
Total nitrogen	1–49	3.7–15.5	35–100	15–35
Total phosphorus	0.1–16	0.5–12.5	10–50	3–10
Orthophosphate	0.1–10	0.1–10	6–35	0.5–10
Total organic carbon	3–384	1–100	100–300	10–120
Chemical oxygen demand	5–3100	2–150	250–1000	25–150
Lead	0.1–10	0.01–2	0.05–1.27	0.0005–0.20
Zinc	0.1–4.6	0.01–1.5	0.03–8.31	0.047–0.02
Cadmium	0–0.1	0–0.05	0.004–0.14	0.0002–0.02
Oils/grease	0–110	0–50	50–150	5–15
Faecal coliform	$55–112 \times 10^6$	$55–112 \times 10^6$	$10^7–10^9$	200
Total coliform	$200–146 \times 10^6$	$200–146 \times 10^6$	$10^9–10^{11}$	1000

[1]all parameters expressed in mg/l, except coliforms (organisms/100 ml)

THE ATMOSPHERE AS A TRANSPORT MECHANISM

Introduction

The flux of pollutants from the atmosphere directly onto the surface of an inland water can, under certain conditions, contribute a significant fraction of the total pollutant input. Such inputs also include the more generally recognized fluxes such as municipal and industrial discharges, rivers and runoff from urban and agricultural areas. This holds true, especially in a rural environment, for large lakes and for components of atmospheric deposition that would otherwise be retained in the catchment.

Wet deposition is a function of contaminants carried to the water surface by precipitation (rain, snow, hail). Over a small lake, the total amount of materials deposited in this fashion is probably not significantly different from that over surrounding land areas. However, for very large lakes (such as the Great Lakes, the Baltic, etc.), the deposition process might be substantially altered by other processes taking place over the lake, an example being the typically lower amounts of precipitation falling onto the lake surface (compared to the surrounding land).

Dry deposition is the common name for all deposition processes other than wet deposition. It includes transfer of gases from the atmosphere to vegetation, soil, lake water, etc., and may involve both biotic and abiotic fixation of the gases. Large particles will fall by gravity, smaller ones must be transferred by atmospheric movements to the surface where they can be captured by impact, interception or molecular diffusion through the layer nearest the surface. As with gases the deposition rate depends on the surface topography and atmospheric turbulence but also on particle size. Dry deposition rates are dependent on particle size distribution.

This section presents a review of measurements of wet and dry deposition with particular emphasis on deposition to inland water surfaces. Some examples in which atmospheric deposition has been measured or estimated and found to be significant compared to other inputs are given in order to provide a basis upon which to assess the importance of atmospheric deposition compared to other known sources in the case of a specific water surface. The examples focus on a few groups of compounds:

(1)   Nutrients, of which nitrogen and/or phosphorus are often a growth limiting factor and, thus, of special concern;
(2)   Trace metals;
(3)   Toxic organic micro-constituents, which include a large number of components (i.e. pesticides, polyaromatic hydrocarbons (PAHs), chlorinated organic compounds such as DDT and PCBs, or volatile organic carbons (VOCs)); and
(4)   Acidifying components (oxides of sulfur and nitrogen).

Nutrients

Nitrogen

Nitrate and ammonium concentration field data have been obtained and published in the United States (Figures 6.1 and 6.2) and Europe (Figures 6.3 and 6.4). National networks may provide more detailed information on the spatial variation of these compounds on a site-specific basis. The concentration fields for nitrate and sulfate (see below) have much in common but the nitrate to sulfate ratio is larger in North America than in Europe, although the ratio increases from central to northern Europe. The maxima in ammonium concentrations are somewhat displaced relative to those of nitrate both in North America and Europe as a consequence of different sources, with ammonia essentially being emitted in farming areas (principally from manure from domestic animals) while nitrate and sulfate emissions in urban areas are related to the combustion of fossil and wood fuels. Concentrations in remote continental regions generally can be seen in Table 6.4.

Organic nitrogen compounds in precipitation have been measured as the difference between 'total nitrogen' and the sum of nitrate and ammonium. Very contradictory results have been published over the years, in which the organic fraction ranges from less than 20 percent to almost equal to the inorganic fraction. Consequently, no firm conclusions can at present be reached on the spatial distribution of these organic nitrogen components.

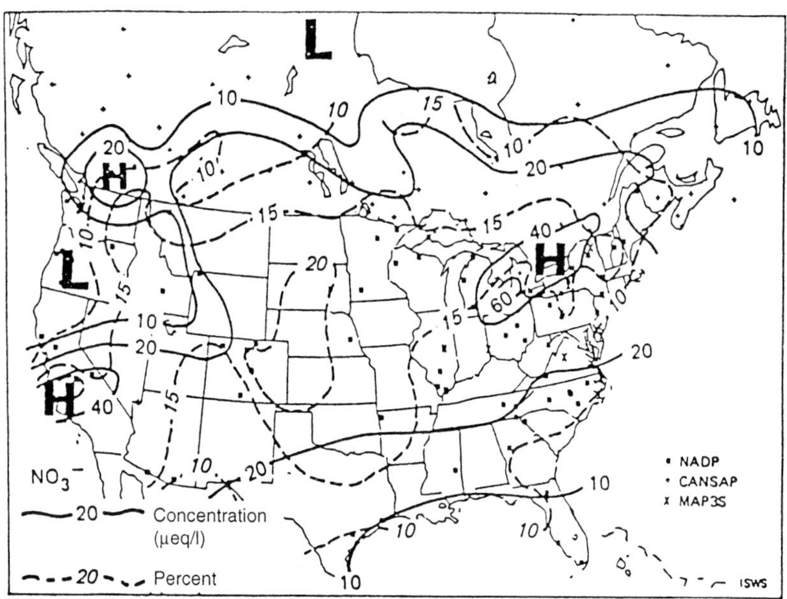

Figure 6.1 Distribution of average monthly nitrate concentrations (in µeq/l) over North America (from Semonin and von Boversox, 1983)

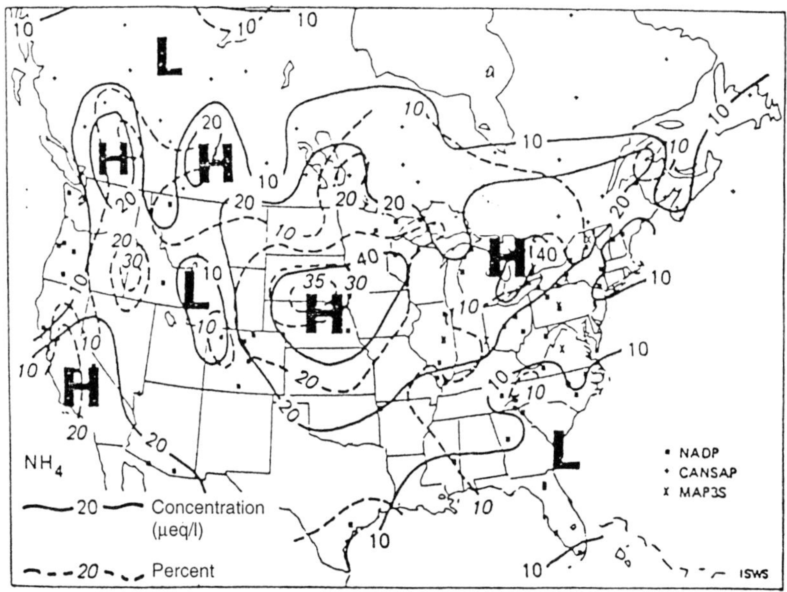

Figure 6.2 Distribution of average monthly ammonium concentrations (in µeq/l) over North America (from Semonin and von Boversox, 1983)

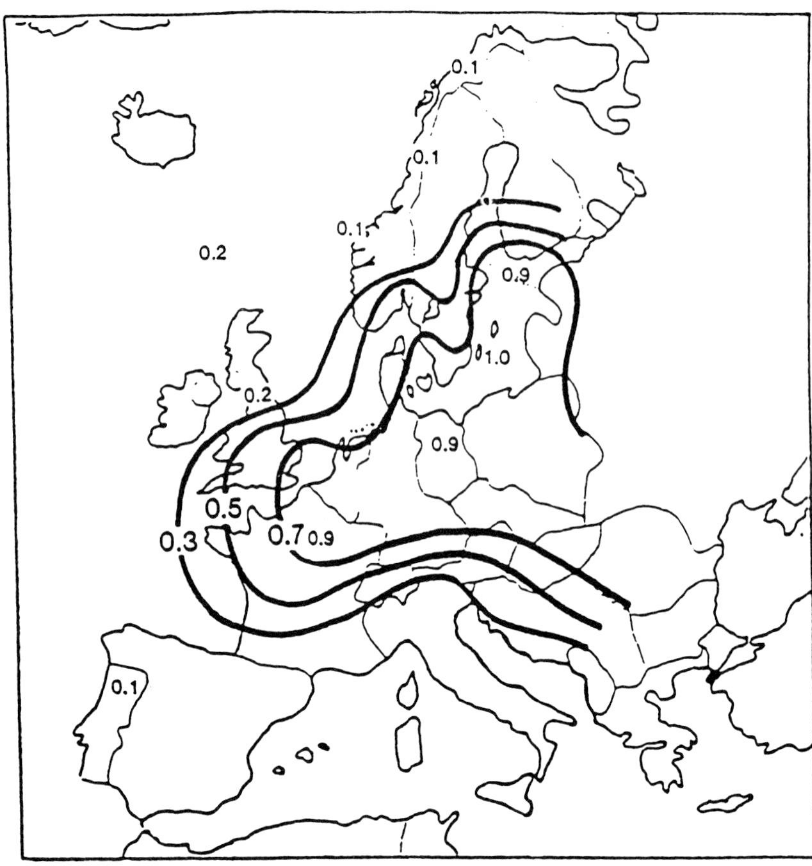

Figure 6.3 Mean concentrations of nitrate (mg NO_3-N/l) in precipitation in Europe (based on EMEP data from 1983)

Components of dry deposition of nitrogen include gases such as N_2, N_2O, NO, NO_2, HNO_3, HNO_2, PAN, NH_3 as well as particulate NO_3, NH_4^+ and organic N. Of these, HNO_3 and NH_3 are very soluble in lake water and the deposition velocity is only limited by the rate of transfer from the gas phase, in a similar manner to SO_2. Only in rare cases one can find a combination of high ammonium concentration and high pH in water and a very low ammonium concentration in the air. Such conditions result in the flux of ammonia from the lake to the atmosphere (Hendry *et al.*, 1981) or a decreased flux to the lake.

The most uncertain part of the flux calculation is presently estimated air concentration levels of the soluble gases mentioned above. Only occasional studies have been performed and reported but indicate, together with model calculations, that the concentration fields of HNO_3 and NH_3 have a quite different structure.

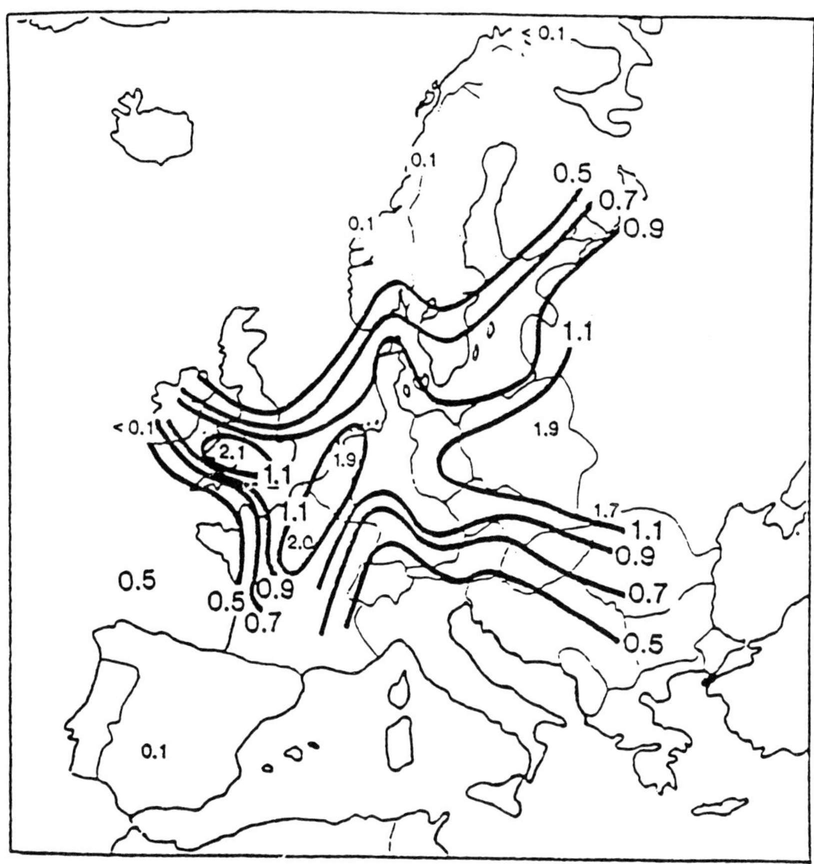

Figure 6.4 Mean concentrations of ammonium (mg NH_4-N/l) in precipitation in Europe (based on EMEP data from 1983)

Ammonia concentrations reported from the US range from about 10 ppb in a farming area to less than 0.5 ppb (Dawson, 1984; US EPA, 1984; Cadle *et al.*, 1982). A similar range was observed in Europe. Assman *et al.* (1988) have mapped the latter and have compiled a contour map of NH_3 concentrations based on estimated emission rates. Values ranged from about 6 ppb in extensive farming regions (The Netherlands) to less than about 0.1 ppb in the forested regions of central Scandinavia. It is to be expected that much higher concentrations occur locally around point sources. It can further be assumed that ammonia has a short residence time in the lower atmosphere both with regard to reaction with acidic aerosols and deposition on vegetation (*cf.* Assman *et al.*, 1988).

For HNO_3, Logan (1983) used 0.8 ppb for the eastern US, 0.07 ppb for the western US, and 0.035 ppb for Canada in his budget calculations; similarly,

Table 6.4 Wet deposition of sulfate, nitrate and ammonium to remote continental areas (from Galloway, 1985)

Location	*Number of samples*	*Precipitation amount* (cm)	SO_4^{2-} *Concentration* (µeq/l)	*Deposition* (gS/m²/y)	*Reference*
Global estimates	–	40	44	0.28 ± 0.08	–
	–	–	–	0.15	
East Africa	61	–	13	0.15	Rodhe *et al.*, 1980
Nigeria	–	–	8	0.11	–
Venezuela					
San Carlos	14	391	2.7	0.17	Galloway *et al.*, 1982
Brazil					
Amazon Basin	28	240	8.8	0.4	–
Lake Calado	59	200	3.8	0.12	Melack unpubl. data
Australia					
Katherine	120	112	4.7	0.064	Likens, 1985

Table 6.4 Continued

| Location | Number of samples | Annual precipitation (cm) | NO_3^- | | NH_4^+ | | Reference |
			Precipitation concentration (μeq/l)	Deposition (gN/m^2/y)	Precipitation concentration (μeq/l)	Deposition (gN/m^2/y)	
Global estimates	–	–	–	0.10–0.23	–	0.23–0.47	Söderlund, 1982
	–	–	–	0.05*	–	0.05*	Logan, 1983
	–	–	–	0.015–0.05	–	0.01–0.06	–
Venezuela							
San Carlos	14	391	2.6	0.14	2.3	0.13	Galloway *et al.*, 1982 Logan, 1983
Brazil							
Amazon Basin	28	391**	2.1	0.11	1.1	0.062	–
Lake Calado	54	200	3.5	0.10	3.1	0.087	Melack, unpubl. data
Australia							
Katherine	40	112	4.3	0.055	2.0	0.033	Galloway *et al.*, 1982 Logan, 1983
	120	112	4.9	0.077	3.4	0.056	Likens, 1985

*Lower estimate; **estimated

Kelly *et al.* (1984) gave levels of a few tenths of a ppb in the Whiteface Mountains of the US. In Europe, Ferm *et al.* (1984) measured about 0.4 ppb as an average for one year on the west coast of Sweden.

In addition to the uncertainties mentioned, it must also be remembered that these measurements of HNO_3 and NH_3 concentrations were made over land with a poorly specified fetch (upwind area). Applying these values of deposition velocties to specific lakes requires, in principle, that the concentrations be measured at a reference height over the water and under such conditions that a proper concentration profile can be developed (see Chapter 7).

For particulate matter, the deposition velocity for ammonium is similar to that of sulfate, with the same uncertainties (see below), but a larger value may be used for NO_3. NO_2 is rather insoluble in water and its deposition velocity therefore small. N_2 is consumed in biological nitrogen fixation by blue–green algae (cyanophytes/cyanobacteria). Messner and Brezonik (1981) quote a typical value of 0.1 kg $N/m^2/y$ in north central Florida (USA). Higher values are also given but are for eutrophic lakes which already receive a substantial N input from other (external) sources.

Based on the numbers given above, dry deposition of NH_3 may be more important than wet deposition near specific sources but probably makes a small contribution to loading in most other cases (for instance, in NW Europe). Thus, where the estimates are most uncertain, other inputs to a lake than from the atmosphere may be more important. In the case of HNO_3, dry deposition to a lake can make a small but significant contribution compared to that of wet deposition.

Phosphorus

Phosphorus is introduced to the atmosphere both with soil dust (such as fertilized soil that is rich in P), from industrial processes, and from vegetation. It may be present in atmospheric aerosols (*cf.* Delumyea and Petel, 1978).

Data on wet deposition of phosphorus, both as orthophosphate or total phosphorus, show great spatial variability when measured with bulk or wet-only collectors. For example, values ranging from 6 to 60 µg P/l have been reported from The Netherlands (KNMI/RIV, 1983), while values ranging 42 to 83 µg/l were measured with bulk collectors and 10 to 94 µg/l with wet-only collectors within a 30 km radius in an area 200 km north of Toronto, Canada (Dillon and Reid, 1981). Other measurements (Granat, unpublished) show a similar variability with a typical mean value of about 10 µg P/l. It might, therefore, be inferred that at least some of the phosphorus in wet deposition is of quite local origin. The larger values quoted above may, not be directly appropriate for the calculation of deposition over a large lake.

Dolske and Sievering (1979) estimated dry deposition of phosphorus from measurements over Lake Michigan and concluded that both soil dust and industrial emissions contributed to the deposition. Dry deposition was equal to the wet deposition. In this connection, it might be interesting to note that, for a lake of 0.5 km width in Poland, about half of the atmospheric input of phosphorus was due to insects transported from the shoreline to the middle of the lake (Kowalczewski and Rybak, 1981). A similar situation was reported from Lake Tahoe (USA) and from various African lakes as a result of locust swarms. These cases, however, may be treated as a special case of dry deposition. Likewise, Nicholls and Cox (1978) report that up to 20 percent of the atmospheric phosphorus load may be due to pollen. Large amounts of phosphorus are also transported from farming areas to lakes as windblown dust. Such transport results in very uneven deposition patterns.

Usually data on phosphorus are insufficient to estimate spatial trends in wet and dry deposition. Moreover, since the transport can include large particles with a short residence time, the load can vary considerably from place to place and within a lake, and thus may require a large number of sampling sites to be assessed properly.

Trace metals

Trace metals are emitted into the atmosphere from a number of sources including natural ones such as resuspension of soil particles, dust, and sea salts, as well as from direct anthropogenic emissions. By comparing the concentrations of a particular metal with a typical reference element in soil and in precipitation, for instance, it is possible to estimate the contribution from various sources, at least in some cases. This is done on a global basis in Table 6.5. A valuable reference is also an emissions inventory from Europe (Table 6.6) which was recently supplemented by results on mercury emissions in Europe which suggested a value of 117 tons/acre (Pacyna, 1984).

The increased deposition rate of trace metals from the atmosphere during the last century has been revealed by analysis of undisturbed sediment cores in remote regions. For instance, the deposition of lead has increased considerably from the middle to the late 19th century (see, for instance, Evans and Dillon, 1982; Evans and Rigler, 1985; Heit *et al.*, 1984; Kahl, 1982). However, these measurements themselves do not give information on how much of the lead was deposited directly to the lake surface.

Galloway *et al.* (1982) reviewed the literature on trace metal concentration in precipitation (Table 6.7). They divided the data into three groups: urban, rural and remote. The last group was essentially reserved for measurements in the Arctic and Antarctic regions. The second group was of special interest in the present context. The concentration ranges for individual elements in this group were very large. Some of the variation can undoubtedly be explained by measurement errors as is obvious from the comparison of data from nearby places obtained by different investigators. However, the range in concentrations for individual elements undoubtedly also reflects a spatial trend in concentration away from major source regions. Few attempts have been made to visualize

Table 6.5 Natural and anthropogenic sources of atmospheric emissions (from Lantzy and Mackenzie, 1979)

Element	Continental dust flux	Volcanic dust flux	Volcanic gas flux	Industrial particulate emissions	Fossil fuel flux	Total emissions (industrial + fossil fuel)	Atmospheric interference factor (%)
Al	356 500	132 750	8.4	40 000	32 000	72 000	15
Ti	23 000	12 000	–	3600	1600	5200	15
Sm	32	9	–	7	5	12	29
Fe	190 000	87 750	3.7	75 000	32 000	107 000	39
Mn	4250	1800	2.1	3000	160	3160	52
Co	40	30	0.04	24	20	44	63
Cr	500	84	0.005	650	290	940	161
V	500	150	0.05	1000	1100	2100	323
Ni	200	83	0.0009	600	380	980	346
Sn	50	2.4	0.005	400	30	430	821
Cu	100	93	0.012	2200	430	2630	1363
Cd	2.5	0.4	0.001	40	15	55	1897
Zn	250	108	0.14	7000	1400	8400	2346
As	25	3	0.1	620	160	780	2786
Se	3	1	0.13	50	90	140	3390
Sb	9.5	0.3	0.013	200	130	380	3878
Mo	10	1.4	0.02	100	410	510	4474
Ag	0.5	0.1	0.0006	40	10	50	8333
Hg	0.3	0.1	0.001	50	50	110	27 500
Pb	50	8.7	0.012	16 000	4300	20 300	34 583

All fluxes are in units of 10 g y
Interference factor = (total emissions continental – volcanic fluxes) × 100

Table 6.6 European emissions (tons/y) of trace elements from various sources during 1979 (from Pacyna, 1984)

Europe	As	Be	Cd	Co	Cr	Cu	Mn	Mo	Ni	Pb	Sb	Se	V	Zn	Zr
1. Conventional thermal power plants	284	21	101	787	1196	1377	1011	352	4580	1138	122	155	12 600	1316	732
2. Industrial commercial and residential combustion of fuels	394	29	155	1214	1580	2038	1378	493	7467	1652	154	218	21 800	1824	939
3. Wood combustion	40		25			1500			375	562				4590	
4. Gasoline combustion			31				92		1330	74 300					
5. Mining			1			192	275		1640	1090		0.2		460	
6. Primary non-ferrous metal production															
6.1 Copper–nickel	4490		595			7850				9250				2500	
6.2 Zinc–cadmium	910		1550			440				7880		13		48 800	
6.3 Lead	300		8			120			140	10 450				180	
7. Secondary non-ferrous metal production															
7.1 Copper			2			61				55	1			660	
7.2 Zinc														2630	
7.3 Lead			1							387				150	
8. Iron, steel ferro-alloys manufacturing			58		15 400	1710	10 770		340	14 660				10 250	
9. Refuse incineration	11		84	4	53	260	114		10	804	100	32	19	5880	
10. Phosphate fertilizers			27			77			77	6		v.s.		230	
11. Cement production			15		663					746					
Total	6500	50	2700	2000	18 900	15 500	17 600	850	16 000	123 000	380	420	34 500	80 000	1700

Table 6.7 (a) Range and (b) median concentrations (in µg/l) of toxic metals in wet deposition (from Galloway *et al.*, 1982)

a.

	Urban		Rural		Remote	
	Range (µg/l)	*Number of references*	*Range (µg/l)*	*Number of references*	*Range (µg/l)*	*Number of references*
Sb	–	–	–	–	0.034	1
As	5.8	1	0.005–4	9	0.019	1
Cd	0.48–2.3	5	0.08–46	23	0.004–0.639	4
Cr	0.51–15	4	< 0.1–30	9	–	–
Co	1.8	1	0.01–1.5	2	–	–
Cu	6.8–120	6	0.4–150	19	0.035–0.85	5
Pb	5.4–147	8	0.59–64	32	0.02–0.41	6
Mn	1.9–80	8	0.2–84	28	0.018–0.32	5
Hg	0.002–3.8	6	0.005–2.2	10	0.011–0.428	4
Mo	0.20	1	–	–	–	–
Ni	2.4–114	6	0.6–48	15	–	–
Ag	3.2	1	0.01–0.48	7	0.006–0.008	2
V	16–68	3	0.13–23	6	0.016–0.32	3
Zn	18–280	9	< 1–311	32	0.007–1.1	8

b.

	Urban	*Rural*	*Remote*
Antimony (Sb)	–	–	0.034
Arsenic (As)	5.8	0.286	0.019
Cadmium (Cd)	0.7	0.5	0.008
Chromium (Cr)	3.2	0.88	–
Cobalt (Co)	1.8	0.75	–
Copper (Cu)	41	5.4	0.060
Lead (Pb)	44	12	0.09
Manganese (Mn)	23	5.7	0.194
Mercury (Hg)	0.745	0.09	0.079
Molybdenum (Mo)	0.20	–	–
Nickel (Ni)	12	2.4	–
Silver (Ag)	3.2	0.54	0.007
Vanadium (V)	42	9	0.163
Zinc (Zn)	34	36	0.22

such spatial trends or to construct contour plots for trace metals. An early attempt was made by Lazrus *et al.* (1970) based on measurements made during nine months in 1966–67. A model calculation based on known sources of lead was claimed to confirm Lazrus' results (Nieman, 1984). In contrast, Barrie and Vet (1984) compared present-day measurements of lead in snow (in southern Canada) to these forecasts and found much lower values. They concluded that the discrepancy can be due both to contamination during the earlier measurements and to actual reductions in lead emissions. At any rate, the old plots may not be adequate for assessing the present-day situation as lead emission rates have changed dramatically (the use of lead in the developed countries has decreased, while emissions from automobile exhausts and other processes may have increased in developing countries). Due to the many problems in sampling and analysis of metallic pollutants, one must realize the possibility of even rather large errors; a now classic example was given by Patterson and Settle (1976). Without a very detailed interpretation on how trace metal data were obtained in each reported study, it is difficult to judge and organize available information, although some generalizations are given below.

In a major source region, large spatial variations can be found. Thus, Nurnberg *et al.* (1984) reported an increase by a factor of 10 from 'rural' to 'problem areas' with respect to concentrations of several trace metals in precipitation. It is then very satisfactory to see that independent measurements at rural locations by Thomas (1984) agreed very well with those of Nurnberg *et al.* (1984). Data from the northern part of their study area also agreed quite well with data from Denmark and southern Sweden (Ross, 1985).

With rather few sampling stations for trace metals, additional information can be obtained by extensive surveys of snow accumulated during the winter (Ross and Granat, 1985; Barrie and Vet, 1984) or from analysis of mosses. Data have been reported from such sources for deposition fields over Scandinavia for arsenic, cadmium, chromium, copper, iron, lead, nickel, vanadium and zinc. Two examples are reproduced here as Figures 6.5 and 6.6, one illustrating large depositions around certain point sources (Cr) and one illustrating depositions dominated by large-scale gradients (Cd). The snow survey of Ross and Granat (1985) showed that the concentrations of cadmium, lead and zinc had a remarkably good relationship to the simultaneously measured sulfate concentrations, except near (< 30 km) a large point source.

Spatial trends in trace metal deposition in North America seemingly have similar characteristics. Data from Chan *et al.* (1987) show the large local influence of emissions from industrial sources in Sudbury, Canada. Subsequently, Gatz *et al.* (1988) evaluated data from the Great Lakes Atmospheric Deposition (GLAD) network and created contour plots of lead and cadmium deposition over the Laurentian Great Lakes. Data from Thornton and Eisenreich (1982) and Lindberg and Turner (1984) show a similar spatial variability over different scales and suggest that the influence of soil dust in the northwestern part of the US (Michigan) decreases towards the west. Their study also indicated that the

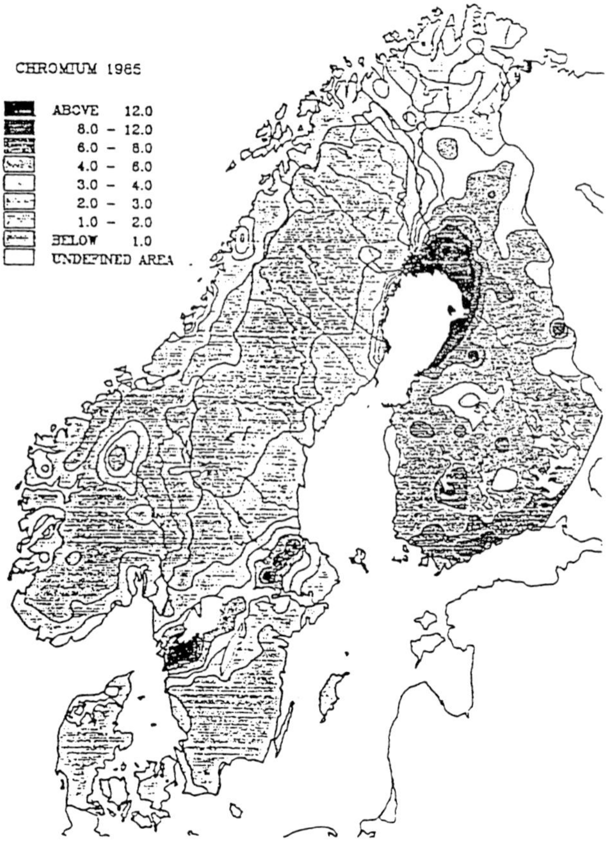

Figure 6.5   Chromium concentrations (in µg/g dry weight) in moss during 1985

anthropogenically derived components had about the same concentration as the dust. It was further observed that spatial trends over distances of about 130 km in the southeastern US (in the southern Appalachian Mountains) were greater than the difference between this location and north-central US (Michigan) for some metals. Finally, comparing data from the east coast of the US (Church *et al.*, 1984) with those obtained in Sweden, there is remarkably good agreement between the absolute values of the atmospheric metals and the use of sulfate as an indicator of anthropogenic influences from the major source regions (in North America and Europe, respectively).

Most trace metals are found in particulate matter. Exceptions include the gaseous components of mercury, lead and selenium. Outside the major source regions, particulate toxic metals of anthropogenic origins are mainly found in the accumulation mode (*cf.* Lannefors *et al.*, 1983).

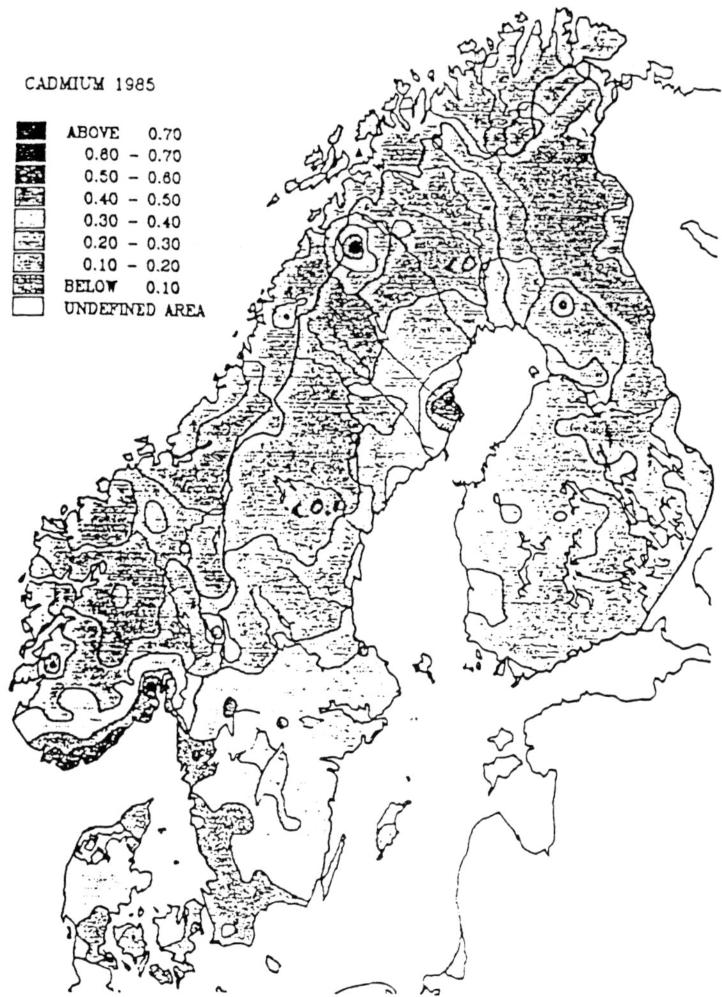

Figure 6.6   Cadmium concentrations (in µg/g dry weight) in moss during 1985

Based on the deposition velocities quoted earlier for sub-micron sized particles, the rate of dry deposition of metals will be small compared to wet deposition (about 10 percent) in rural areas. It should also be noted that trace metals can be released as large particles which have an appreciable settling velocity. In such cases (for example, within some 30 km of the smelters in Sudbury, Canada; Chan *et al.*, 1987), the deposition velocity will be larger and dry deposition more important. Such conditions could, however, not be generalized and must be considered from case to case.

The wet and dry deposition of trace metals of anthropogenic origin seems to be spatially more variable than, for instance, for sulfate (see below). Such a relationship is to be expected with regard to the chemical and physical processes involved in their removal. Mesoscale variations (100 km) seem to be large in source regions while careful measurements in more remote areas have revealed a more coherent picture, similar to that obtained for sulfate. Nevertheless, it appears to be possible to give some typical values for rural areas at various distances from major source regions. Such numbers might be relevant to within a factor of 2 to 4 and could be useful in assessing the order of magnitude of possible inputs to a particular lake (see, for instance, Table 6.6). To obtain a deposition estimate of metals which have not been monitored regularly an analogy might be tried between the component of interest and a 'reference component' which is measured in precipitation and with a 'known' source strength for a whole region, as in Table 6.6. The reference element might be sulfate or nitrate.

Synthetic organic chemicals

Trace organic components of concern include both deliberately distributed organics (herbicides, insecticides, fungicides) and those released during fossil fuel combustion, incineration or other industrial activity. For a detailed list of components of interest, see US EPA (1979). Components which undergo little degradation in the atmosphere and are adsorbed on other particulates will have a residence time which permits transport over hundreds to thousands of kilometers, while those present as, or adsorbed on, larger particles will be removed quite rapidly from the atmosphere (as discussed by Hites, 1981, among others). The concentrations of trace organic components in atmospheric deposition have been reviewed by Galloway *et al.* (1979) and the results are summarized in Table 6.8.

Of the many different synthetic organic chemicals, only a few have been included in regular measurement networks (i.e. the GLAD network around the Laurentian Great Lakes and a network in The Netherlands; KMNI, 1987). Deposition measurements of trace organic components are limited and principally related to studies of deposition on the Great Lakes (see Table 6.9). Table 6.9 is one example (of many) of the more typical, occasional study in which a large number of components were determined, including several types of organic pollutants.

Polycyclic aromatic hydrocarbons (PAHs) have been studied extensively, both in urban and more remote areas, due to the carcinogenic effect of some components belonging to this group. PAHs are produced by combustion and are released to the atmosphere adsorbed on soot or fly ash. Van Vaeck and van Cauwenberghe (1985) measured the particle size distribution of some 60 non-volatile organic constituents present as ambient aerosols. They concluded that most of the PAHs are generated by anthropogenic sources and incorporated in

Table 6.8 Overview of trace organic pollutants in atmospheric precipitation (from Galloway *et al.*, 1979)

Compound	Air concentration		Precipitation concentration	
	Range (ng/m³)	*Mean (ng/m³)*	*Range (ng/l)*	*Mean (ng/l)*
Total PCB	0.1–2	1	15–50	25
Total DDT	0.01–0.05	0.03	1–10	5
Dieldrin	0.01–0.1	0.05	1–4	2
Aldrin	0.1–1	0.5	0.5–3	2
Toxaphene	0.02–2	0.5	1–10	5
α-BHC (Lindane)	1–4	2	1–15	8
Hexachloro-benzene	0.1–0.3	0.2	1–4	2
Chlordane	0.01–0.5	0.1	1–5	2
Heptachlor	1–5	2	1–9	3
Dibutylphthalate	0.5–5	2	4–40	10
Diethylhexylphthalate	0.5–5	2	4–40	10
Total PAHs	10–100	20	50–300	100

Table 6.9 Concentrations of trace organic compounds measured in the Laurentian Great Lakes region (from Strachan and Huneault, 1979; Eisenreich *et al.*, 1981)

a.

	Mean concentrations (ng/l)			
	L. Ontario (16)	*L. Erie (7)*	*L. Huron Georgian Bay (13)*	*L. Superior (14)*
Total PCB	32	9	11	26
Lindane	4.7	6.1	6.0	4.9
α-BHC	19.1	10.3	13.3	4.6
ΣDDT-residues	5.6	3.8	2.7	0.8
α-Endosulfan	3.8	1.6	0.1	0.2
β-Endosulfan	12.0	2.0	2.1	1.0
Dieldrin	1.3	2.6	1.0	0.5
Methoxychlor	8.5	13.1	9.5	1.6
HCB	nd	nd	nd	2.8

nd, not determined

Table 6.9 Continued

b.

Compound	Air		Precipitation	
	Range (ng/m³)	*Mean** (ng/m³)	*Range* (ng/l)	*Mean** (ng/l)
Total PCB	0.4–3	1.0	10–100	30
Total DDT	0.01–0.05	0.03	1–10	5
α-BHC	0.25–0.4	0.3	1–35	15
γ-BHC	1–4	2	1–15	5
Dieldrin	0.01–0.1	0.05	0.5–30	2
HCB	0.1–0.3	0.2	1–4	2
p,p'-Methoxychlor	–	1	1–20	8
α-Endosulfan	–	1	1–10	2
β-Endosulfan	–	1	1–12	3
Total PAH	10–30	20	50–300	100
Anthracene	0.1–1	0.6	1.3–2.3	2
Phenanthrene	0.1–1	0.6	2.0–2.3	2
Pyrene	0.1–4	1.1	1.3–4.5	2
Benzo[a]anthracene	0.1–1	0.5	2.6–3.1	3
Perylene	0.1–2	0.5	–	1
Benzo[a]pyrene	0.1–2	1	0.1–3.1	2
TCC	$2–15 \times 10^3$	9×10^3	$1–5 \times 10^5$	2×10^6
DBP	0.5–5	2	4–10	6
DEHP	0.5–5	2	4–10	6

*concentrations used to calculate dry and wet deposition

particulate matter with a marked preference for the sub-micron sized particles with the largest specific surface areas.

As with trace metals, the stability of the PAHs makes it possible to estimate anthropogenic contributions from the historical record of atmospheric deposition in more remote areas from lake sediment samples. Hites (1981) showed that total PAH concentrations ranged from 14 000 ppb near the sediment surface to less than 120 ppb at the core bottom and concluded that the higher concentrations present in sediments of more recent origin were due to anthropogenic combustion sources. This conclusion is consistent with rapid increase in concentration beginning around 1900.

Inputs to large waterbodies of polychlorinated biphenyls (PCBs) have been of much concern due to their bioaccumulation and toxic effects. The presence of measurable concentrations of these components is found throughout the northern and southern hemispheres. The atmosphere has been shown to be a major pathway for PCB inputs into several large lakes (see further below). Although PCB production has been stopped in the US and reduced to very low levels in Europe, emissions from municipal landfills and incinerators still seems to be of concern as a source of PCB deposition in the Laurentian Great Lakes region (Murphy *et al.*, 1985). The effect of resuspension of already deposited material in the lakes also has increased in relative importance as external contributions are controlled (Baker *et al.*, 1985).

Salts

Sulfur is included in this overview because non-sea salt sulfate (excess sulfate) contributes to acid deposition. Sulfur may also be used (with caution) as a general indicator of dispersion and deposition of other air pollutants, particularly materials such as sub-micron aerosols and moderately reactive gases, emitted in the major industrialized areas of the world.

Contour maps of sulfate wet deposition or concentration in precipitation have been reported both for North America and Europe based on data from large-scale networks. Figures 6.7 and 6.8 are some published examples. Additional information is also available from many national networks in Europe that has not yet been integrated into such maps of large-scale concentration/deposition fields.

The probable deviation between the concentration fields given in Figures 6.7 and 6.8 and the true value for a limited area (e.g. a specific lake) depends very much on where the comparison is made. Rather little experimental evidence is available but we can at least present some hints of the possible magnitude.

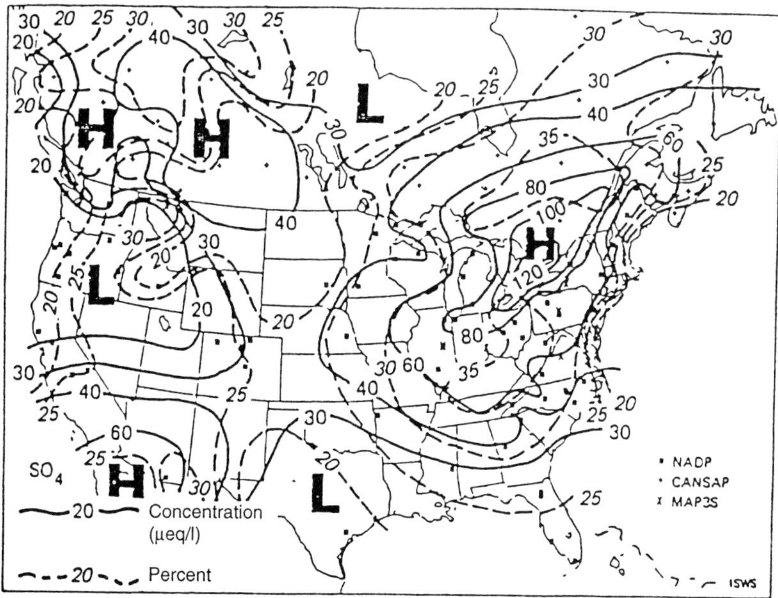

Figure 6.7 Distribution of average monthly sulfate concentrations (in μeq/l) over North America (from Semonin and van Boversox, 1983)

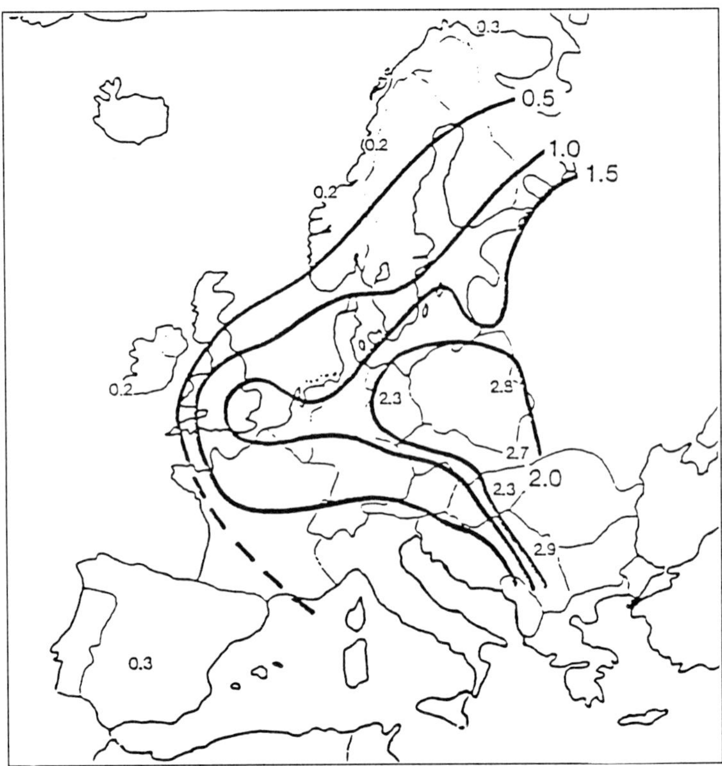

Figure 6.8 Mean concentrations of sulfate (in mg SO_4–S/l) in precipitation in Europe (based on EMEP data from 1985)

Examples from national networks with a high station density may indicate a between-station variability equal to or less than 10 percent. In major source areas the spatial variability between sampling sites is probably larger but detailed information is not readily available.

Measurements in China and Japan indicate that excess sulfate concentrations are of the same order of magnitude as in Europe although few data have been published outside of these countries. In China, excess sulfate has made the precipitation acidic in contrast to a normal pH, in other areas, of above 6. There is some evidence that part of the sulfate contributing to this acidification problem is from windblown dust (Zhao and Sun, 1986; D. Zhao, personal communication). Much lower sulfate concentrations have been reported from rural areas in Africa as indicated in Table 6.10.

As mentioned above, the deposition flux of sulfur dioxide to water surfaces depends on wind speed and atmospheric stability which both vary with season and geographic location. Better estimates of this flux can be made for any given

Table 6.10 Mean concentrations (in mg/l, except for the metals which are in µg/l) of various contaminants measured in precipitation in South Africa during the International Geophysical Years, 1957/66, and 1979/80 (after Bosman and Kempster, 1985)

		1979/80	
Determinand	*1957/66*	*Wet-only*	*Bulk*
pH	6.70	3.9	4.22
Ca	1.75	0.4	1.82
Mg	0.71	0.2	0.57
Na	0.53	0.6	0.86
K	0.36	0.3	0.53
SO_4	3.96	3.7	5.71
Cl	1.12	0.7	1.31
$NO_2 + NO_3\text{-}N$	0.41	0.2	0.48
$NH_4\text{-}N$	0.23	0.3	0.55
Al		34	58
Cr		< 2	< 2
Cu		1	7
Fe		15	34
Mn		11	27
Mo		< 5	< 5
Ni		< 5	7
Pb		< 6	6
Zn		< 1	< 1

place if such climatological data are obtained. Other parameters of relevance to the deposition of sulfate include water and air temperatures. Concentration fields for both sulfur dioxide (see Figure 6.9) and sulfate are known in general terms in Europe and North America and in more detail for some countries (NILU, 1985; KNMI, 1988) and can be used for dry deposition estimates.

Acidic compounds

Hydrogen ion concentrations in precipitation have been measured extensively in many countries in the temperate zone. Concentration ranges from between 80 and 100 µg/l to less than 10 µg/l. In regions with small levels of emissions of soil dust (such as in eastern North America or NW Europe), the spatial gradients in pH are similar to those of excess sulfur. However, in western North America and Eastern Europe, where alkaline soil dust is more abundant, an excess of bicarbonate is found in precipitation (pHs above 6). Dry deposition of gases such as SO_2, NHO_3 and NH_3 and of acidic or alkaline particles also contributes – positively or negatively – to acidification, and is of significance, as can be seen from the deposition estimates for the elements given above.

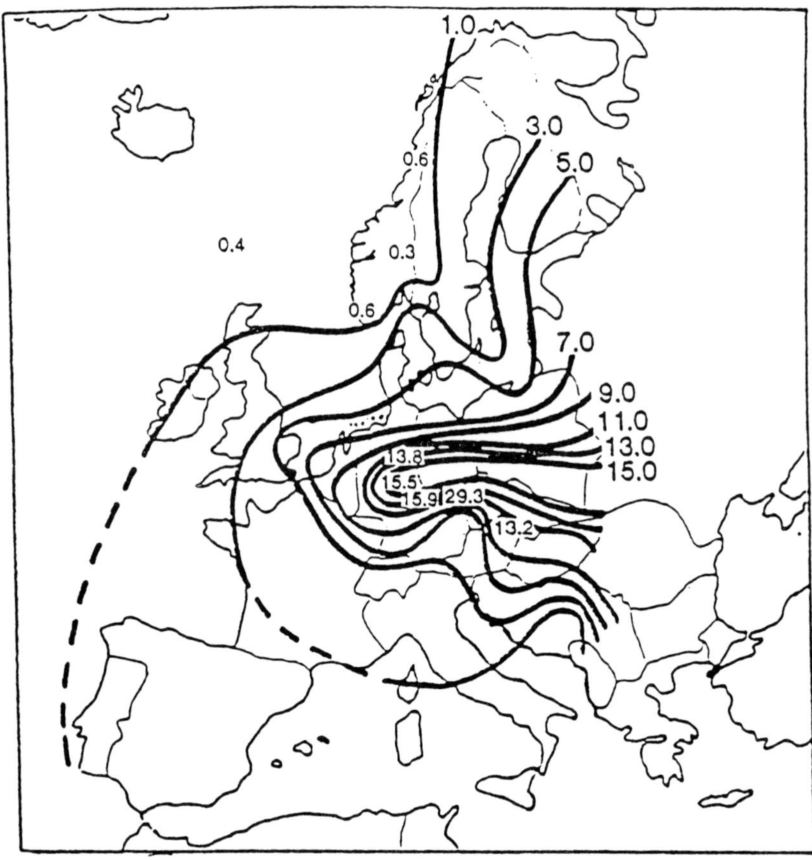

Figure 6.9 Mean concentrations of sulfur dioxide (in µg SO_2–S/l) in Europe (based on EMEP data from 1985)

SILVICULTURE

Introduction

Most pollutants exported from forested lands leave the area with the discharging water draining the basin. The pollutants are either dissolved or suspended, and they may be due to natural sources or residues resulting from human activities in the forests. In this regard, the distinction between natural and human-induced sources is important, but very difficult to make. Many studies throughout the world have shown the pronounced natural variability in chemical composition of stream waters from forested lands (Pierce *et al.*, 1970; Keller and Strobel, 1982; Grip, 1982; Balci *et al.*, 1981; Fuehrer, 1983; Hueser, 1977; Lehnardt *et al.*, 1983). This variability is not only due to the forest environment. It is a combination of effects of meteorological conditions, geologic substratum, geomorphic conditions, and forest type.

An analysis of data from different locations identifies several parameters that help to explain the observed variations. One of the most obvious parameters is streamflow volume or discharge rate. This variable is a good integrator and describes the hydrological conditions at the time of sampling quite well. This variable also lends itself to graphical analysis through plots of flow or discharge versus time. Looking at the flow regime over a year (or longer period) – the duration curve is a suitable means of characterizing flow regimes – the frequency of high and low flow occurrences can readily be seen. Likewise, related water quality patterns (often inverse relationships between the frequency of flow and concentration of a chemical component) can be apparent from similar plots of concentration *versus* time.

Another variable often considered to explain the variability of water quality parameters is water temperature. Many pollutants or dissolved materials are reactive to stream temperature. Some seasonal variation may well be explained using water temperature as variable. For example, water temperature distributions in two neighbouring basins in the pre-Alps of Switzerland differ markedly, despite their similar size and general climatic/geologic environment. However, they have different vegetation, land uses and topography which contribute to the observed temperature difference. The behaviour of nitrate in the two streams could be an example of the influence of temperature on other stream variables, since the differing concentrations of nitrate in the two streams could be due to temperature effects on microbial activity.

While both flow and temperature frequency distributions may be suitable bases for the interpretation of the occurrences of nutrients, minerals and pollutants being discharged in streams from nonpoint sources in forested areas, the immediate environmental conditions prior to sampling for water quality are also important, especially during a particular rainfall event as noted in Chapter 4. Among these conditions are the following:

(1)   Soil moisture conditions;
(2)   The type and intensity of precipitation falling on the forested area;
(3)   The chemical composition of the rain;
(4)   The atmospheric conditions, wind speed and direction, and occurrence of air pollutants;
(5)   The vegetative conditions, both as a result of seasonal variations and as a result of land uses (i.e. logging operations, plantation development, road construction, tree thinning activities, drainage work in wet forests, etc.).

Although these considerations make it difficult to estimate the average values of nonpoint source contaminants, as well as their variabilities, regional and often local sources of information on climate, geology, topography, soil and forest type may be available to permit development of appropriate estimates. In many regions of the world, meteorological and forestry stations are situated in or near forests to minimize anthropogenic disturbances. These stations can provide monitoring data on this most important type of landscape, from which extrapolation into areas with no observations seems to be possible. Indeed, observations from forested watersheds are often used to determine baseline conditions in disturbed watersheds, as noted in Chapter 1. Information from such places is the integrated result of the above-ground forest vegetation, of the underground root and soil compartment with all biological activities, and of the underlying geological parent material. Any long-term change in the behaviour of the various hydrologic and water quality parameters must be considered an effect of changes in the environmental conditions, such as air and precipitation chemistry, land usage, economic activities, and climatic variabilities, and/or, sometimes, changes in observation or monitoring methods (Lerner, 1986; Aubertin, 1977).

Human-induced pollutants are considered to be the difference between the natural baseline and observed concentrations. In forested lands, some activities may, and some may not, result in pollution. Forest management practices which disturb only very small areas probably result in negligible impacts on the export of pollutants. However large-scale forest practices (road construction, clearcutting, logging), particularly when using mechanized harvesting methods, significantly disturb the forest floor – at least temporarily. Erosion is the immediate consequence of such activities and increased exports of water, nutrients and pollutants cannot be excluded (Blackie *et al.*, 1980). Even more critical is the introduction of chemicals to forested lands. These may be fertilizers or wood preservatives used to treat logged timber, pesticides, etc. Once such pollutants are introduced into forest ecosystems, it is very difficult to control their transport.

Undisturbed forests and woodlands represent the best protection of lands from sediment and pollutant losses. Woodlands and forests have low hydrologic activity due to the high level of surface water storage in leaves (interception), ground mulch, and rough terrain (see Chapter 4). Furthermore, forest soils frequently have high permeability, and even lowland forests with a high groundwater table absorb large amounts of precipitation, actively retain water and retard runoff. In addition, the tree canopy, ground cover and increased organic content of forest soils significantly reduce erosion losses. Surface runoff from forested areas is often almost nonexistent (Novotny and Chesters, 1980). Nevertheless, streams draining lowland forests may have elevated organic and nutrient levels caused by leaching from the forest soils by interflow and base flow. Despite this effect, woodlands are often the determinants of background pollution levels against which other land uses are judged.

Reported and simulated sediment loadings from forested lands are typically less than 10 kg/ha/y. Uncontrolled logging operations – clearcutting – often disturb a forest's resistance to erosion. Observations and records indicate that almost all of the sediment reaching waterways from forested lands originates from the construction of logging roads and from clearcuts. Chief among these sources of erosion are roads that disrupt or infringe upon natural drainage channels.

The Hubbard Brook Ecosystem Studies

Much of what is known about the impact of forestry management practices has been derived from a series of 'systems' studies undertaken in the White Mountains of New Hampshire (USA). These studies were specifically designed to investigate the energy and biogeochemical relationships of a northern hardwood forest. The study site was located in the Hubbard Brook Experimental Forest in West Thornton, New Hampshire. The facility there, maintained and operated by the US Forest Service, was established in 1955 and was the major laboratory for research pertaining to management of forested watersheds in New England. The Hubbard Brook site lent itself well to an ecosystem study since small, discrete drainage basins can be readily defined in close proximity to each other (and, hence, could be easily compared with one another).

The Hubbard Brook Experimental Forest has a rugged terrain and is covered by an unbroken woodland of northern hardwoods, with spruce and fir present at higher elevations. The climate is typified by short, cool summers, with an average July temperature of 19°C, and long, cold winters, with an average January temperature of –9°C. Six small watersheds were chosen by F. H. Bormann and Gene Likens for biogeochemical studies in the mid-1960s; each watershed was entirely forested by uneven-aged, second-growth northern hardwoods, primarily sugar maple, beech, and yellow birch, mixed with a few red spruce and balsam fir.

During the late fall and winter of 1965, all of the vegetation on one watershed was cut and subsequently treated with herbicides in an experiment designed to determine the effect on (1) the quantity of stream water flowing out of the watershed, and (2) the fundamental chemical relationships within the forest, including nutrient cycling and enrichment of stream water. The hydrological and chemical characteristics of these watersheds have now been monitored for over three decades at stream gauging stations (Figure 6.10).

The annual distribution of precipitation and runoff for Hubbard Brook Forest is fairly uniform throughout the year, and is typical of moist, temperate locations. Records kept by the Forest Service between 1955 and 1963 indicate that the average annual precipitation into the system was 123 cm, while average annual runoff was 72 cm. Thus, an average value of 51 cm could have been lost to evapotranspiration. The impact of the clearcutting operation on the water cycle was to increase runoff, reducing evapotranspiration by 70 percent and increasing the velocity of stream flow. Between 1969 and 1984, the clearcut watershed was virtually undisturbed, and left to recover. As might be expected, values for stream flow and evapotranspiration in the manipulated watershed have returned

Figure 6.10   Stream gauge used to measure runoff from small watersheds (Photo: M.M. Holland)

to values similar to the other, undisturbed watersheds – runoff has decreased and evapotranspiration has increased. The role of forest vegetation in the hydrologic cycle was dramatically illustrated.

Results from the studies of nutrient cycling at Hubbard Brook were even more dramatic. The natural concentrations of nitrate in stream water from undisturbed forests showed a seasonal cycle, being higher in winter (November to April) than in summer (May to October). The decline in May and low concentration throughout the summer were correlated with the heavy demand for nutrients by the vegetation, as well as with the generally increased biological activity associated with a warming of the soil. Beginning in June 1966, the concentration of stream nitrate in the deforested watershed rose sharply (Figure 6.11) indicating that nitrate was rapidly flushed from the system by runoff, since it was no longer being utilized by terrestrial vegetation. However, by 1971, nitrate concentrations in stream samples from both the forested and reforested clearcut systems were the same. During a mere two years with no disturbance, the forest had recovered sufficiently to begin, once again, accumulating and cycling large amounts of nitrogen (Bormann *et al.*, 1977).

This research also led to broader insights into the annual nitrogen budget of northern hardwood forests. Each year 68 percent of the nitrogen added to the ecosystem came from nitrogen fixation, while 32 percent entered the system from precipitation. Moreover, of the total nitrogen entering a watershed during

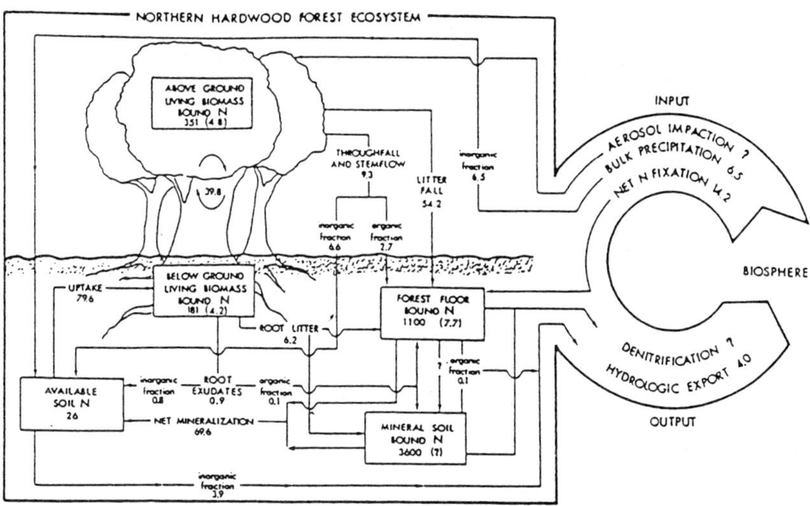

Figure 6.11 Annual nitrogen budget for an undisturbed hardwood forest ecosystem at Hubbard Brook (USA). The pool size is shown in the boxes as kg N/ha with accretion rates in kg N/ha/y shown in parentheses; transfer rates are shown on the arrows as kg N/ha/y (from Bormann *et al.*, 1977)

a given year, about 81 percent was retained within the ecosystem. Further studies suggested that storage of the critical elements, nitrogen and phosphorus, by watershed vegetation greatly exceeded stream output; therefore, the role of forest vegetation and soil in preventing nutrient loss was most important.

One of the initial goals of the clearcutting experiment was to determine the relationship between the removal of vegetation and stream eutrophication. The deforestation experiment resulted in severe pollution of the drainage stream in that watershed. From 1966 to 1970, the nitrate concentration in stream water exceeded, almost continuously, the maximum concentration recommended for drinking water. As a result of increased temperature, light, and nutrient concentrations, and in sharp contrast to the undisturbed watersheds, a dense bloom of algae appeared in the stream draining the deforested watershed during those four summers. The nutrients, previously used for terrestrial vegetative production, were obviously used in the enhanced production in the aquatic ecosystem.

Several important conclusions regarding forestry management practices can be drawn from these studies:

(1) A forest ecosystem is a highly complex natural unit composed of organisms and their inorganic environment;
(2) The parts of the ecosystem are intimately linked by the natural biological, geological, and chemical processes that are part of the ecosystem;
(3) The uptake of water and the storage of nutrients by vegetation is critical to maintaining a healthy balance within a terrestrial ecosystem;

(4) The stability of an ecosystem is linked to the orderly flow of nutrients between the living and the non-living components of the system;

(5) Individual ecosystems are linked to their surrounding land–water ecosystems, as well as to the biosphere in general, by the water and nutrients that move through them in their individual biogeochemical cycles.

Related forest hydrology research results

Streamside vegetation not only reduces sediment and nutrient transport from the terrestrial to aquatic environment, as noted above, but also has potential to modify stream water temperature to enhance the oxygen-carrying capacity of streams (Karr and Schlosser, 1978). Forest hydrology research provides a number of generally applicable observations:

(1) Natural near-stream vegetation can reduce the nutrient and sediment transport from the terrestrial to aquatic component of ecosystems;

(2) Near-stream vegetation can be used to reduce temperature-associated water quality problems in agricultural watersheds (by shading and modifying the stream microclimate through altered evapotranspiration rates);

(3) Natural stream morphology reduces the unit stream power, bank erosion, and suspended sediment concentrations (Table 6.11 and Figure 6.12) (Karr and Schlosser, 1978).

In an undisturbed, natural watershed, suspended solids will be very low because erodible sources of sediment will be minimal in both terrestrial and aquatic areas. When land is cleared (by clear-cutting forests or for row-crop agriculture), sources of erodible sediment will be increased. Water quality will decline because of the increased availability of sediment and surface runoff (Karr and Schlosser, 1978). Sedimentation commonly results in declining productivity across all trophic levels in aquatic ecosystems by altering the structure and productivity of plant, invertebrate, and vertebrate communities. Therefore, reductions in sediment loadings will have a beneficial effect on stream biota. However, efforts to reduce sediment loads must be accompanied by the more informed

Table 6.11 Percentage of change in sediment and nitrate loads over a 15-year period under varying forestry practices (from Karr and Schlosser, 1978)

Forestry practice	*Suspended sediments*	*Nitrates*
Clear-cut	205	400
Clear-cut with buffer strip along streams	54	0
Control	0.1	0

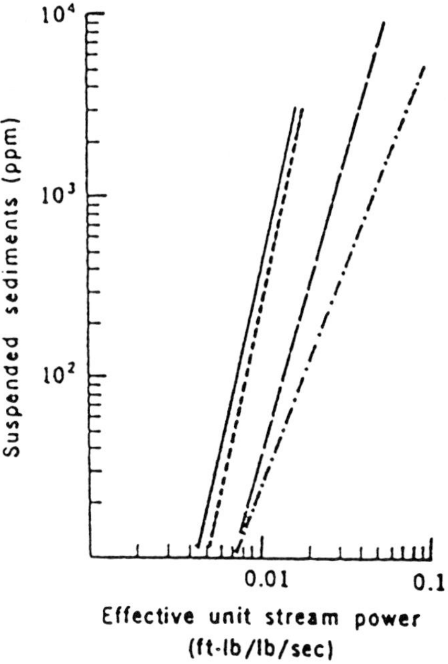

Figure 6.12 Relation between effective unit stream power and measured suspended sediment concentrations for four streams (after Karr and Schlosser, 1978)

management of other stream attributes for optimal effectiveness in remediating water quality and rehabilitating stream biological communities (Karr and Schlosser, 1978).

AGRICULTURE

Introduction

Water is an absolutely essential prerequisite for agricultural production. At the same time, however, water usage, and its drainage from agricultural lands, constitutes one of the major causes of water quality degradation. Fortunately, the general goals of sustainable agriculture and environmental protection of aquatic systems are similar; farmers and environmental regulatory officials both desire (1) the maintenance or enhancement of agricultural production, and (2) water resources of sufficient quantity and quality to ensure this production.

It is clear that agricultural practices (including water management) can have fundamental impacts on specific components of the hydrologic cycle, thereby impacting both agricultural soil and water resources. Many large-scale impacts on surface and groundwater resources are dominated by agricultural activities. This is especially the case in regions where (1) the larger part of arable land is in intensive agricultural use and (2) water quality problems originate with, or are triggered by, agricultural activities. The areal extent of agricultural activities, as compared to other human-dominated land uses, is so great that agricultural impacts are often more significant than any other land use impacts, at least in regard to soil conditions, near-surface groundwater quality, and surface receiving water quality (rivers, lakes, reservoirs). In regions with intensive agriculture, environmental impacts related to agricultural activities (and agricultural water management practices) are associated with:

(1)   Erosion and soil degradation;
(2)   Nutrient enrichment (eutrophication) of rivers, lakes, and reservoirs and nitrate contamination of groundwater;
(3)   Biodegradable organic matter;
(4)   Heavy metals;
(5)   Synthetic organic chemicals (pesticides, herbicides);
(6)   Salts (salinity);
(7)   Bacteria.

Among the above-noted environmental impacts, sediment erosion from agricultural land is of special significance because it both (1) affects the fertility of agricultural lands, and (2) can be a transport mechanism for nonpoint source contaminants that can readily sorb onto sediment particles (nutrients, heavy metals, some synthetic organic chemicals). It is further noted that nitrate contamination of groundwater appears to be a water quality degradation problem uniquely associated with agricultural activities, specifically excessive usage of fertilizers, and its subsequent leaching from the land surface. It also should be recognized that, regardless of the specific nonpoint source contamination problem, water is both (1) the main transport mechanism for agricultural contaminants, and (2) the main agent of agricultural chemical and biological processes. Thus, almost all agricultural practices affect the hydrologic properties of the agricultural soil–plant system and, consequently, the quantity and quality of agricultural drainage to surface and groundwaters. Examples of the specific impacts of the major agricultural pollutants are described in more detail below.

Sediment

Sediment is by far the most important agricultural nonpoint source contaminant on a global scale. Cropland soil erosion is the major source. Sediment, as a pollutant, is defined as solid materials which have been transported from their place of origin, *via* water or wind erosion. It is the original pollutant affecting, and being affected by, agricultural activities. To begin with, soil is essential for the support and growth of plants and crops used by humans for food and fiber, and by domestic and wild herbivores for grazing. Consequently, protection of soil is vital to the economic well-being of every nation. Of most concern to farmers, therefore, is the loss of productive topsoil and associated organic matter and nutrients. This is particularly true for small subsistence farms in developing countries, where the farmer is most reliant on the inherent fertility of the soil for sustained production. In developed countries, although a source of water quality degradation, high levels of production have been maintained in the face of accelerated erosion by continued application of artificial fertilizers (although at the cost of deterioration of the physical and chemical quality of the soils).

In the United States, about one-third of the top soil has been washed away, and the natural productivity of the land has decreased by 10 to 15 percent since the advent of industrial agriculture. In the Russian Federation, erosion-susceptible soils occupy at least 120 million ha, or about one-half of the total area of cropland of the member countries (Golubev, 1980). Although there is only sparse information on the extent of soil erosion in developing countries, there is no reason to believe that conditions are any less severe in these locations than in the developed countries.

Using a simple model and considerable literature, Golubev (1980) has made an assessment of the worldwide conditions of water erosion of soils, which is summarized in Table 6.12. These rough estimates highlight several important conclusions:

(1)   Soil erosion currently is five times greater than during the pre-agricultural period;

(2)   The main reserves of tillable land are in sub-tropical and tropical regions – thus, it is in these regions that a considerable increase in soil erosion can be expected;

(3)   The highest increment of soil erosion presently is in the humid regions of the world.

As previously noted, agricultural soil erosion has a major impact, both on (1) soil productivity and (2) water quality. Decreased soil productivity occurs as a result of (1) loss of storage capacity for plant-available water, (2) loss of plant nutrients, (3) degradation of soil structure, and (4) decreased uniformity of soil conditions within a field. Subsequent changes in land use can occur, because higher production costs reduce the profitability of farming eroded land, and because gully erosion can make field operations difficult. The second major environmental impact of erosion is off-site sedimentation, and transport of sediment-associated contaminants (i.e. those with high K_d values; see above) to receiving waters.

Erosion is the detachment and movement of soil or rock fragments by water, wind, ice or gravity. It occurs when erosive agents (rainfall) are sufficiently intense to both detach and transport soil particles. The five principal types of water erosion are (1) splash, (2) sheet, (3) rill, (4) gully, and (5) natural erosion. Agricultural erosion (sediment yield) frequently is termed 'sheet erosion'. This form of erosion occurs in small rills or larger gullies. Generally, this type of soil erosion results from the exposure of bare soil to raindrop energy and the convective forces of sheet-flow and 'micro-channel' flow. The detachment,

Table 6.12   Assessment (in tons/km^2/y) of water erosion of soils (from Golubev, 1980)

Zone and province	Lowlands			Mountains			Total		
	Past	*Present*	*Future*	*Past*	*Present*	*Future*	*Past*	*Present*	*Future*
TROPICAL									
Humid	100	810	2540	190	1840	2950	110	950	2600
Dry	150	330	640	280	500	780	170	360	660
Arid	200	210	340	360	360	360	210	220	350
Subtotal	140	520	1430	250	1150	1810	150	600	1480
SUB-TROPICAL									
Humid	80	2220	3730	210	2280	2280	140	2240	3100
Dry	200	720	1040	390	1120	1120	270	860	1070
Arid	100	160	190	190	310	560	110	180	250
Subtotal	130	750	1170	280	1380	1440	170	930	1250
SUB-BOREAL									
Humid	80	2930	3120	250	6440	6440	150	4330	4440
Dry	220	1020	1140	790	1220	1220	340	1060	1160
Arid	60	90	110	300	300	300	160	180	190
Subtotal	140	1200	1310	390	2500	2500	220	1640	1710
BOREAL									
Forest	50	560	860	160	160	540	80	460	780
Permafrost	30	30	100	90	90	90	60	60	100
Subtotal	40	420	660	120	120	300	70	320	540
WORLD	120	620	1180	250	1220	1430	150	760	1240

transport, deposition, resuspension, further transport, etc., of soil particles on the land surface depends on a number of hydrological processes and factors previously discussed in Chapter 4. Soils can be eroded by exposure to wind, as noted in Chapter 5.

The major agricultural factors that can affect the extent of water-induced erosion include:

(1)   Farming on long slopes without terraces or runoff diversions;
(2)   Farming (especially row crops) in fall-line directions, and especially on steep slopes;
(3)   Improper contour tillage;
(4)   Crop seeding on bare soils;
(5)   Leaving bare soils between crop harvests and the establishment of new crop canopies;
(6)   Intensive runoff from upslope pasture or rangeland traversing areas of row crops;
(7)   Intensive cultivation close to streams;
(8)   Poor crop stands;

(9)   Creation of erosion-sensitive areas by ill-sited agricultural access roads, such as dirt roads leading to agricultural plots along fall-lines which frequently turn into deep, canyon-like gorges – in the valley bottoms, the same roads will be blocked by excessive mud deposits.

The erosion process can occur unabated, unless the soil surface is protected by vegetation, a surface mulch, or some other type of ground cover. When the water runoff concentrates into small, natural depressions or in marks left by farm equipment, eroding channels or rills can form. Once a soil is tilled, this channel-erosion process can increase rapidly as small channels are deepened and widened. It is more severe on steep or long slopes than on gentle or short slopes. In fact, rill erosion can be the main source of soil loss from a field, and can result in loss of production as normal farm machinery is often unable to cross deep rills. Generally, some of this eroded soil is deposited at the base of the slope, or in depressions or areas of increased vegetative cover; thus, the amount of soil actually entering streams can be less than the amount initially eroded from the land surface.

The steeper the slope of the land surface, and the finer and less cohesive the soil particle, the more vulnerable bare soil is to erosion. Nevertheless, water action, especially rainfall and snowmelt, and the subsequent water runoff over the land surface, is the direct cause of water-induced soil erosion. This means that agricultural water-management practices can be used to attempt to alter the sediment export rate from farms. It is noted that, by weight, the sediment load in surface waters in the United States is 500- to 700-times greater than the sewage load discharged to streams. However, by weight, the pollution potential of sediment is much less than that of municipal sewage loads.

At least 30 years is required to develop one-inch of good topsoil (about 340 tons/ha). Further, at least 150 years is required to develop the minimum five-inches of topsoil needed for agricultural production when the soil is supplemented by adequate subsoils and an appropriate climate. In developing countries, where extended periods of extreme climatic conditions occur, this period of development can be appreciably longer.

In the United States, the average, annual soil loss is about 11 tons/ha/y (about 340 tons/ha over a 30-year period). Under the very best conditions, therefore, soil losses generally are occurring about as fast as soil is being formed. However, these soil loss rates are conservative estimates; soil losses as great at 20 tons/ha/y are common on croplands. Therefore, in areas where soil erosion exceeds the rate of topsoil formation, there is likely to be a decrease in potential crop production (although this effect can be masked by increased applications of agricultural chemicals and other technological inputs which cause other water quality problems – see below).

It is noted that eroded sediment in agricultural drainage can be very different from that of the original surface soil from which it was eroded. In the process of particle detachment and transport in overland flow (water erosion) or suspension by wind (wind erosion), there is a tendency for larger particles (sand and aggregates) to settle out preferentially as the transport energy dissipates. This usually enriches the runoff or wind-blown sediment load with finer particles, primarily silts and clays. This can be described mathematically by an enrichment ratio, defined as the ratio of the concentration of a soil constituent in surface soil to the concentration of the same constituent in the sediment. Pollutant enrichment ratios greater than one indicate selective erosion and transport of that pollutant. Some researchers have attempted to relate sediment characteristics (e.g. clay content) to those of surface soils, using appropriate enrichment ratios. However, this is difficult to do in a watershed of diverse soil types because it is not easy to determine the relative contributions of the different watershed regions to sediment generation.

As with most human-induced interventions, soil erosion tends to increase as the degree of cultivation or tillage of agricultural land increases. This is exemplified by a detailed study of the quality of water from 20 watersheds in Oklahoma and Texas (USA) over a ten-year period (Smith *et al.*, 1983). The annual, average soil loss in surface runoff from these watersheds for the period 1977–1984 is presented in Figure 6.13, and related to intensity of cultivation. As expected, a dramatic increase in soil loss occurs with increasing cultivation and reduced vegetative soil-cover.

The preceding discussion focuses on cropland, primarily because this land generally is the most susceptible to soil erosion. However, significant soil erosion can also occur with other agricultural land uses. For example, soil erosion rates

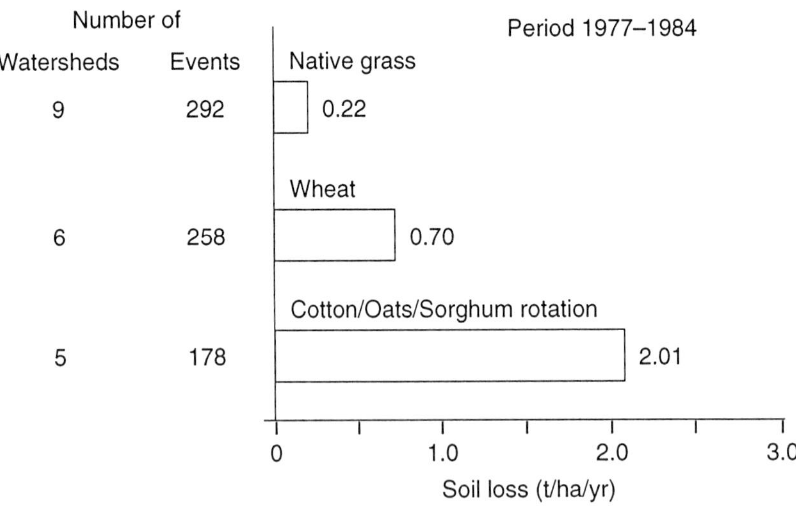

Figure 6.13 Annual soil loss in runoff from drainage basins in Oklahoma and Texas (USA), averaged for 1977–1984, as a function of cultivation of selected crops (From Smith *et al.*, 1983)

in excess of 11 tons/ha/y are also reported for eight million hectares of steep and erodible pasture land in the United States. Furthermore, sheet and rill erosion from forested lands can vary from 1 to 22 tons/ha/y (see above). Soil erosion losses can also increase substantially when forested land is grazed. Therefore, as native range- or forested-lands are brought under agricultural production, there is an immediate risk of soil erosion, unless adequate control measures are carefully planned ahead of time and put into practice. Measures for controlling sediment loss from agriculture, and other types of land use, are discussed in more detail in Chapter 9.

Nutrients

Major agricultural sources of plant nutrients (phosphorus and nitrogen) to waterbodies include both the sediment and water components of agricultural runoff. Nutrient runoff represents an economic loss of fertility to the farmer as eroded nutrients are no longer available for (terrestrial) plant growth. In fact, croplands represent a very large (possibly the largest) nonpoint source of phosphorus and nitrogen to surface waters (Bailey and Waddell, 1979). Major agricultural nonpoint nutrient sources include:

(1)   The natural nutrient content of the soil which is washed away in dissolved or particulate forms;

(2)   Losses of inorganic fertilizers applied to agricultural lands *via* surface and sub-surface runoff;

(3)   Exports of organic fertilizers applied to lands as solid and liquid manures, sewage, sludges, etc.

Phosphorus

Crops commonly require greater quantities of phosphorus than is typically available in natural soils. Therefore, phosphorus fertilization is an integral part of crop production, especially in developed countries. Phosphorus exists in nature in a number of chemical and mineralogical forms (Table 6.13). It has no important gaseous forms, and does not enter into as many biologically-mediated reactions as nitrogen. The various chemical forms of phosphorus all involve the orthophosphate molecule (PO_4), which combines with a wide range of metallic cations to form various salts and minerals. In soils and sediments, PO_4 is strongly adsorbed to hydroxylated metal surfaces (M-OH). In this compound, the metal ion (M) can represent iron, aluminium, calcium, magnesium, manganese, or a trace metal. The bond between the metal cation and the orthophosphate molecule is not easily broken. Because of the strong affinity of orthophosphate for metal surfaces, phosphorus has a high partition coefficient, K_d, and is almost entirely associated with sediment in runoff. Movement of phosphorus in leachate to groundwater is therefore small, compared to its movement into surface waters.

Table 6.13   Important environmental forms of phosphorus (after Logan, 1983)

P fraction		*Chemical characteristics*	*Percent bioavailability*
Total P	*Total dissolved P*	Dissolved inorganic P	> 90
		Dissolved organic P	< 50
		Labile P Includes adsorbed, exchangeable, easily dissolved, easily hydrolyzed	Potentially – 100 Immediately – ?
	Total particulate P	Inorganic P Relatively stable, primary and secondary minerals with Fe, Al, Ca	Potentially – ? Immediately – 0
		Organic P Relatively stable humus compounds such as inasirais	Potentially – ? (< 50) Immediately – 0

Analysis of agricultural sources of phosphorus is complicated by the fact that different sources and transport mechanisms are dominant under various flow conditions. Davenport (1983) reported, for example, that phosphorus runoff from a predominantly agricultural watershed in west-central Illinois (USA) consisted largely of insoluble sediment-associated phosphorus derived from cropland. During high-flow and baseflow conditions, dissolved phosphorus was the dominant form. During low-flow conditions, dissolved soluble phosphorus from livestock wastes, on-site disposal systems and natural geochemical weathering, *via* groundwater inputs, was likely to dominate.

Major sources of agricultural phosphorus are discussed below. Representative losses of phosphorus measured in agricultural runoff in the United States are presented in Table 6.14. Jolankai (1983) and Johnson (1983) have also documented examples of average nutrient concentrations in runoff both in the United States and Europe (Table 6.15).

Sediment-associated phosphorus constitutes the principal particulate phosphorus component of land runoff. Soils contain between 200 and 2000 mg/kg of total phosphorus. Young soils of high clay content tend to have the highest phosphorus levels. In contrast, the highly-weathered soils of the tropics, as well

Table 6.14 Representative losses of phosphorus in runoff from agricultural watersheds in the United States (from Nelson and Logan, 1983)

Watershed			P Transported (kg/ha/y)	
Location	Size (ha)	Land use	Diss. P	Part. P
Ohio				
Maumee River Basin[†]	1.54×10^4	Mixed	0.29	1.53
Portage River Basin[†]	1.11×10^5	Mixed	0.30	0.24
Pier III[†]	3.2	Soybeans	0.13	1.09
Ohio	123	Pasture and forest	0.07	–
Indiana	5×10^3	Mixed	0.15	1.90
Michigan				
Avg. of plots[†]	0.3	Row crops	0.71	–
Mill Creek	–	Mixed	0.2	0.2
Illinois				
Kaskaskia River Basin	1.3×10^4	Mixed	0.1	–
Wisconsin				
Tributary to Lake Kegonsa	–	–	0.11	0.46
Dairy farming	546	Mixed	0.53	0.77
Iowa				
Watershed 2[†]	3.3	Corn	0.09	–
Connecticut	8.5×10^2	Forested	–	0.22
Arkansas	3.1×10^3	Mixed	–	2.3
Maryland				
Potomac River Basin	2.3×10^4	Mixed	–	0.27
North Carolina				
Pigeon River	3.5×10^4	Mixed	–	0.17
Watershed 2	1.5	Rotation	0.27	–
Oklahoma				
Watershed C	17.9	Cotton	1.1	5.5
Maine				
Stetson River	7.4×10^3	–	–	0.04
Agricultural watershed	–	Mixed	0.5–0.9	0.3–0.4

Predominantly cropland with some forest and pasture
[†]Average data

as soils derived from coarse-grained parent materials, tend to be low in phosphorus content. Further, because organic matter contains a fairly constant ratio of carbon:nitrogen:phosphorus, the soil total phosphorus content increases with increasing organic matter. Phosphorus inputs, in the form of fertilizers or wastes, will increase the total phosphorus content of soil, even where applications are designed to just meet a crop's phosphorus requirements. This is because only about 10 to 20 percent of the particulate soil phosphorus is readily available for uptake and utilization by crops.

Table 6.15 Concentrations and loading rates of selected chemical contaminants in rainfall

Parameter	Concentration (mg/l)	Area loading rate (kg/ha/y)	Remark
NH_4–N	0.6–2.7	0.7–15.0	
NO_3–N	0.1–1.2	0.9–5.0	
Total–N	1.0–7.0	1.2–28	
PO_4–P	0.0–0.3	0.0–2.0	
Total–P	0.0–0.5	–	
TSS	1.3–11*	100*	*only these data
COD	9.0–16*	124*	
SO_4	0.0–28	2.0–114	
Ca	0.2–20.0	1.0–52	
Cl	0.2–7.0	1.5–29	
K	0.1–6.0	1.4–12	
Na	0.1–2.5	0.4–35	
Mg	–	1.1–17	

Soil phosphorus is composed of a complex mixture of humus-associated organic forms, minerals (apatite, octacalcium phosphate, variscite), and phosphorus adsorbed to various mineral surfaces. Most of these forms do not undergo rapid transformations and are not easily available for biological uptake. The environmentally-important component of soil phosphorus, the so-called 'labile-phosphorus' pool, consists of a fraction of the adsorbed phosphorus, some soluble mineral phosphorus (dicalcium phosphate), and a small fraction of the organically-bound phosphorus. This pool, ranging between 5 to 50 percent of the total phosphorus (Sohzogni *et al.*, 1982), can undergo chemical and biological transformations as conditions (e.g. pH, redox potential, biological demand) change. Other, finer soil particles (i.e. clays and organic matter) also tend to be enriched in phosphorus, compared to the original surface soil. Because of the association of phosphorus with clay minerals and soil humus, eroded sediments tend to be phosphorus-rich. This appears as an increase in labile phosphorus.

In addition, dissolved phosphorus is also transported from agricultural lands. Under normal conditions, the smaller soil pores are partially or completely filled with soil water containing solutes. This soil water usually contains low concentrations (usually less than 0.5 mg/l) of orthophosphorus, and small quantities of organic phosphorus. The adsorption/desorption reactions between soil particles and pore water is very rapid. This equilibrium process regulates the concentration of dissolved phosphorus in land runoff. Labile soil phosphorus will be desorbed until a new equilibrium is reached (although such an equilibrium state is unlikely in land runoff because the water volume is large, compared to the mass of sediment), or until the labile pool has been depleted.

Other phosphorus sources from agricultural lands include washoff of some portion of the inorganic fertilizer phosphorus commonly applied to croplands. Application rates commonly range between 20 to 100 kg P/ha, with most fertilizers used today being in the form of soluble calcium or ammonium orthophosphate salts. These are either (1) surface applied and incorporated by tillage, (2) sub-surface injected with the seed at time of planting, or (3) surface applied without incorporation (the latter may be the case when no-till farming methods are used). When these soluble salts contact the soil, they dissolve in soil water and react immediately (*via* adsorption or precipitation processes) with the metallic cations in the soil. The result is that the soil labile-phosphorus fraction is increased, as well as the equilibrium concentrations of inorganic phosphorus in soil-water solution. (In addition to increasing the soil phosphorus content, phosphorus fertilizer applied to the land surface, but not subsequently incorporated into the soil, can be lost in runoff if rainfall occurs prior to completion of the reaction of the fertilizer with the soil – this is particularly true of granular fertilizers, which react more slowly with the soil.) Even after fertilizer phosphorus has reacted with the soil, if it has been surface applied, it can cause a high localized concentration of labile and dissolved phosphorus at the soil surface. Sediment subsequently eroded from such a surface also will have a higher phosphorus content.

Finally, the decay of crop residues on the soil surface can release nutrients which are either adsorbed at the soil surface, or washed off the land in runoff. The contribution of decaying crop residues to phosphorus runoff is greatest when conservation tillage cropping systems – in which the crop residues remain on the soil surface without incorporation into the soil (see Chapter 9) – are employed.

Nitrogen

Nitrogen is a ubiquitous element in nature, as well as an essential constituent of proteins and other important macromolecules in all organisms. The environmental chemistry of nitrogen is dominated by biological oxidation and reduction reactions, most of which are microbiologically mediated. Nitrogen exists in nature in a number of important chemical forms (Table 6.16), only a few of which are environmentally significant. Ammonia (NH_4), nitrate (NO_3), nitrite (NO_2), and organic nitrogen are the principal nitrogen forms of interest in regard to nonpoint source pollution. Organic nitrogen is primarily formed and degraded by biological action. Fertilizers, biological fixation, mineralization, and nitrification processes contribute nitrate to aquatic ecosystems. Denitrification, plant uptake, immobilization in soil organic matter, and leaching are the primary processes by which nitrate is removed from the environment.

Ammonia can hydrolyze in water to form the ammonium cation (NH_4^+), which can be adsorbed onto clays. This is a completely reversible association; thus, adsorbed ammonium can be bioavailable to terrestrial plants, crops,

Table 6.16   Important environmental forms of nitrogen

Form	Oxidation state gaseous forms	Environmental importance
N_2	0	79% of Earth's atmosphere
N_2O	+1	Implicated in ozone depletion; source of acid rain
NO	+2	Source of acid rain
NO_2	+4	Source of acid rain
NH_3	–3	Toxic to fish and other biota
	Dissolved ions	
NO_2^-	+3	Toxic to plants, humans and livestock
NO_3^-	+5	Most common form in soil and water; toxic to humans and livestock when reduced to NO_2^-
NH_4^+	–3	Common form in soils and sediments; in equilibrium with NH_3
	Soils	
Exchangeable and fixed NH_4^+	–3	–
Organic–N	Primarily –3	Various biomolecular including amino acids, proteins and nucleic acids

phytoplankton and aquatic plants. Fertilizers generally contain highly soluble ammonia and ammonium salts. When the concentrations of such compounds exceed plant requirements, precipitation or irrigation waters can transport these nitrogen compounds to aquatic systems. Ammonia associated with clay minerals enters the aquatic environment *via* soil erosion, although soils containing large amounts of clay-type minerals can immobilize ammonia.

On the other hand, the nitrate anion (NO_3^-) is repelled by soil and sediment particles which normally have net negative charges. Nitrate often is associated with groundwater contamination (see below), although surface runoff also contains nitrate from seepage and from underground tile drains. Nitrate can have negative effects on human and animal respiratory systems. Livestock, particularly bovine animals, are susceptible to nitrate poisoning, especially when the nitrate concentration in water is of the order of 50 mg/l. Human drinking water standards generally specify nitrate concentration of less than 5 mg/l (see Chapter 5).

In natural systems, nitrate is in a steady-state equilibrium with major inputs from aquatic or terrestrial nitrogen fixation processes, and from the oxidation of nitrogen gas (N_2) to nitrate during thunderstorms. Most of this nitrogen is stored as soil humus, microbial and plant biomass, and in higher life forms. There are some natural losses of nitrogen from the terrestrial system in the form of runoff and erosion of soil nitrogen, volatilization of ammonia and denitrification products (N_2, N_2O, NO), and leaching of nitrate to the groundwater.

In these natural ecosystems, runoff and leaching of nitrate and ammonia is generally low, and stream concentrations seldom exceed 1 to 2 mg NO_3-N/l and 0.1 mg NH_4-N/l. Human-induced increases in ammonia and nitrate in runoff are commonly attributable to disruptions of the natural nitrogen cycle, in which nitrogen inputs far exceed the environment's assimilation capacity. As discussed below, major nitrogen sources include (1) soil nitrogen, (2) nitrogen fertilizer, (3) livestock wastes, and (4) municipal sludge and wastewater.

Terrestrial systems generally favour the mineralization of organic nitrogen to ammonia, with subsequent rapid nitrification to nitrate. Concentrations of water-soluble nitrogen in surface runoff usually are highest at the beginning of the cropping season, declining progressively during the remainder of the year. This seasonal effect is attributable to the progressive removal of nitrogen by crops and leaching to groundwater, as well as the establishment of a condition of equilibrium between nitrogen and soil organic matter. Johnson (1982) and Jolánkai (1983) reported average nitrogen concentrations in runoff, based on United States and European data.

Surface soils contain between 0.1 to 0.25 percent nitrogen, the vast majority being inorganic nitrogen in the form of humus. Relative to this fraction, the mineral nitrogen content of soil is low. Soil nitrogen is transported in the watershed primarily as eroded sediment, in which most of the nitrogen is organic. This nitrogen is relatively stable, with an annual mineralization rate of less than 0.2 percent/year. Soils also contribute nitrate as percolation waters resulting from the mineralization of organic nitrogen. In a natural ecosystem, most mineralized soil nitrogen is recycled by biological assimilation. In soils receiving fertilizer or organic wastes in excess of crop assimilation, there is a net mineralization of soil nitrogen, resulting in a pool of nitrate that can potentially leach or run off the land surface. This pool can be significantly affected by tillage practices.

Nitrogenous fertilizer is commonly used on non-leguminous annual crops such as feed grains (maize, rice, wheat), and other staple crops (potatoes). Large amounts of nitrogen are also used on high-yield pasture grasses and turf. Most crops have low efficiencies of mineral nitrogen utilization, and uptake rarely exceeds 50 percent of applied nitrogen. Nitrate losses from fertilizer use are, therefore, relatively high, but can be minimized by (1) limiting application rates, (2) properly timing applications, and (3) using slow-release fertilizers. Nevertheless, there will be some increase in nitrate loads in runoff above natural levels whenever lands are fertilized. The greatest losses are usually associated with maize and potato crops, while much lower losses are associated with crops such as wheat.

Biodegradable organic matter

Agricultural substances whose microbially-mediated decomposition in natural waters results in consumption of oxygen include row-crop and other vegetative residues. Waters with BOD values greater than 10 mg/l are considered polluted since they contain large amounts of organic matter requiring oxygen-consuming bacterial decomposition. Davenport (1983) found BOD and COD values in agricultural runoff, measured over a two-and-a-half-year period, to be highest following late winter and spring rainfall runoff events. It should be noted that non-rainfall runoff conditions characterized Davenport's study aquatic system over more than 90 percent of the annual cycle.

Heavy metals

Most heavy metals occur naturally in the environment, and are present in small amounts in virtually all surface and ground waters. In contrast to nutrients and sediments, however, the loss of heavy metals from agricultural lands is usually small. The losses that do occur generally result from residues of materials (pesticides, herbicides, fertilizers) applied to agricultural lands, and subsequently washed off as a component of land runoff. However, certain heavy metals can negatively impact the structure and stability of aquatic ecosystems, including biological communities and overall water quality. Extensive use of metal-containing substances on agricultural lands might, therefore, constitute an environmental problem in some specific circumstances. Commonly-used agricultural chemicals that can be sources of heavy metals in runoff waters include some fertilizers and pesticides that contain zinc, copper, cadmium and mercury. Insecticides can be a source of lead, which has the ability to accumulate in the skeletal structure of humans and animals *via* food chain relationships.

Synthetic organic chemicals

Included in this category of nonpoint source contaminants are pesticides and herbicides, which are synthetically-produced organic chemicals used to control the occurrence of weeds, insects, and pest-mediated diseases. The agricultural community has used an ever-increasing array of such chemicals for these purposes since the 1940s. Both the numbers of pesticides, and the quantities used worldwide, have increased dramatically since that time. The four major categories of pesticides important in agriculture are (1) insecticides, (2) fungicides, (3) herbicides and (4) rodenticides. Their agricultural benefits are many, including (1) promotion of higher crop yields and improved products, (2) aiding in the mechanization of agricultural production (with concomitant reductions in labour requirements), and (3) improving the utilization and management of agricultural lands. While most agricultural chemicals are used in the developed countries, the use of pesticides is increasing in the developing nations. It is estimated that, for example, agricultural products would be likely to cost consumers two to three times the present levels if the use of such chemicals was eliminated.

In evaluating nonpoint source pesticide pollution, the persistence and toxicity of the biocide should be considered together (Leng *et al.*, 1995). This is because a rapidly-degraded toxic compound may pose only temporary threat when its residues are transported in land runoff from treated areas. Some highly soluble pesticides (especially salt formulations of acid herbicides) are mobile in soil, and, while they are less available for removal in surface runoff, can migrate downward with percolating rainwater. Fortunately, most mobile herbicides do not normally contaminate groundwaters, primarily because the adsorption capacity of the soils above the water table is sufficient to retain the chemicals until they decompose, although pesticide contamination of groundwater is known.

Pesticides applied to the soil can either be (1) absorbed by the soil, (2) degraded, (3) volatilized into the atmosphere, or (4) percolated downward through the soil. Maas *et al.* (1984) reported five primary routes of pesticide transport to aquatic systems: (1) by direct application, (2) *via* land runoff (dissolved, granular or adsorbed on particulates), (3) through aerial drift, (4) as a result of volatilization to, and subsequent deposition from, the atmosphere, and (5) through uptake by biota (with subsequent movement into food chains). Studies of large drainage basins indicate that the greatest fluxes of agricultural pesticides are in (1) tributary loads associated with peak runoff at the time of pesticide application (i.e. *via* land runoff), (2) precipitation (i.e. as a result of volatilization), and (3) resuspension of previously-deposited sediments (important for long-lived compounds such as DDT). Generally, pesticides in groundwater are less of a water quality problem than in surface waters, except in areas with highly-permeable soils and high chemical use.

Additional factors affecting the transport of pesticides in runoff following their application to agricultural lands include (1) pesticide placement, (2) soil moisture prior to pesticide application, (3) time of pesticide application, and (4) proximity of pesticide application to streams. Pesticides incorporated into the soil will not be lost in land runoff to as great an extent as pesticides applied and left on the land surface, or sprayed directly onto foliage. Some pesticides will exhibit greater runoff losses when applied to wet soil than to dry soil, particularly if runoff occurs soon after application. Pesticide losses are generally greatest in the first runoff occurring after its application. The losses then decrease as the time between application and runoff increases.

Past studies on pesticide transport in land runoff have suggested that, except when heavy rainfall occurs shortly after application, measured concentrations of pesticides in runoff waters are very low. The total quantity of pesticides in agricultural runoff from croplands during the year is less (often much less) than five percent of the amount initially applied to croplands. Nevertheless, some chemicals are highly toxic to fish, other aquatic fauna, and humans, primarily because they can persist and accumulate in the aquatic environment. Even very low levels of pesticides in runoff, therefore, can be of environmental concern. On the other hand, agricultural chemicals that (1) are not acutely toxic to animal life, (2) do not persist from one crop season to the next, and (3) do not accumulate in food chain organisms can usually be used at normal application rates without causing unacceptable environmental damage. The impacts of pesticides on water quality depend primarily on their (1) persistence and formulation, (2) rate and method of application, and (3) mobility.

The pollution potentials of some pesticides commonly used on croplands are presented in Table 6.17. Their transport mechanisms are highly dependent on these physical and chemical properties. The predominant transport modes (indicating the partitioning of the chemical between water and sediment) were derived from field experiments reported in the literature, and from the chemicals' water solubilities. The persistence of the pesticides in soil is the time required for 90 percent of the mass applied to disappear from the soil (Table 6.17). The approximate values can vary by $\pm$ 50 percent due to the climate (exposure to UV radiation) and the soil texture, moisture content, acidity, temperature and microbial activity. As noted in Table 6.17, organochlorine insecticides break down more slowly than the other types; they also tend to be toxic at lower concentrations.

The effect of elapsed time is particularly noticeable with short-lived pesticides, and pesticides not incorporated into the soil. Concentrations of the chemical in subsequent runoff events decrease at a rate dependent largely on the persistence of the pesticide in the soil. For example, field experiments with the carbamate insecticides, carbaryl and carbofuran (Table 6.18), showed that pesticide concentrations in both runoff water and sediments in the third runoff (which occurred within one to two months after application) were less than five percent of the concentrations in the first runoff (Caro *et al.*, 1974). In contrast, concentrations of the persistent organochlorine insecticide, dieldrin, in the third runoff (three to four months after application) were about 15 percent in the water phase and over 30 percent in the particulate phase of those in the first runoff (Caro *et al.*, 1974). Because sloping croplands are rarely adjacent to continuous streams, runoff containing pesticides must usually traverse some untreated land before reaching the water. This intervening area can trap some of the pesticides, resulting in reduced levels of stream contamination.

Since the primary mode of pesticide transport from agricultural lands generally is associated with sediment (Table 6.17), movement of sediment-associated pesticide is affected by an enrichment process in much the same way as nutrients (see above). Pesticides are adsorbed primarily on organic soil colloids which are easily removed from the soil surface by runoff. They also remain in suspension longer than coarser soil particles. Thus, while erosion control can reduce the mass of synthetic organic contaminants transported to water courses, a reduction in erosion does not always produce a proportional reduction in pesticide contamination.Consequently, effective pesticide control practices must reduce both water and soil losses.

Herbicides account for the greatest proportion of synthetic organic chemical usage in agricultural activities in the United States, with most of these chemicals being applied to a single crop, maize. In Europe, fungicides are widely used on

Table 6.17 Toxicity, persistence in soil and primary mode of transportation of several pesticides applied to agricultural lands

Pesticide	Toxicity* (mg/l)	Persistence[†] in soil (months)	Transport[‡] mode
Organochlorine insecticides			
Aldrin	0.003	24	S
Benzene hexachloride	0.79	36	S
DDT	0.002	36	S
Dieldrin	0.003	36	S
Endosulfan	0.001	24	S
Endrin	0.002	24	S
Heptachlor	0.009	24	S
Methoxychlor	0.007	24	S
Urea and triazine herbicides			
Atrazine	12.6	12	SW
Cyanazine	4.9	12	SW
Diuron	760	10	S
Fluometuron	760	4	SW
Linuron	16	4	S
Metribuzin	7100	5	W
Prometryn	71.0	2	S
Propazine	7100	12	S
Simazine	5.0	12	S
Carbamate insecticides			
Carbaryl	1.0	1	SW
Carbofuran	8.0	1	W
Phenoxy herbicides			
2, 4-D (acid)	750	1	W
2, 4-D (salt)	715	1	W
2, 4-D (ester)	4.5	1	S
2, 4, 5-T (acid)	0.5–16.7	4	W
2, 4, 5-T (salt)		6	W
2, 4, 5-T (ester)		4	W
Organo phosphorus insecticides			
Diazinon	0.030	3	SW
Methyl parathion	1.9	1	SW
Parathion	0.047	1	S
Phorate	0.0055	1	SW

*Toxicity expressed as a lethal concentration to 50% of test fish – bluegills or rainbow trout; [†]approximate time for 90% disappearance from soil, these values may vary ± 50%; [‡]movement most likely to occur primarily with sediment, S, water, W, appreciable movement with both sediment and water, SW

Table 6.18 Partition coefficients (K_{oc}), half-lives and water solubilities of selected pesticides

Compound	K_{oc}	Half-life (days)	Water solubility (mg/l)
Atrazine	172		35
Alachlor	20		242
Bromacid	72		–
Cyanazine	190		171
Carbofuran	29		700
Chlorpyrifos	13 490		–
Diuron	339		42
DDT	243 118		0.001
Glyphosate	–		12 000
Linuron	841		75
Lindane	1081		–
Malathion	1773		–
Metribuzin	–		1220
Metolachlor	–		530
Parathion	7161		–
Paraquat	–		> 10% by wt.
Picloram	26		430
Simazine	158		5
Trifluralin	–		0.05
2, 4-D	20		900

wheat, while, in tropical countries, insecticides are the most widely used chemicals for control of pests that afflict crops, livestock and humans. Not only are there regional differences in global pesticide use, but there have been significant changes in the types of chemicals used. Concerns over the ecological and human health effects of persistent chlorinated hydrocarbon insecticides (DDT, dieldrin, heptachlor, chlordane, 2,4,5-T, DHC, toxaphene, picloram) have led to partial or complete bans on their use in North America and Europe since the 1970s. These compounds are characterized by (1) long half-lives, (2) low water solubilities, and (3) high affinities for organic materials. Affinity is a measure of a compound's chemical character (e.g. 'lipophilicity' = fat-loving; 'hydrophobicity' = water-hating). The more nonpolar a compound, the more hydrophobic and lipophilic it is. This characteristic can be expressed by the propensity of the compound to partition between water and octanol (a nonpolar organic solvent). The octanol/water partition coefficient is a measure of a compound's tendency to associated with soils and sediments, rather than with water, as well as of its affinity for organic materials (plants, animals). The partition coefficient is corrected for use with organic compounds (e.g. pesticides) by expressing the partitioning on the basis of organic carbon content of the soil or sediment. It is termed K_{oc}. Table 6.18 presents half-lives, water solubilities and K_{oc} values for a number of commonly-used agricultural pesticides. Most have partition coefficients low enough to suggest that pesticides in runoff are associated with the water, rather than the particulate, phase. This is particularly true of most widely used herbicides. Risks associated with pesticide usage are given in terms of a chemical's acute toxicity to humans and biota, and its ability to bioaccumulate (increase in concentration as it moves up a food chain). The major human health concerns are related to the chronic (rather than acute) toxicity of the compound, and, especially, its longer-term carcinogenic, teratogenic and mutagenic effects. These effects typically are identified through animal-testing procedures and expressed in terms of a lethal dosage, the LC_{50} (see Chapter 5).

Pesticide movement to waterbodies *via* atmospheric transport (precipitation, fallout) is a particularly difficult problem to solve, and is especially severe for compounds like DDT and PCBs, which appear to have been distributed globally by such means.

Salts

Salts can be transported in runoff waters as dissolved solids, or as suspended solids attached to sediments. In the United States, salt (salinity) problems are primarily found in the waters of the arid- and semi-arid west. Salinity in surface and groundwaters results from both natural and anthropogenic causes. Human-induced salinity problems can result from the (1) salt concentrating effects of agricultural consumptive water use, (2) accelerated erosion, and (3) long-term irrigation. For example, irrigation tends to increase water salinity by (1) accelerating the leaching of salts through soils, (2) inducing the erosion and transport of salty sediments, and (3) concentrating salts in surface waters and groundwaters.

A widespread salt problem occurred in 1985 in the United States, when the US Bureau of Reclamation closed the Kesterson National Wildlife Refuge in California as a result of extensive contamination of surface and groundwaters with selenium, boron, arsenic and chromium. The US Bureau of Reclamation ultimately determined that the selenium and other elements were transported to the reservoir in the refuge by drainage water from irrigated farmlands in the central portion of the State.

In a similar case, irrigation of marginal lands in semi-arid and arid central Asia resulted in the catastrophic salination of the Aral Sea. In this case, the almost complete diversion of the river waters to the irrigation schemes compounded the problem by cutting off the inflows to the Sea, which promoted further salination as a result of evaporative losses of water vapour from the lake surface. At this time, the Aral Sea has lost over 50 percent of its former surface area and increased in salinity from less than 5 ppt to more than 35 ppt. This rapid salination (between 1960 and 1990) has extirpated the endemic fauna from the lake and is alleged to have seriously threatened human health in the lower portions of the watershed (UNEP, *s.d.*).

Bacteria

Bacterial contamination of surface waters can also affect the health of humans and livestock. For example, many infectious diseases are transmitted *via* animal manure, a common fertilizer for row crops throughout the world. Daniel *et al.* (1978) reported that such diseases include anthrax, encephalitis, tetanus, histoplasmosis, leptospirosis, bronchitis and salmonellosis. In addition to these potential health hazards, bacteria can impair the recreational use of surface waters and groundwater. They also can increase the cost of subsequent water treatments to meet other beneficial uses. In many developing countries, the construction of man-made lakes to supply water to irrigation schemes has additional undesirable consequences, such as promoting the spread of schistosomiasis, onchocerciasis, malaria, and dracontiasis.

Measurement of the faecal coliform:faecal streptococcus ratio (FC:FS) can provide an indication of whether bacteria in aquatic systems are of animal or human origin. A summary of FC:FS ratios indicative of human and various livestock species is presented in Table 6.19.

Nitrate contamination of groundwater

Nitrogen contamination of groundwater, as previously mentioned, is a special concern in agricultural settings. In densely-farmed agricultural regions of the world, nitrate pollution of groundwater was triggered by a dramatic increase in global fertilizer application rates during the 1960s and 1970s, and presently is considered to be a severe water quality problem in many countries. Most shallow, phreatic aquifers of Europe, for example, are contaminated with nitrates, and the nitrate levels in such aquifers continue to show a steeply-rising trend. Moreover, the problem is likely to get even more serious in the future, as the downward migration of nitrate has not yet reached the aquifers in many regions.

Near-surface groundwater resources are the primary drinking water sources in many regions. The primary concern relative to nitrate-enriched groundwater relates to potential health effects such as methaemoglobinaemia, a serious illness that occurs in infants when the nitrate concentration exceeds 10 mg NO_3-N/l. While there are economically-feasible methods for removing nitrate from surface waters, there are no practical means of purifying contaminated aquifers, or of even stopping the continuing downward migration of nitrates to groundwater. Furthermore, the natural self-purification capacity of certain underground formations (especially the presence of reducing agents) is nearly exhausted in some regions of Europe. A significant proportion (5 to 15 percent of the total number of wells) of European water supply and observation wells contain waters with nitrate concentrations already exceeding the EEC maximum-allowable level of 50 mg NO_3-N/l (Jolánkai and Roberts, 1987).

The ploughing of grassland enhances aerobic microbial mineralization processes, and converts organic nitrogen forms into leachable nitrate. Ploughing

Table 6.19 Representative FC:FS ratios (from American Public Health Association, 1980)

Organism	FC:FS ratio
Humans	4.4
Ducks	0.6
Sheep	0.4
Chickens	0.4
Pigs	0.4
Cows	0.2
Turkeys	0.1

adds plant material to the soil, and promotes its eventual conversion to inorganic nitrogen. Nitrate levels exceeding 200 mg/l were frequently found in the groundwaters under freshly-ploughed grasslands. Similarly, leaving land cropless through the autumn and winter seasons increases the probability that nitrate released after ploughing crop residues into the soil will be susceptible to leaching before it can be utilized by crops in the following spring. Drainage by surface ditches or subsoil drains may increase oxygen availability, thereby reducing denitrification, and increasing nitrification reactions. It also may increase the nitrate leaching rate by allowing more water to pass through the cultivated zone.

It is virtually impossible to quantitatively determine the contribution of nitrate from different sources, or to identify the affecting factors or their dependence on human activities (i.e. controlability). Nevertheless, nitrate transport and transformation processes are closely interrelated. The latter processes determine, or at least influence, the quantity of nitrogen available for overland transport. A general conclusion is that the main transportation routes are defined by water movement, surface runoff and leaching. Surface runoff in temperate regions occurs mostly during fall–winter and early spring. If all other conditions are equal, the leftover nitrogen on the land during the dormant season will define the magnitude of its transport, especially if the soil is left bare (without winter crops). Reported runoff values run as high as 170 kg/ha/y, although the characteristic range for agricultural watersheds is 10 to 40 kg/ha/y.

Only the soluble forms of nitrogen (primarily nitrate) are leached to groundwaters. Since the nitrogen pool in the upper soil layers is roughly an order of magnitude greater than the sum of all of the inputs to the soil, the amount of nitrate made available for leaching in the soil determines the extent of downward nitrate transport. Thus, processes favouring mineralization of organic nitrogen can significantly affect nitrogen leaching, and can easily exceed the effect of nitrogen fertilizer inputs. This also explains the wide variation of reported leaching rates, even under similar cropping, watering and fertilization conditions. Reported leaching rates are as high as 150 kg/ha/y, while a characteristic 'inner range' is 30 to 50 kg/ha/y. Nitrate leached from the soil zone will have to pass the unsaturated portion of the soil horizon before entering groundwater. Downward migration of the nitrate front depends on water movement, oxygen levels, and the presence or absence of reducing agents. Water movement, on the other hand, depends mostly on the soil and rock structure (permeability and fissuring of the unsaturated zone), and the rate of water recharge from the surface. Under conditions of predominantly intergranular water movement, the average characteristic rate of travel will be in the order of 1 m/y. This means that many (presently incipient) impacts still remain to be seen, and that further increases in the nitrate concentrations of groundwaters can be expected.

Of the nitrogen applied to agricultural lands, it is estimated that arable crops recover about 40 to 60 percent (as crop material) under present cultivation practices. The value for grasslands exceeds 80 percent. The additional uptake from the soil's nitrogen pool typically varies between 10 and 30 kg/ha/y. This means that roughly 50 percent of the nitrogen applied in the form of fertilizer will remain in the soil system, thereby being available for export in land runoff, leaching to groundwaters, or other loss processes, such as denitrification, addition to the soil pool, and immobilization in organic materials.

Certain microorganisms use nitrate as a nitrogen source, producing elemental nitrogen and nitrous oxide which are lost to the atmosphere. The extent of these denitrification losses depend on nitrate concentration, carbon supply, moisture, and the content of ferrous materials in the soils. A characteristic range of denitrification losses may be 10 to 30 kg/ha/y. In contrast, other microorganisms nitrify ammonia to nitrite and nitrate at rates which characteristically range between 40 and 120 kg/ha/y. However, a large fraction of the ammonia content of applied manure may be volatilized; reported characteristic values of ammonia losses range between 3 and 10 kg/ha/y. Nitrogen accumulation in the soil pool is not thought to be significant, and inorganic nitrogen appears to be only temporarily immobilized by microorganisms – plant residues and microorganisms ploughed into the soil decay and return the inorganic nitrogen to the organic pool in the soil.

In summary, the risk of groundwater contamination by nitrate is very high. There is, and will continue to be, a substantial external input of easily leachable, inorganic nitrogen fertilizers. In addition, conventional farming practices favour mineralization of organic nitrogen in the soil, which leads to a nitrate input of the same order of magnitude as the external load. Techniques known to increase the overall cumulative uptake rates and reduce the rates of mineralization include splitting applications of fertilizers to meet crop demands, growing intercrops ('scavenger' crops), applying appropriate tillage (including no-till) techniques to reduce mineralization, etc. (see Chapter 9). Nevertheless, the present state of knowledge regarding the processes and factors governing the fate of nitrogen in the soil–plant–water system (especially on the farm-scale) is inadequate.

TRANSPORTATION

Modern social development seems to imply an increased need to transport people. At least for the foreseeable future, therefore, the continued development of societies will lead to the further growth of the transportation network. It has been envisioned that the transportation sector could show as large an increase as about 7 percent annually during the coming years. If this trend continues, the global vehicle stock is expected to approximately double in the next 30 to 40 years. Aviation also has shown gradually increased transport volumes which have encouraged the growth of aircraft fleets. For these reasons, transportation is commonly regarded as posing one of the more significant environmental and health threats known in densely populated areas; these threats consist of the large-scale effects of air pollutants on vegetation, the high concentrations of pollutants – that are potentially dangerous to human health – (e.g. in the vicinity of arterial roads and in certain congested street environments), and disturbances from noise.

The expansion of car traffic in cities imposes a considerable strain on the environment. It is especially the increasing incidences of air pollution that are giving cause for alarm. In the cities, exhaust fumes account for a significant share of this pollution. The emission of particles from diesel-powered vehicles is of considerable importance in the context of human health effects. The various contaminants emitted to air, soil and water from the different transportation sectors (vehicle, rail, ship, and air) differ in significance from country to country, depending on the standard of living, population density, and the pattern of infrastructure. It is beyond any doubt that car traffic is a major source of air pollution in almost all parts of the world. Usually direct pollution to soil and water are of minor importance but can be a problem on the local scale.

Exhaust gases from road vehicles contain several air pollutants. Carbon dioxide makes up about 10 percent of the gases, with the major part consisting of molecular nitrogen and water vapour. Most of the hazardous compounds are found in the 3 to 4 percent of the gases that include carbon monoxide, nitrogen oxides, hydrocarbons and particulates. Carbon monoxide is the dominant pollutant in exhaust gases. Especially high concentrations of carbon monoxide are created during idling and when driving at low speed. Nitrogen oxides are created as a component of combustion and originate from the molecular nitrogen in the air. High concentrations are emitted during acceleration and when driving at high speed. Hydrocarbons (PAHs) in the exhaust gases are formed as a result of incomplete combustion. Petroleum consists of relatively volatile hydrocarbons, while diesel fuels consist of heavier, less volatile hydrocarbons. Particulates in exhaust gases are mostly made up of small lead particles from gasoline-powered vehicles, and coal and soot particles from diesel-powered vehicles. One tenth of the lead emitted from gasoline-powered vehicles is in the form of organic lead which is the more toxic form (compared to inorganic lead). However, the emissions of lead from cars have been reduced to a great extent in many countries through restrictions on the lead content of gasoline. In addition, chlorofluorocarbons (CFCs) are commonly used in many air-conditioning systems. Considerable amounts of these CFCs can be used during a life-span of a vehicle if the system needs to be repeatedly recharged over time. On top of this, about 500 thousand people die every year in traffic accidents, and about 45 million are injured, one-third so seriously that they need hospital treatment.

Pollution from rail traffic is considered by many to be connected with old and badly handled wood preservation facilities, the use of embankment materials contaminated with CFCs, and disposal of locomotive oils containing PCBs. In addition, however, environmental problems associated with this mode of transportation also include pollution from the use of cadmium and asbestos, discharges from train toilets and the use of chemicals for weed control along the rights of way. Further, the action of the bogeys against the tracks can disperse considerable quantities of fine metallic particulates, principally iron, up to 100 m from the track bed.

Shipping imposes various forms of environmental pollution which are usually associated with emissions to the air from the use of bunker oils. The quality of internationally available bunker oils has been deteriorating in direct parallel to improvements in refinery techniques, although a desulfurization technique is being further developed within the petrochemical industry. Another major environmental problem linked to transoceanic shipping is the discharge of oils to the water. Historically, accidental oil spills have been the environmental issue most associated with shipping, but the regular discharge of bunker washings and ballast waters are now recognized as being the more significant contributors of hydrocarbons to seas and harbours. The dumping of chemicals and garbage (and other macro-pollutants) at sea are also environmental problems associated with shipping. These latter have been restricted in recent years by a number of international and regional maritime agreements (*cf.* UN, 1983).

Environmental concerns over air traffic can be separated into pollution of the air and water, and noise problems. The most important air pollutants include jet engine emissions of nitrogen oxides, hydrocarbons, carbon monoxide, benzene, ethylene, polyaromatic hydrocarbons (PAHs) and soot. About 20, 30 and 50 percent of the total emissions of nitrogen oxides, hydrocarbons and carbon monoxides, respectively, are emitted at high altitudes (at cruising level) during a typical 400 km flight. In addition to these direct impacts, various indirect impacts associated with the terrestrial aspects of aircraft operations can occur. For example, aircraft fuels can be spilled or can evaporate during the refuelling process, releasing PAHs to the atmosphere or soil, while water pollutants (e.g. glucose, urea) can be released during de-icing and as a result of fire-fighting practices.

These forms of transportation account for a considerable part of the total emissions of air pollutants. The environmental debate so far has been focused on air pollution from vehicles and aircraft. Comparatively little information is therefore available for the significance of environmental pollution from railways and shipping. In very general terms, transportation can be regarded as responsible for the following pollutant contributions on a global scale: 80 percent of the carbon monoxide emissions, 70 percent of the nitrogen oxide emissions, 50 percent of the hydrocarbon emissions, 5 to 10 percent of the sulfur dioxide emissions, and 5 percent of the particulate matter emissions. Of these, vehicle emissions constitute the key sources of pollution from the transportation sector, accounting for the following proportions of the total emissions: 90 percent of the carbon monoxide emissions, 80 percent of the nitrogen oxide emissions, and 80 percent of the hydrocarbon emissions. Road traffic makes up almost the entire amount of these emissions (over 90 percent). Road traffic also accounts for about one-third of the total carbon dioxide emissions, with about 5 to 6 percent being car-generated. This latter source is the fastest growing source of carbon dioxide to the atmosphere. Emissions of lead to the air continue to originate from road traffic due to the use of lead additives in gasolines, primarily in developing countries; in the developed nations, vitreous oxide is a by-product of the catalytic converters used to control such impurities in exhaust emissions. These emissions may contribute to global climate change ('the greenhouse effect'). Aircraft emissions contribute a relatively small amount of the total air pollutants: about 4 percent of the carbon dioxide, 2 percent of the nitrogen oxides, and 1 percent each of carbon monoxide and total hydrocarbons.

RESIDENTIAL AND CONSTRUCTION SITES

Hydrologic effects of urbanization

In an undeveloped area, stormwater runoff (water draining from the land surface during and immediately following a rainstorm) is circulated throughout natural systems (see Chapter 4). When undeveloped lands are converted to other land uses, however, stormwater runoff can become a significant water quality problem. Of all land use changes affecting the hydrology of an area (both quantity and quality of runoff water), urbanization is probably the most significant. Urbanization typically results in the creation of extensive areas of impervious surface (streets, roofs, parking lots). As a result, urban stormwater runoff cannot soak or percolate into the soil as readily as it could in undeveloped areas (where there are large areas of non-paved surfaces). As an area becomes urbanized, the peak water flow and the rate of storm runoff increases and runoff is concentrated in sharper, faster, and higher peaks. As a result, the runoff is rapidly transported to receiving waters (rivers, lakes, reservoirs). Nonpoint source contaminants also more readily accumulate on the impervious surfaces of urban areas, leading to a dramatic change in both quantity and quality (Herricks and Jenkins, 1995).

The nature (quantity and quality) of urban stormwater runoff is the result of several factors, including:

(1)   Reductions in the infiltration, evaporation, transpiration and storage of urban stormwater runoff;

(2)   Increased impervious surface area;

(3)   Modifications of surface drainage patterns (including development of stormwater management facilities).

Historically, urbanization has resulted in the development of stormwater drainage systems. These systems were built to lessen the risk of flood damage to property and minimize the loss of life. Thus, urban stormwater drainage systems (storm sewers) initially were designed to promote public safety and convenience, without recognition of the potential water quality impacts of urban runoff on receiving waters. In recent years, however, particularly in developed nations, the potentially devastating impacts of urban stormwater runoff have become a major focus of nonpoint source control programmes.

The streets and highways of urban areas are covered with the debris of modern society. More obvious examples include paper, plastic containers, beverage cans, cigarette butts, animal excrement and other macro-pollutants. Less obvious, but potentially more serious, contaminants include fertilizers and pesticides used in residential areas, motor oils, etc. An assessment of stormwater impacts in urban areas in Florida (USA) found that:

(1)   Stormwater supplied virtually all the sediment deposited in the State's natural waters;
(2)   The urban stormwater load of biodegradable (oxygen-consuming) substances was nine times greater than that of the municipal wastewater treatment plant;
(3)   Nutrient loads from urban stormwater were comparable to municipal wastewater treatment plant discharges;
(4)   Urban stormwater was the source of 80 to 95 percent of the heavy metals entering the State's natural waters.

Urban hydrograph and the first flush

Peak water flows and contaminant loads in urban stormwater may not coincide. Rather, the first portion of storm runoff (sometimes called 'first flush') typically has a higher concentration of contaminants than the remainder of the stormwater runoff. The first flush describes the washing action that stormwater has on accumulated contaminants on the land surface. In the early stages of a rainstorm, impervious surfaces are flushed clean by the stormwater runoff. This can result in an initial large pulse of contaminants to receiving waters. The Florida studies have shown the first flush equates to approximately the first one inch of rainfall, which carries up to 90 percent of the pollution load from a rainfall event (US Geological Survey, 1984). In an Australian study, Cordery (1977) reported that concentrations of suspended sediments in the first flush were higher than the sediment concentrations found in raw sewage. The quality of the runoff in even the later stages of the flush was equivalent to that of secondary effluent.

Although the contaminant concentration typically is higher in the first flush, the first flush may not carry the largest portion of the total nonpoint contaminant load. This latter aspect of the first flush phenomenon depends on the total volume of stormwater. Studies in Austin, Texas (USA) have shown that the first flush accounts for about 40 percent of the total nonpoint source contaminant load in a typical rainfall event. Overall, first-flush effects generally diminish as the area of the drainage basin increases, and the amount of impervious surface area decreases. Nevertheless, treatment of the first flush of urban stormwater runoff will help minimize its water-quality impacts (see Chapter 9).

The movement of urban nonpoint contaminants also is a function of the type of urban drainage system. Combined sewer systems (stormwater runoff is collected by the municipal wastewater treatment plant collection system) offer the possibility of attempting to treat stormwater runoff before it is discharged to receiving waters. The noteworthy disadvantage of these systems is that municipal wastewater treatment plants typically lack the ability to store large volumes of stormwater, and, as a safety feature, allow the overflow to bypass the plant and drain directly to the receiving waters. Because the stormwater and municipal wastewaters are mixed together in such a system, combined sewer overflows (CSOs) mean that essentially raw sewage can be discharged into receiving waters. This situation is common in older cities. In contrast, separated sewer systems allow for the essentially unrestricted movement of stormwater runoff to receiving waters. These sewer systems do not result in municipal wastewater treatment plant overflows, but, rather, convey runoff and any associated urban contaminants directly to receiving waters without the benefit of any type of treatment.

Stormwater runoff contamination is highly variable and strongly dependent on local conditions. It also is a function of (1) the random occurrences of storms and (2) the widely varying accumulations of polluting materials on catchment surfaces. Table 6.20 illustrates the average concentrations of several traditional water quality parameters measured in both separated and combined sewer systems in some industrialized countries. These data were taken primarily from European and North American studies, and illustrate the effects of diverse land uses, different seasons, and various types of rainfall events. The methodologies also varied between studies. (Although the values may be exceeded under some conditions, the tabulated results nevertheless provide a reasonable indication of the concentrations of some urban nonpoint source pollutants (UNESCO, 1987).)

Another method of reporting storm runoff data is in the form of annual unit area loads for various land uses. These also are known as nutrient export coefficients (Rast and Lee, 1983). Examples of unit area loads, from studies conducted in the United Kingdom (Ellis, 1986), Canada (Dick and Marsalek, 1979 and Finland (Menanen, 1981), are presented in Table 6.21 (see also Chapters 4 and 5).

Nonpoint contaminants associated with urban areas

Uunk (1983) categorized nonpoint source contamination problems inherent in urban stormwater runoff in two main groups:

Table 6.20 Ranges of pollutant concentrations (mg/l) in combined sewer overflows (CSOs) and stormwater runoff (from UNESCO, 1987)

| | *Land use* | | | | | |
| | *Residential* | | *Commercial* | | *Industrial* | |
Parameter	*CSOs*	*Separate*	*CSOs*	*Separate*	*CSOs*	*Separate*
BOD$_5$	2–600	1–145	4–600	0.5–173	82–685	0.5–88
	(23–114)	(8–55)	(46–95)	(17–38)	(86–153)	(9–28)
COD	33–1762	4–1740	41–626	3–610		0.7–605
	(138–209)	(28–213)	(138–145)	(46–170)		(86–343)
Settleable	0.1–656	0.1–4500		0.1–440		0.1–1270
solids	(165–238)	(50–435)		(76–160)		(151–374)
Total	24–1260	1–12 000	20–1800	1–4803	124–1000	1–11 900
suspended	(177–271)	(28–736)	(90–391)	(56–275)	(274–637)	(114–1220)
solids						

The first line for each water quality parameter displays the minimum and maximum values from all studies; the values in parentheses are the range of the average concentrations from all studies

(1) Problems relating to the quantity of runoff which include (a) increased peak flows in receiving streams (which may result in flooding and erosion), (b) decreased base flows in receiving streams (endangering aquatic life and sometimes water supply), and (c) depletion of groundwater aquifer storage (which may result in damage to vegetation, contamination of groundwater and reduction of water supply);

(2) Problems relating to the quality of runoff including impacts on aquatic ecosystems, such as (a) depletion of the oxygen content of receiving waters (due to the oxygen demand of the stormwater runoff), (b) eutrophication of receiving waters due to nutrient loads in urban runoff, (c) toxic effects due to heavy metals and synthetic organic chemicals in stormwater runoff, and (d) interference with human uses of receiving waters (including recreation and drinking water supply).

On a site-specific basis, several types of water quality impacts are associated with urban runoff (US Environmental Protection Agency, 1983):

(1) Rapid, short-term changes in water quality during and shortly after storm events;

(2) Long-term water quality impacts associated with suspended solids and in-place pollutants, and with nutrients, which enter aquatic ecosystems with long water retention times (slow flushing rates);

(3) Short-term water quality impacts caused by scour and remobilization of pollutants previously deposited in the sediments.

Urban nonpoint source pollutants enter aquatic ecosystems through pipes,

Table 6.21 Annual unit area loads for stormwater and combined sewer overflows (from Ellis, 1986; Dick and Marsalek, 1979; Melanen, 1981)

	Annual pollutant loadings (kg/ha/y)				
Source	*Total suspended solids*	*BOD*	*COD*	*Total nitrogen*	*Total phosphorus*
Runoff in storm sewers	100–6300	5–170	20–1000	2–12	0.2–2.2
Residential area runoff	600–2300	5–100	20–800	2–12	0.2–2.2
Commercial area runoff	100–800	40–90	100–1000	5–12	1.2–2.2
Industrial area runoff	400–1700	10–90	200–1000	5–10	1.0–2.1
Highway runoff	120–6300	90–170	180–3900	–	–
Combined sewer overflows	1200–5000	500–1300	500–3300	15–40	4–8

channel drains, or direct surface runoff. Urban drainage basins are comprised of transportation corridors, housing estates, recreational areas, and business and commercial districts. Although the surfaces associated with these land uses vary in permeability, most are essentially impervious (US Environmental Protection Agency, 1983). Accordingly, the major contaminants associated with urban land uses and stormwater runoff vary, but typically include:

(1)   Suspended solids (sediments);
(2)   Plant nutrients (phosphorus, nitrogen);
(3)   Biodegradable organic material;
(4)   Heavy metals;
(5)   Bacteria;
(6)   Road salts and grit;
(7)   Macro-pollutants (organic and inorganic litter);
(8)   Synthetic organic chemicals and hydrocarbons (from vehicular and industrial emissions and leakages, and the degradation of street surfaces).

The major nonpoint source contaminants typically associated with short-term changes in water quality are toxic substances (heavy metals) and biodegradable organic materials. Longer-term impacts are associated with toxic substances (heavy metals) and nutrients. The third type of impact, that associated with sediment scour and remobilization, involves primarily toxic substances (heavy metals). Several of these contaminants are discussed further below.

Sediment

In terms of volume, sediments (suspended solids) are the primary nonpoint contaminant found in urban stormwater runoff; this is similar to the situation found in agricultural areas (Daniel *et al.*, 1978), and, likewise, is a result of soil erosion by wind and/or water (Knox, 1977). One long-term effect of sediment transport is the accompanying movement of sediment-associated pollutants to receiving waters. The total suspended solids concentration in urban stormwater runoff is relatively high, compared to municipal wastewater treatment plant discharges (US Environmental Protection Agency, 1983). This is especially true for construction sites. In a study of urbanization effects, conducted near Washington DC, Guy (1965), for example, reported that 423 tons/ha (189 tons/acre) of sediment were discharged from a 23 ha (58 acres) area during three years of development and construction. In another development near Washington DC, Guy (1965) reported that 311 tons/ha (139 tons/acre) of sediment were transported in runoff from a construction site. This is double the 156 tons/ha (70 tons/acre) measured prior to the commencement of construction activity.

Nutrients

Plant nutrients (phosphorus, nitrogen) are common constituents of urban stormwater runoff. These nutrients have a primary role in the eutrophication process. Nutrients in stormwater runoff exist in both dissolved and particulate form, with the particulate forms accounting for about 60 percent of the total. Most nutrients in urban runoff were derived from precipitation, dust fall, leaching from vegetation, street litter, lawn fertilizer, and petrochemical combustion. Leaf deposition is also a significant source of phosphorus in urban runoff (Cowen and Lee, 1973).

Biodegradable organic matter

Biodegradable organic matter can cause oxygen depletion in receiving waters, primarily as a result of the oxygen requirements of bacteria in the water column that decompose the organic materials. The majority of the biodegradable organic matter in urban runoff is associated with very small (< 45 μm) particles. On an annual basis, oxygen-consuming material in stormwaters in the United States is comparable to that of secondarily-treated municipal wastewater discharges. The major urban contributors of oxygen-consuming materials to receiving waters, in terms of increasing oxygen demand, are light industrial, heavy industrial, residential, and commercial areas (US Environmental Protection Agency, 1983; Daniel *et al.*, 1978).

Heavy metals

Common heavy metals of concern in urban runoff are lead, zinc, copper, chromium, cadmium, nickel and mercury. Lead, zinc and copper account for about 90 percent of the dissolved heavy metals, and 90 to 98 percent of the total metal concentrations in many urban stormwaters (Harper, 1985). Heavy metals in highway runoff originate primarily from motor vehicles, direct atmospheric fallout and the degradation of highway materials. Although lead is the principal metal in urban runoff (because of its use in automotive fuels), zinc and other heavy metals are also found at deleterious levels (Porcella and Sorensen, 1980). Automobile tyre wear is a primary source of zinc in urban runoff. Daniel *et al.* (1978) reported that street materials from commercial areas had a higher concentration of toxic metals than materials from industrial or residential areas. By volume, however, the contribution of heavy metals was highest from industrial areas, and lowest from commercial areas. Studies in the United States suggest that copper is the most significant toxic metal in urban stormwater runoff (US Environmental Protection Agency, 1983).

Microorganisms

Microorganisms in urban stormwater runoff can pose potential human health hazards, and increase drinking water treatment costs. The US Environmental Protection Agency (1983) reported high levels of coliform bacteria in urban runoff, even where high degrees of dilution occurred. These same studies also documented substantial seasonal differences, with higher levels occurring during the summer. In several urban lagoons and estuaries in Cape Town (South Africa), an opposite, yet consistent, result has been reported; bacteria levels peaked during winter, in response to the greater inflows from the urban watersheds during the rainy season.

Major sources of urban nonpoint source pollutants

Atmospheric deposition

Pollutants present in the air originate from other sources in, or adjacent to, a drainage basin (traffic, industry, etc.). Airborne pollutants can reach the land surface as both wet and dry deposition. However, most atmospheric contaminants are washed from the air during the early stages of rainfall events. Washout of atmospheric pollutants by rainfall droplets is very effective (i.e. 'first-flush effect'), and contributes to the high pollutant concentrations in the first part of stormwater runoff (Goettle, 1978; Randall *et al.*, 1978).

Nonpoint source inputs, particularly from vehicular and smokestack emissions, also can cause acid rain. Rainfall acidity varies greatly because of local conditions (e.g. numbers and types of industries present, prevailing wind direction). In urban areas, the pH of rainfall is often lowered by sulfur dioxide and nitrogen oxides attributed to stack emissions from energy production processes (Junge, 1960). However, Novotny and Kincaid (1982) report that this acidic precipitation can be buffered in some urban drainage systems by the dissolution of calcium and magnesium carbonates from pavements. Some reported mean values of atmospheric loading rates are presented in Table 6.22.

Compared to raindrops, snowfall has an increased pollutant scavenging efficiency, primarily because of the large snowflake surface area. The increased ionic strength of the subsequent melt water (particularly associated with salting activities) can enhance metal exchange capabilities and facilitate movement of

Table 6.22 Mean values of atmospheric loadings in urban drainage basins (from Ellis, 1986)

Pollutant	Total deposition rate ($g/m^2/yr$)	Wet deposition rate (mg/l)	Snowmelt (mg/l)	Contribution to runoff (percent)
Suspended solids	8.4–36.2	5–70	263–690	10–25
COD	0.44–31.6	8–27	15–25	15–30
Sulfate (SO_4)	6–15	4.8–46.1		31–100
Total phosphorus	0.021–0.204	0.02–0.37		17–140
Nitrate-nitrogen	1.8–8.2	0.05–4.4	4.1–5.7	30–94
Lead	0.04–4.0	0.03–0.12	0.3–0.4	15–54
Zinc	0.1–1.3	0.05–0.38	0.35–0.41	20–62

Table 6.23 Mean concentrations of pollutants in snow in several urban areas in Sweden (from Malmqvist, 1978; Hogland, 1979)

City	Population density (persons/ha)	Concentration (mg/l)				
		Sulfate	Total phosphorus	Copper	Zinc	Lead
Gothenburg	250	< 5	0.41	0.05	0.36	0.25
„	115	< 5	0.11	0.01	0.05	0.04
„	22	< 5	0.09	0.01	0.06	0.04
Lund	110	8.8	0.07	0.05	0.14	0.20
„	32	5.9	0.05	0.10	0.07	0.15
„	*	5.3	0.02	0.23	0.06	0.04

*Rural area just outside the city

metal ions into an aqueous phase (Morrison *et al.*, 1984). Malmqvist (1978) and Hogland (1979) reported such enhanced concentration values among pollutants in 'untouched' urban snow on lawns and in parks in Swedish cities (Table 6.23). The concentration of pollutants in the snow increased with the age of the snow, due primarily to this scavenging ability.

Dry fallout is usually a relatively minor component of atmospheric-derived pollution, but can vary considerably as a function of local climatological conditions, land-use activities, etc. Dustfall originates mostly from unpaved roads and parking lots, uncovered material storage sites, construction and demolition sites, industrial and domestic refuse, and from biological sources (pollen, spores, etc.). Annual rates of urban dustfall range from 0.01 to 13 tons/ km^2, depending on local conditions. For most metropolitan areas in the United States, deposition rates range between 1.7 and 3.2 t/ha/y (American Public Works Association, 1969), and are associated with dust particles < 30 μm in size.

Particles in this size range make up 99 percent of the total dry fall (Novotny *et al.*, 1985).

Vehicular traffic and related activities

As noted above, traffic-associated pollution consists primarily of vehicle exhaust emissions, losses of fuels and lubricants, particle emissions from vehicles and asphalt wear, and road-deicing chemicals. Shaheen (1975) reported approximately 1300 mg/vehicle/km of solids could be attributed to vehicular traffic. Direct traffic emissions were reported to be 200 mg/vehicle/km (Anonymous, 1977) and 120 mg/vehicle/km from abraded tyres (Brunner, 1975). In addition, deicing compounds can contribute significant salt loads to water courses, depending on local climatological conditions and legislative requirements – similar salinity impacts could also result from chloride-based dust suppressants used in many countries, including those not subject to snowfalls. The most widely applied deicing chemical is generally sodium chloride which results in salt loads of 1 to 3 tons/km (Nelissen, 1982; Brunner, 1975). However, road salt application recently has been restricted in several countries, with sand or stone chips being applied for deicing purposes. Where salts are necessary, calcium chloride provides a competitive alternative to sodium chloride.

Residential activities

Natural and anthropogenic residential nonpoint pollutants include (1) street litter (trash), (2) animal wastes and carcasses; and (3) household chemicals. Litter includes such discarded items as paper, bottles, plastics, building materials, tobacco residues, yard waste (e.g. grass clippings, garden residues, parts of vegetation), food leftovers, etc. A litter accumulation rate of 1.8 kg/person/y was reported in the United States (Proctor and Redfern Ltd., 1981). Due to its large average size, however, litter represents an aesthetic problem and usually is not a major water quality problem unless it accumulates in runoff channels, storm sewers, and other areas. In such cases, flooding and other water-related problems can occur.

Animal wastes (faeces) can cause serious water quality problems in countries where pets are numerous. For example, 1.5 million dogs produce approximately 22.5 million kg faeces/y in The Netherlands (Uunk, 1983). This faecal load causes bacteriological contamination and oxygen depletion in receiving waters. Dead animals and animal excreta also are present in urban milieu.

The use of household chemicals (detergents, fertilizers, insecticides and herbicides, paints, oils, etc.) also can be locally important sources of urban nonpoint pollution. This is due primarily to the solubility of these materials, and the ease with which they can be washed from the land surface by stormwater runoff. In many instances, these substances are introduced directly to stormwater drainage systems as a (illegal) disposal mechanism.

Soil erosion

Pervious surfaces in urban areas are generally well protected by vegetation, at least in most developed countries where urban lawns and gardens are culturally desirable. Urban soil erosion rates in these countries are usually much lower than in agricultural (rural) areas. However, urban areas which do not have protective vegetative surfaces can be major contributors of suspended solids to receiving waters. Such areas include construction sites, unpaved roads and unmaintained lots. In developing countries, depending on climatic factors, vegetated homesteads are less the cultural norm. However, even in such situations, the major source of urban soil erosion is generally the fast-growing urban fringe. In such countries, erosion is often evident along the banks of depressed urban stream channels. In addition, in many developing countries, economic exigencies often mean that homestead gardens are cropped to supplement food supplies and are, thus, more susceptible to soil washoff (see the discussions on agricultural soil losses, above). As in agricultural regions, soil erosion rates are a function of geomorphology, climate, soil types, and even social conditions. As a result, reported ranges of erosion rates from urban areas are extremely large. For example, Heaney *et al.* (1975) reported that erosion rates in cities in the United States ranged between 500 and 8000 kg/ha/y, with extreme values as high as 350 000 kg/ha/y at construction sites. In urban settings, sediment-associated pollutants can be computed as a fraction of the sediment concentration in runoff (Huber, 1986).

Vegetation inputs

Leaf fallout can be an important source of organic matter, nutrients and oxygen-consuming substances in urban areas (Cowen and Lee, 1973). The quantity contributed to receiving waters varies with land usage, and vegetation type and density. During autumn, for example, a mature tree can produce between 15 and 25 kg of organic leaf litter. Further, trees can also enrich the rainwater penetrating the tree canopy with nutrients and organic materials exuded from both living and senescent cells.

Corrosion

The effects of the corrosion of industrial products exposed to the weather in urban areas has not yet been well-quantified. Certainly, acidic precipitation has caused considerable damage to monuments and structures made of limestone and/or cement. In turn, the rusting of re-bar and other metals can potentially contribute metals to stormwater runoff, while creating the need for extensive structural repair or replacement (and the contingent nonpoint source pollution consequences that result from construction activities).

INDUSTRIAL AND COMMERCIAL ACTIVITIES

Introduction

Industrial activities cover a multitude of human production activities, ranging from primary industries such as ore extraction and processing to final assembly of finished goods. Commercial activities transfer such goods from the production process to the consumer sector. The scales of these activities range from one-person, home operations to large, multi-national public corporations. Industrial activities tend to be consumptive activities, making use of primary resources, energy, and human labour and creativity, while commercial activities, although less resource-intensive, also consume energy, and human labour and creativity in the process of transferring industrial goods to consumers. While industrialization has led to great boons for society, including far-reaching advances in sanitation, transportation and standard of living, it has also led to equally significant impacts on the environment and quality of life. Urbanization, in the modern sense, is a direct result of industrialization – manufactories require a concentrated supply of labour, consumers and the centralization of materials in stockpiles. In turn, urbanization has created a need for industrial farms to supply the needs of the urban populace, extensive transportation networks to meet the need for efficient transfers of materials between plants and to markets, and waste disposal systems to carry away excess materials and refuse. Each of these activities has an actual or potential nonpoint source contaminant impact, which is dealt with in this chapter. This sub-section covers those impacts related to industrial and commercial enterprises.

While many industrial and commercial impacts are in the form of point source discharges, runoff from the factory yards, impervious surfaces and material stockpiles can and do contribute to nonpoint source contamination of aquatic systems. Virtually every form of impact is possible – while many impacts are industry-specific, a great many are common among all industries. Hence, we have adopted a composite approach based on a cross-section of commercial and industrial enterprises, including the following:

(1)   Brewing and beverages;
(2)   Edible oils;
(3)   Sugar;
(4)   Pulp and paper;
(5)   Dairy;
(6)   Meat packing and processing;
(7)   Leather and textiles;
(8)   Laundry;
(9)   Metal finishing and processing.

Industrial and commercial nonpoint source contaminants

As has been noted, the contaminants generated as the result of industrial and commercial activities can be either point sources, where process-, waste- and storm-waters are discharged *via* pipelines or sewerage systems, or nonpoint sources of contamination where waters and/or wastes are simply discarded. While process wastes and waste materials are usually considered to be point source contaminants in most developed nations, poor infrastructure or inappropriate technologies employed in less developed nations can lead to these same by-products being nonpoint source contaminants. In addition, even in developed nations, product or material spills and/or poor housekeeping practices can lead to a nonpoint source contamination component of the total waste load generated.

To gain some perspective of the nature of nonpoint (and point) source contaminants that can be anticipated from industrial and commercial operations, Tables 6.24 and 6.25 summarize a large volume of data gathered during the 1980s as part of the National Survey of Water Use conducted by the South African Water Research Commission. These tables suggest that the most common nonpoint source contaminants to be expected from industrial and commercial operations include:

(1)   Sediments;
(2)   Nutrients;
(3)   Metals;
(4)   Oxygen-consuming substances and organic materials;
(5)   Salts;
(6)   Microorganisms;
(7)   Acids;
(8)   Macropollutants.

In other words, commercial and industrial operations can produce the entire spectrum of nonpoint source contaminants defined in Chapter 5. Obviously, the magnitude of the potential contamination as well as its precise composition varies widely between industries. Thus, few industries contribute the full range of pollutants, while most contribute more than one form. Macropollutants, in the form of litter and trash, are the most common class of contaminant common to virtually all commercial and industrial operations.

While the shortcomings of Tables 6.24 and 6.25 have been partially noted above, reference to the extensive compilations of van der Leeden *et al.* (1990), which relate primarily to the United States and North America, suggests that the range of conditions reflected in the Tables is probably widely applicable. Water use, in terms of gross volume of water consumed and discharged, is significantly higher in the United States than in South Africa, but the percentage of waters recycled and the types of contaminants – and magnitude of contamination – remain comparable in both countries. Hence, while there is potential for significant variations in these parameters between countries and continents, the following discussion will have some general relevance.

Table 6.24   Composition of industrial effluents from various industries in South Africa (Binnie and Partners, 1986, 1987a; Steffen, Robertson and Kirsten, 1989a, 1989b, 1989f, 1990b). Concentrations given as kg per unit of product produced

Industry effluent	*COD*	*OA*[a]	*TSS*	*TDS*	*TOC*	*pH*	*SRP*	*SO$_4$*	*Cr*	*SOG*[b]
Brewing and bottling (kg/m³)										
Malt beer	10.4	–	2.9	7.9	–	–	–	–	–	–
Sorghum beer	5.2	0.5	1.7	2.3	1.0	–	0.01	–	–	–
Soft drinks										
bottle wash	3.8	–	0.5	5.3	–	–	–	–	–	–
no wash	3.5	–	0.1	2.1	–	–	–	–	–	–
Edible oils (kg/ton)										
Oils	9.8	–	–	15.4	–	5.9	–	7.6	–	1.6
Pulp and Paper (kg/ton)										
Pulp and paper	57.8	–	48.2	84.2	–	–	–	–	–	–
Paper	6.9	–	12.1	37.6	–	–	–	–	–	–
Leather and tanning (kg/ton)										
Full tanning	166.5	–	18.0	346.5	–	–	–	–	4.5	–
Wet-blue tanning	130.6	–	31.0	225.1[c]	–	–	–	–	4.0	–
limeyard[d]	96.8	–	27.0	91.4[c]	–	–	–	–	0	–
chromeyard[e]	33.8	–	4.0	133.7[c]	–	–	–	–	4.0	–

[a]Oxygen absorbed = permanganate value
[b]SOG = soaps, oil and grease
[c]Unsalted hides; salted hides would have significantly higher TDS loads
[d]Area where hides are soaked and de-haired
[e]Area where hides are pickled in acids, salt and chrome tanning salts

Table 6.25 Composition of industrial effluents from various industries in South Africa (Binnie and Partners, 1987b; Steffen, Robertson and Kirsten, 1989b, 1989c, 1989d, 1989e, 1989f, 1989g, 1990a, 1990b). Concentrations given in mg/l of effluent discharged

Industry effluent	COD	TSS	TDS	pH	SRP	TKN	SO$_4$	Cr	Al	TotMet[a]	CN
Sugar refining											
Sugar	1154	2760	–	8.5	12	–	–	–	–	–	–
Pulp and paper											
Kraft pulp	1400	–	–	2.8	1.4	> 0.6	229	0.1	0.1	> 2.3	–
Dairy											
Milk											
Pasteurized	2500	–	1800	–	–	17.9	–	–	–	–	–
Skimmed	1000	–									
Cream: 30% fat	8600	–									
Buttermilk	1100	–									
Whey	750	–									
Fruit juice-milk drinks	1500	–	1100	–	–	10.0	–	–	–	–	–
Other milk products	2750	–	1500	–	–	8.0	–	–	–	–	–
Butter	4000	–	2000	–	–	20.0	–	–	–	–	–
Cheese	3000	–	3000	–	–	25.0	–	–	–	–	–
Meat packing											
Red meat	5000	1500	1530	7.4	–	–	–	–	–	–	–
White meat	3410	730	1250	7.2	–	11.0	–	–	–	–	–
Tanning and leather											
Leather	9700	1970	19 600	–	–	–	–	120	–	–	–

Table 6.25 Continued

Industry effluent	COD	TSS	TDS	pH	SRP	TKN	SO_4	Cr	Al	TotMet[a]	CN
Commercial laundering											
Wash water	890	225	2100	11.1	69	–	–	–	–	–	–
Rinse water	60	54	115	8.6	11	–	–	–	–	–	–
Final waste	625	184	1400	9.1	51	–	–	–	–	–	–
Metal finishing[b]											
Electroplating	483	–	2337	7.5	–	–	–	–	–	94	178
Anodizing	248	3392	3122	7.4	96	–	968	–	215	–	–
Phosphating	727	–	1069	7.0	182	–	–	–	–	–	–

[a]Total metals include Cr, Ni, Cu, Zn, Cd, Sn, Ag, and Au
[b]Electroplating deposits a thin coating of a valuable metal over a base of a less valuable metal; anodizing provides a protective oxidizing layer over the surface of a metal; and phosphating (and chromating) provides a base onto which paints will adhere more effectively

Sediments

Sediment or particulate loads generated from commercial and industrial operations typically consist of materials eroded from factory or commercial premises by stormwaters, or of materials eroded or discharged from material stockpiles or waste dumps (excluding litter and trash, see below). Many production or extraction operations generate large quantities of waste materials, whether spent grains, coagulants, flocculants, or fibers. Many of these materials can accumulate on hardened surfaces as dust, such as clay dust or other residues, that is washed off the surfaces by storms and shown as total suspended solids (TSS) in the Tables. Other particulates are contributed through non- or poorly-treated process waters discharged overland or otherwise indirectly to natural water courses. Many of these substances also contribute to the load of oxygen-consuming substances (shown as COD). A major source of sediment from commercial and industrial areas comes from soils eroded from factory yards, city streets and parking areas. At times, industrial processes can contribute to the dearth of vegetation in industrial areas which further enhances the runoff of suspended solids – sites downwind of acidic stack discharges often suffer from erosion due to this factor. Vehicular traffic, pedestrian traffic and other frequent disturbances also enhance soil disturbance which, in turn, leads to enhanced erosion from these unprotected surfaces, especially in the absence of paving or at the margins of impervious surfaces (where runoff velocities and volumes are high due to the effect of the impervious surface). In the Tables, the highest values of TSS are primarily (but not exclusively) associated with agri-industrial operations; e.g. the sugar cane/sugar beet fibres produced as organic waste after the sugar-bearing juices are extracted, the bone chips and other organic fragments produced during the butchering process, and the hair and other particulate matter left after hides are initially processed into leather. Many of these potential contaminants can be re-used as protein supplements in cattle feeds if accumulated, de-watered and formed into cakes or pellets.

The supply of particulate matter from industrial and commercial operations is extremely susceptible to process manipulations, with various types of processes producing greater or lesser quantities of particulates as in the case of the metal finishing industry (Table 6.25). Many of these particulates are also amenable to treatment and retention by on-site wastewater treatment facilities, although some industries do require fairly large-scale operations before such treatment is affordable or practicable (hence, many small-scale operations in developing countries, especially in the rural areas, are generally without post-production treatment).

Nutrients

The principle source of nutrients from industrial and commercial operations is wash waters containing detergent residues. On the whole, nutrient enrichment from industrial and commercial nonpoint sources is minimal compared to the other types of contamination that can arise from these operations. Granted, some industries do produce significant quantities of nutrient-rich process waters which are sent to waste – the dairy and textile industries are prime examples (Table 6.25) – but these are generally in the minority. Nevertheless, nutrients arising from industrial operations such as fertilizer production, packaging and transfer can be significant in some cases. Landscaping and transportation activities, ancillary to the industrial or commercial processes, can also result in significant nonpoint source nutrient loads.

Metals

Lead and cadmium are two of the most ubiquitous metals present in the biosphere in artificially high concentrations. Motor exhausts (see above) and other human activities related to urban living – including industrial and commercial activities – are responsible for such introductions, which commonly find their way into water courses in the form of street wash off. In addition to these contaminants, a great many other minerals and metals find their way into industrial effluents. Obviously metal finishing and processing is a major source of metal contamination both in the form of discharged plating bath liquids and in the form of rinse waters where excess plating materials are washed from the plated products – usually by an acid bath (Table 6.25). The leather industry, which uses chromium in its tanning process, can also discharge significant quantities of metals (Tables 6.24 and 6.25). Again, while many of these sources are amenable to treatment, such treatments are expensive and usually not employed by small-scale operations.

Organic matter

The release of oxygen-consuming substances and organic materials, including synthetic organic compounds, is probably the greatest single class of water contaminant released by industrial and commercial enterprises – the only possible exception being the macropollutants which are generated by all sectors of society. These substances are both particulate and dissolved and together form the high chemical oxygen demands (COD) recorded in Tables 6.24 and 6.25. These values can range up to and exceed 8600 mg/l. Without treatment, such high COD values can rapidly reduce oxygen concentrations in receiving waters to the point where anaerobiosis can occur. The presence of H_2S and CH_4 in the urban waterways of industrializing Europe in the 19th Century gave olfactory evidence of the consequences of discharging untreated (textile?) effluents, and gave such centres the characteristic smells mentioned frequently in the classical literature. Some suggestion of this remains to this day as those who live and work in the vicinity of pulp mills, for example, can attest. Fortunately, many of these oxygen-consuming substances are amenable to treatment and re-aeration using settling pond technologies that are usually not prohibitively expensive.

The range of synthetic organic materials released in industrial effluents and occurring from industrial wastes is legion. Probably the best known of these contaminants are the PCBs (polychlorinated biphenyls) and PAHs (polynuclear aromatic hydrocarbons) that are only just being recognized as carcinogenic. The PCBs are intentionally manufactured for use in the electrical industry to lubricate and insulate transformers and other electrical equipment, and have been released both intentionally in wash down waters and accidentally through equipment failures and fires. In addition, a range of synthetic organics can occur as by-products of many industrial and commercial operations, such as the THMs (trihalomethanes) that can occur in waterworks as the result of over-chlorination of algal-laden raw waters. Similarly, in the pulp and paper industry, toxic and non-biodegradable chlorphenolics result as by-products of the pulping process. These substances are more intractable and are not readily amenable to common treatment processes; hence, synthetic organic substances continue to reach watercourses *via* the effluent stream. Further, such substances can lie within the sediments for extended periods and continue to threaten the public health. Typically, longer terms management of many of these substances is associated with their replacement by less toxic substances and phasing out of service over time. Unfortunately, this phasing out process usually includes transporting such substances to developing nations where their use can continue to decades (e.g. DDT; Bowonder, 1987).

Salts

Most industries for which data exist contribute large quantities of dissolved solids to the environment (see TDS in Tables 6.24 and 6.25). Those industries that produce saline effluents, such as the tanning industry – where hides are routinely preserved with salts which serve as temporary preservatives (although there is a move toward freezing the hides which saves time and costs) – and the metal plating industry, can contribute directly to the salt load of receiving waters, other industries also contribute; e.g. the dairy industry where brines are used in the cheese-making process, the oil industry where brines are regularly used to extract additional oil from wells (and brines are often extracted from wells), and the transportation industry where salts are regularly used to slake the generation of dust during road construction and maintenance and where, in temperate climates, salts are applied during winter to control ice formation on roadways. Many of these salts reach the environment in diffuse form – carried by runoff from salt stockpiles, washed off road surfaces, or re-solubilized from evaporation ponds or over-irrigated agricultural fields (salination). The latter is a major problem in semi-arid environments and constitutes a significant threat to continued agricultural production in some areas (e.g. India, northern and western China, the southwestern United States, Australia and southern Africa). Nonpoint source contamination by salts can be reduced through the use of properly managed evaporation ponds, and many brines are amenable to re-use even by smaller-scale industries. While alternative practices such as refrigeration of hides are being introduced in some industries, reclamation of salts from brines is also providing cost-effective solutions to the problem of salination of watercourses. While not all brines or saline effluents are so amenable to treatment at low cost, re-use does provide a reasonable alternative in many cases.

Microorganisms

Bacterial contamination of waters can be both direct, as in the case of the brewing industry where 'spent' yeasts are routinely disposed of, and indirect as in the case of the abattoir industry where such contaminants are discharged *via* faecal matter and waste blood. Microbes can also form on the rich organic substrates provided as particulate waste from many industries (see above, much particulate waste generated by industry is organic in nature). While few data are available, TOC can be used as a diagnostic in some cases – if this determinant is adopted as a diagnostic indicator, then bacterial contamination of waters is less of a concern than the organic, oxygen-consuming substance loads with which they may be associated. In terms of those microorganisms released directly into the environment, few are dangerous to human health and many are amenable to reuse as cattle feed supplements (e.g. brewer's mash is often 'recycled' in this manner). On the other hand, human health impacts can develop from the release of contaminated waste, especially from the poultry industry where *Salmonella* sp. is a common and persistent (surviving in soils for up to 9 months) contaminant capable of infecting humans. While little attention is commonly paid to such impacts – either as a point or as a nonpoint source contaminant – disinfection of wastes should be considered as an adjunct to the hygiene standards already practiced throughout much of the food industry.

Acidic substances

Industrial and commercial enterprises have been identified as primary sources of acidic contamination of the water environment. While many of the impacts are indirect, *via* the atmosphere (see above), significant contamination can occur from effluent discharges and wash off from waste dumps and material stockpiles. The latter are very common in the mining and related industries, especially where the minerals are extracted from matrices within sulfide rocks. Both initially, at the point of extraction where groundwater seepage is often highly acidic, and subsequently, during various processing stages, where spoils and extractants can both contain significant residual concentrations of acidic substances, direct discharge and stormwater runoff can introduce acids to watercourses, while spills of finished products, including sulfuric acid, can also contribute to pH depression within the aquatic environment.

While these contaminants are usually acidic in nature, similar contamination can at times result from the discharge of alkaline substances to the environment. Extreme pHs in either direction can cause environmental damage.

Table 6.25 suggests that both types of environmental contamination are possible from industrial sources. pH values as low as 2.8 are common in the pulp industry, while pH values of over 11 have been recorded in commercial laundering operations. Fortunately, extreme pH values are easily adjusted by addition of the appropriate acids or bases, but many times this may be prohibitively expensive for smaller-scale industries. Dilution or waste stream combination – where acidic substances discharged by one industry may neutralize alkaline substances discharged by another – do provide some alternatives. Generally, the release of process waters with extreme pHs is accidental or results from carry-over of extreme pH solutions between portions of a process. Thus, acidic contamination exclusive of that created within the atmosphere is not a major nonpoint·source problem within the industrial and commercial sectors.

Macro-pollutants

By far the most common nonpoint source contaminant of the general environment is litter and trash, both *via* direct disposal and/or poor housekeeping practices (i.e. wind-distributed solid waste) and indirect means. Both are of concern here. Direct disposal of solid wastes in or along watercourses has been standard practice in most parts of the world for decades. Wetlands, especially, have suffered from deposition of 'clean fills' and solid wastes in both permitted and illegal (and possibly unknown) landfills and dumps. The latter have been implicated in many of the most startling and tragic human impacts yet known (e.g. Love Canal). The full extent of the impacts from waste dumps (not many of which are illegal simply because the law was silent on the siting of disposal facilities until only relatively recently; *cf.* Mandelker, 1992) is yet to be experienced. In the interim, some of the impacts of solid waste contamination of watercourses are immediate, such as flooding caused by blocked storm drains and stormwater channels and the associated aesthetic degradation of the environment – few industrial areas are well-known tourist destinations, for example, and many commercial centres (especially in Europe and North America) are approaching this state. While these types of contamination can be rectified by conducting river/lake clean-ups and through promotion of good housekeeping practices (e.g. by matching environmental aspirations with recycling and anti-littering campaigns, see Chapter 9), the more insidious forms of contamination are generally less readily controlled. These are the indirect contaminant loads resulting from the decomposition of solid wastes or the release of chemicals into the environment from ruptured waste containers. Table 6.25 presents some representative analyses of seepage waters arising from solid waste landfills serving municipal areas in the United States. Control of these contaminants usually requires extensive treatment.

Commercial nonpoint source contaminants

Nonpoint source pollution impacts from commercial and other business operations are generally associated with macropollutants and generally occur within the ambit of urban or business park environments. These impacts are typically related to waste disposal, transportation or to the general process of urbanization. These potential impacts are covered elsewhere in this chapter. For example, urbanization and transportation impacts are dealt with above. As noted, macropollutants are generally amenable to control through recycling and litter control measures, although their frequent disposal in landfills or dumps can also lead to more intractable problems (see below).

Summary

While most industrial and commercial impacts on water quality are in the form of point source discharges of contaminants, these same contaminants can form nonpoint source pollutants, especially in developing nations and the United States where urban infrastructure is absent or in a poor condition. Seepage, leaks and spills of process waters, illegal discharges (often encouraged by stringent regulations that leave small producers with no other means of disposing of waste liquids – they cannot discharge to wastewater sewers nor afford the services of waste hauliers – and uncontrolled discharges and overflows will continue to form a nonpoint source component of this waste stream. Good maintenance, sewer extensions, and provision of disposal alternatives (e.g. through co-operatives, etc.) can reduce many of these sources to manageable proportions, although accidental spills, etc., will probably never be eliminated. This reinforces the need for a good spill management and control plan as part of any industrial operation, and for the installation of good stormwater management practices which provide for at least some degree of detention prior to the waters entering natural watercourses. The elaborateness of these schemes will obviously depend on the size of the operation and the presence or absence of municipal treatment systems. Nonpoint source controls are discussed in Chapter 9, while point source controls, insofar as they relate to eutrophication management, are discussed in the companion volume, *The Control of Eutrophication of Lakes and Reservoirs* (Ryding and Rast, 1989).

WASTE MANAGEMENT

Land applications

As used in this book, the term 'land application' refers to the spreading of human wastes (sewage effluents and/or sludges), livestock wastes (cattle, pig, poultry and horse manures and barnyard wash waters) onto the land surface as a method of waste disposal. These wastes may be subjected to treatment or containment prior to their application to the land surface. In reality, land application is a method of recycling such wastes back into the soil, whereby they can potentially be (re)utilized as nutrient sources by plants, including crops. In this manner, a basic cycle of production and consumption of crops by humans is enhanced. This cycle, however, can have both positive and negative impacts on water quality.

The general advantages of land application of wastes, identified by Vermes (1987), include:

(1)   Combining waste treatment with plant production processes;
(2)   Using primarily natural, renewable energy sources for the decomposition of organic matter, in contrast to artificial treatments which often require fossil fuel energy for the same purpose;
(3)   Allowing for the natural recycling of different materials, and the multiple reuse of already-utilized water before returning it to the hydrological cycle;
(4)   Increasing the agro-ecological potential of a given location by supplying cultivated plants with water and nutrients;
(5)   Reducing the gap between a community's drinking water supply and sanitation requirements, by filling the need for a harmless and relatively economic system of waste disposal.

Because virtually all essential nutrients, including nitrogen, phosphorus, potassium, calcium, trace nutrients and humus colloids, are found in human and livestock wastewaters, they may be expected to produce enhanced crop yields. Sopper and Kardos (1975) and McKendrick (1982) provide evidence of positive responses to spray irrigation with municipal wastewaters from forests, old fields, pastures and croplands. The wastewater was relieved of its nutrients and colloidal organic load as it passed through the organic litter and the upper 15 to 30 cm of soil. This zone is rich in soil organisms and has been called a 'living filter'. The earthworm populations were significantly increased on the study sites, and added to the positive effects of the living filter by increasing the potential for incorporation of organic matter and recycling of nutrients within the soil system. The increased earthworm population also increased the soil porosity and water-holding capacity. Crop yields were increased – maize by 50 percent, while hay yields increased by as much as 300 percent.

The currently-used management systems for recycling human and livestock waste products can be classified into two broad groups, (1) aquatic (water–flora–fauna) and (2) terrestrial (soil–flora–fauna). The basic schemes are illustrated in Figures 6.14 and 6.15. After any required pretreatment, the wastewaters are discharged to (a) reedy filter fields or marshlands, (b) existing forests or other lands with natural vegetation, (c) special tree plantations or pastures, or (d) cultivated crop fields (arable land). Attention usually is directed toward terrestrial systems offering the highest economic benefits, however. Essential to development of such a rational approach to land application of wastes is the concept of 'carrying capacity'. This concept includes the capacity of (1) the heterotrophic bacteria in the soil to decompose biodegradable materials

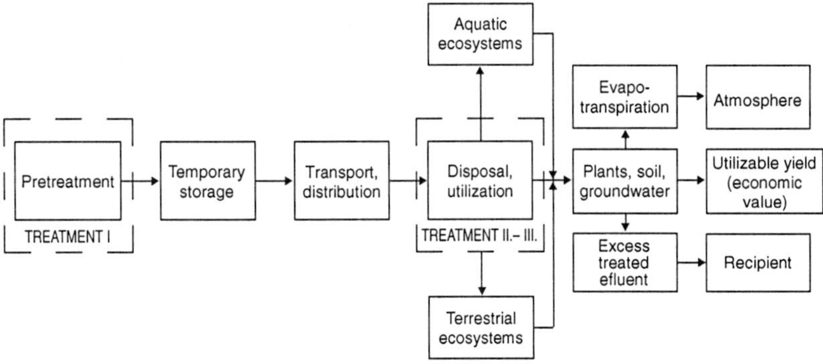

Figure 6.14 Wastewater treatment, disposal and utilization in aquatic and terrestrial systems using a conventional engineered approach (after Vermes, 1987)

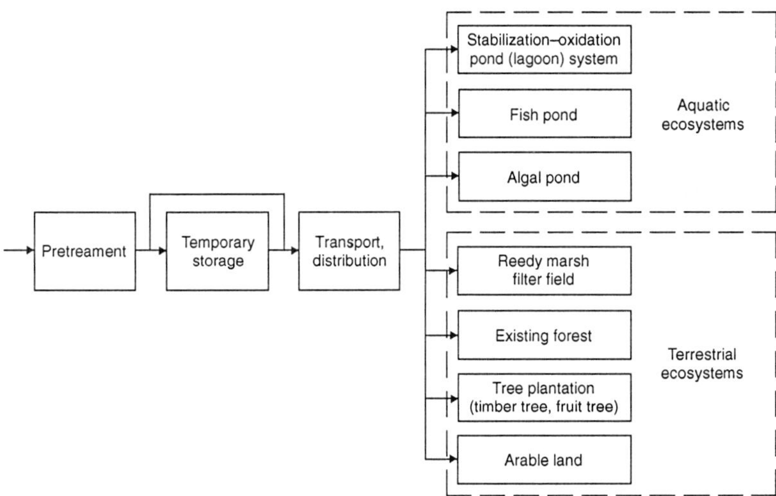

Figure 6.15 Wastewater treatment, disposal and utilization in aquatic and terrestrial systems using biotechnology (after Vermes, 1987)

without the soil becoming septic (i.e. depleted of oxygen), (2) the vegetation (crops, trees or grass) to utilize nutrients released from decomposition of the wastes, and (3) the soil minerals to bind toxic trace elements and prevent their uptake by crops or their movement to groundwater.

There are also disadvantages associated with the land application of wastes which must be considered, such as the nonpoint source pollution of aquatic ecosystems often associated with such schemes. In contrast to previous eras, the modern era has generally produced waste volumes so great that the natural cycle of its production and disposal has become disrupted. As an example, crops produced over a large area are often processed at large centrally-located plants, producing large amounts of solid and liquid wastes, without adequate nearby land on which to dispose of them. In addition, city populations consume large amounts of food and fiber. The resulting sewage and garbage is concentrated in the urban setting with no land suitable for waste disposal. The large agro-industrial feedlots result in a similar 'over-load' of wastes concentrated in a relatively confined area.

A major nonpoint source pollution problem related to land application of wastes is the nature of their contamination. The soil system has little difficulty degrading naturally-occurring organic waste products. However, the wastes of modern society often contain a variety of contaminants, including disease-causing organisms, heavy metals and persistent synthetic organic chemicals. For example, sludges derived from municipal wastewater treatment plants in cities with combined sewers can have high concentrations of heavy metals and synthetic organic compounds (e.g. pesticides). These contaminants have the potential to interfere with the soil's natural recycling processes, and to contaminate water and crops *via* nonpoint source runoff processes. They may also be the source of potential human health effects due to their bioaccumulation in agricultural crops fertilized with the contaminant-laden sludges. Of more recent concern, and as yet unquantified, is the potential effect of growth hormones and other agro-industrial chemicals used in animal husbandry that are released into the environment through excretory processes. These, and other nonpoint source pollution concerns are discussed further below.

Nutrients

Sewage sludge is a good nitrogen and phosphorus source (Table 6.26). The nitrogen content of sludge is about two-thirds organic nitrogen and one-third ammonia-nitrogen (US Environmental Protection Agency, 1985). Most sludges contain very little nitrate. About 20 to 30 percent of the organic nitrogen will be biologically available in the year following the land application of digested sludges. The corresponding value is about ten percent for composted sludge. Nitrogen mineralization decreases to about ten percent or less in subsequent years. Ammonia loss from land-applied sludge is about 20 to 50 percent if the sludge is not immediately incorporated into the soil. The losses can be much higher for lime-treated sludges generated by tertiary wastewater treatment plants using chemical phosphorus precipitation.

Phosphorus exists in sludge in a variety of complexed forms, although it is primarily associated with iron and aluminium, either as precipitates or adsorbed to sludge solids. Some phosphorus is also contained in the organic phosphorus fraction. Field studies indicate that about 50 percent of sludge phosphorus is as biologically available as the phosphorus from inorganic fertilizer.

Table 6.26 Representative quantities of selected components in municipal sewage sludges

Element	Median concentration (mg/kg)	Element	Median concentration (mg/kg)
Nitrogen	39 000	Iron	17 000
Phosphorus	25 000	Mercury	6
Potassium	4000	Manganese	260
Aluminium	16 000	Molybdenum	4
Arsenic	10	Nickel	80
Cadmium	10	Lead	500
Cobalt	30	Tin	14
Copper	800	Selenium	5
Chromium	500	Zinc	1700
Fluoride	260		

As previously noted, most sewage sludge land application programmes are based on meeting crop needs for either nitrogen or phosphorus. Nitrogen-based rates are normally higher than phosphorus-based rates, and are usually used where land for sludge disposal is scarce. Continued use of nitrogen-based application rates, however, usually results in a buildup of biologically available phosphorus in soils far in excess of crop needs. This increases the potential for phosphorus loss in runoff. Therefore, for watersheds with eutrophication problems, and with available land for sludge application, the application rates should be based on supplying crop phosphorus requirements.

Similarly, manure application rates to soils are normally calculated on the basis of the nitrogen or phosphorus requirements of the crop(s) to be grown. Application rates based on nitrogen requirements are usually much higher than rates based on phosphorus. The exception is weathered soils of the tropics and sub-tropics, where phosphorus requirements are very high. Nitrogen-based manure application rates are typically based on actual chemical analysis of the manure, or on average literature values. Depending on how manure is applied and incorporated into the soil, an ammonia volatilization loss of 5 to 50 percent is assumed. Organic nitrogen is assumed to mineralize at a rate of about 30 percent for the first year after application, about 10 percent for the second year, and 2 to 5 percent per year thereafter.

In contrast to phosphorus and nitrogen, all other manure nutrients are assumed to be completely soluble, and behave as if they were applied as chemical fertilizer. Some manures (Table 6.27) contain high levels of calcium, magnesium, or potassium. High manure application rates over many years, therefore, can shift the balance of these elements in the soil, affecting plant uptake. This condition can be monitored by soil tests for elemental content. In dry areas, a particular problem with manures is their high salt content, and its buildup in the soil should

Table 6.27 Representative quantities (%, by weight) of selected components in livestock wastes

Source	Nitrogen	Phosphorus	Potassium	Calcium	Magnesium	Sodium
Beef	1.3	0.5	1.5	1.3	0.5	0.7
Dairy	2.0	0.4	1.7	1.9	0.9	0.4
Hog	2.8	1.0	1.5	1.6	0.4	0.4
Poultry	2.2	1.5	1.8	3.0	0.5	0.5
Horse	1.7	0.3	1.5	2.9	0.5	–

Table 6.28 Representative quantities of selected components in municipal wastewater effluents

Constituent	Typical value (mg/l)	Constituent	Typical value (mg/l)
Suspended solids	25	Arsenic	< 0.005
BOD	25	Cobalt	< 0.02
COD	70	Chromium	< 0.05
Nitrogen	20	Copper	0.10
Phosphorus	10	Mercury	0.0009
Potassium	14	Molybdenum	0.007
Calcium	20	Nickel	< 0.10
Magnesium	17	Lead	< 0.2
Sodium	50	Selenium	< 0.005
Chloride	45	Zinc	0.12
Boron	1.0		

be carefully monitored. This is more of a problem with liquid manures than with solid manures.

Cattle and horse manures are usually handled as solids, while pig manure is handled as a liquid. Dairy and poultry manures can be handled either as a liquid or solid. The method of manure storage and handling can also affect its nutrient content. Solid manure is susceptible to loss of ammonia by volatilization, while manure stored as liquid can have significant denitrification losses of nitrogen. The liquid fraction of manure contains nitrogen (nitrate and ammonia), phosphorus and potassium. These components will be lost if the liquid fraction (primarily urine) is not recovered. A particular concern related to poultry manures is the likely presence of excessive quantities of copper which are typically used as dietary supplements in chicken feeds.

The water remaining after the removal of waste solids also contains large quantities of nutrients, especially phosphorus (Table 6.28). Except in areas where hydrologic conditions are limiting (e.g. regions with high net rainfall, periods of frozen soil, or soils of low hydraulic conductivity), wastewater effluents have been successfully applied to pastures, forests and croplands (Sopper and Kerr,

1979). Plants with high assimilative capacities (e.g. grasses) have been able to effectively remove wastewater nitrogen. In addition, most surface and subsoil soils have adequate capacity to accept large amounts of wastewater phosphorus. In dry areas, where water is at a premium, wastewater irrigation is a well-established practice.

Biodegradable organic materials

It is also recognized that manures and unstabilized sludges are undecomposed organic wastes. They can have high BOD values, and often contain high levels of bacteria. The direct runoff of manures to streams and other receiving waters can, therefore, pose potential hazards to drinking water supplies. It should be noted that some sludge is digested, either in the presence of air or in sealed tanks, to produce aerobic or anaerobic sludge; these materials are highly stabilized, have low levels of pathogens, and are ideal for land application. Lime stabilization is another method of sewage sludge treatment. Calcium hydroxide ($Ca(OH)_2$) is used to raise the sludge pH to a value of about 12. This effectively kills all the sludge pathogens, even though the sludge solids remain unstabilized. This material has been successfully applied to land, although there may be odour problems if the sludge is not immediately incorporated into the soil. There is also considerable sludge ammonia loss due to volatilization. To overcome these problems, raw, partially dewatered sludge can be mixed with a carbonaceous material (e.g. wood chips, tree bark, shredded newspaper, municipal garbage) and allowed to incubate (compost) with (or without) turning or forced aeration for periods of 2 to 12 weeks. Heterotrophic bacteria decompose part of the sludge solids and the added carbonaceous materials. The sludge temperatures increase to $50\,°C$ to $70\,°C$, with a high pathogen mortality. The final product is dry, high in organic matter and suitable as a soil conditioner.

Cannery wastes, whey, sugar beet pulp, meat packing waste, sugar cane fermentation mash, fish offal, brewery yeast, and other food processing wastes present unique disposal problems, although most can be successfully applied to land. Most food processing wastes contain high levels of biodegradable organic matter and nutrients (Table 6.29). Thus, the principle of 'carrying capacity', typically applied to manures and sludges, should be equally applicable to these materials. A thorough knowledge of the physical, chemical and biological nature of each material is required for its safe land disposal. This can be obtained from a complete analysis of the waste, as well as an understanding of the food-processing methodologies. Disposal of food-processing wastes often is more a problem of necessary land availability, than of human health concern. Crops are usually grown some distance from the processing plant, and are transported to the plant, in most agro-industrial operations. This results in a large concentration of nutrients and BOD in the plant area. Where this is the case, digestion of the raw wastes prior to land disposal can reduce some of the BOD and solids, and minimize land requirements.

Table 6.29 Representative quantities (%, by weight) of selected components in food-processing wastes

Waste	Nitrogen	Phosphorus	Potassium
Brewer's waste	0.9–3.2	0.04–0.18	0.83
Tannery waste	0.1–14	–	–
Coffee waste	0.7–3.1	0.01–5.5	0.01–0.75
Municipal refuse	0.6–0.8	0.08–0.25	0.31
Paper mill sludge	0.15–2.33	0.16–0.50	0.44–0.85
Cannery waste	0.97	0.14	0.13

Heavy metals

Almost all chemical elements can be found in sewage sludge (Table 6.26). However, only a few of these elements are toxic to plants (phytotoxic), or are of concern in regard to animal or human health. Phytotoxic metals include copper, zinc, nickel, and barium, while metals of veterinary concern include copper, cobalt, chromium, cadmium, lead, vanadium and arsenic. Those metals having public health implications include cadmium and lead. A recent evaluation of land application risks of wastewaters suggests that only cadmium and lead normally require special attention, however. Nevertheless, individual wastes can have very high contents of particular heavy metals, depending on source (e.g. municipal wastes of industrial origin). Therefore, no sludge land application programmes should be permitted without an extensive chemical analysis of the metal content of the sludge. Logan (1983) made an extensive review of trace elements in sewage sludge, while the US Environmental Protection Agency (1985) manual on sludge disposal also provides guidance concerning sludge land application programmes.

Synthetic organic chemicals

Sewage sludge and manures may contain a wide array of synthetic organic compounds most of which occur in concentrations of a few milligrams per kilogram or less. There is considerable debate as to the risk to human and animal food chains from such organic chemicals. Part of the uncertainty occurs because the health effects of various synthetic organic chemicals are presently not well known, and because the ultimate environmental fate of these organic chemicals is also unclear. It has been suggested however, that the risk from sludge- and manure-applied organic chemicals low, compared to risks from other contaminants.

Microorganisms

Municipal sewage sludges and livestock wastes can contain all the disease elements that the human and animal populations discharge to the disposal system, including viruses, dysentery-causing bacteria, and parasites (Table 6.30). Most bacterial pathogens are normally readily killed, even when raw wastes are applied directly to the land surface. This is due to pathogen desiccation, photo-decomposition, nutrient starvation, and predation. In contrast, viruses and parasites can be long-lived in soil, and have the potential to contaminate crops and water supplies. When wastes are digested at high temperature (or are pasteurized; see Morrison, 1986), pathogen mortality is very high (though not complete) for both aerobic and anaerobic sludge digestion processes, and can be essentially complete when using composting and lime treatment techniques. Data from studies of pathogen transmission from land-applied wastes suggest that the pathogen risk usually is low for such treated wastes.

Acidic compounds

Few commonly applied wastes contain acidic compounds or acid producing substances. However, fly ash, the residual solid from the combustion of coal, is often produced in enormous quantities by large coal-fired (electric generating) plants, stockpiled, and occasionally land-applied. Similarly, scrubber sludge, formed by the neutralization with lime of sulfur dioxide (SO_2) produced as stack gas in coal-fired power plants, is also disposed of by stockpiling and land-application. It is primarily composed of gypsum ($CaSO_4.2H_2O$). In addition, scrubber sludge can contain fluoride and sulfuric acid. Nevertheless, these materials have been used to reclaim surface-mined areas. In this context, their primary benefit is physical, providing bulk to supplement the remaining mine spoils which are left after extraction of ores. These sludges are, however, low in nutrient content and generally require the addition of sewage sludges, compost, lime or fertilizers before they can sustain vegetative growth. The main environmental hazards in fly ash are the phytotoxic element, boron, which can adversely affect the regrowth of the vegetation on the reclaimed mine spoils; the potentially biotoxic element, fluoride; and the acidic compound, gypsum.

Other considerations

Vermes (1987) has summarized the major obstacles to large-scale land application of wastes, as follows:

(1) A dominant engineering philosophy regarding waste treatment which has resulted in the belief that wastewater problems are best solved by technological waste treatment methodologies;
(2) A concern over the public health and potential environmental impacts of toxic substances, as well as a lack of regulations and guidelines for providing adequate protection to humans and the environment;
(3) A lack of interest from the agricultural community regarding the use of wastes;

Table 6.30 Pathogens in sewage sludge

Group	Pathogen	Disease caused
Bacteria	*Salmonella* (1700 types)	Typhoid, paratyphoid, salmonellosis
	Shigella (4 spp.)	Bacillary dysentery
	Enteropathogenic *Escherichia coli*	Gastroenteritis
	Yersinia enterocolitica	Gastroenteritis
	Camoylobacter jejuni	Gastroenteritis
	Vibrio cholerae	Cholera
	Leptospira sp.	Weil's disease
Protozoa	*Entamoeba histolytica*	Amebic dysentery, liver abcess, colonoid ulceration
	Giardia lamblia	Diarrhea, malabsorption
	Balantidium coli	Mild diarrhea, colonic ulceration
Helminths	*Ascaris lumbricoides* (Round worm)	Ascariasis
	Ancyclostoma duodenale (Hook worm)	Anemia
	Necator americanus (Hook worm)	Anemia
	Taenia saginata	Taeniasis

Virus	Number of types	Diseases caused
Enteroviruses		
Poliovirus	3	Meningitis, paralysis, fever
Echovirus	31	Meningitis, diarrhea, rash, fever, respiratory disease
Coxackievirus	23	Meningitis, herpangina, fever, respiratory disease
Coxackievirus	6	Myocarditis, congenital heart anomalies, pleurodynia, respiratory disease, fever, rash, meningitis
New enteroviruses (Types 68-71)	4	Meningitis, encephalitis, acute hemorrhagic conjunctivitis, fever, respiratory disease
Hepatitis Type A (enterovirus 72?)	1	Infectious hepatitis
Norwalk virus	1	Diarrhea, vomiting, fever
Calicivirus	1	Gastroenteritis
Astrovirus	1	Gastroenteritis
Reovirus	3	Not clearly established
Rotavirus	2	Diarrhea, vomiting
Adenovirus	40	Respiratory disease, eye infections

(4) Institutional and organizational problems that hamper practical solutions, as well as the lack of sufficient, practical demonstration systems.

Thus, the current challenge to the land disposal and treatment of manure and sewage wastes is to achieve maximum degradation of the waste, with minimum contamination of other environmental components (soil, plants, water).

Waste management: landfills, dumps and hazardous wastes

Mankind has disposed of unwanted materials, refuse and other solid wastes by depositing them on the land surface for millennia. Much of what we now know of ancient cultures has come from excavating their dumps or middens which contained the castoff evidence of their meals, wares and worldly goods. Today, the concept of terrestrial or subterranean disposal of waste continues, with the variety of goods and substances being disposed of in dumps and landfills becoming ever more diverse and dangerous. For this reason, there are real differences in the construction, operation and contamination potential of modern sanitary landfills, secure landfills and dumps compared to their historic counterparts (see Chapter 9 for a more detailed discussion of the various disposal options currently employed). The term 'dumps and landfills' is used in this chapter to refer to any terrestrial solid waste disposal site.

Some of the more infamous modern waste disposal sites have become synonymous with public health hazards; to wit, Love Canal in upstate New York. Although many types of materials are commonly disposed of in waste dumps and landfills, and many substances leach or infiltrate from these dumps and landfills into the surface and ground waters, the most persistent and deadly are those commonly referred to as 'hazardous waste'. Hazardous waste is surprisingly difficult to define. It includes a broad range of substances, produced daily in thousands of different waste streams, ranging from toxics, carcinogens and pathogens to flammables, explosives, etc. It is also surprisingly difficult to quantify, as the mere quantity is not always a good measure of the hazardous nature of these substances.

Since the Second World War, the number of products and processes contributing to this waste stream has increased at an accelerating rate (> 1000/y), with increasing amounts of this waste being comprised of synthetic chemicals. That is not to say that the more than 7000 synthetic chemicals currently on the market are 'bad' *per se* (or that naturally occurring chemicals are 'good') because many of these chemicals have led to significant improvements in quality of life, but there is an added element of risk associated with these compounds that relates to their 'newness'. Because of their lack of history, their long-term impacts are poorly known, even despite the extensive testing that has been mandated in recent years.

Estimates of the mass of hazardous waste produced range from 300 to 800 million tons in the OECD countries of Europe to about 240 million tons in the United States. In Europe, the principal generators of hazardous wastes are

Germany, France, the United Kingdom and Italy. It is assumed that the developed countries generate over 90 percent of the world's hazardous waste, although data on hazardous waste generation in the developing nations is scarce. Nevertheless, Brazil, India, South Korea and China are believed to generate substantial volumes of such waste. A possible explanation for the vagueness of these estimates could lie in the fact that there is no generally accepted international definition of what constitutes hazardous waste; the WHO defines these wastes as substances 'having physical, chemical and biological characteristics which require special handling and disposal procedures to avoid risk to health and/or other adverse environmental effects'. At this time there is, also, no global standard on hazardous waste (although such standards are being mooted in Europe; Ryding, 1992).

In addition to these more notorious contaminants, dumps and landfills also contain a range of more common constituents which are also capable of leaching or infiltrating into ground and surface waters. These include a range of nutrients and oxygen consuming substances that, while not overly harmful in and of themselves, can cause environmental damage or create public health concerns if present in high enough concentrations (e.g. nitrogen which in its nitrate form can cause methaemoglobinaemia in infants and lead to the nutrient enrichment – eutrophication – of surface waters). The contaminants arising from solid waste management include:

(1)   Nutrients;
(2)   Oxygen-consuming substances;
(3)   Heavy metals and radionuclides;
(4)   Synthetic organic chemicals;
(5)   Salts;
(6)   Microorganisms and metabolic products;
(7)   Acidic compounds;
(8)   Macropollutants, litter and trash.

While only about half of these – heavy metals and radionuclides, synthetic organic chemicals, microorganisms, and acidic compounds – may be considered hazardous in terms of the categories set forth in the US Resource Conservation and Recovery Act (ReCRA), all eight of these potential nonpoint source contaminant groups may be present in leachate and/or runoff from waste disposal sites and can adversely affect water quality in streams, lakes or groundwaters (Rhyner *et al.*, 1995; Landreth and Rebers, 1996).

Table 6.31 presents representative concentration values of some of the various contaminants found in leachates from landfills and dumps. These concentrations may match or exceed the levels of these contaminants contained in municipal wastewaters (van der Leeden *et al.*, 1990). Typically, fresh leachates are more 'active' than older, stabilized leachates, but nonetheless significant environmental contamination can occur, even from well-constructed sites. As these contaminants are primarily a function of the materials disposed of, the specific

Table 6.31 Characteristics of landfill leachates; concentrations in mg/l of leachate (after van der Leeden *et al.*, 1990)

Constituent	United States		Sweden (Municipal)	Coal ash
	'Fresh'	'Old'		
P	7.35	4.96	1.1	–
Total N	989	7.51	80	–
Conductivity (mS/m)	920	140	340	–
TDS	12 620	1144	–	–
TSS	327	266	–	–
Ca	2136	254	–	–
Mg	277	81	–	–
Fe	500	1.5	30	10
Mn	49	–	2.5	10
Zn	45	0.16	0.6	0.3
Cl	742	197	500	1000
SO_4	–	500	–	1000
Cu	0.5	0.1	0.05	0.1
Cd	–	0.4	0.005	0.02
Cr	–	–	0.05	0.2
Ni	–	–	0.05	0.2
Pb	–	1.6	0.04	0.2
Hg	–	–	0.0003	500
pH	5.2	7.3	7.1	–
BOD_5	14 950	–	600[a]	–
COD	22 650	81	800	–

[a] BOD_7

contaminant load from any given facility may exceed the tabulated figures by up to three orders of magnitude or more. In terms of total annual contaminant loads, van der Leeden *et al.* (1990) cite US Geological Survey and Environmental Protection Agency data reporting BOD and nitrogen nonpoint source pollutant loads from landfills in the United States as 300 000 tons/y and 26 000 tons/y, respectively.

Leachate flows from sanitary landfills are primarily controlled by the percolation rate (i.e. the precipitation–runoff–evapotranspiration change in water storage). This can vary considerably, often being between 15 and 50 percent of precipitation. The pollution load from a landfill also depends on the leachate quality; e.g. the concentrations of present pollutants. The leachate flow, however, is often the most important factor controlling the pollution load (e.g. the mass flow). The primary concentrations of pollutants are mainly controlled by physicochemical and biochemical processes, such as solubilization, sorption, ion exchange or biological degradation. Physicochemical processes act as sinks for pollutants, resulting in a substantially decreased pollutant mobility. The apparent effect of this phenomenon is lower concentrations of pollutants in the leachate (compare the 'fresh' and 'old' leachate concentration values in Table 6.31).

Nutrients

Nutrients are typically released from dumps and landfills as the result of decomposition processes affecting organic matter disposed of in the solid waste disposal sites. These decomposition processes are typically aerobic early in the cycle and anaerobic thereafter (see below), with the result that, while nitrates are often released during the early stages of decay, ammonia is the dominant nutrient fraction released from these sites. This has been clearly shown from some Swedish landfills where ammonia values exceeded nitrate values by almost two orders of magnitude (23 mg NH_4-N/l versus 0.6 mg NO_3-N/l). Phosphorus release is less extreme and more constant over time, averaging about 1 mg/l in the Swedish study.

Oxygen consuming substances

Leachate quality in sanitary landfills is, to a very high degree, dependent on biological degradation. It obviously will control the biochemical oxygen demand (BOD) and chemical oxygen demand (COD) of the leachate as well as metal and sulfate concentrations. Any landfill containing biodegradable material will undergo separate degradation phases, although the necessary time might differ substantially from one case to another. When the 'landfill' passes through these phases, the leachate quality changes from a high pollution level to a rather low pollution level. These phases could be described as follows:

(1)   The initial phase, when aerobic degradation of easily degradable material begins, oxygen is consumed, and carbon dioxide production will prevent replacement by fresh air; redox potential and pH might be high;

(2)   The acid phase, when anaerobic conditions dominate, nitrates and sulfates are reduced to ammonia and sulfides, and organic matter is bio-transformed to fatty acids, resulting in a decrease in pH and high levels of BOD, COD and metals;

(3)   The methane phase, when the redox potential is low, anaerobic conditions dominate, fatty acids are biotransformed to methane and carbon dioxide (inducing low BOD and COD levels), metals precipitate as sulfides and sorption (ion exchange) to the organic matrix occurs.

A sanitary landfill in the 'methane phase' is in a stable condition, only generating a low pollution load (with the exception of ammonia). A stable methane phase is indicated by a pH greater than 7 and BOD and COD levels less than 0.2 mg/l in the leachate.

The critical step in the waste degradation process is the transfer from the acid phase to the methane phase. This can depend on conditions extending over various periods of time: toxic organic compounds and high concentrations of some inorganic compounds may inhibit this transfer. The pH is perhaps the most important factor (values below pH 6 to 7 can have an inhibitory effect) because it is controlled initially by the production of fatty acids. The rate of production of fatty acids should be balanced, and, at times, may require the addition of alkaline materials to the landfill. A high proportion of easily degradable material in the waste, such as food waste, might cause a rapid production of fatty acids that can prevent a transfer to the methane phase. It is, for example, a known fact that household waste containing food wastes produces a leachate with a higher content of metals than waste without this highly degradable fraction. Shredding (creating small particle sizes) or a high initial water content can have the same effect (e.g. a prolonged acid phase). It is not possible, however, to give any general guidelines for the practical management of this phenomenon.

Heavy metals

As noted above, the metals content of landfill leachate is a function of both pH and landfill content. Where sources of metals are present, a low pH will mobilize these substances into the aqueous phase and increase the content of elemental metals in the leachate. The most common heavy metals found in municipal leachates are copper, cadmium, lead, mercury and zinc. Other metals may be more common at industrial waste disposal sites, ash dumps, or in mine tailings and slimes dams; these dumps typically do have a low pH due to the presence of high concentrations of sulfides in many mineral assemblages containing minable quantities of metallic ores (Thompson, 1980). Metal concentrations in leachate from metal processing wastes, for example, will be controlled by the solubility of metal hydroxides, carbonates and sulfates. After an initial phase (the length of which is dependent on the degree to which the cap and foundation are truly impermeable), leachate will begin to be produced at a relatively steady rate and have metal concentrations that are likely to become equally constant; for the metal processing wastes in a range of between 1 and 10 mg/l total metals. Similarly, metals in fly ash are also controlled by solubility and dissolution rates and reach an equilibrium at between 0.1 and 0.3 mg/l total metals. In this case, the release of leachates is governed by the water storage capacity of the ash and the permeability of the landfill. Assuming a 10 m depth of ash has been emplaced in the landfill, the low permeability of the waste (10^{-7} to 10^{-8} m/s = 50 mm/y, infiltration rate) will probably retain any leachates produced for up to a century after the disposal of the ash has been discontinued – and only after that time will the landfill begin to produce a leachate with the equilibrium metal concentrations mentioned above! (Obviously, other nonpoint source contamination from this landfill may occur in the intervening period; the foregoing example relates solely to the production of contaminated leachates from the disposal site.)

Synthetic organic chemicals

Not all of these chemicals are exotic; hazardous wastes include many common compounds such as pesticides, metal finishing, pickles, paints, solvents and other process residues, which, when properly disposed of, pose little or no hazard to the public or the environment. Industry, including the agricultural industry, is the primary source of these contaminants, with the chemical, metals and petroleum industries accounting for nearly 80 percent of the total hazardous waste generated (based on data from the United States; Carol and Rubin, 1985; van der Leeden *et al.*, 1990).

The fate of persistent or recalcitrant organic compounds in landfill environments is controlled by complex processes that are not fully known. On a strictly qualitative basis, one might conclude that release of persistent organic compounds with a high bioconcentration potential will theoretically be strongly retarded by sorption to the organic matrix. The prolonged retention time in the landfill, however, could result in their degradation or biotransformation to some extent. This can change rapidly if organic solvents or complexing agents are present in the wastes. These organic compounds are also sorbed to colloids which might have a high mobility.

Despite a good theoretical assessment of the hazards posed by synthetic organic compounds disposed of in dumps and landfills, few quantitative data seem to be available on these substances, of which chlorinated organics and phthalates would seem to be the most common. Data given by Robinson (1986) suggest that methylene chloride and *trans*-1,2-dichloroethene are the most significant contaminants present in solid waste leachate (in concentrations of 2.65 and 1.3 mg/l, respectively), while 1,1-dichloroethane, toluene, ethyl benzene and chloroform (all volatile organics); phenol (an acid organic); and diethylphthalate, bis(2-ethylhexyl)phthalate and dibutylphthalate (base-neutral organics) are also present (Table 6.32).

Many environmental discharges of these substances are due to regulatory failures. Small producers, especially, often fall between the private producers (usually households) serviced by local authorities and the large producers who can contract with waste disposal companies. The US Environmental Protection Agency (1985) has estimated that these small producers comprise up to 98 percent of the hazardous waste generators in the United States, but account for only 0.5 percent of the waste disposed of in licensed facilities. The United States holds the record for abandoned hazardous waste dumps (about 75 000, mostly unlined) and for hazardous waste dumps in urgent need of treatment (about 10 000). Nearly 2 percent of the groundwater resources of that nation have been contaminated by hazardous wastes; the State of Michigan has estimated that up to 15 percent of all groundwater contamination in the state is

Table 6.32 Organic substances detected in landfill leachates; concentrations in µg/l of leachate (after Robinson, 1986). Number in parentheses reflects the total number of organic compounds in the category that were detected

Constituent	*Concentration* (µg/l)
Acid organics (3)	
Phenol	293
Volatile organics (23)	
Methylene chloride	2650
Toluene	420
1,1-Dichloroethane	570
trans-1,2-Dichloroethene	1300
Ethylbenzene	150
Chloroform	71
Base-neutral organics (8)	
Bis(2-ethylhexyl)phthalate	110
Diethylphthalate	175
Dibutylphthalate	100
Chlorinated pesticides (1)	–
PCBs (1)	–

due to hazardous waste disposal by small generators of these materials. While there have been significant declines in the number of hazardous waste sites in the United States since 1985, over 20 000 active, permitted sites continue to operate across that country. In Europe, the numbers of active hazardous waste disposal sites range from between 500 and 900 in Scandinavia to over 4300 in The Netherlands; while over 3000 abandoned sites exist in the United Kingdom, The Netherlands and Denmark; 35 000 abandoned sites exist in Germany, one of the larger European waste producers.

Salts

Leachates from landfills and dumps tend to be rather saline, with TDS values in excess of 10 000 mg/l being recorded from 'fresh' discharges (Table 6.31). While these concentrations typically diminish as the landfills and dumps stabilize over time, the release of leachates with TDS concentrations in excess of 1000 mg/l is not uncommon. The principle saline constituents of the leachate are chlorides and sulfates; these elements are even more highly concentrated in leachates from ash dumps and similar solid waste disposal sites – especially brine disposal facilities.

Microorganisms and metabolic products

Few data exist on the magnitude of microorganisms and their metabolic products released from landfills. While microbial processes form the dominant means of organic decomposition within the dumps and landfills, common measures of their presence, such as TOC (total organic carbon) and bacterial plate assays, do not appear to have been tabulated in the literature. Generally, it is accepted that few of these organisms will escape a well-designed and operated solid waste facility. Evidence from sludge composting and pasteurization operations, where the number of pathogens could be expected to be significantly higher, would support this contention (Morrison, 1986).

Acidic compounds

While most waste disposal operations quickly achieve a stable pH that approximates neutrality, Robinson (1986) reports a range of pH values measured in leachates from North American landfills that spans the gamut from extremely acidic (pH = 1.5) to relatively alkaline (pH = 9.5). While these extremes are just that, the release of acidic compounds may occur in some instances – especially from mine dumps and related industrial disposal areas. Generally, however, the pH range experienced in landfill leachate is from 5.7 to 7.7 (Robinson, 1986).

Macro-pollutants

As noted many times, the generation of macro-pollutants – litter, debris, odours and other aesthetic impairments – from well designed and operated solid waste disposal areas is minimal. Paper and plastics are especially prone to windborne redistribution. In this regard, the effects of wind erosion (as well as water erosion) on the landfill cap or cover should not be overlooked, as this can impact even well-constructed facilities both during operation and after closure (G. Boddington, personal communication). Quantification of the magnitude of such debris lost from dumps and landfills is almost totally site and process dependent. No estimates of the magnitude of such losses are readily available in the literature.

INTERNAL LOADING AND RECEIVING-WATER MANAGEMENT

Introduction

Major fractions of the pollutant loadings to waterbodies ultimately accumulate in the sediments. The proportions of the external load retained in streams and lakes varies with the water flow rate and hydraulic detention time (water residence time), the physicochemical properties of the pollutant and the macro composition of the sedimenting material, the productivity of the waters, their mixing behaviours and several other factors. Although no simple relationships can be presented, mass balance studies have shown retention coefficients for lakes (the fraction retained) that are well over 50 percent and frequently closer to 100 percent. This is true for:

(1)   Sediments;
(2)   Nutrients (OECD, 1982);
(3)   Heavy metals (and iron);
(4)   Organic chemicals (and micropollutants);

(unless the areal hydraulic loading is high in combination with a low partition coefficient, in the latter cases; an uncommon combination of high flushing rate and poor adsorption onto the particles settling in the lakes: Ali *et al.*, 1984). Thus, most lakes and streams are efficient traps for pollutants, and their sediments can be described as the 'dustbins' of the watershed. Unfortunately, substances 'lost' to the sediments generally remain available for subsequent release into the water column.

The mass of pollutants present in the top few centimeters of the sediment is frequently two to three orders of magnitude higher than that present in the overlying water column. Concomitantly the rate of change of the concentration of pollutants in the sediments due to changes in inputs from the watershed is rather slow compared to the response of the water column. Considering a mixing depth of 5 to 10 cm and a rate of accumulation of 5 to 10 mm, which are typical values in many lakes, the rate of dilution of the active sediment is in the order of a decade. This is an average, and large spatial variations will occur in lakes and streams due to the existence of zones with net erosion or net accumulation, and transport zones. Sill sediments, especially, are large buffers in the stream–lake ecotone that equilibrate slowly to new conditions, and can retard the response of a waterbody to both point and nonpoint contaminant control measures.

It was in connection with such a delayed response that phosphorus (P) liberation from sediments (known as 'internal loading') under aerobic conditions in shallow eutrophic lakes was first recorded in the mid-1970s by Ryding and Forsberg (1977) and Lee, *et al.* (1977). Their observations of high phosphorus release rates during periods of high oxygen concentration in the water mass were extraordinary and contradicted the existing theory of anoxic phosphorus release formulated by Mortimer (1941, 1942), who had shown that phosphorus was associated with oxidized ferric iron compounds (Fe^{3+}) in the sediment and released when the ferric iron was reduced to ferrous iron (Fe^{2+}) during anoxic conditions. Subsequently, other researchers have shown similar releases of metals and organic contaminants, although in both cases the rates were generally higher under anaerobic conditions than under aerobic conditions (see reviews by Bostrom *et al.*, 1982; Ryding, 1985; Löfgren, 1987; Enell and Löfgren, 1987; Persson and Jansson, 1988; Thornton *et al.*, 1993). Given this commonality in behaviour, all four of the contaminants identified above will be discussed simultaneously.

Occurrence of substances in sediments

The physical and chemical form in which pollutants enter the sediments is generally different from the form in which they are present in the sediments. Both in the water phase and in the sediment phase, transformations by physical, chemical and biological processes occur. As pollutants enter a watercourse from terrestrial sources, each with their own composition, various transformations may occur that affect the distribution of chemicals; processes such as adsorption, desorption, dissolution and precipitation. The growth of algae produces particles which transport the incorporated nutrients downward by settling, although an important fraction can be mineralized again before the material reaches the bottom. Algae can also transport heavy metals by surface complexation. Sigg (1987) reported metal:carbon ratios in settling materials on the order of 2×10^{-4} for Al, 6×10^{-5} for Cu, 4×10^{-5} for Pb and Cr, and 2 to 3×10^{-2} for Fe. Algal growth also affects pH, which in turn has an influence on the adsorption of phosphate and heavy metals onto iron and manganese (hydr)oxides. These (hydr)oxides are also known as scavengers of these elements. Another important autochthonous particulate material is calcite. This material will form upon supersaturation at low Mg:Ca ratios (Stabel, 1985). At high pH, calcite is known to absorb phosphates (Stumm and Leckie, 1971), but its role in transporting the nutrient to the sediments is of minor importance as compared with particulate organics (Sigg, 1987). Calcite in sediments, however, is important in controlling pH. Finally, the particulate organic fraction of the settling material is mainly responsible for the transport of organic micropollutants.

Physical processes further influence the distribution of settling materials and their associated pollutants. Shear forces due to wind-induced waves and currents prevent fine and light materials from settling and, by resuspension during events of high shear stress, remove such materials if they have settled. Through this type of physical sorting, finer sediments accumulate preferentially in the deeper areas and the coarser materials in the shallower. Shallow lakes, therefore, are often subject to more or less permanent internal loading. Pollutants are also concentrated in the deeper areas and are often subject to exposure in hypolimnetic (possibly deoxygenated) waters during periods of stratification.

Upon accumulation in the sediment, be it permanent or temporary, a further redistribution of pollutants may occur due to biochemical, physical and chemical processes. As these transformations also modify the internal loading they will be reviewed in the following sections.

Transport processes contributing to internal loading

Pollutants can be returned to the water phase in two forms: dissolved, in which case diffusion and advection are sometimes active, and particulate, associated with particles during resuspension. Some transport may be caused by organisms leaving the sediments or by pumping actions (bioturbation), but generally the role of organisms is considered to be an indirect one in that they actually enhance diffusional transport through bioturbation. The macro-fauna, by its burrowing activity, enlarges the water–sediment contact area, and it has been shown that the flux of a conservative material can be amplified by a factor of at least two by Chironomidae (Brinkman and van Raaphorst, 1986). The enhanced oxygenation related to the larger contact area also affects the distribution of non-conservative substances in the sediments, and no simple predictions of the effects of macrofauna on the release rates can be given.

Diffusion and advection of dissolved material within and out of the sediments interact with the (bio)chemical processes; neither can be treated as fully independent. Transport affects the concentration gradients, and hence reaction rates, and *vice versa*. The interactions will be discussed later. The molecular diffusion within the sediments is attenuated by the porosity and the tortuosity, which are interrelated (Krom and Berner, 1980). Based on this relationship, these authors have presented an empirical formula to define the effective coefficient of diffusion, D_e:

$$D_e = 0.77 \, p \, D \tag{6.1}$$

where:　　p = porosity; and
　　　　　D = the molecular diffusion coefficient

The porosity generally increases towards the surface and decreases in deeper strata due to compaction. This compaction causes an upflow of pore water, but in most situations this advective transport is of only minor importance to internal loading. Only in systems where pressure gradients in the subsoil cause seepage or infiltration does advective flow become important. The relative importance of advection and diffusion in transport is expressed by the dimensionless Peclet number, P_e:

$$P_e = (u \, L) / D_e \tag{6.2}$$

where:　　u = the advective velocity within the sediment, corrected for
　　　　　　porosity;
　　　　　L = a characteristic length; and
　　　　　D_e = the effective coefficient of diffusion defined above

Assuming the length, L, is equivalent to the depth of the near-surface sediment concentration gradient (frequently the thickness of the aerobic surficial layer), the order of magnitude will be about 1 cm. Only for Peclet values of $P_e > 1$ will advection be important for transport through the interstitial water. With D_e about 5×10^{-10} m²/s, this means that, for infiltration or seepage rates of less than 0.5 cm/d, advective transport does not contribute significantly to internal loading.

Resuspension will occur if erosive (or critical) shear stresses at the sediment–water interface exceed 0.05 to 0.5 N/cm² (Partheniades, 1972; Terwindt, 1977; Fukuda and Lick, 1980; Lick, 1982; Kusuda *et al.*, 1984; Luettich, 1987). Erosive shear stresses are a function of the grain size distribution, porosity, density of sediment particles and cohesion; cohesion being a complex function of chemical and physical properties of the particles and environmental conditions such as pH and redox potential. Reduced conditions promote erosion. Erosive shear stresses in lakes are generally due to wind action which induces waves and currents (flows); in streams, currents are the dominant forcing mechanism. The patterns are complex, but theoretical and semi-empirical relationships to predict wave-induced resuspension in shallow lakes have been developed (e.g. Banks, 1975; Groen and Dorresteyn, 1976; Aalderink *et al.*, 1984; Shanahan *et al.*, 1986; Luettich, 1987). Two approaches can be discerned in the attempt to describe resuspension: the first is based on a hydrodynamic lake transport model in which erosive shear stress must be measured (Sheng and Lick, 1979; Lick, 1982), and the second, more popular method is based on estimation of sedimentation rate parameters directly from field data on suspended solid concentrations and wind (Somlyody, 1980; Aalderink *et al.*, 1984; Brinkman and van Raaphorst, 1986) or wave height (Luettich, 1987). There is a tendency to refine the description by sub-division of the solids into size classes. The major problem, however, with generalized models lies in the site-specific values of the parameters for entrainment. Even within a specific lake, these will vary spatially and it is difficult to relate them to sediment properties across the entire lake basin. A promising concept in the further development of these models is the notion that the erosive shear stress necessary to cause resuspension at the lake/stream bed is different from that required at the shallow boundary layer at the top of the lake sediments or along the stream banks (resulting in a range of conditions within the waterbody/watercourse under which shear stress-induced erosion can occur). This may explain the hysteresis effects observed in the field and laboratory in systems with increasing and decreasing shear stresses (Partheniades, 1965; Umita *et al.*, 1984; Luettich, 1987).

The relative importance of these processes is not well documented. For phosphorus, resuspension may be potentially significant, especially during summer storms, when particles loaded with phosphorus are likely to desorb phosphate into the relatively phosphorus-poor overlying water – the higher pH in the water column may also favour such a release. Generally, however, diffusion of dissolved pollutants from the sediments is more important; the process is more continuous and not restricted to shallow lakes or littoral zones, being driven by a deeper layer of sediment than the very thin layer involved in resuspension.

Transformations within the sediments

The most important processes causing a redistribution of pollutants among the different sediment pools are mineralization of organic matter, adsorption/ desorption, and precipitation/ dissolution. The latter two categories of processes cannot always be discerned properly. Again, the processes can be described individually but their extent and consequences are indissolubly coupled to the transport processes within the sediments. The decay of detritus by microorganisms starts in the water phase and is continued in the top layer of the sediment. Not all components of the detritus are equally available to degradation; e.g. cell walls of algae are comparatively difficult to mineralize (Gunnison and Alexander, 1975), green algae decay less readily than blue-greens (Rodgers and DePinto, 1983), up to 50 percent of algal cell material is refractory (Jewell and McCarthy, 1971), and decomposition is controlled by the area available for colonization by bacteria. Hydrolysis is the rate limiting step under anaerobic conditions (Cappenberg *et al.*, 1979; Foree and McCarthy, 1970). In addition, some 40 percent of the detritus deposited at the sediment surface cannot be mineralized anaerobically or condensed into humics having virtually a zero decay rate. The biodegradable fractions of this detritus contain more nitrogen and phosphorus than the refractory material, which means that a proportionally greater amount of P and N will be mineralized than carbon. As part of the ammonia can be nitrified in the aerobic zone of the sediment, and subsequently denitrified in the anoxic zone, nitrogen is lost, and the proportion of P to N in the sediments is generally much lower (2–3:1) than in the material settling in the water column (5–10:1). Some other products of mineralization are also of great importance for the mobilization of pollutants. These are dissolved organic carbon (DOC), reduced manganese, and reduced iron. DOC has a strong capacity for complexation with heavy metals, while Fe^{3+}-hydroxides and oxides also adsorb metals, either directly or through organic intermediates. Upon reduction of these (hydr)oxides the metals can be solubilized, together with phosphate (Lijklema, 1977).

Interactions between hydrous oxides and hydrogen, hydroxide, and metal cations and anions can be interpreted to a large extent in terms of co-ordination chemistry or as competitive complex formation between these ions (Schindler and Stumm, 1987). This explains why metals are adsorbed effectively on hydrous and mineral oxides at high pH while phosphorus is not. Iron and manganese oxides have been identified as playing important roles in controlling the concentrations of trace metals (Jenne, 1968) and phosphate (Einsele, 1936): the redox cycles of iron and manganese have a pronounced effect upon the fate of these pollutants (Davison, 1985), with pronounced hysteresis effects occurring in relation to aging and oxidation effects in Fe^{3+}-hydroxide formations (Lijklema, 1980). Humics, also, have rather strong complexing capacities for different metal cations – in particular for Cu^{2+} but also for Fe^{3+} – and, accordingly, phosphorus can be bound by adsorption/complexation in a humic–Fe–PO_4 association which seems to be rather stable. Close to a neutral pH, clay minerals, such as illite, can bind up to 2 mg P/g (Edzwald and Toensing, 1976), although, in most sediments, the contribution of clays will be of limited importance. Nitrate and ammonia are also adsorbed onto clays but are generally more mobile than phosphorus. All of these adsorption/desorption reactions are, to a large extent, reversible.

Finally, pollutants can (co-)precipitate in sediments as heavy metal sulfides, calcium phosphate compounds, and vivianite (combining both phosphate and iron). Precipitation of sulfides is a dominant process in anaerobic marine sediments, but, in freshwater systems, the concentration of sulfate is limited and the rate of production of sulfides in anaerobic sediments is limited by diffusion. Generally however, the sulfides produced will either bind iron, forming the well-known black sediments, or compete with iron for phosphate. The mobility of phosphate under such conditions – which typically occur in calcareous lakes with naturally high sulfate concentrations – may be enhanced considerably, especially where the proportion of sulfide is high in comparison with the available iron (Hasler and Einsele, 1948). In contrast, heavy metals will probably be effectively immobilized in such systems unless high concentrations of free hydrogen sulfide are formed.

Interactions between transport and transformation

Internal loading is the result of the combined effects of the diffusion, mineralization and redistribution processes discussed above. While the rates of these individual processes vary widely (mineralization and diffusion are slow compared to most adsorption and desorption reactions), the net effect is that, frequently, an (chemical) equilibrium appears to exist between the solutes and solids which is driven by the decay of detritus and diffusion. The mineralization reaction exhibits seasonal variations due to temperature changes and variations in the supply of organic matter settling from the overlying water (which is much higher during the growing season). Accordingly, oxygen consumption rates will be higher during the summer and early autumn, leading to a decrease in the thickness of the aerobic top layer of the sediment. Eventually the sediments may become anaerobic throughout and this condition may extend into the overlying water, especially when stratification occurs and the oxygenation of the hypolimnion is restricted by a thermocline (or pycnocline, in tropical systems). This situation causes steep concentration gradients to develop at the sediment–water interface with respect to phosphorus, ammonia and heavy metals, while, at the same time, decreasing the binding capacity of the freshly precipitated hydroxides due to the (much) higher pH in the overlying water as compared to the pore water. For redox-sensitive systems, these phenomena contribute to the release of phosphorus, heavy metals and ammonia, sometimes of considerable magnitude (Davison, 1985).

This mechanism also applies, to a certain extent, to the daily variation in oxygen concentration at the sediment–water interface arising from the production–respiration cycle in the water column. The concomitant variation in the oxygen penetration depth in the sediments, and the establishment of an oxygen gradient by a combination of diffusion and reaction, over the daily cycle has a smaller time constant than the adjustment of the iron profile by diffusion. Hence, iron, heavy metals and phosphorus will still be concentrated near the

Table 6.33 Measured sediment–water releases of contaminants as reported in the literature

Study location	Element	Rate (mg/m²/d)	Reference
Peel–Harvey Inlet (Australia)	P	4–22	McComb and Lukatelich (1990)
	N	100–200	
Potomac Estuary (USA)	P	1–73	Callender (1982)
Lake McIlwaine (Zimbabwe)	P	3–14	Thornton (1979)
	N	–14–5	Thornton (1986)

interface, but in a more diffuse manner. The extremely high flux rates that may result from the processes described above on a seasonal basis will be attenuated on a daily basis, but may still result in a higher contaminant release rate, to the degree that variations in the oxygen concentrations occur, during the darkness (especially before dawn) than during daylight hours.

Finally, the rate at which sediment–water exchange of contaminants occurs can be modified by the physicochemical climate of the sediments themselves. In sediments where contaminants are associated with organic (e.g. humics) or calcareous materials, for example, the magnitude of phosphorus release to the overlying waters on a seasonal basis may be greatly reduced (Petterson, 1984). In contrast, in the case of nitrogen, the presence of reduced conditions which promote ammonification can enhance nitrogen release, even though ammonia has more opportunities for adsorption within the sediments than does nitrate (Klapwijk and Snodgrass, 1984). Similarly, the solubility and mobility of sediment-bound heavy metals is a complex function of pH, redox conditions, the availability of complexing organic matter and their relative ionic strengths, and the quantity and character of potential adsorbents (such as humic materials and oxides) within the sediments themselves. While a general discussion of these interactions affecting metals in lake sediments can be found in Förstner *et al.* (1984), the evidence suggests that, unless acidification plays a role, the release of metals from sediments does not seem to be a major problem in freshwater environments. Table 6.33 provides a few examples of sediment release rates reported in the literature.

CHAPTER SUMMARY

This chapter has briefly reviewed some of the pertinent literature on nonpoint sources of pollution from an extremely large bibliographic data base. Many of the citations are from studies conducted in temperate climates, although reference has been made, where appropriate, to studies conducted outside of this setting. While non-temperate studies are not numerous, some work has been done in Australia and southern Africa. Generally, these studies have confirmed the similarities between pollutant sources in at least two climatic zones – the semi-arid zone and temperate zone. Variations lie primarily in the transport mechanism (i.e. the seasonality of rainfall in the non-temperate zones), although some inherent seasonal variation controlled by temperature may also be perceived. For example, data from both South Africa and the United States suggest that bacterial contamination of recreational waters is greatest in summer, which reflects, perhaps, the role of temperature in exacerbating water quality problems. In addition, summer was the period in which the greatest number of users were present, and shortly after the spring rains had flushed the watersheds. (Such multiple correlations demonstrate, yet again, the complex inter-relationships that must be considered in nonpoint source management, involving not only the availability of pollutants as in point source pollution control but also a variable transport mechanism, the hydrological cycle.)

In some respects, this chapter has reiterated information presented in Chapter 5. The basic pollutant groups discussed in the latter chapter provide the source materials of nonpoint source pollution. Combining these pollutant groups with the specific human land-use activities reviewed in this chapter, and the hydrological cycle discussed in Chapter 4, has resulted in the identification of the degree to which these activities contribute to nonpoint source pollution. This permits (1) the identification of the greatest potential problem areas – usually those activities which result in the greatest disturbance of the land surface (such as mining, agriculture and construction), and (2) their initial quantification. This information forms the basis for conducting a nonpoint source pollution control inventory which is discussed further in Chapter 7.

CHAPTER 7

MEASUREMENT OF NONPOINT SOURCE POLLUTION

J. Gayer, G. Jolánkai, L. Vermes, W. Rast and J.A. Thornton

LAND USE INVENTORY

Introduction

The concept of land use, as related to determination of unit area loads (UALs) for different nonpoint source pollutants, was identified previously (see Chapter 6). Implicit in the utilization of information about land use in a drainage basin is the assumption that one can accurately identify specific land uses within the basin. Accordingly, the purpose of this section is to discuss the techniques for identifying specific land uses, and to provide some distinguishing characteristics of these land-use activities.

Determination of land-cover features and resultant land-use activities

Identification of the use to which land is being put requires identification of the natural (e.g. vegetative cover, terrain features) and anthropogenic (e.g. buildings, roads, cultivated areas) features of the land surface. The assumption is that specific types of land uses can be identified by the presence of specific features or structures on the land surface. Ideally, such information is already available in the form of land-use/land-cover maps that can be obtained from appropriate governmental or municipal planning agencies. This approach, obviously, assumes that the land use information is accurate and up-to-date. This can be substantiated, to some degree, by noting the period when the land-use mapping was prepared. Maps that are many years old may contain obsolete or inaccurate information for assessing the extent of nonpoint source pollution, and for devising control programmes to alleviate the pollution problem(s).

If such land-use information is not available, one must collect detailed surface data from the parcel (or larger area) of land of interest. For small watersheds, it is possible to walk the land on foot, or drive, and identify specific features of the land which will allow identification of its primary use(s). However, for larger drainage basins (and for purposes of expediency), this task is more readily accomplished with the use of photographic images (photographs) of the land surface. Indeed, this is also the approach typically used by governmental agencies to develop the above-noted land-use maps.

There are a variety of ways to obtain the necessary images, including satellite photos ('Landsat' imagery) and/or aerial photographs with a plane-mounted or

217

hand-held camera. The necessary detail of the images depends primarily on the purpose for which they are being collected. Information obtained in this manner typically is known as remote-sensing data (so-called because the detection device is not in physical contact with the object being studied). Regardless of the manner in which they are obtained, the photographic images are then subjected to photointerpretation, with the goal of identifying land surface features or structures unique to specific land uses. A post-interpretation survey to obtain 'ground truth' information or data that confirm the results of the photointerpretation effort is usually useful for substantiating the results, and ensuring their accuracy for planning and other purposes.

There are many types of potential land uses. There are also many features or structures on the land surface that should be considered in identifying specific land uses. Thus, a reliable 'land use classification' system must be used to identify specific land uses in a comprehensive and consistent manner. One such system, typically used in conjunction with remote-sensing data, is that developed by Anderson *et al.* (1976). This classification scheme can be divided into several levels of detail. An example of the concept of land-use levels is provided in Table 7.1. As shown, the first level is the most coarse delineation of land use (e.g. agricultural *versus* residential areas). The second level provides a more precise delineation of these coarse categories (e.g. agricultural land is sub-divided into cropland *versus* orchards). A further delineation (Level III) adds a more specific characterization of the land usage (e.g. in regard to the types of crops grown). Experience suggests that a delineation of land usage to at least the degree inherent in Level II is required to produce land-use maps of sufficient detail to assess nonpoint source pollution.

To illustrate the degree of detail required for delineating specific land-use types, Table 7.2 provides information on Levels I, II and III for major land uses, based primarily on the classification scheme of Anderson *et al.* (1976). This classification scheme identifies some of the features and structures on the land surface which can be used to delineate specific land uses.

Additional information on remote-sensing techniques and possibilities are provided by Lintz and Simonett (1976). In addition, details on photointerpretation of photographs (or photographic images) obtained by remote sensing and other techniques are provided by Loelkes *et al.* (1976).

ANALYTICAL PROCEDURES

Introduction

A majority of the nonpoint source contaminants defined in Chapter 5 occur in the form of dissolved substances and/or microscopic particulates in natural waters. Few lend themselves to direct visual detection and quantification. Even those that are macroscopic tend to be difficult to quantify without the adoption of standardized methodology to ensure consistency between observers. This section is not intended to be a detailed analytical guide, however, but, rather, it

Table 7.1   An illustration of land-use classification for a watershed with several land uses

Level I	Level II	Level III
URBAN	Residential	Single-family dwellings (60%) Apartments (40%)
	Commercial	Shopping mall (95%) Parking lot (5%)
AGRICULTURE	Cropland	Row crops [corn] (80%) Non-row crops [barley] (20%)
	Orchard	Apple trees (40%) Orange trees (40%) Nut trees (20%)
	Feedlot	Beef cattle (100%)

Table 7.2   An illustration of land-use information, based on the classification scheme of Anderson *et al.* (1976) (modified from Loelkes *et al.* (1976))

Level I	Urban	Land characterized by surface predominantly covered with structures associated with cities, towns, industry, etc.
Level II	Residential	Land containing structures used as residences
Level III		Single-family residences, apartments, condominiums, mobile homes, recreational homes, and any other building or structure used as a residence
Level II	Commercial	Areas with structures predominantly used for the sale of products or services classified as commercial
Level III		Central business districts, shopping centres, commercial strip developments, and junkyards
Level II	Industrial	A wide variety of land uses, delineated primarily as light industry and heavy industry
Level III	Light industry	Activities focusing on design, assembly, finishing, and packaging of raw materials. The materials used generally have been processed at least once
Level III	Heavy industry	Activities devoted to heavy fabrication, making and assembling, parts which are 'dirty', large and/or heavy, or to the processing of basic raw materials (including ore processing operations). Examples include steel mills, cement plants, chemical plants, electric power generators, oil refineries and underground mining operations

Continued

Table 7.2 Continued

Level II	Transportation, communications and utilities	Major transportation and/or utility routes
Level III		Highways (including rights-of-way and interchanges); railroads (including track rights-of-way, stations, and repair and switching yards); airports (including runways, terminal areas, fuel storage areas, parking lots, etc.), seaports (including docks, shipyards, locks, etc.), and communication and utility areas involved with processing, treatment, and transportation of water, gas, oil, electricity, and radio, radar, and television communications facilities
Level II	Industrial and commercial complexes	Industrial and commercial land uses that typically occur together or in close proximity
Level III		Industrial parks and accompanying parking lots, etc.
Level II	Other urban or built-up land	Urban land used for other purposes
Level III		Parks, cemeteries, waste dumps, water control structures and spillways, golf courses, and undeveloped lands
Level I	Agriculture	Land used primarily for production of food and/or fibre, or for built-up service areas associated with agricultural activities
Level II	Cropland and pasture	Cropped land and land used for pasture
Level III		Harvested croplands, summer-fallow and idle croplands, croplands in soil-improvement grasses and legumes, croplands used for pasture in rotating crops, and improved and idle pastures
Level II	Orchards, groves, vineyards, nurseries and ornamental horticultural areas	Land used for tree cultivation
Level III		Land used for production of various fruit and nut crops; also includes floricultural and seed-and-sod areas, some greenhouses, and tree nurseries
Level II	Confined feeding operations	Land used for livestock production
Level III		Large, specialized livestock production enterprises, including beef cattle feedlots, dairy operations with confined feeding, large poultry farms, and hog feedlots, and related buildings, fences, waste-disposal areas, etc.
Level II	Other agricultural land	Agricultural land used for other purposes
Level III		Farmsteads, livestock-holding areas (e.g. corrals), small farm ponds, fish farms/hatcheries, horse farms, etc., and related supporting facilities

Table 7.2 Continued

Level I	Rangeland	Land on which the potential natural vegetation is predominantly grasses, grasslike plants or shrubs, or where natural herbivory is an important influence in its precivilization state
Level II	Herbaceous rangeland	Land dominated by naturally-occurring grasses and forbs, or land modified to include this vegetation when managed for rangeland purposes
Level II	Shrub and brush rangeland	Land with shrub or scrub vegetation; characteristic of arid or semi-arid regions
Level I	Forests	Land containing trees capable of producing timber and other wood products, which have a tree-crown areal density of at least ten percent, and which exert an effect on the climate or water regime
Level II	Deciduous forest land	Forested land with a predominance of trees that lose their leaves at the end of the frost-free season or the beginning of the dry season
Level II	Evergreen forest land	Forested land with a predominance of trees that remain green throughout the year
Level II	Mixed forest land	Forested lands with more than one-third intermixture of either deciduous or evergreen tree species in a specific area
Level I	Wetland	Land areas where the water table is at, near, or above the land surface for a significant portion of the year
Level II	Forested wetlands	Wetlands dominated by woody vegetation (e.g. bottomland oaks, shingle oaks, black gum, cypress, etc.)
Level II	Nonforested wetlands	Wetlands dominated by herbaceous wetland vegetation, or unvegetated

is intended to direct the reader to those resources that are available and which have been accepted by many regulatory authorities as providing standardized, reproducible results when used to examine waters and wastewaters. Sample collection methodologies are described later in this chapter.

> ***Caution*** *For regulatory purposes, the reader is referred to the requirements for sample collection, analysis and reporting established by the responsible government agency in their country; many agencies have strict guidelines concerning the timing and manner of sample collection and preservation, the number of samples and replicates taken, and the qualifications of the laboratories involved in their analysis, not the least concern of which is the need to maintain a documented 'chain of custody' for each sample to avoid its disqualification as evidence in a court of law. Samples taken and analyzed by regulatory agencies should*

always be collected, handled and analyzed in the manner prescribed by law as they may be required for evidentiary purposes at some future time.

Sample handling and preservation

Water samples are collected for a variety of purposes. Some typical purposes for which samples are collected include environmental monitoring (samples tend to be collected continuously from fixed stations; e.g. air quality or water quality monitoring programmes); compliance monitoring (samples are collected in a prescribed manner at a fixed sampling point within a fairly well-defined period of time (e.g. water pollution control monitoring programmes); and scientific research (samples can be collected at discrete intervals or in an event-related manner, usually from fixed sampling points).

Samples collected for environmental monitoring purposes are usually analyzed for a broad range of contaminants and constituents; for example, weather monitoring stations usually record wet- and dry-bulb temperatures, relative humidity, barometric pressure, precipitation volumes, evaporation rate, cloud cover, wind velocity and direction, and other parameters (such as insolation, or cloud base height, or visibility). These data are collected hourly throughout the day, every day, at a set location. They may be used for weather forecasting purposes, but may be used for other purposes such as establishing maxima and minima, climatological analysis, etc. For these data to be comparable from station to station, the manner and form in which they are collected have been standardized; any weather observer collecting these data at any time will follow the same procedures (sometimes referred to as a sampling protocol) and use the same or very similar instruments. In contrast, samples obtained for regulatory purposes may be collected at random intervals from various premises, but will follow similar prescribed methods of collection, transmittal and analysis to ensure their reproducibility and the equitable treatment of each activity being inspected. The range of parameters analyzed will be prescribed by law. Scientific samples, likewise, must be reproducible and station-specific. The range of parameters sampled and analyzed, however, will be determined by the purpose of the investigation and may be collected in response to specific natural events (such as rainfall, river stage, etc.) or other stimuli (e.g. during daylight hours, etc.). The samples may be obtained as instantaneous, discrete samples ('grab' samples) or as samples integrated over time, flow volumes, depth or other spatial or temporal period. Environmental samples will be obtained on an on-going basis, regulatory samples on an *ad hoc* (but on-going) basis, and scientific samples for the duration of a specific project. At times, it may be advantageous from a cost and manpower efficiency standpoint to combine environmental monitoring programmes with scientific field programmes to minimize duplication; generally, however, each of these types of programmes should stand alone as their purpose, intensity and frequency of conduct are not often compatible – scientific samples usually requiring closer interval sampling than environmental monitoring samples, for example.

Sample collection methods are described in subsequent sections of this chapter. However, most samples are collected either as records (flow gauge charts, rainfall and temperature record charts, etc.) or as physical specimens. Records are usually in the form of paper documents or magnetic media; many records that were formerly placed on paper forms are now commonly stored electronically, a process that facilitates automatic or long-distance information gathering techniques. For example, stream flow measurements (see below) may be obtained by having an observer read a value from a gauge plate, abstracting a value from a paper chart or interrogating a stream-side microprocessor unit. The choice of the method employed will depend on cost (electronic information, while becoming more cost-effective, is relatively expensive), equipment availability (isolated stations may not have the power and telephone linkages necessary for remote microprocessor operation), and the purpose of the study. Unless otherwise specified or required, the choice of sampling equipment should usually err in favour of proven, robust, simple techniques and technologies.

Records require relatively little in the way of handling and preservation. It is usually a good idea to back-up or duplicate the information, especially if it is in a magnetic format, and, in the case of computer records, to retain a 'hard copy' of the information. Duplicate copies can easily be made using various means; photocopiers, microphotography, 'disk copy' procedures, etc. Digitizing of paper chart records and storage as magnetic media permit further analysis of these records; for example, stream flow records can be analyzed for seasonality and other types of periodicity. The use of 'working copies' also ensures the preservation of the original records. As in any type of sample handling, a methodical, common-sense approach to record keeping is a major advantage when it comes to data manipulation, retrieval and analysis. Whether the information is kept in filing cabinets or on computer files, attention to storage climate is an important part of sample preservation – exposure to climatic extremes should be avoided. Archivists and government librarians can provide specific instructions on the preservation of historic or other valuable records which may require special handling and storage.

The handling and preservation of samples is somewhat more problematic than the treatment accorded records, primarily as the nature of the specimens varies so widely. No single procedure will apply universally to biological specimens, water samples and sediment cores, and the treatments that are applied can vary with the constituents being analyzed; e.g. it is not useful to preserve samples with mercuric chloride if one is analyzing for mercury or chloride, although this fact is occasionally overlooked! Nevertheless, there are types of preservation techniques that can be applied to various groups of specimens, and these are discussed briefly below. Further information on recommended sampling and analytical procedures for water quality constituents in streams, lakes and reservoirs can be found in Mancy and Allen (1982), Wilson (1982), Hunt and Wilson (1986), Chow *et al.* (1988), Ryding and Rast (1989), Wells *et al.* (1990), and Hauer and Lamberti (1996).

Thermal preservation

Both drying and freezing are accepted preservation techniques; indeed, freeze-drying of plant materials is also recommended in some cases. While thermal preservation does not adulterate the specimens, it normally results in a change in the state or character of the specimens.

Plant specimens are usually dried – mounted and dried typological specimens are common in herbaria, for example. Guidelines for mounting, drying and preserving plant specimens are given in many botany texts (e.g. Porter, 1967; Harrington and Durrell, 1957). Sediment samples, likewise, have commonly been stored in a dried and crushed state, although recent evidence (Twinch, 1987) suggests that such treatment can reduce the nutrient exchange potentials of the sediments and may alter the grain size distribution due to aggregation of fine particles.

Specimens, such as water samples and chlorophyll *a* samples, can be preserved in a frozen state prior to analysis. While chlorophyll *a* samples have been preserved at $-10°C$ for up to 1 year without significant deterioration, water samples should be kept for no more than 3 months – less if being analyzed for nitrogen fractions. Freezing may cause algal cells to lyse or rupture, so care should be exercised if element speciation is the objective of the particular study – this is not a problem if 'total' element content is being analyzed, however. To overcome this limitation, water samples may be chilled to $4°C$, at which temperature most biological activity ceases. Filtered samples can be kept for up to a week without significant deterioration at this temperature.

Finally, samples of blue-green algae, being assayed for toxicity, have been preserved in a freeze-dried state for a year or longer without significant deterioration.

Poisoning

The use of preservatives such as mercuric chloride, sulfuric acid, nitric acid, alcohol, formaldehyde, etc., overcomes the problems of change of state of the sample mentioned above but introduces other complications. The need to avoid contaminating the sample has been alluded to above; while standard addition techniques may be useful in measuring low concentrations of some elements, addition of preservatives rarely meets the level of accuracy necessary to satisfy the requirements of this analytical procedure. Hence, choice of preservative is important. Choice of preservative is also important from the point of view of effectivity; mercuric chloride may be adequate for nutrient samples but does not suffice for metal samples, which should be preserved in an acidic medium, for example. Again, reference to most limnological field guides and laboratory manuals will suggest which preservatives to use for specific analyses (see Lind, 1974; Strickland and Parsons, 1972; Golterman and Clymo, 1971). For biological specimens, the use of preservatives is commonplace, although formaldehyde is generally being replaced with alcohol for many applications due to the potentially carcinogenic nature of the former. All preservatives are poisonous and should be treated with due care.

No preservation

Some applications will require that specimens remain fresh or viable following collection; for example, fish specimens to be used in behavioral studies or for breeding purposes, or algal samples to be used in bioassay studies. Such specimens are generally transported at as close to ambient water temperature as possible; living organisms such as fish may require aeration if the journey from the field station to laboratory is prolonged.

> ***Special note pertaining to the humane treatment of biological specimens*** *As a general rule, every effort should be made to avoid undue suffering when obtaining samples of or from living creatures. The use of anaesthetics such as sodium benzocaine has proven beneficial in fisheries studies in minimizing damage to specimens. By definition, many field and laboratory techniques are destructive, in that tissue samples or otoliths are analyzed, but the methodologies need not be cruel. Good sample design will also maximize the information obtained from the fewest possible samples.*

Finally, as a general rule, the period between sample collection and analysis should be kept to a minimum. Field filtration of chemistry samples and field preservation will greatly enhance the accuracy of the laboratory analyses (i.e. the closer in time and space one can get to field analysis, the closer the result obtained will be to the actual condition within the environment). In this same vein, collection of as much information on site as possible is usually advisable; for example, good inexpensive temperature and conductivity probes can be purchased or made in most nations, field pH/Eh meters are usually available, and ion-specific electrodes are becoming more readily available (and more reliable). Some of the more technologically advanced nations are experimenting with remote data collection platforms permanently anchored in lakes and reservoirs that can be interrogated by radio or other means (satellite telemetry or land-line). Notwithstanding these technical advances, however, the reader should bear in mind that selection of the least complex and most robust methods, based on the purpose behind their sample collection, will generally ensure delivery of results more often than highly complex but extremely sensitive methods – appropriate methodology does not mean inferior methodology!

Physicochemical methods

The use of standard (= comparable) methods has already been mentioned elsewhere in this manual. A variety of handbooks exist which document both field and laboratory analytical techniques. The more common compilations of standard methods include the following:

(1)   APHA/AWWA/Water Environment Federation (1992). *Standard Methods for the Examination of Water and Wastewater, 18th Edition*, (Greenberg, *et al.*, eds.), Water Environment Federation, Alexandria VA. 1042 pp. – probably the most commonly used methods handbook throughout the world, in its various editions; a most comprehensive text covering all aspects of water chemistry and biology;

(2)   Golterman, H.L. and Clymo, R.S. (1971). *Methods for Chemical Analysis of Fresh Waters*. International Biological Programme Handbook No. 8. Blackwell Scientific Publications, Edinburgh. 166 pp. ISBN 0-632-05540-5 – covers the common limnological parameters as well as many not routinely analyzed; one of a series of standard reference manuals used in the International Biological Programme (see below for other methods manuals in this series);

(3)   US EPA [United States Environmental Protection Agency] (1983). *Methods for Chemical Analysis of Water and Wastes*, Report No. EPA-600/4-79-020, US Environmental Protection Agency, Cincinnati – the official chemical analytical guide covering the methodologies tested and approved by the US EPA for environmental monitoring and regulation in the United States; methods contained in this guide have been abstracted by many other water resources agencies throughout the world as standard practice in the laboratory [see also US EPA (1972), *Biological Field and Laboratory Methods for Measuring the Quality of Surface Waters and Effluents*, Report No. EPA-670/4-73-001, US Environmental Protection Agency, Washington; US EPA (1985a), *Methods for Measuring Acute Toxicity of Effluents to Freshwater and Marine Organisms*, Report No. EPA-600/4-85-013, US Environmental Protection Agency, Washington; and US EPA (1991), *Methods for the Determination of Metals in Environmental Samples*, Report No. EPA-600/4-91-010, US Environmental Protection Agency, Cincinnati, for further related methodologies];

(4)   Strickland, J.D.H. and T.R. Parsons (1972). *A Practical Handbook of Seawater Analysis*. Bulletin No. 167. Fisheries Research Board of Canada [Canadian Board of Fisheries and Aquatic Sciences], Ottawa. 310 pp. – the standard wet chemical reference for oceanographic studies, most of these methods have applicability in freshwater and estuarine systems as well;

(5)   Wetzel, R.G. and Likens, G. (1979). *Limnological Analysis*. Saunders, Philadelphia. 357 pp. – a companion volume to the popular treatise on

limnology (Wetzel, 1975); a thorough treatment of the topics covered in the text and practical guide to field techniques; and

(6)   Mackereth, F.J.H. (1963). *Some Methods of Water Analysis for Limnologists*. Freshwater Biological Association Scientific Publication No. 21. FBA, The Ferry House, Windermere, UK. 70 pp. – a brief but thorough treatment of methods of chemical analysis applicable to freshwater ecosystems; although many methods have been superseded by revised methodologies given in the more recent compilations of chemical methods, this volume remains a useful analytical guide for wet chemistry laboratories.

While the foregoing references are commonly available in many laboratories throughout the world, many countries have also published compilations of standard analytical methods adopted for regulatory and other purposes within their borders. Many of these methods have been selected for their specific applicability under the climatic and other conditions pertaining in a specific region. Such compilations are usually available from the state health authorities or national laboratories. Where possible, readers should obtain copies of these guides, even though they may later opt to use one of the texts cited above. The use of nationally-accepted methodologies will usually be required for samples analyzed for regulatory purposes, and may be required if the analyst or researcher is performing studies with funding provided by the state.

Biological methods

The following represent some of the more common compilations of analytical methodologies relating to the collection and evaluation of biological specimens:

(1)   Smith, G.M. (1950). *The Fresh-Water Algae of the United States*. McGraw-Hill, New York. 719 pp. ISBN 07-058809-0 – a comprehensive guide to phytoplankton with discussions of their evolution, distribution and collection;

(2)   Vollenweider, R.A. (1974). *A Manual on Methods for Measuring Primary Production in Aquatic Environments*. International Biological Programme Handbook No. 12. Blackwell Scientific Publications, Edinburgh. 225 pp. ISBN 0-632-00531-9 – the standard methodological handbook of the International Biological Programme;

(3)   Ward, H.B. and Wipple, G.C. [W.T. Edmondson, ed.] (1959). *Freshwater Biology, 2nd Edition*. Wiley, New York. 1248 pp. – probably the most comprehensive account of the biology of freshwaters yet produced, this guide to the taxonomy and ecology of freshwater animals remains one of the most referred to texts in its field;

(4)   Edmondson, W.T. and Winberg, G.G. (1971). *A Manual on Methods for the Assessment of Secondary Productivity in Fresh Waters*. International Biological Programme Handbook No. 17. Blackwell Scientific

Publications, Edinburgh. 358 pp. ISBN 0-632-06430-7 – another in the excellent series of field method guides produced for the International Biological Programme; this manual covers all aspects of sampling and measuring zooplankton-based, secondary productivity;

(5) Holme, N.A. and McIntyre, A.D. (1971). *Methods for the Study of Marine Benthos*. International Biological Programme Handbook No. 16. Blackwell Scientific Publications, Edinburgh. 334 pp. ISBN 0-632-06420-X – although designed for marine studies conducted in association with the International Biological Programme, this volume provides a useful methodological overview of techniques and methods that have equal applicability in freshwater systems;

(6) Sorokin, Y.I. and Kadota, H. (1972). *Techniques for the Assessment of Microbial Production and Decomposition in Fresh Waters*. International Biological Programme Handbook No. 23. Blackwell Scientific Publications, Edinburgh. 112 pp. ISBN 0-632-08850-8 – although lacking some of the most recent advances in microbial production assessment (i.e. the use of tritiated thymidine assays), this International Biological Programme handbook covers all of the basic methodologies necessary to quantify the bacterial component of freshwater ecosystems; and

(7) Ricker, W.E. (1971). *Methods for Assessment of Fish Production in Fresh Waters*. International Biological Programme Handbook No. 3. Blackwell Scientific Publications, Edinburgh. 348 pp. ISBN 0-632-08490-1 – this volume remains the standard reference work on fisheries methods with global applicability.

The reader is again referred to the note given in the preceding section regarding the use of nationally-recognized standard methods. Similar compilations of standard methods may also be available for analyzing biological samples in many countries. In addition, there are numerous regional taxonomic keys, similar to Smith's guide to algae, that may be more appropriate guides in areas outside of North America. Nevertheless, the foregoing handbooks represent the more commonly used references in the field of biological sample collection and preservation that have international applicability. In addition to these texts, readers are encouraged to scan the scientific literature and/or join an appropriate professional body with interests in water resources management in their region.

General texts

The reader may also wish to consult a number of texts and supplementary reading materials which provide basic introductions to the many subjects – limnological, ecological, geological and hydrological – that comprise the multi-disciplinary foundations of nonpoint source contaminant control. Some relevant materials include:

(1)   Wetzel, R.G. (1975). *Limnology*. Saunders, Philadelphia PA. 743 pp. ISBN 0-7216-9240-0 – a concise, thorough-going treatment of the physics, chemistry and biology of freshwaters; one of the standard textbooks in this field and an excellent source-book for further readings in limnology;

(2)   Hutchinson, G.E. (1957, 1967, 1975, 1993). *A Treatise on Limnology, Volumes I–IV*. Wiley, New York – the culmination of a lifetime's study of freshwater ecosystems, this multi-volume treatise is the most in-depth account available of the study of lentic systems, their physics, chemistry and biology; the classic textbooks of limnological sciences;

(3)   Hynes, H.B.N. (1972). *The Ecology of Running Waters*. University of Toronto Press, Toronto. 555 pp. ISBN 0-8020-1689-8 – the most comprehensive account of the fluvial environment and its biology written, and a standard textbook in the field of lotic ecology;

(4)   Stumm, W. and Morgan, J.J. (1970). *Aquatic Chemistry: An Introduction Emphasizing Chemical Equilibria in Natural Waters*. Wiley, New York. 583 pp. ISBN 0-471-83496-3 – for readers wishing a more in-depth account of chemical interactions in aquatic environments, Stumm and Morgan provide a detailed account of the chemistry behind the measurements and observations commonly obtained from freshwater systems;

(5)   Gregory, K.J. and Walling, D.E. (1985). *Drainage Basin Form and Process: A Geomorphological Approach*. Edward Arnold, London. 458 pp. ISBN 0-7131-5923-5. – a standard textbook of stream hydrology including the dynamics of erosion and watershed-based stream system analysis;

(6)   Ryding, S.-O. and Rast, W. (1989). *The Control of Eutrophication of Lakes and Reservoirs*. UNESCO MAB Series Vol. 1. Parthenon Press, London. 314 pp. ISBN 0-929858-13-1 – a review of the processes involved in, and control of, the enrichment of waters by nitrogen and phosphorus; a companion volume to this manual;

(7)   Stahre, P. and Urbonas, B. (1990). *Storm-water Detention for Drainage, Water Quality, and CSO Management*. Prentice Hall, Englewood Cliffs NJ. 338 pp. ISBN 0-13-849837-7 – a useful account of the determination of rainfall–runoff relationships and the design and construction of stormwater management systems, including calculation examples;

(8)   Novotny, N. and Olem, H. (1994). *Water Quality: Prevention, Identification and Management of Diffuse Pollution*. Van Nostrand-Reinhold, New York. 1054 pp. ISBN 0-442-00559-8 – probably the most thorough and comprehensive overview of the assessment and control of nonpoint source pollution in the temperate zone yet written, this volume extends and up-dates the text of Novotny and Chesters (1981) by introducing the ecological basis of environmental engineering and management, and concepts of sustainable use of water resources;

(9)   Chapman, D. (1992). *Water Quality Assessments: A Guide to the Use of Biota, Sediments and Water in Environmental Monitoring.* Chapman and Hall, London. 585 pp. ISBN 0-412-44840-8 – a practical guide to the design of water quality monitoring programmes and the interpretation of water quality data; a second edition of this work was published in 1996;

(10) Hauer, F.R. and Lamberti, G.A. (1996). *Methods in Stream Ecology.* Academic Press, New York. 674 pp. ISBN 0-12-332905-1 – a comprehensive course book and manual of current methods in stream ecology, covering hydrology and hydraulics, water chemistry, and stream flora and fauna, and their interactions at the community and ecosystem levels;

(11) Ryding, S.-O. (1992). *Environmental Management Handbook: The Holistic Approach – from Problems to Strategies.* Lewis Publishers, Boca Raton. 777 pp. ISBN 0-87371-753-8 – sets forth the underlying rationale to environmental management for government, business and the public; and

(12) Cooke, G.D., Welch, E.B., Peterson, S.A. and Newroth, P.R. (1993). *Restoration and Management of Lakes and Reservoirs.* Lewis Publishers, Boca Raton. 548 pp. ISBN 0-87371-397-4 – a second, expanded edition of the popular book on lake management in the northern hemisphere; this text summarizes the range of, and some of the representative experience with, remedial measures that can be applied *in situ* to lakes.

As noted previously, this list is not intended to be a comprehensive bibliography of water resources management texts, but rather an indication of those materials of relevance to a broad spectrum of water resources professionals dealing with nonpoint source management issues. These books represent only a small fraction of an extensive water resources literature. These books will provide the reader with a sound working knowledge of the principles and concepts underlying the multi-facetted aspects of nonpoint source contaminant management. Again, there may be local or regional texts of greater relevance to specific climatic zones or management situations. These should be consulted in addition (or in preference) to the abovementioned literature.

ATMOSPHERIC DEPOSITION

Wet deposition

Regardless of where atmospheric deposition may fall (i.e. on the land surface or directly onto a waterbody), the contaminant load in wet deposition usually is estimated as the product of the contaminant concentration in the collected precipitation (rainfall, snow, hail) and the quantity of precipitation. It is noted that atmospheric deposition (wet or dry) on to the land surface should not be calculated as a separate component of the total contaminant load to a lake or reservoir, primarily because it already would be included as a component of the tributary load. However, atmospheric deposition directly on to the surface of a waterbody should be measured or estimated as a separate component of the total contaminant load to the waterbody. It is especially significant for relatively pristine lakes or reservoirs, where it can constitute a substantial portion of the total contaminant load (PLUARG, 1978; Ryding and Rast, 1989).

A wet-only precipitation collector (a collector covered with a lid between periods of precipitation) is only exposed a fraction of the time. Therefore, it is somewhat protected from contaminant artifacts, such as insects, bird droppings, vegetation debris, and dust from the ground. Bulk collectors (open all the time) also can give reliable samples of wet deposition in regions where the sampling site is protected from local emissions of large particles. An example would be a small clearing in a forest setting (Granat, 1987). However, when bulk collectors are used, it is advisable to use two or three units spaced some distance apart (several hundred meters) at each sampling site, in order to reveal possible contamination from non-precipitation sources.

One can obtain supplementary information about spatial trends in wet deposition by collecting a large number of samples from accumulated snow – in areas where snow accumulates – over a period of several months. (This will provide an appropriate time integration.) In these situations, however, it is important to compare the results to data from routine sampling sites in order to account for losses due to partial melting of the snow (Johannessen and Henrikssen, 1977), and to put the results in perspective with regard to seasonal variations.

One can also obtain semi-quantitative estimates of atmospheric deposition by analysis of moss samples. Mosses retain many heavy metals and organic micro-constituents from both wet and dry deposition and are not particularly influenced by the type of surface on which they grow. Analysis of moss samples can be especially valuable when evaluating small-scale variability in contaminants within a watershed; an example would be point sources within a large-scale concentration field obtained from a network of precipitation collectors.

On a global scale, analysis of ice-core samples from the Alps, Greenland, and Antarctica provide long and interesting records on atmospheric deposition. However the findings from such cores are often difficult to extrapolate to polluted areas as a result of an imprecise understanding of global circulation patterns. This is especially true of samples obtained in the polar areas.

Errors associated with measurements

Although calculation of wet deposition loads to a waterbody is simple in theory, there are many potential sources of error associated with the measurement of rainwater composition, and in establishing the spatial variation in concentration or deposition. Errors in point estimates (i.e. the difference between the true value and the measured value at a certain sampling site) may be due to: (1) contamination either from handling the collected sample, from the collector, or from the local surroundings; (2) loss of sample integrity due to changes in contaminant concentrations before the sample is analyzed in the laboratory; (3) poor precipitation collection efficiency; and/or (4) analytical errors. Errors involved in establishing areal trends may be due to: (1) a non-representative sampling site which introduces errors in the extrapolation of results from inland and shoreline sampling sites to pelagic areas in large waterbodies – both with respect to the quantity of precipitation and contaminant concentration; (2) an inadequate sampling network which is not sufficient to provide accurate information on spatial variations in contaminant concentration and/or quantity of precipitation; and (3) inappropriate calculation procedures used to calculate atmospheric deposition loads, examples being smoothing, use of correlations between contaminant concentration and quantity of precipitation, etc.

Errors in point measurements depend largely on: (1) the local surroundings of the sampling site, and (2) the type of sampling equipment and its maintenance. These potential errors should be evaluated throughout a sampling network, and even in sub-components of the network, in order to adequately assess the actual conditions. Such errors can result in quite different results being obtained when various sampling techniques have been considered. Although dictated to some degree by the environmental conditions at the sampling sites, these techniques include: (1) the use of bulk *versus* wet-only collectors, (2) differing sample collection times, and (3) the necessity of using preservatives, cooling, etc. Examples of differing results from such variables can be seen in the work of Galloway and Likens (1978), Whelpdale and Berry (1980), Söderlund (1982), Slanina *et al.* (1987), and Granat (1987). The sample collection efficiency of wet deposition collectors has been studied in connection with meteorological networks. Depending on the exposure to wind at a particular location, the collection efficiency can vary from nearly 100 percent for rain collected at low wind speed, to less than 50 percent for snow samples.

Laboratory analytical procedures also can be a source of error in determining wet deposition contaminant loads. However, it is noted that such errors can typically be assessed by intercalibration between laboratories and/or appropriate quality control and quality assurance programmes. Furthermore, such errors may often be small, in comparison to sampling and sample handling errors.

The establishment of large-scale, atmospheric deposition collection fields is often evaluated first by considering the concentration field of a particular contaminant in the wet deposition, and the quantity of precipitation obtained at many points in the proposed network. This information can sometimes be obtained from national meteorological networks. Of particular importance is the degree of co-variation between contaminant concentration and quantity of precipitation. The contaminant concentration might increase, decrease, and be independent, of the precipitation quantity. Increases and decreases in concentration are related primarily to the rather extreme conditions at high elevations with, and without, capping clouds. Contaminant concentration independence from precipitation quantity might be the most typical finding, although it has only been verified in a few cases (see Granat, 1988).

A meaningful, large-scale contaminant concentration field (contour map) can be constructed if (1) measurements of sufficient number and reliability are made, and (2) the small-scale variability in concentrations is low. The spatial variability of wet (and dry) deposition depends, in principle, on the configuration of the contaminant source and the contaminant atmospheric residence time.

It is noted that, in addition to meteorological conditions (e.g. windspeed, turbulence, occurrence of rain and/or snow), the reactivity (of gases) and size (of particles) are the key factors determining the atmospheric residence time. An approximation of particle size distribution, residence time and terminology is summarized in Figure 7.1. Wet deposition or removal of particulate matter in the size range that can be deposited in precipitation is estimated to have a mean residence time in the lower atmosphere of about three days in Europe and North America (Rodhe, 1978). Highly reactive or soluble gases and particles can be deposited more rapidly by dry deposition processes. Less soluble gases (including some organic particles not typically adsorbed onto particulate matter or vegetation) can have a greater atmospheric residence time – longer than a few days. As an example of residence time variations, large particles typical of smelter emissions may be deposited within kilometers from the smelter, while, in contrast, sub-micron particles can be transported over thousands of kilometers and undergo gradual deposition (primarily with precipitation). In addition, some atmospheric contaminants may fall into only one of these groups, while others can belong to both groups, with respect to residence time. For example, compare the rather smooth concentration field of sulfate in precipitation outside major source regions (Figure 6.8) with that of chromium (Figure 6.5).

Measurement of deposition onto water surfaces

With regard to measurements of wet deposition directly on to lake surfaces, the procedures are similar to those used for the land surface. Typically, contaminant concentration and precipitation quantity samples are collected at sampling sites in the vicinity of the waterbody. Sometimes, samples also may be collected in samplers placed on platforms on the lake surface. Such a technique should be used on very large waterbodies, in which the deposition pattern may vary over the large surface area (e.g. North American Great Lakes). In both cases (large and small waterbodies) it is assumed that contaminant concentration samples collected along the shoreline of a waterbody are better for estimating the atmospheric deposition load directly to the waterbody than samples collected a large distance from the waterbody. The concentration over the lake can be estimated by interpolation of near-lake measurements (Granat, 1987). In contrast, the quantity of precipitation directly onto a waterbody is usually lower (by up

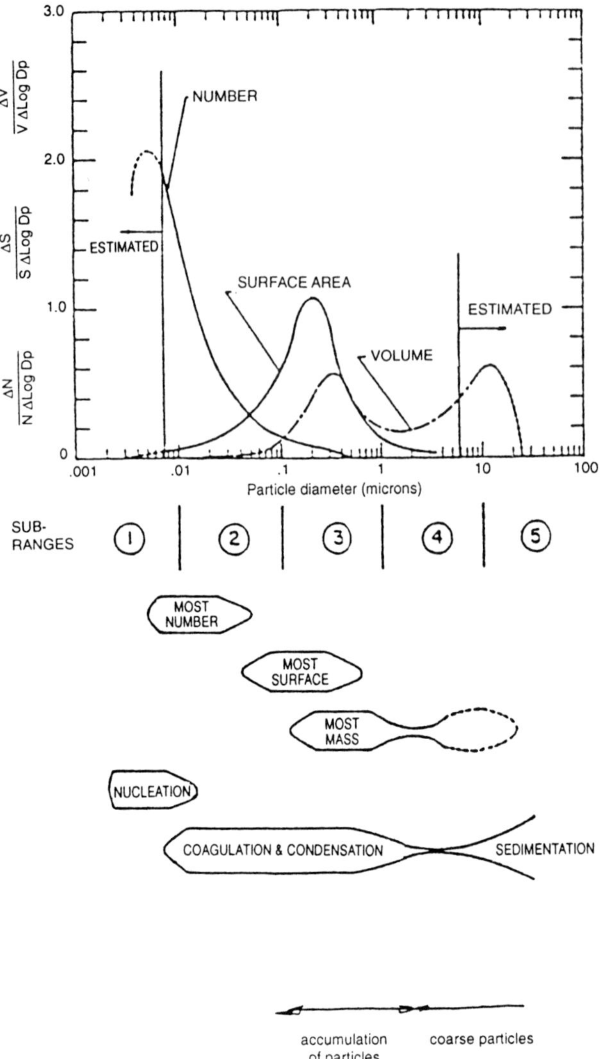

Figure 7.1   Characteristic properties of atmospheric aerosols (after Whitby, 1978)

to 20 to 30 percent) than that over the land surface. These facts should be taken into account in calculation of the atmospheric deposition load directly onto the lake surface.

Dry deposition

Dry deposition comprises all deposition processes other than wet deposition. Accordingly, it includes transfer of gases from the atmosphere to terrestrial

234

vegetation, soils, lake water, etc. It can involve both the biotic and abiotic fixation of the gases as well as the movement of particulates. Large particulates will fall by gravity to the land and waterbody surface, while smaller particles must be transferred by atmospheric processes to the surfaces, where they can be captured by impaction, interception, or molecular diffusion through the layer nearest the surface. As with gases, the particulate deposition rate depends on the surface structure of the land, atmospheric turbulence, and particle size.

In contrast to wet deposition, dry deposition is usually not measured directly with an instrument. Attempts to estimate the dry deposition flux onto a waterbody as the difference between bulk and wet-only collectors, or from the dry side of a wet and dry collector, will often fail because they do not represent sampling conditions at the waterbody surface. Accordingly, the usual procedure is to measure the concentration of a contaminant in the air (gas or particulate), and then calculate the flux based on an understanding of the various processes involved in the transfer from the atmosphere to the surface. This can be expressed as:

$$F = v(g).\Delta C \qquad (7.1)$$

where: $\quad$ F = flux downwards;
$\qquad \Delta C$ = contaminant concentration difference between reference levels in the air and the collection medium; and
$\qquad$ v(g) = transfer coefficient velocity (often called dry deposition velocity)

An overview of methods for measuring or estimating the flux and the dry deposition velocity in Equation 7.1 is given by Liss and Slinn (1983), who provide an extensive review and discussion of this topic applied to deposition on water surfaces.

When the transfer of an atmospheric contaminant is controlled by the gas phase, Equation 7.1 can be simplified to:

$$F = v(g).C(z) \qquad (7.2)$$

where: C(z) = contaminant concentration at the reference height in air

The dry deposition velocity can be related to wind speed, atmospheric stability, and the structure of the water surface. Hicks and Liss (1976) suggested that it can be calculated as:

$$v(g) = Q.U(z) \qquad (7.3)$$

where: U(z) = windspeed at a particular reference height; and
$\qquad$ Q = factor that depends both on wind speed and atmospheric stability (temperature profile); the value for Q is given in Figure 7.2

Estimates of the liquid phase transfer velocity are more uncertain than those of the gas phase transfer velocity. An example is given in Figure 7.3. Correction of these transfer velocities for gases other than carbon dioxide and oxygen is

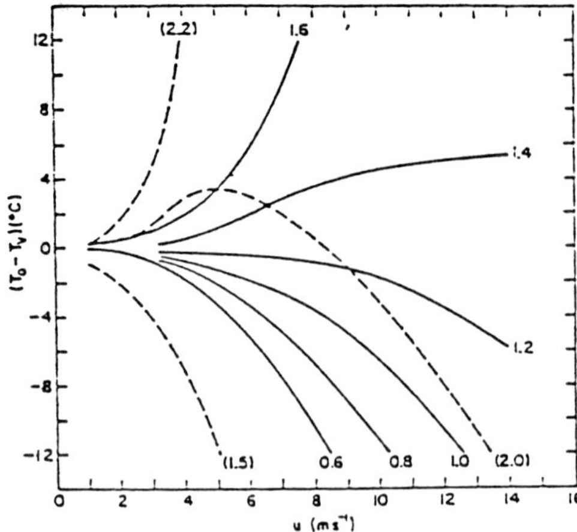

Figure 7.2 Parameter ($Q = v_g/u$) for estimation of deposition velocity during a gas phase controlled transfer to a water surface (from Hicks and Liss, 1976)

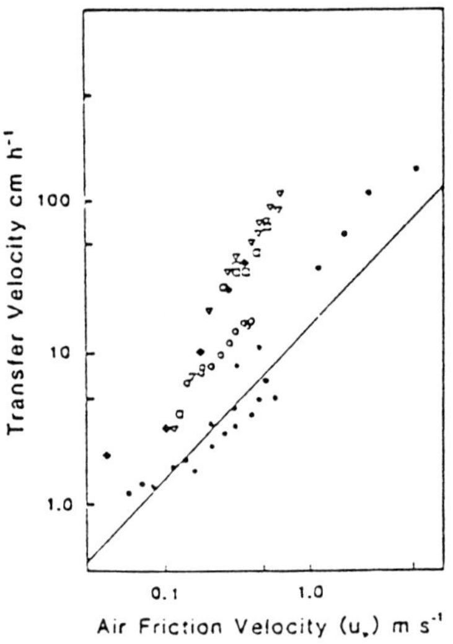

Figure 7.3 Transfer velocity during a liquid phase controlled transfer to a water surface (from Liss, 1983)

discussed by Liss (1983). Calculation of the deposition velocity for particles is also considerably more difficult than for gases; specific problems are outlined in Figure 7.4. Further guidance on the calculation of this parameter is given by Slinn and Slinn (1980) and Williams (1982). Figure 7.4 provides some deposition velocity values calculated by Williams (1982). Although higher than many other estimates, their use is suggested in order to avoid underestimation of the dry deposition flux to water.

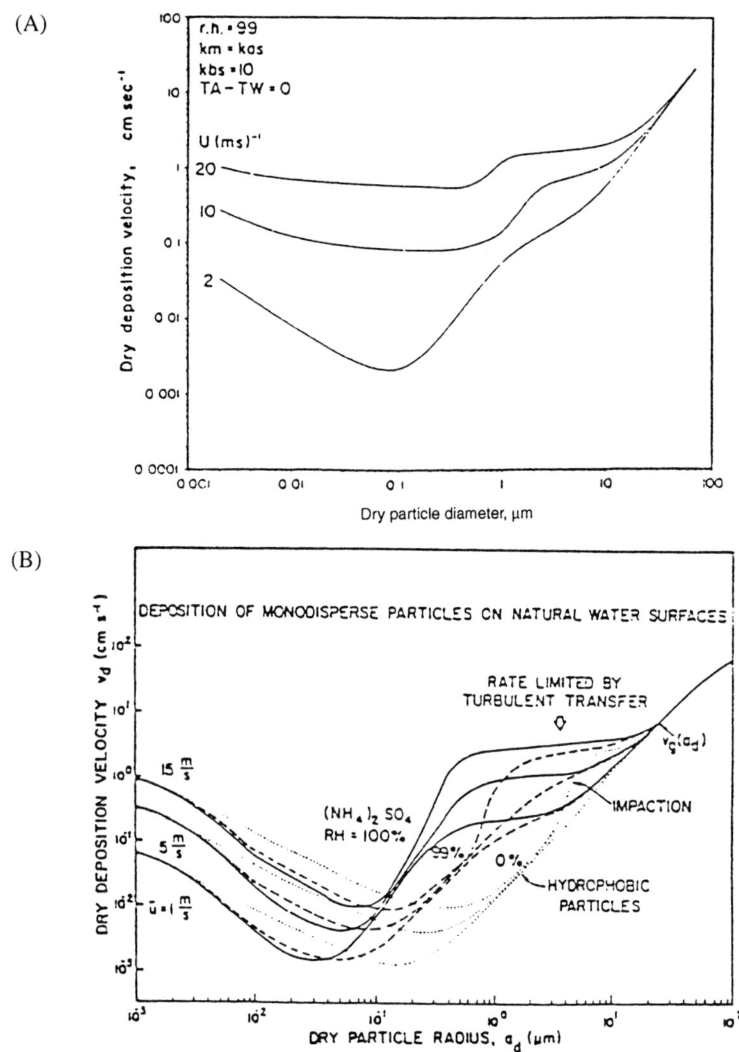

Figure 7.4 (A) Proposed most likely variation of deposition velocity with wind speed and particle diameter (from Williams, 1982), and (B) dry deposition velocity as a function of dry particle size and equilibrium size at 0, 99 and 100 percent relative humidity (from Slinn and Slinn, 1980)

Several other investigators have discussed the occurrence of an organic micro-layer at the water surface and its effects on atmospheric deposition (e.g. see Meyers *et al.*, 1983). In principle, this layer presents special problems in estimating the resistance of the water surface to gases, especially water-soluble gases. However, Liss (1983) has concluded that the film would only substantially reduce the deposition flux when the density of the film was so large that it affected light reflectance (i.e. it could be observed as a slick). The slicks would also be broken up by wind. Furthermore, deposition of gases onto water surfaces in calm weather is typically very small under most conditions.

STREAM GAUGING

Site selection

Selection of a stream or tributary sampling site depends on (1) the specific data needs, and (2) the nature of the stream flow and other conditions. Ideally, a stream sampling site should be located at or near a gauging station. This will allow water quality and stream contaminant loads to be related more easily to stream flows. Furthermore, the stream sampling site should ideally have uniform depths and water velocities across the stream channel. As a practical matter, stream cross-sections used to make stream flow measurements can usually serve also as locations for collecting water quality (e.g. nutrient) samples (Wells *et al.*, 1990).

The presence of other tributaries entering the stream being measured must also be considered. For example, a stream sampling site should not be located immediately downstream from either (1) the confluence of two streams, or (2) a specific pollutant source. This is because the water in the stream immediately downstream of these areas may not be well mixed, thereby increasing the possibility of collecting a biased water sample. The possibility of backwater effects also dictates that sampling sites not be located immediately upstream of the confluence of two streams. In either case, once a sampling site has been selected, one should try to continue to (1) make measurements of stream flows, and (2) collect samples, at the same location throughout the period of record.

Uniform water quality across a stream will provide the most accurate estimates of contaminant loads to be made. Accordingly, one should make a series of measurements across the stream (a minimum of 10 transects is suggested) to determine if the values of the water quality parameters vary significantly across the stream cross-section. If the values do vary, then the sampling effort should be tailored to taking water quality samples proportional to their distribution, and to the stream water velocity, across the stream cross-section.

Methods for measuring stream flows

This section generally refers to measurement of water and contaminant loads in streams and tributaries draining watershed-sized areas. In contrast to plot-sized measurements described below, watershed-scale monitoring should include any channels that have a base flow (i.e. large drainage ditches, irrigation channels, storm sewers, and larger stream systems).

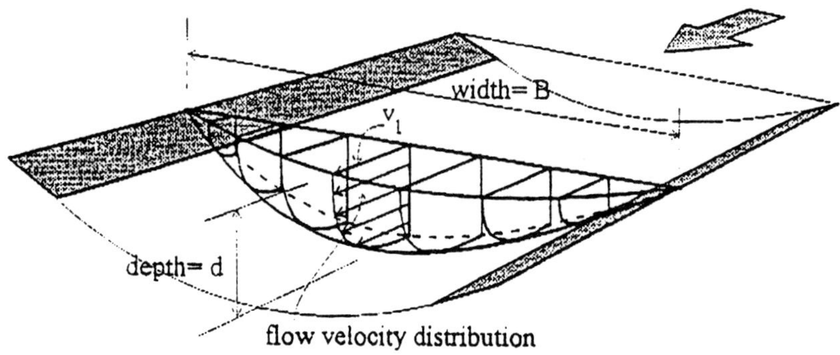

Figure 7.5 The concept of measuring the rate of flow in open channels (from Starosolszky, 1970)

Streamflow measurement

The basic principle of streamflow measurement in an open channel is the determination of the volume of water that is defined by the plane of the cross-sectional area and a surface which corresponds to the end points of the velocity vector components perpendicular to the plane of the cross-section, as shown in Figure 7.5 (Starosolszky, 1970). Theoretically, it involves the solution of the following inetgration:

$$V = \int_0^d \int_0^B v \, dh \, dw = Q \ (m^3/s) \tag{7.4}$$

where: V = volume of water passing the cross-sectional area per unit time (sec);

v = velocity of the flow perpendicular to the plane of the cross-sectional area at a given point (m/sec);

h = depth of water (m) at a given point in the cross-section (varying between zero and d, where $h = 0.00 - d$); and

w = width (m) of the water surface (varying between zero and B, where $w = 0.00 - B$)

In practice, the streamflow measurement means the measurement of the wetted cross-sectional area (the area bordered by the channel surface and the water surface) and the measurement of the cross-sectionally averaged rate of flow or water discharge:

$$Q = v.A \tag{7.5}$$

where: Q = streamflow (m³/sec);

V = velocity (m/sec); and

A = cross-sectional area (m²)

This is sometimes known as the area–velocity method.

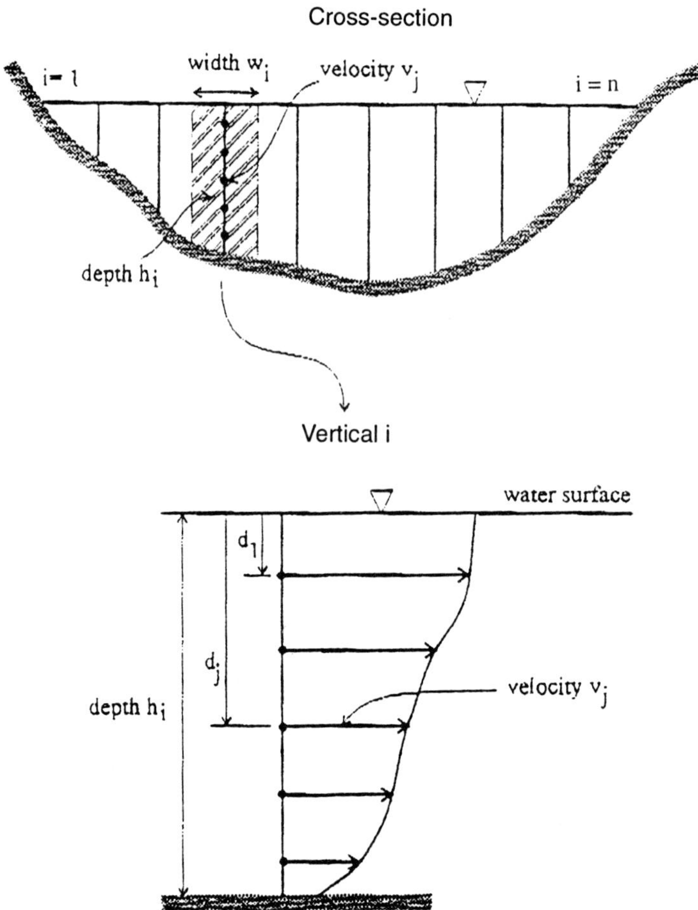

Figure 7.6 Illustration of the measurement and calculation of water discharge in open channels (from Chow *et al.*, 1988)

Measurement of the area is carried out by measuring discrete water depths, h_i, corresponding to discrete water surface widths, w_i, across the stream profile (Figure 7.6).

The flow velocity, v_i, which corresponds to each elemental cross-sectional surface area ($A_i = h_i.w_i$) is then determined by measuring flow velocities at discrete points, j, along a vertical line in the middle of each discrete width, w_i. These velocity measurements, v_j, are averaged over the entire water depth.

Having determined the discrete values of width (w_i), depth (h_i), and velocity (v_i), the theoretical integration can be approximated by the following summation:

$$Q = \sum_{i=1}^{n} h_i w_i v_i \quad (m^3/s) \tag{7.6}$$

where: n = number of discrete widths/depths measured

The method of calculating the average velocity in a vertical profile also depends on similar measurements taken within the vertical plane. Sometimes, certain rules for depth distribution are used for which the mean velocity is obtained as the arithmetic mean of the measured values. In the general case of an uneven distribution of points (Figure 7.6), the calculation is made as follows:

$$v_i = \frac{1}{h_i} \sum_{j=1}^{m} \frac{v_j + v_{j-1}}{2} (d_j - d_{j-1}) \tag{7.7}$$

where: m = number of measurement points in the vertical plus 1.0 (e.g. d_m = h_i and $v_m = 0.5\, v_{m-1}$ is the velocity measured near the bottom)

Equation 7.7 calculates the depth-weighted average velocity as the total depth-velocity area divided by the total depth, h_i. Note that surface flow velocity is considered to equal the first measured velocity below the surface, while the bottom flow velocity is half of that of the nearest measured velocity.

Considerable additional detail on streamflow measurements, including available measurement equipment, and streamflow calculation methods, is provided in the well-known hydrologic text of Chow *et al.* (1988).

Stage measurements

The simplest method of measuring stream stage is with a staff gauge. This is a graduated pole erected vertically along the stream bank. If a bridge abutment is the site of the streamflow measurement, the staff gauge, or gauge plate, can be attached to it. Such a gauge can be read manually by an observer. Similar types of manual gauges include chain, tape, and wire gauges. These latter devices are often placed on bridges. The stage is determined by lowering the weight or other device to the water surface, and reading the corresponding stage from a counter or other graduated device on the apparatus. An advantage of these latter devices is that they can be read under virtually all flow conditions.

This process can also be automated. The simplest automated method involves measurement of the rise and fall of a float in a cylinder (called a 'stilling well'). The stilling well is typically connected by a horizontal pipe set into the stream to allow entry of the stream water into the well. This ensures that the water level in the stilling well is at the same level as the water in the stream channel. The stilling well reduces the influence of streamflow (such as the vibration of the float by water movement in the stream) on the gauged height by isolating the float from the stream. The pipe connecting the stilling well to the stream is generally placed well into the stream bed to allow measurement of low flows but should be protected from clogging with debris during both high and low flows. The float within the stilling well is connected to a wheel and drum chart recorder (Figure 7.7), or paper tape device, which traces the rise and fall of the stream level over time. The wheel and drum

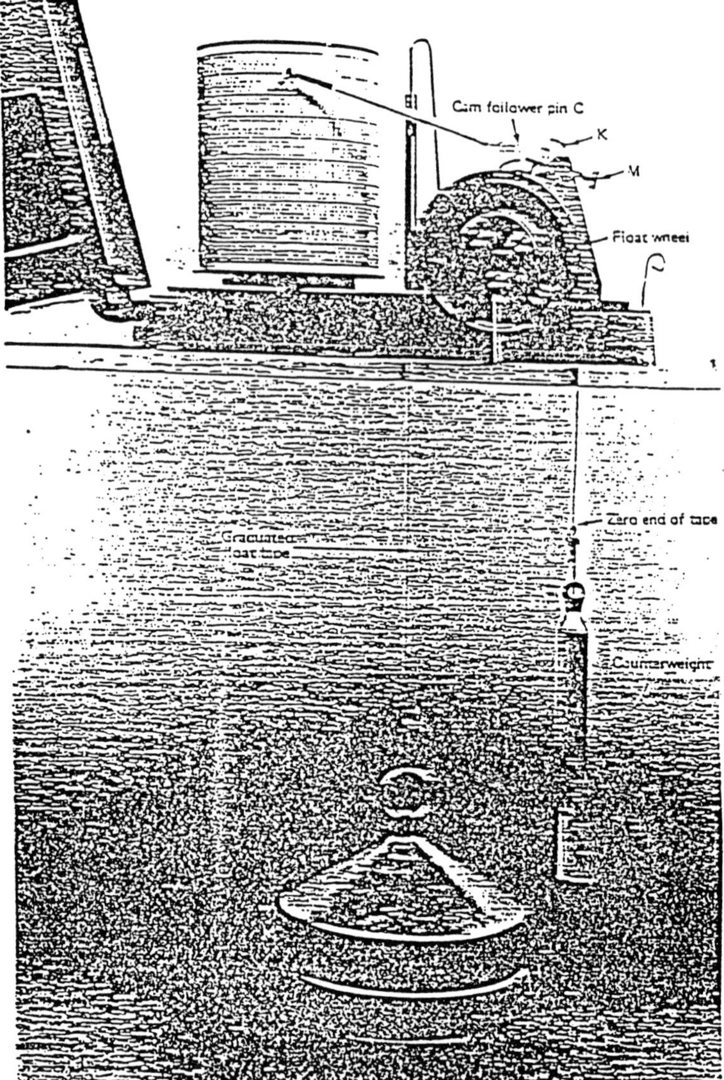

Figure 7.7 Water level recorder showing float, counterweight and recording drum (from USDA, 1979)

recorder traces the hydrograph onto a paper strip on a rotating drum; the rotating drum can be powered by a spring mechanism or a battery. The punched tape device works on the same principle, except that it records the stage as a series of punched holes (Chow *et al.*, 1988). Both devices are relatively inexpensive to use, reliable, easy to maintain, and require no electricity. Their primary disadvantage is that the data, often voluminous, must be transferred by hand to a computer for analysis.

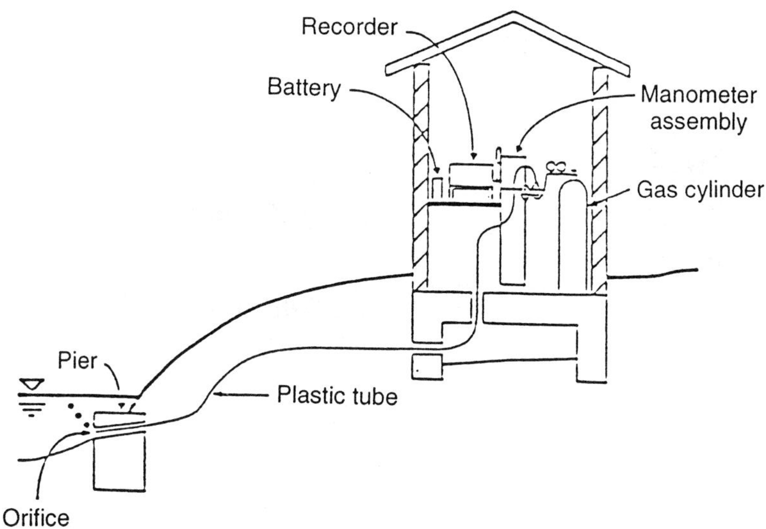

Figure 7.8 Water level measurement using a bubble gauge recorder (from Chow *et al.*, 1988)

Two automatic stage-measuring devices operate on the principle of hydraulic pressure, in which the stage is proportional to the pressure exerted by the height of the stream water column above the stream bed. These devices are the bubble gauge and the electronic pressure transducer. The bubble gauge (Figure 7.8) continuously emits a stream of gas bubbles (usually carbon dioxide) through a tube leading from a pressurized cylinder. The orifice of the tube is placed on the stream bottom. In this position, the pressure required to emit the gas bubbles, measured most commonly with a mercury manometer, is proportional to the stage. Data are recorded onto a punched tape or electronic media. In warm climates, bubble gauges have been known to suffer from a build up of microbial and/or algal growth at the outlet orifice.

The pressure transducer device, though a more recent development, operates on the same general principle as the bubble gauge. The transducer is connected to a digital recording device. The advantage of this device is that the digital data can be transferred directly to a computer disk or tape for subsequent processing. Thus, this system is very efficient in terms of data management; however, it is also much more costly and requires electricity at the stream measuring site.

To augment the operation of these stream gauges, structures such as flumes or weirs can be constructed to more accurately define the cross-section of the channel. As a result, the rating curve is reduced to a simple mathematical calculation which does not require time-consuming calibration. Flumes and weirs vary in their geometric configuration, depending on the type of flow control required in a given situation (Figure 7.9). Common examples include the Parshall

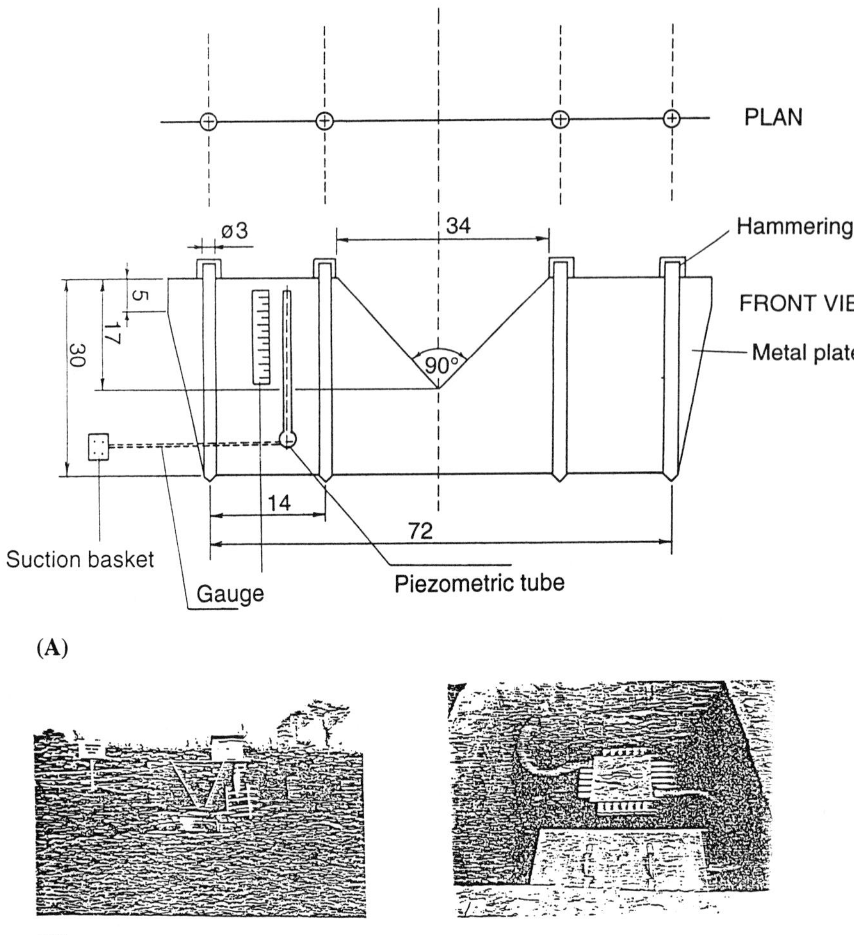

Figure 7.9 (A) Thompson-type discharge measuring weir, and (B) an H-flume with water stage recorder and Cochocton wheel (left) and multi-slot divisor and storage tank (right) (from Lal and Stewart, 1994)

flume, broad-created weirs, H-flumes and V-notched weirs. In their use, the stage is measured as described previously, and used to directly calculate water discharge. Considerable additional detail on stream stage measurements, including available measurement equipment, and methods of data analysis, is also provided in Chow *et al.* (1988).

Sediment load measurement

Sediment is transported in streams during periods of high or flood flows, when the turbulent energy of the stream is sufficient to suspend and carry the sediment particles. As the streamflows abate, the suspended sediment load is again deposited on the stream bed, usually downstream from its origin. This process of resuspension and deposition produces a 'ratchet-like effect' in terms of which particles are carried progressively downstream from their point of origin.

The concentration of sediment in stream water is proportional to the energy of the stream, which is proportional to streamflow. Sediment loads are measured by sampling the streamflow, and determining the sediment concentrations over a range of flows. A sediment load curve, analogous to a rating curve, can be established for the stream site by plotting the logarithm of the instantaneous sediment load *versus* the logarithm of the instantaneous streamflow. An obvious advantage of developing a sediment discharge curve is that sediment loads can then be estimated without the necessity of monitoring sediment concentrations.

A sediment load curve can be improved by segregating the data according to the type of discharge. For example, summer storm data should be separated from snowmelt runoff data, since erosion conditions in a watershed differ considerably under these two conditions. Also, data from the rising portion of the hydrograph should be separated from the falling portion of the hydrograph; the rising portion tends to have a higher sediment concentration for the same flow than does the falling portion of the curve as the rising portion of the curve is more closely related in time to the erosive event that resulted in the movement of the particulates. Another method of preparing more accurate sediment load–streamflow relationships is to divide the double logarithmic plot into flow segments and to establish separate regression equations for each segment.

It is noted that some contaminants, particularly those with high partition coefficients, are able to bind readily to sediment particles. As a result, their concentration–discharge relationships will be similar to those for sediment. Examples of such contaminants include phosphate, some pesticides, and heavy metals. It is also noted, however, that if the sediment concentration is low, the concentration of such contaminants (even with high partition coefficients, K_ds, as defined in Chapter 6) can be primarily associated with the water phase of streamflows. Obviously, contaminants with low partition coefficients (e.g. nitrate, chloride, some pesticides) are not correlated with sediments loads, and may not even be correlated with streamflow. For example, nitrate may show a concentration decrease with increasing streamflow. Likewise, the concentration of a pesticide may be correlated not with streamflow, but instead, to time of application to agricultural land in the watershed.

Methods for collecting water samples in streams

The compilation and manipulation of streamflow and water quality data, in order to determine annual or seasonal nonpoint source contaminant loads to receiving waters, is discussed elsewhere (see Chapter 8). This section presents guidance on the sampling of streams and tributaries for water quality parameters.

The simplest stream sampling method is the manual 'grab sample'. This technique can be as simple as lowering a bucket or bottle into a stream, and removing ('grabbing') a near-surface water sample. For dissolved contaminants (e.g. nitrate), this method may be adequate. For sediment or sediment-bound contaminants, however, this method is inadequate primarily because sediment concentrations can vary both vertically and horizontally within a stream cross-section. Studies involving the sediment bedload (the gravel and larger particles moving along the bottom of a watercourse by saltation) can be done with special bedload samplers (USDA, 1979). These samplers comprise a lead-weighted, bomb-shaped device suspended from a cable, and containing a sample container which is open to the stream flow. The water flow forces the bedload sediment into the sampler, which is then closed to obtain the sample. For sediment-associated contaminants, however, this method is of little use, primarily because bedload particles typically have low surface area to volume ratios and are thus relatively unreactive; unlike suspended particulates, such as clay and silts, they generally absorb insignificant amounts of such contaminants.

The US Geological Survey recommends two general methods of collecting water samples: (1) the equal discharge increment method, and (2) the equal width increment method. Although both methods yield the same results, if properly applied, the relative advantages of the two sample collection methods are summarized below in Table 7.3. Valid reasons for deviating from one of these techniques would include the fact that a stream is too shallow or narrow to collect a depth-integrated, multivertical water quality sample. In these latter cases, it is recommended that a sample be collected from the centroid of the stream channel and composed in a churn splitter (see Wells *et al.*, 1990). The method used in a given situation will depend on whether or not the stream is transporting sand particles. This can be determined by taking a single water sample at the fastest-flowing area of the stream cross-section, and allowing it to sit still. If sand is present, it will settle to the bottom within a few minutes; in this case, water samples should be collected with a suspended-sediment sampler (see Wells *et al.*, 1990). Otherwise, one can use a sample bottle to collect the water sample. Regardless of the sample collection vessel used, either sample can be collected at the same location.

Equal discharge increment method

This water quality sample method involves collecting water from the centroids of equal discharge increments (Figure 7.10). A limitation of this method is that it requires prior knowledge of the distribution of streamflow across the stream cross-section. This knowledge may be based on a long period of record, or on a streamflow measurement made immediately prior to selecting the sampling intervals. An advantage of this method is that fewer vertical samples are required to make an accurate measurement than for the equal width increment method.

Table 7.3 Comparison of the equal discharge increment and equal width increment water-sample collection methods (modified from Wells *et al.*, 1990)

Equal discharge increment method	*Equal width increment method*
Fewer vertical samples required, resulting in shorter collection time	Prior knowledge of streamflow in stream cross-section is not required
Sampling during rapidly changing stream levels is facilitated by shorter sampling time	The method is easily learned
Variable transit rates can be used in stream cross-sections	Less sample-collection time usually is required if no streamflow measurement is required, and the stream cross-section is stable

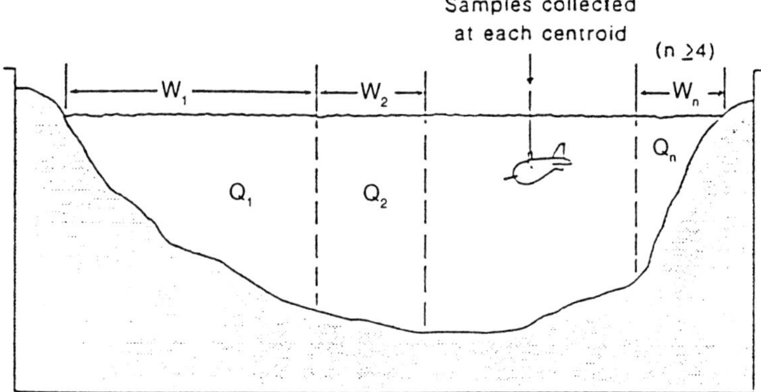

Figure 7.10 Equal discharge increment samples collected at the centroid of flow of each increment (from Wells *et al.*, 1990)

A minimum of four, and a maximum of nine, vertical samples are recommended for this method (Wells *et al.*, 1990). It is assumed with this method that a sample collected at the centroid of each discharge subsection represents the mean concentration of a water quality parameter for the subsection. If this is not the case, as shown by lateral variations in values of field parameters (e.g. temperature, specific conductance, pH, or dissolved oxygen), then (1) the number of vertical subsections can be increased, or (2)

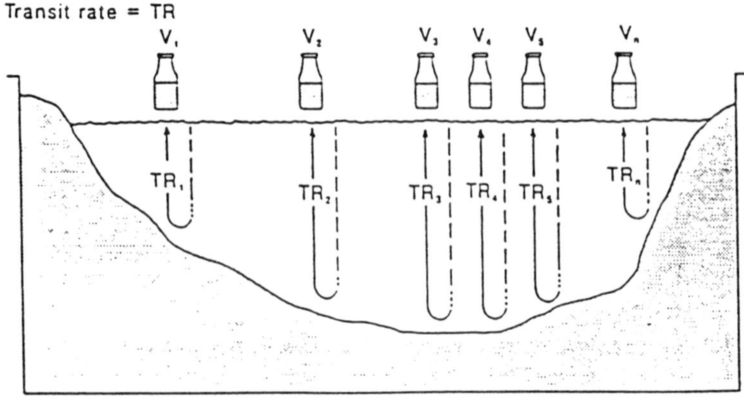

Figure 7.11 Vertical transit rate relative to sample volume collected at each equal discharge increment centroid (from Wells *et al.*, 1990)

the equal width increment method can be used. Wells *et al.* (1990) provide guidance on choosing between the latter two options.

It is noted that equal sample volumes are of primary importance when using the equal discharge increment method of sampling a stream. This means that an individual using this method of collecting water quality samples must submerge and raise the sample collection bottle at a faster rate (i.e. the transit rate) in the deeper subsections of the stream cross-section than in the shallower subsections of the stream (Figure 7.11).

Equal width increment method

This method differs from the equal discharge increment method in that it requires a water sample volume proportional to the amount of streamflow at each of several equally spaced vertical subsections in the stream cross-section. This method is most often used in (1) shallow streams which can be waded, (2) sand-bed streams where the distribution of streamflow in the cross-section may vary between sampling trips, and (3) dendritic streams where tributary inflows have not completely mixed with the main streamflow.

The width of the vertical increments to be sampled can be determined by dividing the width of the stream by the number of vertical samples necessary to collect a sample representative of sediment and streamflow in the cross-section (Figure 7.12). A minimum of 10 vertical samples is recommended for streams

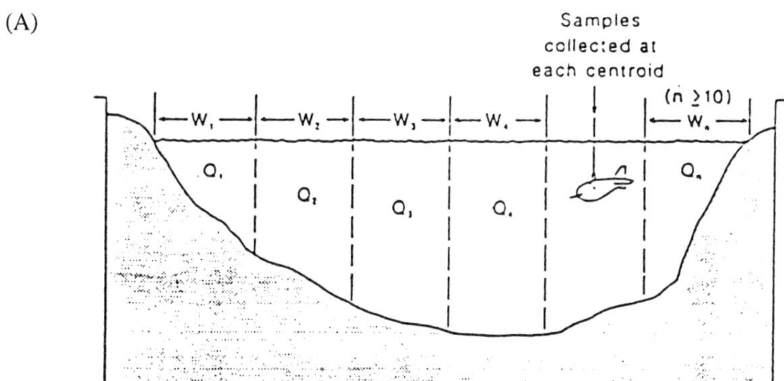

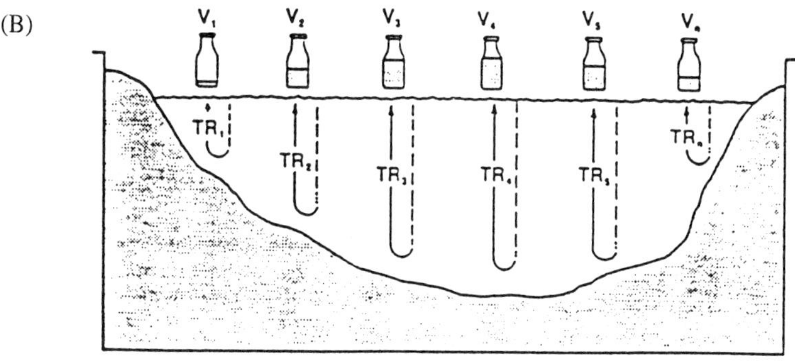

Figure **7.12** (A) Equal width increment sampling technique, and (B) equal width increment vertical transit rate relative to sample volume that is proportional to water discharge at each vertical (from Wells *et al.*, 1990)

more than 1.5 m (5 feet) in width. A maximum of 20 vertical samples is usually adequate, except for very wide and shallow streams. For streams of less than 1.5 m (5 feet) in width, as many vertical samples as possible, spaced a minimum of 1 cm (0.5 inches) apart, is recommended.

In contrast to the equal discharge increment method which required equal sample volumes, the equal discharge method requires equal sample collection rates. That is, the transit rate (the speed at which the sample collection vessel is lowered to the stream bottom, and subsequently raised to the water surface) must be as constant as possible. In this manner, a volume of water proportional to the discharge in the vertical subsection will be collected. A practical method of assuring constant sample collection speed is for an individual to select a reference point (e.g. chest level), and begin and end the vertical sweep of the water sample collection vessel at this point (even if part of the descending or ascending phase of the sample collection effort is in the air). An equal volume of water will *not* be collected at each vertical subsection. Rather, larger water volumes will be collected in fast flowing streams and smaller volumes will be collected in slower streams (Figure 7.12). Wells *et al.* (1990) also provide further guidance on this sample collection method.

Collection of point water samples

The use of depth-integrating samplers is not possible in streams where the depth exceeds approximately 5 m (Wells *et al.*, 1990). In these situations, it is necessary to use a 'point sampler' to collect representative water quality samples.

However, the criteria for choosing between the equal discharge increment or equal width increment methods of sample collection also apply here. For streams with depths between approximately 5 and 10 m, the procedure is to lower the sampler, containing a clean, closed sample-collection bottle, to the streambed. The sampler is then raised to the stream surface at a constant transit rate, opening the sample bottle at the time the upward movement of the sampler commences. The sample bottle is left open until the sampler clears the water surface. A new sample collection bottle is then inserted into the sampler, opened, and then lowered to the streambed at a constant transit rate. The sample collection bottle is then closed and raised to the surface. The two water samples can then be mixed to form a single composite sample.

If the stream depth is greater than approximately 10 m, the same procedure is applied, except that the stream is segmented into vertical subsections no greater than 10 m in depth. An example of point sampling in a stream with a depth of about 20 m is illustrated in Figure 7.13. Note that the transit rate used in the downward direction (TR_1 and TR_2) is not equal to the transit rate used in the upward direction (TR_3 and TR_4). However, TR_1 and TR_2 are equal, as are TR_3 and TR_4. Wells *et al.* (1990) and Chow *et al.* (1988) provide further details on different types of sampling equipment, sample collection and treatment, and field procedures and measurements.

Automatic water sample collectors

In contrast to the manual collection of samples, automatic sampling equipment is also available. Such equipment can be especially convenient when sample

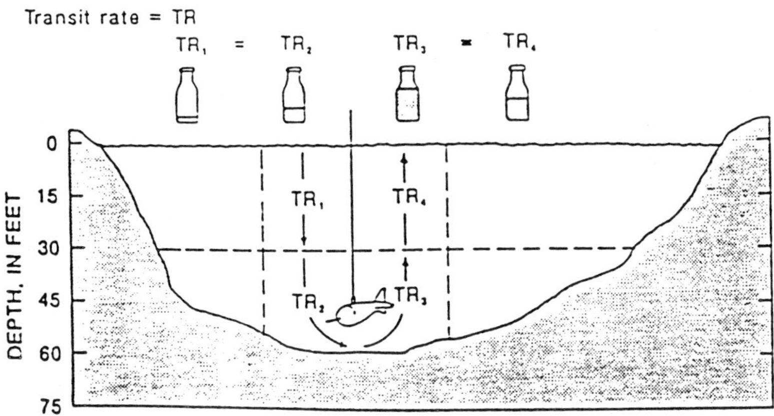

Figure 7.13   Use of a point-integrating sampler for depth integration of a deep stream (from Wells *et al.*, 1990)

sites are located at great distances, or when it is inconvenient or impossible to rapidly respond to storm event runoff. Furthermore, for contaminants only discharged during short time intervals (e.g. pesticides), grab samples (see above section) are often inadequate. In such cases, automatic sampling systems may have to be used to obtain accurate contaminant loading estimates.

All automatic sampling systems involve pumping a water sample from one or more points in a stream, at predetermined intervals. The time of sampling is also noted so that the appropriate instantaneous flow volumes can be used to calculate the contaminant flux. The total flux, or load, for the runoff event is determined by summing the products of the sample concentration and instantaneous flow intervals, as follows:

$$\text{Contaminant load (kg/storm event)} = \sum C_i . Q_i . t_i \tag{7.8}$$

where: C_i = contaminant concentration for time interval, $t(i)$;
Q_i = average streamflow for time interval, t_i; and
t_i = time interval between samples (e.g. every 15 min)

The water samples pumped from the stream are automatically discharged into an array of sample containers of appropriate size. The container size is determined by the amount of sample required for analysis, while the container material is determined by the nature of the contaminant. For example, pesticides can only be collected in glass containers; they adsorb to plastic. Most commercial pump samplers (e.g. ISCO samplers) offer a choice of sample container types, composition, size and number. They can also be programmed to take water samples at predetermined time or flow volume intervals. For example, during low flow periods, water samples can be obtained every 6 h. This level of sampling effort may adequately characterize the quality of these flows, where the flow rate is relatively constant. Time-based sampling would be appropriate, for example, for monitoring wastewater discharges where most of the contaminant load occurs at low flows. Many samplers can also be programmed for flow-based sampling, or to switch from time-based sampling to hydrograph-based sampling when the stream stage exceeds a certain value (i.e. taking flow-proportional samples over a range of stream stages). For high flow conditions, the sampling frequency can be adjusted on the basis of the shape of the hydrograph.

Frequency of stream sampling

The required sampling frequency depends on the techniques used, the objectives of the measurement, the precision and accuracy desired, and the accessibility and logistic support available at the sampling site. For field-scale sampling, the frequency may range for several measurements within a single storm event to one measurement per year (sampling is often based on a hydrological or water year, which typically runs from October to September). In addition, the sampling frequency is a function of the duration of the period of measurement and the duration of the phenomenon being investigated. Short-term studies generally require a more intensive monitoring programme than long-term studies. Similarly, short-lived phenomena may require more intensive sampling over a shorter period of time than long-lived phenomena. For example, describing cause–effect relationships on a field scale, or for studying chemical and/or biological processes, may require several measurements per storm.

The nature of watershed scale studies has changed over the years. Prior to the 1960s, much water quality sampling was based on sampling ambient conditions during low-flows for the relatively traditional parameters typically associated with municipal wastewater treatment plant effluents (e.g. pH, temperature, dissolved oxygen, turbidity, BOD). Stream sampling was often undertaken with a frequency as low as once per month. With the advent of sampling programmes designed to monitor nonpoint source contaminants during the 1960s, however, it became apparent that a large percentage of contaminants generated from nonpoint sources and discharged into lakes and reservoirs by their tributary streams occurred during a few large storm events. For these types of contaminant loads, fixed interval, low-flow sampling was clearly inadequate. Instead, in order to determine a more accurate nonpoint source contaminant load, it was necessary to sample the large runoff events generated by storm events. This is not to say that low-flow sampling was no longer needed to monitor nonpoint source contaminants, but rather that the sampling effort had to be augmented to include high-flow events, regardless of their timing. It already has been noted that the currently available automatic samplers can be programmed to operate in both a low-flow and high-flow mode. The switch between these modes is usually triggered by the hydrograph exceeding a pre-determined value.

Where such technology is not available, flow-proportional sampling can be done manually. Because it generally costs less to collect water quality samples than to analyze them, the manual method of flow-proportional sampling involves the collection of more samples than may be needed. Then, based on a subsequent analysis of the hydrograph for the period during which the sample was collected (but prior to the samples being analyzed), a number of samples are discarded. The number of samples to be analyzed or discarded is generally determined using stream-specific criteria. In this way, a more accurate estimate of the load, over the curve of the hydrograph, is possible. It should be noted that the frequency of sampling over the hydrograph depends on its steepness, and on the change in contaminant concentration with change in the hydrograph. Generally, larger watersheds produce broader hydrographs than smaller watersheds. Furthermore, the rising portion of the hydrograph is typically steeper than the receding portion of the hydrograph.

Calculation of the tributary pollutant load

The pollutant load contributed by a tributary is the product of the water discharge and waterborne pollutant concentration over a specific time interval. Experience has suggested that changes in stream discharge are much more significant to the calculation of an accurate pollutant load than are changes in the concentrations of waterborne pollutants. Thus, because storm events can dramatically affect the tributary discharge volume (and the total pollutant load reaching a waterbody), it is critical such events be specifically sampled. This usually requires the use of continuous streamflow monitoring equipment. In fact, Walker (1987) indicated that, in situations where the pollutant concentration does not vary greatly with flow, the product of the flow-weighted pollutant concentration and average streamflow provides a reasonably accurate estimate of the pollutant load.

A reality is that, in many situations, the pollutant load to a waterbody is calculated as the product of the arithmetic average pollutant (or sediment) concentration and arithmetic average streamflow. However, a significant problem with this approach is that it typically underestimates the pollutant load in areas where storm events are a characteristic or important hydrological phenomenon. As alternatives, the US Environmental Protection Agency (1990) has provided several alternative methods of calculating the pollutant load, as discussed below in further detail.

Mid-interval technique

This method involves assigning representative pollutant concentration data to corresponding streamflow data. The measured pollutant concentration is combined with the discharge measurement (made at the time of sample collection) to calculate an instantaneous pollutant loading rate. This rate is assumed to characterize the tributary load over a specific time interval associated with the collected water sample, as follows:

$$\text{Pollutant load} = \sum_i C_i.Q_i.T_i \tag{7.9}$$

where: C_i = pollutant concentration in i^{th} sample;
Q_i = instantaneous discharge at time of sample collection; and
T_i = time interval associated with i^{th} sample

One typically uses a time interval equivalent to one-half the time interval between the collected water sample and the previously collected sample, plus one half the time interval between the collected sample and the sample to be collected on the next sampling trip. The product of the instantaneous load for each sample and the associated time interval provides an estimate of the total load. This method is especially useful when samples are collected on a uniform, fixed time interval basis; the 'average' pollutant concentration is determined by directly averaging concentrations, mainly because each sample characterizes a stream for the same time interval. However, a disadvantage of this approach is that its accuracy is very sensitive to storm-event loadings.

Time-weighted mean pollutant concentration

For situations in which water quality samples are not collected on a regular basis and/or are collected during storm events, each collected sample does not represent an equal time interval. Thus, estimating the 'average' pollutant concentration requires one to 'weight' the water quality sample according to the length of time it is used to represent a stream. The resultant 'time-weighted mean concentrations' (TWMC) are calculated as follows:

$$\text{TWMC} = (\sum C_i T_i)/\sum T_i \tag{7.10}$$

where: C_i = pollutant concentration of ith sample;
T_i = time interval for which the ith sample characterizes the stream pollutant concentration (one half the time interval between the samples collected immediately preceding and following the ith sample is typically used for this value)

Flow-weighted mean pollutant concentration

Of most value in calculating the total pollutant load to a waterbody is the estimation of the 'average' pollutant concentration on the basis of weighting the individual water samples in accordance with their associated streamflows. This weighted average is calculated by dividing the total pollutant load by the total discharge for the period of interest. Termed the 'flow-weighted mean concentration' (FWMC), it is calculated as follows:

$$\text{FWMC} = (\sum C_i T_i Q_i)/(\sum T_i Q_i) \tag{7.11}$$

where: C_i = pollutant concentration of the ith sample;

T_i = time period for which the ith sample characterizes the stream pollutant concentration (typically represented by one half the time interval between the samples immediately preceding and following the ith sample); and

Q_i = instantaneous discharge at time i^{th} sample was collected

Of analytical interest is use of the ratio of the FWMC and TWMC loading values to identify major pollutant sources. As indicated by the US Environmental Protection Agency (1990), for a given stream site, an FWMC:TWMC ratio greater than 1.0 suggests pollutant concentrations are increasing with increasing discharge, an indication that nonpoint pollutant sources are important. Conversely, an FWMC:TWMC ratio of less than 1.0 would suggest that pollutant concentrations are decreasing with increased streamflow. This indicates the pollutants are being diluted with increasing flow, an indication of point source inputs.

RURAL NONPOINT SOURCE POLLUTION

General considerations

Before selecting the specific experimental plots to be studied, an investigation of the hydrology (outflow) of the entire drainage basin should be performed. This will provide information about the general hydrologic and pollutant loading characteristics of the watershed (i.e. it will allow identification of the water-quality parameters of concern). The results of this larger scale investigation also will highlight the seriousness of the nonpoint source pollution problems, especially when comparing the total annual load of the drainage basin to the point source load. Thus, the magnitude of the 'basin-scale' unit area load (UAL) for a given contaminant can be estimated as follows:

$$Y_B = 1/(T\,A)\,(L_T - L_P) \tag{7.12}$$

where: Y_B = estimate of the UAL for the whole drainage basin (kg/ha/y);

L_T = total annual load (kg) over the period T (y) at the most downstream outlet stream site;

L_P = total annual load (kg) of point-source discharges over time period, T; and

A = catchment basin area (ha)

The above calculation assumes the availability of multiple year flow and concentration records. If the investigation period is less than a year, UAL rates must be expressed in other units (e.g. g/ha/day or g/ha/mm runoff).

In Equation 7.12, Y_B provides only a rough estimate of the UAL, because L_T and L_P are not additive variables. That is, the actual contribution of the point and nonpoint sources to the final outlet load can be assessed only by more complex mathematical models that also take watershed processes into consideration (unless they can be directly and accurately measured). Information on plot-sized UALs is useful for more exact quantification of basin-scale contaminant loading patterns. Thus, an objective of setting up plot-scale experiments is to identify and quantify the most likely nonpoint source areas in the watershed.

Selection of plot-scale field experiments can be more accurately achieved when (1) the records of more than one river sampling site are available, and (2) the product of Equation 7.12 can be calculated for more than one sub-basin of the river network. The basin-scale UALs help identify the likely pollutant source areas within the basin (i.e. the larger the value of Y_B, the more likely the sub-basin of concern is the primary nonpoint source pollutant contributor).

The next step in selecting a study plot is to review land use and topographic maps of the watershed in an attempt to identify nonpoint source areas. Some general types of nonpoint sources (mostly of plant nutrients) include the following:

(1) Sites susceptible to erosion – including large-scale construction sites and/or sites where the soil is left bare for any purpose;
(2) Arable land/crop land – especially row crops, or crops without conservation and/ or contour-line tillage;
(3) Orchards and vineyards – especially new plantations (because of extra fertilizer dosages);
(4) Grasslands and pastures – especially those used for manure spreading;
(5) Densely-grazed pastures – especially those close to streams;
(6) Rural settlements – especially those without canalization;
(7) Animal feedlots; and
(8) Landfills and waste-disposal sites.

Site selection for plot-scale experiments

Ideally, field experiments would be conducted in the susceptible source areas using all possible combinations of soil, slope, tillage, and crop types. However, the more realistic solution is to select 'characteristic' sites, which are sites that generally represent the type of land use for which the experiment is being made (at least in terms of slope and soil type).

If there is insufficient time to repeat experiments and measurements with different crops and tillage/conservation techniques, the logical alternative is to select the most characteristic crop and land cultivation techniques. Even in this situation, however, it would be advisable to split the experimental plot into two parts (i.e. run dual experiments). With dual experiments, one site should be used as a reference or control plot, in which the conditions representative of the larger area could be documented (in terms of tillage, fertilizer usage, conservation techniques, etc.). On the other plot, the effects of alternative control techniques can be quantified (e.g. reduced and/or time-managed fertilization, contour tillage, overwinter 'catch crops'). The results of such dual plot experiments provide the only realistic basis for running control scenarios with some full-scale catchment basin models. One could also try to make use of the cultivation and conservation factors (i.e. the P and C factors) of the Universal Soil Loss Equation (USLE; Wischmeier and Smith, 1978). However, the data and information gained in real experimental settings is typically much more accurate.

After selecting representative sites, the next step is to decide the size of the experimental plot(s). In selecting the size, the decisive factor is whether or not one wishes to use rainfall simulators, or rely on natural rainfall. In the latter case, the plot size can be larger (up to several hectares), while in the former case, the plot size must be rather smaller (of the order of a few hundred square meters). Such a decision would depend upon the availability and type of simulator equipment (e.g. pump capacity), and the availability of water.

Experiments with natural precipitation inputs

Ideally, an experimental plot should form a natural, small sub-catchment, with natural watershed boundaries. In many cases, however, the natural topography will not allow this possibility. In such cases, the plot should be bordered by a small earthen dyke to prevent runoff from outside the plot entering the experimental area. The dyke will also direct runoff toward the outlet structure, which contains a device for measuring discharge.

The design of such an experimental plot should begin with an accurate survey; that is, development of a contour map (Figure 7.14). The density of the grid of the leveling survey should be such that it allows detection of the micro-topography of the terrain. This is to insure that a plot with a relatively uniform slope (without surface recessions) is selected.

The inner side-slopes of the small earthen dykes bordering the plot should be lined with plastic sheets at their 'toe', in order to inhibit exfiltration, and to allow for diversion of runoff towards the outlet. The plot should also be of a uniform soil type, characteristic of the larger region.

Rainfall measurements

Rainfall measurements should be made with rain gauges, at least one of which should be of the recording type (i.e. pulviograph). Figure 7.15 illustrates a conventional device, highlighting the standard mounting structure used in Hungary, while Figure 7.16 illustrates a pulviograph. The resultant record of a pulviograph is illustrated in Figure 7.17. The instrument used to generate this record was of the 'siphon' type; after a rainfall of 10 mm, the water content of the upper collector vessel drains automatically into the lower vessel, and the record begins again at zero. It is advisable to use several rain gauges, depending on the size of the plot, in order to allow for detection of an uneven spatial distribution of rainfall.

257

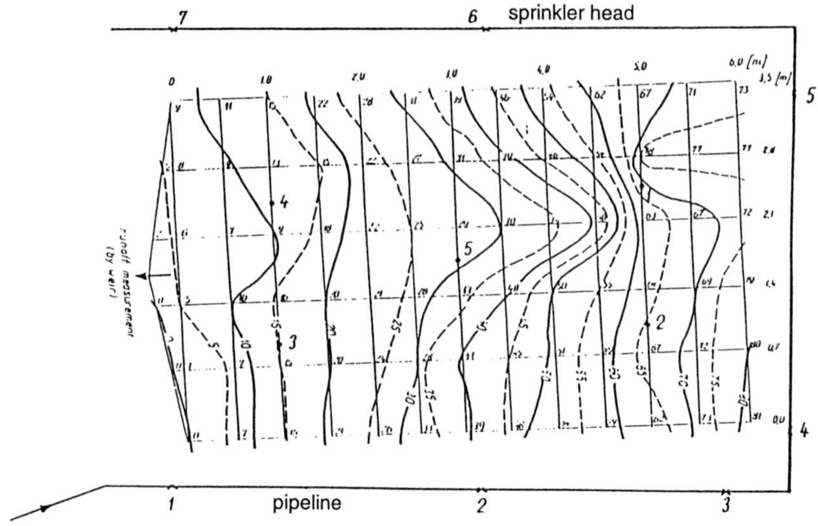

Figure 7.14   Contour map of an experimental plot

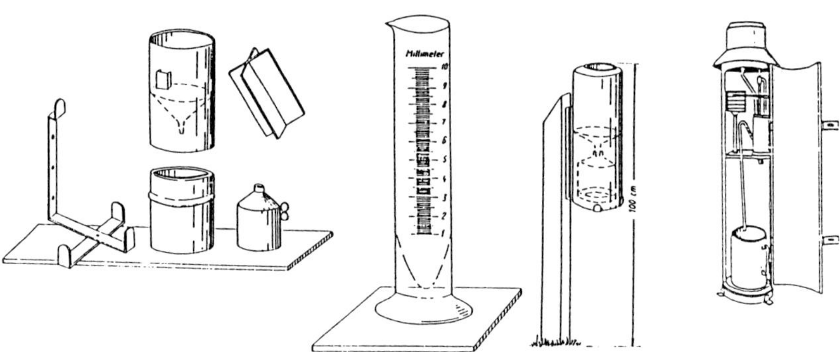

Figure 7.15   Rain gauges

Discharge measurements

Discharge (runoff) measurements can be made in several ways depending on
the magnitude of the expected runoff from the plot. The most usual type of
measuring instrument is a portable measuring weir made of metal plate (Figure
7.9). It can be easily installed in any earth channel, simply by driving it into the
channel bed, and sealing the sides with clay. Written or graphic calibration (flow-
rating) curves are available for these 'Thomson' weirs, and discharge is
calculated as a function of the head (h) over the weir, as:

$$Q = c\, h^{5/2} \tag{7.13}$$

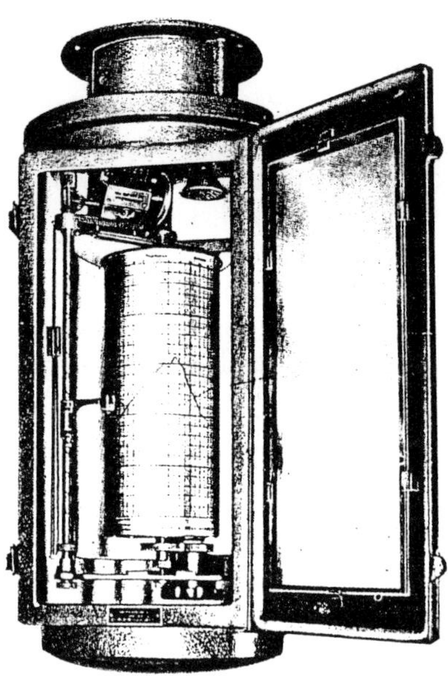

Figure 7.16 Illustration of a pulviograph

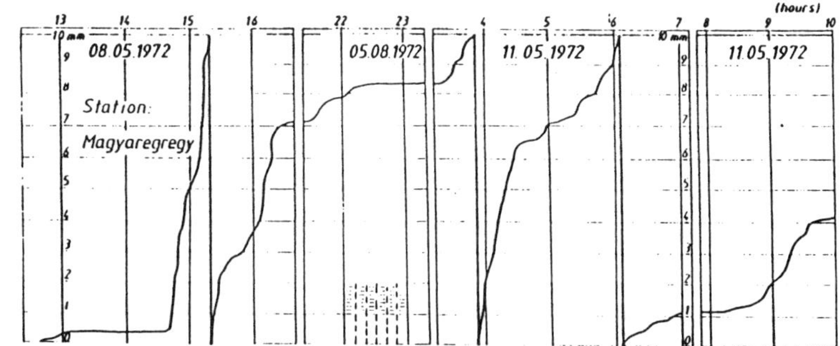

Figure 7.17 Example of a pulviograph chart

where: $c = (2g)^{1/2}$; and

h = head of water passing over the weir

Several other discharge measurement devices are also available commercially, examples being the Venturi and Parshall flumes and various measuring orifices. Alternatively, similar devices can be constructed on the site.

The major problem with measurements of discharge is related to the recording or manual measurement of the overflow depth (h). The ideal solution is to use a water-level recorder. This type of recorder records water depth, and allows for the accurate determination of the total runoff volume, as well as for variations in runoff intensity. However, a very small runoff volume may result in a space too small to install the recorder.

There are many different types of commercially-available water-level recorders, some examples of which are illustrated in Figure 7.18. The two basic types are the mechanical, float-based recorder, and recorders with electrical sensors. Although the installation and operation of these recording devices is both cost- and labour-intensive, they are a fundamental requirement for any accurate plot-sized studies. Indeed, the only other option (manually reading a gauge) is virtually impossible, unless an observer literally lives at the measuring site. (Even with an on-site observer, there is always the possibility of a peak water flow occurring unobserved [e.g. during the middle of the night].)

Water quality measurements

For simultaneously taking water quality samples from runoff waters, the best approach is to use automatic water sampling equipment. The operation of such equipment has been described above; when used in this type of application, the equipment is typically triggered by a critical water-flow condition. Manual collection of water quality samples is possible but often results in the loss of some water quality information. Further, it is often difficult to get sufficient dedicated personnel to collect samples during the nights, or on weekends.

No general rules regarding sampling frequency can be given, primarily because sample collection depends on the duration and intensity of runoff events which can vary from site to site, and between dates at the same site. However, care should be taken to insure that the collection of water quality samples coincides with the runoff peaks, and also temporally over the rising and descending portions of the hydrograph curve. The guidelines for water quality sample preservation should also be observed.

Other measurements

Other relevant measurements include soil moisture before and after runoff events, as well as at regular intervals during the study. The most easily applied instruments are isotope-based moisture probes. Hydrometeorological data (e.g. temperature, wind, evaporation) should also be obtained from the nearest meteorological station.

Information and data should also be collected on: (1) fertilizer application rates (including their actual phosphorus and nitrogen content, and changes in fertilization application patterns over time); (2) tillage and cultivation methods; (3) cropping and harvesting methods (including the nutrient content of the harvested crops and the nutrient content removed from soils by the crops but retained in the field in the form of plant biomass); and (4) the nutrient content of the soil.

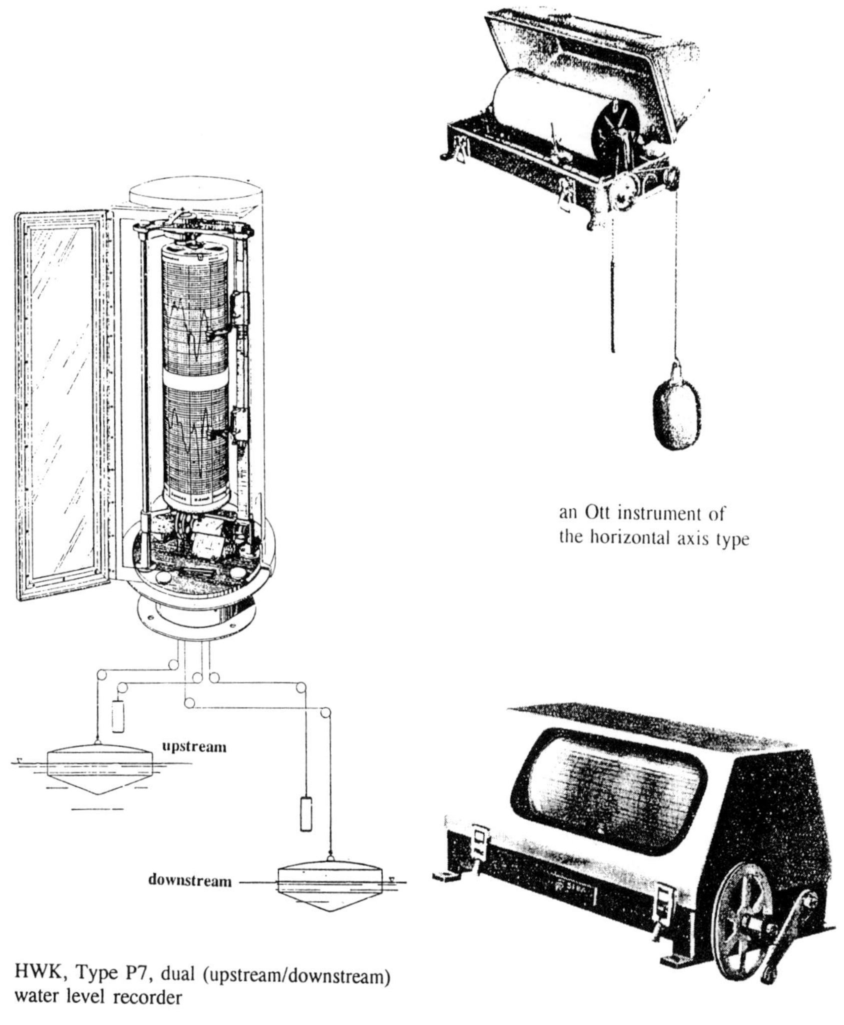

Figure 7.18 Examples of various types of water level recorders (after Starosolszky *et al.*, 1971)

Processing measured data

The processing of the data basically has two objectives: (1) determination of unit area loads (UALs) for various time periods (e.g. monthly, seasonal or annual UALs), and (2) quantification of relationships between the unit area loading rates (and concentrations) and the various factors which affect changes in these rates.

The determination of UALs begins with the calculation of the total load over a certain time period (often a year). This is simply the calculation of the area below the load curve (Figure 7.19), or the product of the water flow (Q) and pollutant concentration (C). Based on these variables, there are several basic cases to be considered in the processing of data. These are discussed briefly below.

    (1) *Continuous flow and time-based quality data*

In this case, the flow record is continuous, and water quality samples are available throughout the entire rainfall event (or for all events) at uniform, discrete time steps; e.g. flow data are extracted from the flow record at discrete intervals with the same time steps. The load time series is calculated as the product of values of C_i and Q_i (Figure 7.19), while the total pollutant load is calculated as:

$$L = \int_{T1}^{T2} Q(t)C(t)dt = \sum_{i=1}^{n} Q_i C_i \Delta t \tag{7.14}$$

where: n = number of time steps (i.e. number of samples) within the period $T = T_2 - T_1$ examined

The unit area load for the period $T = T_2 - T_1$ is calculated as:

$$Y = L/A \text{ (in M.L}^{-2}) \tag{7.15}$$

This corresponds to the mass, M, exported from a unit area, A, during the time period, T. This UAL value then can be converted into any other desired UAL unit (e.g. kg/ha/y; g/km²/day).

    (2) *Continuous flow and noncontinuous quality data*

In this case, the flow record is continuous, but water quality samples do not follow the runoff hydrograph (although the latter is available); two options may be employed to 'complete' the data base:

        (i)   If there is a close correlation between runoff (Q) and concentration (C), then missing C_i values can be calculated from the $C = f(Q)$ relationship (as illustrated in Figure 7.19), and the concentration record completed. The calculation then would proceed as above. It is noted, however, that even close correlations can produce distorted values.

        (ii)   Another possibility is to interpolate between consecutive concentration measurement data, either in a linear manner (e.g. $C_i = (1/2)(C_{i-1} + C_{i+1})$, or by other means (e.g. fitting a nonlinear curve to the concentration measurement data).

    (3) *Noncontinuous flow and quality data*

In this case both the flow and water quality records are noncontinuous (e.g. due to malfunctioning equipment or other problems), unevenly scattered and/or do not follow the assumed runoff pattern; there are again four primary (although uncertain) means of estimating the total pollutant load:

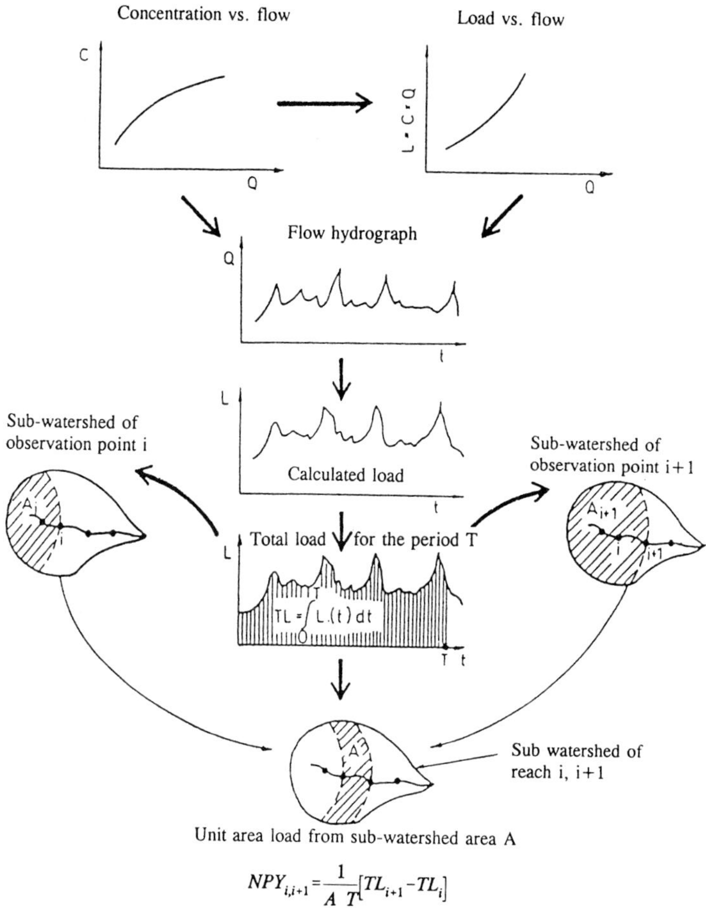

$$NPY_{i,i+1} = \frac{1}{A\,T}\left[TL_{i+1} - TL_i\right]$$

Figure 7.19   Basic types of UAL calculations from measurement data

(i)   If rainfall records are available, some hydrologic models (e.g. the Unit Hydrograph method) may be used to simulate the runoff hydrograph and, subsequently, calculate the load in the same manner as in (2) above.

(ii)   The total load can be estimated as the product of the averages of discharge (Q) and concentration (C) data, and the total time of the measurements, as follows:

$$L = Q \times C \times T \tag{7.16}$$

or as the average of the partial loads (when corresponding data pairs are available in sufficient numbers), as follows:

$$L = (T/n) \sum Q_i C_i \tag{7.17}$$

or, alternatively, as:

$$L = (1/n) \sum Q_i C_i T_i \qquad (7.18)$$

where: T_i = time period(s) over which no sampling data are available.

(iii) When the rainfall records are complete, one can estimate the total pollutant load by correlating the load data with data on rainfall. With this approach, the time lag between the rainfall and runoff events must be estimated (although in a small plot, this lag time may be negligible). The types of relationships that can be used include:

$$L_r = f(I) \qquad (7.19)$$

$$L_T = f(P, M) \qquad (7.20)$$

where: L_r = loading rate (mass/time);
I = rainfall intensity (mm/time; the time unit must be the same as that of the loading rate);
L_T = total load corresponding to a runoff event (mass); $L = f(I, API)$;
P = total rainfall depth corresponding to a runoff event (mm);
M = soil moisture content (or index) preceding the runoff event of concern; and
API = antecedent precipitation index (a measure of antecedent precipitation conditions and, thus, also soil moisture)

To elaborate such relationships, the availability of fairly good records for at least some runoff export events is assumed.

(iv) When there are sufficiently long records (even if incomplete or faulty) available, various statistical methods may be used to estimate the loads produced by unobserved runoff events. These include (a) construction of experimental distribution functions such as a probabilistic simulation model and sampling therefrom, and (b) development and use of flow–load and concentration–load correlations.

Relationships expressing variations in unit area loading rates can also be used in simulation, screening and planning models for larger regions. Such expressions can be elaborated on the basis of the records of the experimental plots:

$$Y = f(R) \qquad (7.21)$$

where: Y = the UAL rate (mass/area); and
R = runoff during the period over which the export was measured (mm, l/ha/sec, etc.) for various crop–soil–slope combinations

and

$$Y = f(P, I) \tag{7.22}$$

where: P = precipitation depth resulting in the unit area export (Y); and
   I = peak rainfall intensity causing the runoff event

An additional, potentially significant variable is the soil moisture or its substitute, the antecedent precipitation index (API). As the soil moisture conditions largely determine runoff (hence, the runoff export), successful event-based simulation of the runoff export process (especially that of the combined events of several sequences) must include a consideration of this variable.

Experiments with rainfall simulators

The advantage of using rainfall simulators in plot-sized agricultural runoff experiments is that runoff events can be 'controlled'. Thus, they can be easily and accurately adjusted for the desired purpose. Even so, there are some shortcomings with these types of experiments, including (1) simulated raindrops (whatever equipment is used) have energy and impact properties that differ from natural raindrops, resulting in simulated loads that only approximate the real situation; (2) potential sites of rainfall and rainfall quantities are limited by the availability of the water supplying the simulator equipment (e.g. by the capacities of the pumps); and (3) long-term UAL rates cannot be easily approximated due to the obvious inability of the simulation to reproduce long-term cycles of rainfall and drought. Rainfall simulators are available commercially, although conventional sprinkler irrigation equipment also may be used for this purpose.

The experiments (including the measurement of rainfall, runoff, soil moisture, etc.) are carried out in a manner similar to that described above in connection with the natural catchments. However, there is rarely need for automatic sampling equipment since experimental conditions are closely controlled; manual sampling and measurement procedures can usually closely follow the changes in both rainfall and runoff. Rain simulator experiments are best suited to the conduct of several simultaneous experiments, with different cultivation, crop management, and fertilization strategies, for example.

URBAN NONPOINT SOURCE POLLUTION

Introduction

As previously noted (see Chapter 6), urban land uses differ from rural and agricultural land uses in that they contain large areas of impervious surface. This consideration modifies the effective delivery of the nonpoint source pollutant load from the urban land surface. The other determinants of nonpoint source pollutant loading rates (the types and quantities of materials used in the watershed, the frequency and duration of rainfall, and the length of time since the previous rainfall) generally influence the pollutant load from all nonpoint sources regardless of origin.

Because of the extensive impervious surface, urban areas are characterized by significantly more rapid hydrologic responses (drainage) to runoff from storm events than other types of land uses. The impervious surface areas also hinder the potential for contact between stormwater pollutants and vegetative cover and the ability of the precipitation to percolate into the soil – characteristics which can ameliorate, to some degree, the impacts of waterborne pollutants, primarily by slowing the rate of water drainage and/or allowing water to move vertically into the soil instead of rapidly draining into receiving waters as surface overflow. Thus, pollutants are less likely to undergo physical and/or biological transformations in urban areas because of their isolation from the soil and its vegetative cover. This affects the manner in which pollutants accumulate in urban areas, and permits a more straightforward (although by no means unequivocal) means of estimating nonpoint source pollutant load from urban areas.

Calculation of urban nonpoint pollutant loads

The accumulation of pollutants in urban regions

Nonpoint source pollutant accumulation in urban areas incorporates all the dry weather processes that can generate or remove pollutants from impervious surfaces. These sources have been discussed in Chapter 6. Dry weather transport of these pollutants may occur as a result of street sweeping, wind erosion (deflation), extraneous water flows, etc. Various models of urban pollutant accumulation have been developed, with most data implying a linear build-up of urban nonpoint pollutants, often expressed in units such as kg/ curb km/day. However, there is ample evidence that urban pollutant build-up can also be nonlinear in some situations. The data of Sartor and Boyd (1972) are often cited as examples of surface pollutant accumulation that tends to approach asymptotic values as a function of the elapsed time since previous street sweeping or storm washoff (Figure 7.20). Ammon (1979) subsequently summarized several studies, suggesting urban pollutant build-up relationships generally fall into one of four functional forms, including (1) linear, (2) power function, (3) exponential, or (4) Michaelis–Menton forms. Examples of these models are as follows:

(1) Linear: $\qquad M = a.t$ $\qquad\qquad$ (7.23)

(2) Power: $\qquad M = a.t^b$ $\qquad\qquad$ (7.24)

(3) Exponential: $\qquad M = M_L.(1 - e^{-bt})$ $\qquad\qquad$ (7.25)

(4) Michaelis–Menton: $\qquad M = ML.t/(a + t)$ $\qquad\qquad$ (7.26)

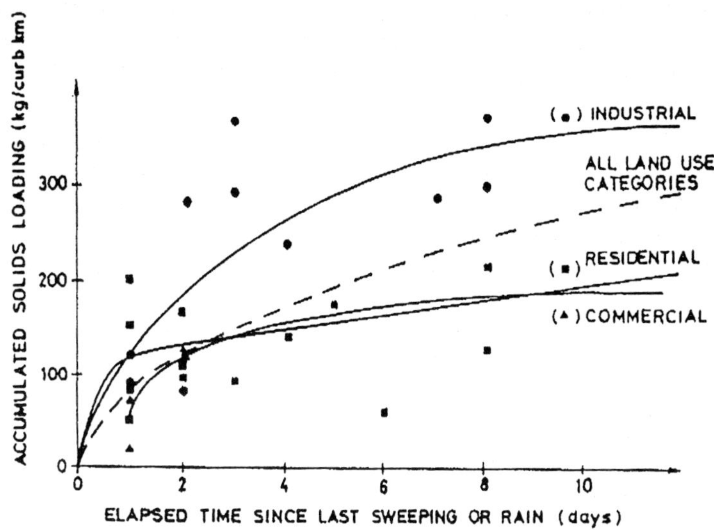

Figure 7.20   Solid accumulation rates (from Sartor and Boyd, 1972)

where: M = mass of pollutant on surface;
   a = coefficient;
   t = time since last cleaning or storm runoff event;
   b = exponent; and
   M_L = limiting (asymptotic) surface load

Hypothetical examples of these four urban pollutant accumulation models are illustrated in Figure 7.21. This example illustrates dust and dirt (or DD) build-up, using arbitrary numerical values for the coefficients (Huber *et al.*, 1981). Although the linear and power equations do not include an upper boundary, one can be imposed arbitrarily. Average pollutant accumulation values for urban watersheds in Chicago (USA) are given in Table 7.4. Most dirt accumulates close to the curb in urban areas. According to Shaheen (1975), dust and dirt accumulation increases with curb height, up to a height of approximately 50 cm. Average amounts of dust and dirt present on street surfaces are presented in Table 7.5.

Urban stormwater pollutant loadings can also be described in terms of determinant factors (Heaney *et al.*, 1976). Table 7.6 illustrates these factors for BOD_5, total suspended solids, volatile solids, total phosphate and total nitrogen. The pollutant loading rates for the two types of storm sewer systems can be expressed as:

(1) Separate sewer networks: $M = A_{(i,j)}.P.F(PD)y$ (7.27)

(2) Combined sewer networks: $M = B_{(i,j)}.P.F(PD)y$ (7.28)

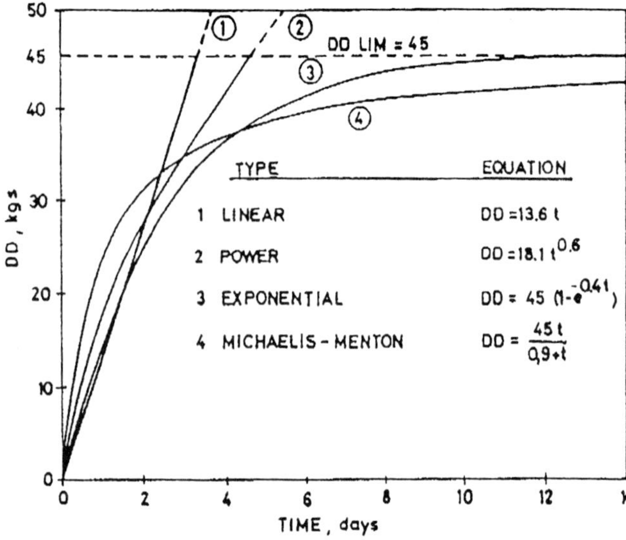

Figure 7.21 Hypothetical build-up rates (from Huber, 1986)

where: M = annual loading of pollutant j from land use i (kg/ha/y);
   P = annual rainfall (mm/y);
   PD = population density (persons/ha);
   y = street sweeping efficiency, defined as: $N_s/20$ where $0 < N_s$
      < 20 days and $y = 1.0$ for $N_s > 20$ days;
   N_s = number of days between street sweeping; and
   $F(PD)$ = function of population density, dependent on land use, as
      follows:
      residential: $F(PD) = 0.142 + 0.134.PD^{0.54}$
      commercial and industrial: $F(PD) = 1.0$
      others: $F(PD) = 0.142$

Runoff associated with storm-event rainfall

On a practical level, the nonpoint pollutant load from urban areas can be
calculated in the same manner as described above for tributaries and streams. A
notable difference between urban settings and rural tributaries or other natural
drainage channels is that the urban drainage network often consists of artificially
constructed storm sewers, concrete-lined channels and storm culverts. An
advantage of these latter systems, in terms of calculating accurate water flow
rates and volumes, is that they typically have uniform dimensions and known
cross-sections. Thus, calculations requiring data on these parameters are likely
to be more accurate than for the more irregular surface features of the natural
drainage networks. In addition, accessibility to sampling sites is typically much

Table 7.4 Pollutant accumulation rates for urban watersheds in Chicago (USA) (from American Public Works Association, 1969)

| | *Dust and dirt (DD) accumulation* | *Pollution rate (mg/g of DD)* | | | | | |
Land use	*(kg/day/curb km)*	*Total suspended solids*	*BOD$_5$*	*COD*	*Phosphate*	*Nitrogen*	*Coliforms**
Residential, single family housing	10.4	1000	5.0	40	0.05	0.48	1.3
Residential, multiple family housing	34.2	1000	3.6	40	0.05	0.61	2.7
Commercial	49.1	1000	7.7	39	0.07	0.41	1.7
Industrial	68.5	1000	3.0	40	0.03	0.43	1.0

*$10^6 \times$ most probable number/g

Table 7.5 Mean values of dust and dirt present on street surfaces (from Sartor and Boyd, 1972)

Location	*Dust and dirt* (kg/curb km)
Residential areas:	
–low activity, old, single-family	240
–low activity, old, multiple-family	250
–average activity, new, single-family	120
–average activity, old, multiple-family	390
Commercial area	80

Table 7.6 Pollutant loading factors (from Heaney *et al.*, 1976)

| | | *Pollutant (j)* | | | | |
Parameter	*Land use (i)*	*BOD$_5$*	*Total suspended solids*	*Volatile solids*	*Total phosphate*	*Total nitrogen*
A (Separate)	1. Residential	0.0353	0.721	0.417	0.00149	0.00579
	2. Commercial	0.1410	0.981	0.617	0.00335	0.01310
	3. Industrial	0.0535	1.290	0.631	0.00312	0.01220
	4. Others	0.0050	0.119	0.115	0.00044	0.00267
B (Combined)	1. Residential	0.1450	2.970	1.720	0.00614	0.02390
	2. Commercial	0.5830	4.060	2.550	0.00138	0.05390
	3. Industrial	0.2210	5.300	2.610	0.01290	0.05040
	4. Others	0.0206	0.491	0.476	0.00182	0.01110

easier in urban regions than in rural regions. Although there are generally more security concerns introduced in the urban setting, the often-enclosed drainage pipes or channels provide for easier temporary or permanent installation of water quantity and quality monitoring equipment. Thus, while these factors suggest (though do not always ensure) that calculations of pollutant loads from urban areas should be more accurate than for rural areas, there is an undiminished need for diligent sample collection that is not offset by the presence of a 'better' measuring situation.

Having pointed this out, consideration must also be given to the factors unique to urban areas that can affect the nonpoint pollutant load. The export of nonpoint source pollutants from the impervious surfaces in urban areas in stormwater runoff also depends on several other factors:

(1) *The type and condition of the street surface*   More dirt and other pollutants are washed off smooth, worn surfaces than off rough, new asphalt surfaces;

(2) *The rainfall intensity and volume*   Particle entrainment, through the impacts of raindrops and the energy of surface runoff water, provides the main pollutant removal mechanism. However, the necessary threshold pollutant movement forces are higher on porous surfaces – low rainfall intensities and volumes enhance particle aggregation and adsorption of soluble materials;

(3) *The size of pollutant particles*   There is little significant difference in the removal rates of pollutant particles of different sizes. This may reflect the 'light' organic nature of many large particles, compared to the 'heavy', inert character of small, clay-sized particles (Ellis *et al.*, 1982); and

(4) *The acidity of the precipitation*   Initial low pH values may lead to dissolution of heavy metals. However, as noted above, the runoff is also subject to the buffering capacity of the street surface.

Under the assumption that the quantity of pollutants washed off impervious surfaces is proportional to the total mass of pollutants on the surface, the nonpoint source pollutant load associated with precipitation and resultant urban runoff can generally be modelled as:

$$M - M_0 = M_0.(1 - e^{-krt}) \qquad (7.29)$$

where: $M - M_0$ = removed mass (kg);
$\qquad\qquad M_0$ = initial amount of pollutants (kg);
$\qquad\qquad t$ = time elapsed (h);
$\qquad\qquad k$ = constant; and
$\qquad\qquad r$ = runoff rate (mm/h)

The constant (k) is a function of rainfall or runoff excess. To determine the value of k, it was assumed that a uniform runoff of 13 mm/h will wash off 90 percent of the initial pollutant load in one hour (Wanielista, 1978). This assumption produces a value of $k_{(i)} = 0.181$ (for impervious surfaces) and $k_{(p)} =$

0.0551 (for pervious surfaces). This basic formulation is still used in many nonpoint source runoff models (e.g. SWMM, STORM; see also Chapter 8).

The major limitation of Equation 7.29 is that it can be applied only in situations where the rate of stormwater runoff is constant. Therefore, discretion is needed in selecting the particular time interval (Δt), during which a constant runoff rate can be assumed. (The initial quantity of surface pollutants can be determined using the equations for separated or combined sewer networks (see Equations 7.27 and 7.28).)

Urban pollutant loads associated with street sweeping

In contrast to rainfall-mediated runoff, removal of urban nonpoint source pollutants under dry conditions is done primarily by street sweeping. Street sweeping removes accumulated dust and dirt from street surfaces (usually for aesthetic purposes). On rarer occasions, it is also used to avoid clogging sewer inlets. The effectiveness of street sweeping in removing urban nonpoint source pollutants (Uunk, 1983) is a function of:

(1)   Frequency of street sweeping;
(2)   Number of passes of street sweeper;
(3)   Speed of street sweeping equipment;
(4)   Type of street sweeping equipment;
(5)   Pavement conditions; and
(6)   Presence of obstacles (e.g. vehicles) along the curb.

The removal rate of conventional equipment (depending on the particle size of the street dirt) is given in Table 7.7. The particle-removal efficiency of broom-type sweepers depends on particle size (Table 7.7). In contrast, vacuum sweepers can achieve good removal rates through the full range of particle sizes. For example, vacuum sweepers can attain a 95 percent removal efficiency of fine particles. However, they are more expensive to operate than broom-type sweepers.

Sartor and Boyd (1972) described street dirt removal with street sweeping equipment, as follows:

$$M = M^* + (M_0 - M^*).e^{-KET} \tag{7.30}$$

where: M = remaining dirt (g/m^2);
M_0 = dirt initially present (g/m^2);
M^* = non-removable dirt (g/m^2);
K,E = removal constants (equipment-dependent); and
T = 'sweep minutes' per surface unit (minutes/m^2)

Calibration of Equation 7.30 for a concrete surface of 2.5 m width, and a speed of 10 km/h, yielded good results for particles larger than 104 µm, with M_0 = 0.45 g/m^2; M = 0.05 g/m^2; K = 300; and, E = 2.5 × 10^{-3} m^2/minute.

The effect of sweeping interval on BOD and suspended solids removal from urban streets is shown in Figures 7.22 and 7.23. Pitt and Shawley (1981)

Table 7.7 Street-sweeper removal rate versus particle size (one pass, broom-type sweeper) (from Sartor and Boyd, 1972)

Particle size (μm)	*Removal* (percent)
2000	79
840–2000	66
246–840	60
104–246	48
43–104	20
<43	15
Overall	50

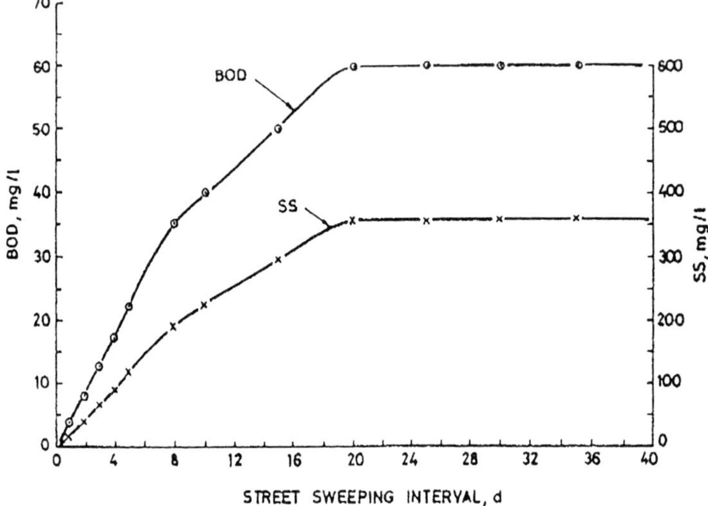

Figure 7.22 The effect of sweeping interval on BOD and SS (from Heaney *et al.*, 1976)

suggested that the cleaning interval was the most important factor in regard to particle-removal effectiveness. Studies in the United States, however, suggested that sweeping generally was an ineffective technique for improving the quality of urban stormwater runoff (US Environmental Protection Agency, 1983); Novotny *et al.* (1985) indicated that a statistical comparison of pollutant concentrations in urban runoff from test (swept) and control (unswept) sites showed only a very small (statistically insignificant) effect of street sweeping on reducing urban nonpoint pollution loads, primarily because the efficiency of street sweepers was generally low, often being negative with low street solids loads. A computer model simulation confirmed these findings. Notwithstanding this lack of effectiveness in improving the quality of urban stormwater runoff

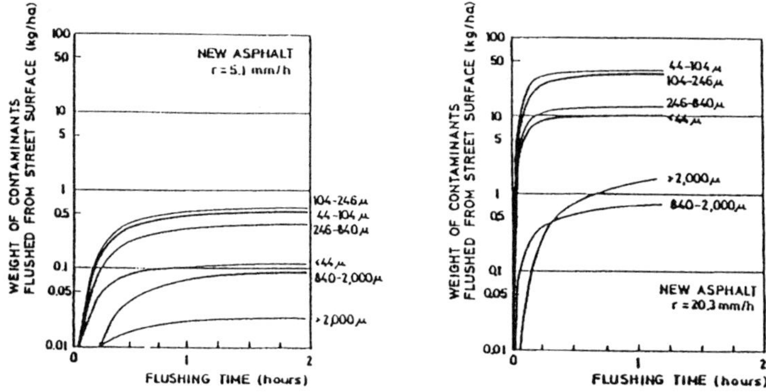

Figure 7.23   Results of wash-off experiments (from Sartor and Boyd, 1972)

(or for reducing the biologically-significant urban nonpoint pollutant loads to receiving waters; see further discussion in Chapter 9), street sweeping retains its usefulness as a means of removing the larger, aesthetically offensive particles and litter from urban regions.

GROUNDWATER NONPOINT SOURCE POLLUTION

Introduction

Groundwater, as indicated by its name, flows under the land surface (in contrast to the surface waters discussed heretofore which flow over the land surface). For this reason, assessment of groundwater pollution from nonpoint sources is typically more difficult than assessing surface water contamination. This is because of several factors including: (1) the slow movement of water through the ground, (2) the difficulty of clearly defining groundwater aquifer boundaries and/or water movements, and (3) the differing chemical conditions often existing in the subsurface environment. In addition, contaminated groundwaters are considerably more difficult to treat than surface waters (see Chapter 9).

Surface and groundwaters are inseparable parts of the hydrological cycle (see Chapter 4). Groundwaters can be contaminated from the same pollutant sources as surface waters, as well as from sources unique to the ground system (e.g. injection wells). At the same time, groundwater is a fundamentally important component of the usable global water resources, amounting to about 90 percent of the world's freshwater resources and supplying about half of the drinking water supply of the United States (Schalf *et al.*, 1981). Further, the hydrological connections between surface streams and groundwater aquifers fundamentally affect each component (see Todd, 1961; Schalf *et al.*, 1981; US EPA, 1987). As described by the US EPA (1987a), there are three basic types of streams; namely, ephemeral, intermittent, and perennial. Ephemeral streams, which receive their entire flow from surface runoff, are generally distinct from groundwater flows.

However, intermittent and perennial streams are fundamentally influenced by the relationship between the depth of the water table and the stream channel bottom. Although intermittent streams flow only for part of the year, during the dry season, their flows are comprised entirely of groundwater discharged into the stream channels. This occurs because the depth of the groundwater water table is above the bottom of the stream bed. Perennial streams are simply an enhancement of this phenomenon, with the water table remaining above the stream bottom throughout the year. As a result, these streams flow throughout the annual cycle.

Basic sampling procedure

The basic sampling procedure for groundwater typically involves siting and drilling observation wells (including lining of the well channel), and extraction and analysis of water samples.

Siting and drilling of wells

Proper spatial and vertical locating of observation wells is of fundamental importance to obtaining accurate groundwater samples. The number of sampling wells in any monitoring programme is dictated by the purpose of the programme. Dissolved constituents in surface waters typically descend vertically through the unsaturated soil layer immediately below the land surface until they meet the saturated zone or water table. Within the water table, the dissolved constituents will move horizontally in the direction of prevailing groundwater flow. The logical placement of sampling wells will therefore be downgradient of the first permeable, water-bearing underground soil layer. Consideration should also be given to seasonal and/or artificial fluctuations in the level of the water table when siting the observation well. If several water-bearing layers exist at different depths, then multiple vertical depth wells may be necessary to adequately characterize the groundwater resources. In addition, other factors (e.g. the pollutant of concern, the existing underground chemical conditions, the regional hydrogeology) can be important in the proper placement of groundwater monitoring wells, and should be considered in a site-specific manner.

Various methods of drilling monitoring wells exist, including mud and air rotary drills, chisel-type pounding rigs, and augers. These are discussed in detail by Schalf *et al.* (1981). Operational concerns associated with the drilling of groundwater observation wells include: (1) the diameter of the well or well casing, (2) the depth of the well, and (3) the nature of the intake structure of the well (well screen). The appropriate diameter for a groundwater monitoring well depends primarily on the diameter of the sampling device, if any, to be lowered to the water-bearing layer(s). A 4- to 6-inch (10–15 cm) diameter well is often sufficient to obtain necessary water quality samples for analysis. Larger diameter wells are typically only useful in situations where remedial measures are contemplated, or where special requirements (e.g. great depths) must be met.

Because of the potential for soil layers of different permeabilities, as well as other hydrogeological considerations, monitoring wells should be constructed to be discrete-depth; i.e. to sample only one specific water layer of concern, without hydrologic connection to other underground water layers. This often means that some type of material, such as cement grout, dry bentonite, or bentonite slurry, must be placed in the well casing structure at a depth above and below the water layer of concern to restrict water flows into the intake portion of the monitoring well. Monitoring wells are typically sealed both within or near the saturated soil zone to isolate the water intake opening and at the ground surface to inhibit downward migration of surface contaminants. The lining of wells with well casings, as well as provision of an intake screening structure through which water enters the well casing, also contribute to the discrete-depth operation of the observation well. The casing and the screen materials must retain their structural integrity and be non-reactive with the water quality parameter(s) of concern being sampled from the well. For example, a preliminary ranking, from best to worst, of commonly used materials for establishing monitoring wells near hazardous waste sites includes Teflon, stainless steel, polyvinyl chloride (PVC), low carbon steel, galvanized steel, and carbon steel (Barcelona *et al.*, 1984). Such factors as expected pollutants, hydrogeological conditions, and expense must be factored into the selection of an appropriate construction material from this list.

Collection of groundwater water quality samples

The accurate collection of groundwater samples for water quality analysis from groundwater monitoring wells may be fraught with difficulties (*cf.* Barcelona *et al.*, 1984; Stolzenburg and Nichols, 1985; Ho, 1983). However, there are three fundamental steps that, if employed, can minimize these difficulties; namely (1) measurement of the well water level before sampling, (2) purging the well before sampling, and (3) collecting the sample in accordance with an established groundwater sampling protocol. Measurement of water depth can be done with a flexible tape measure or other graduated device. Knowledge of the water depth can allow a more accurate assessment of the purge volume, or volume of water necessary to be pumped from the well prior to taking a sample at the surface. Purging or flushing is done primarily to remove stagnant water from the monitoring well prior to collection of a sample. Upon completion of well purging, the well can be expected to contain fresh water from the aquifer water layer of interest (e.g. greater than 90 percent aquifer water). The recommended pumping or bailing time is related to (1) aquifer hydrogeology, (2) the type of sampling equipment, and (3) the water quality parameters being sampled. A common procedure is to purge a volume of water equal to between four and ten well-bore volumes. However, others, such as the US EPA (1987), caution that the purge volume should be more precisely calculated from an analysis of the field hydraulic conductivity measurement which will allow the well's individual hydraulic characteristics to be the guiding factor.

Water samples are most commonly obtained from monitoring wells with bailers or pumps. A bailer can be a collection container such as a Kemmerer sampler or other container that is lowered into a well, typically by hand, to the water-bearing layer. Once filled, the collection container can be pulled back to the surface along with the enclosed water sample. Advantages of bailers include their low cost, ease of construction (from a variety of materials) and lack of requirement for an external power supply. Disadvantages include their general clumsiness and the possibility of contamination during the deployment and retrieval process. Pumps, in contrast, can be either suction-lift pumps or submersible pumps. If the water layer of interest is within the suction depth (less than about 20 feet or 6 m) of the land surface, a centrifugal or peristaltic pump may be used to obtain a water sample. Hand-operated diaphragm pumps can also be used in some situations. Advantages of the suction-lift pump include their wide availability, relatively low cost, and portability. Noteworthy disadvantages include their requirement for an external power supply and restriction of use to within 20 feet of the sampling point. Portable submersible pumps overcome this latter disadvantage by being lowered into the monitoring well to the depth of the water layer of interest. These pumps are effective for withdrawing water at depths of several hundred feet (up to about 100 m). Advantages of submersible pumps include their portability and high rates of pumpage; disadvantages include the need for a typically larger well casing diameter, greater logistical difficulties associated with hoses and cables, and their unsuitability for obtaining samples of organic substances. An alternative to pumping water from the ground is to force it to the surface by air pressure applied to the monitoring well. This is typically done with an air compressor or high pressure air pump. Advantages of this method include the possibility of using either portable or permanently installed pumps, and the ease of both purging the well and collecting water samples. Disadvantages include the possibility of chemical reactions mediated by the forced aeration/oxygenation of the sample. These and other water sample collection techniques are discussed further by the US EPA (1987).

IN-STREAM AND IN-LAKE NONPOINT SOURCE POLLUTION

Introduction

In Chapter 6, the role of the sediments as sources and sinks of contaminants was noted. In particular, the deposition–resuspension of sediments and associated contaminants from within stream beds and the more complex processes at work in lentic waters were reviewed. It was observed that these internal processes can substantially affect and even alter the response of these watercourses to management actions, although the effects were more pronounced in terms of sediments and nutrients than for metals and other contaminants. Estimation of these processes is less easy than estimation of external contaminant loads; for this reason, estimation of internal loading in lakes is usually carried out by difference in mass-balance calculations, while in-stream processes are usually lumped with the external load to lakes (see above). Neither of these methods is particularly satisfactory if the sources of the contamination are subject to pollution control mandates or other legal proceedings – it would be too easy for the litigant to point out the potential for error in these rather coarse estimations. For this reason, more precise measurements may be required. Some common measurements techniques are, therefore, presented below.

In-stream load measurement

It has previously been noted that many contaminants are adsorbed onto particulates, the level of contamination being a function of sediment grain size (i.e. surface area available for adsorption) and mineralogy (i.e. the nature and intensity of the particle charge). For this reason, Walling and Moorehead (1989) make a persuasive case for conducting a more detailed investigation of the nature of the material being transported in streams than is usually done when estimating sediment transport to waterbodies (e.g. UNESCO, 1987). They also note that these characteristics are also indicative of the recent geological history of the particulates.

While there is an obvious connection between the particle size distribution recorded in a particular river system and the regional soils and geology, Walling and Moorehead (1989) suggest that, within these regional level constraints, two further processes are at work; namely, selective erosion and selective deposition. Selective erosion may be a significant process in certain areas – Meyer *et al.* (1975) and Lal and Stewart (1994) report enrichment of the suspended load with fine particulates (silts and clays) in preference to sand in the US mid-west and Nigeria, respectively – but it rarely results in a deposition product that is indistinguishable from the parent material. In other words, the eroded material still bears a recognizable relationship to the original soil from which it was eroded. Where this has not been the case, the erosion site was generally on disturbed soils or steep slopes. In short, the process of erosion generally results in the delivery to the stream of an identifiable group of particulates, the product of slope erosion.

This distribution can be greatly altered once the particulate load enters a water course. In contrast to the general lack of selective erosion, selective delivery can greatly modify the particle size distribution between the point of entry of the erosion products and the point at which the particulates ultimately leave the system. Employing no other process than the application of Stokes' Law (i.e. independent of flow; see Krumbein and Sloss, 1963), the heavier particles will progressively sink from suspension leaving the finer particulates to be carried from the system. These fine particulates thus represent the expected result of terrestrial or slope erosion. Where there is a high concentration of coarse particulates in the stream sediment load, identification of the stream bed as the source of these materials (channel erosion or scour) is likely. Confirmation of the origin of the particulates can be gained from examination of the particle size distribution under high and low flow:

(1) Where there is no change in the particle size distribution with increasing flow, the sediment load is probably source-limited and independent of flow;

(2) Where there is an increase in fine particulates within the suspended sediment particle size distribution with increasing flow, the sediment load is probably dominated by slope erosion; and

(3) Where there is an increase in coarse particulates within the suspended sediment particle size distribution with increasing flow, the sediment load is probably dominated by channel erosion (or internal stream bed sediment and contaminant loading).

For purposes of operational definition, fine particulates may be considered silts and clays, while coarse particulates may be considered sands and large (> 63 µm) aggregates; aggregates, which alter the settling characteristics of the particulates, are most common in systems where clays and organic materials predominate. Walling and Moorehead (1989) cite examples of these processes from various rivers throughout the world and across most climatic zones, including Africa, Asia, Australia, Europe, the Middle East, and North and South America.

The other common technique used to determine the origin of sediments and contaminants is to identify a unique tracer occurring in the area of concern; saline tracers are most commonly used, although isotopes or other elements may also be used for this purpose (Boyle *et al.*, 1974). Where there are no sources or sinks within a system, the dilution curve for these elements (Figure 7.24) is a straight line. Should a source be present in the system, whether a tributary or in-stream source, the curve will become concave, while a sink will have the opposite effect. A tributary will result in a curve having two distinct slopes. In the example shown in Figure 7.24, the concentration of the substance of concern is plotted against the concentration of a conservative substance, usually salinity or conductivity. Obviously, these curves work best where there is a significant concentration gradient, such as in estuaries.

In-lake load measurement

As noted above, many estimates of in-lake loading, or internal loading as it is more commonly known, are derived by difference from mass balance calculations, where input, output and in-lake element masses are measured – inputs are commonly external [riverine] loads, although some include atmospheric loading which may be significant for some elements (see Chapter

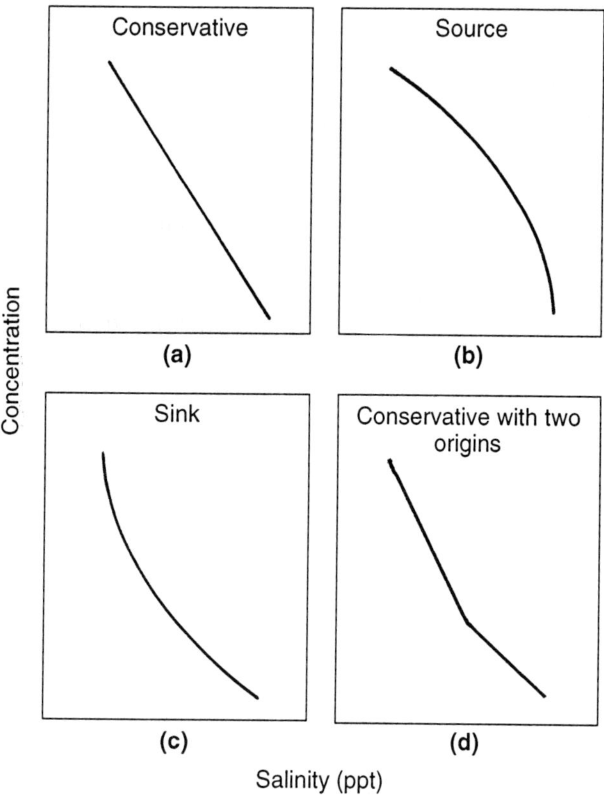

Figure 7.24 Salinity diagrams depicting conservative and non-conservative mixing processes as referred to in the text (after Boyle *et al.*, 1974)

6). Where the in-lake element mass exceeds the sum of the inputs, the difference is rightly assumed to have been generated within the waterbody (Ryding, 1982), primarily from the sediments (see Thornton *et al.*, 1995, for a recent review of the processes involved; while some of the element may be regenerated through the lysing and/or decay of organic matter in the water column, the largest portion of the internal load is released from the sediments through resuspension and remobilization: see also Chapter 6).

More direct measurement of the sediment–water exchange of elements (and contaminants) is possible. Several variants of the standard 'bell jar' technique have been published in the literature (see Glibert, 1976; Lee *et al.*, 1977; Thornton, 1979). These chambers enclose a discrete area of sediment and volume of overlying water for a period of time, isolating the effects of sediment–water exchange and removing the effects of hypolimnetic mixing which tend to dilute the effect of sediment–water exchange given the typical existence of a strong concentration gradient between the interstitial or pore

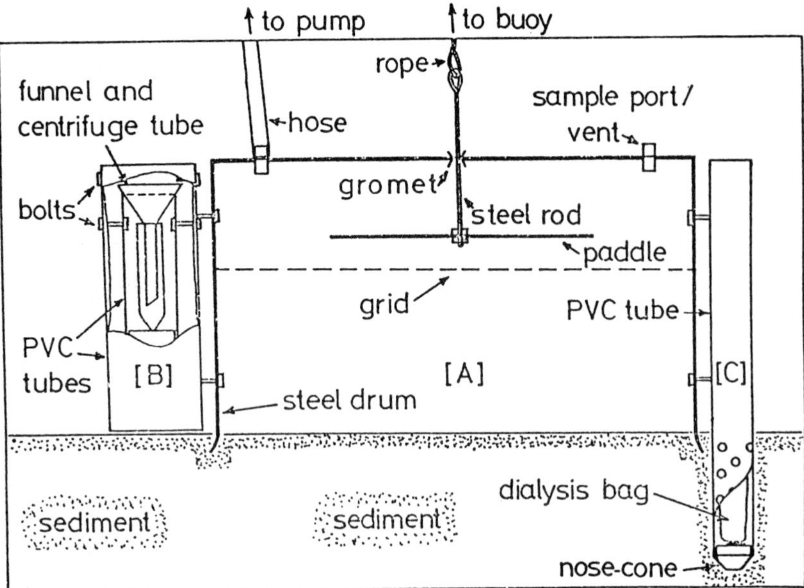

Figure 7.25 Cut-away diagram showing the construction and disposition of the bottom sampling chamber [A] and related experiments; [B] sediment trap and [C] modified corer for determining interstitial water phosphorus concentrations (from Thornton, 1979)

waters within the sediment and the overlying waters of the lake. A typical bell jar is illustrated in Figure 7.25. These chambers may be constructed of various materials, such as plexiglass [perspex] or steel, but, where constructed of a material that may adulterate, any samples subsequently obtained (i.e. in the case of a steel chamber), should be epoxy-coated with at least two or three coats of a good quality epoxy paint to minimize the chance of contamination. A stirring mechanism is usually included to homogenize the water within the chamber prior to sampling. While such chambers can be used to measure a variety of contaminant exchanges, they are normally employed in the measurement of nutrient exchanges which are likely to be the major sediment–water exchange process (Förstner *et al.*, 1984).

Other methods proposed for the measurement of sediment–water exchanges include the use of conservative tracers injected into surfacial aquifers or other porous strata through small-scale injection wells located on the lake shore. Water samples are then collected at close intervals along transects across the littoral zone and the quantity of dye or tracer measured. Using standard dilution formulae, the rate of transfer from the shoreland to the lake bed is calculated. These techniques are normally used in septic seepage surveys seeking to identify malfunctioning private wastewater systems. Tracers range from rhodamine dye to ammonium concentrations to conductivity values (i.e. based on salts such as bromide or chloride). Twinch and Breen (1982) also used radio-tracers to measure ^{32}P uptake and release from sediment cores. Measurements of sediment release or uptake (exchange) are normally reported in either mg/l of overlying water – which can be multiplied by the hypolimnetic volume to determine the relative mass of element released from the sediments – or mg/m^2 of lake bottom – which can be multiplied by the area of lake bottom exposed to the conditions pertaining in the bell jar to determine the mass of element released (or taken up).

The methods described above measure sediment–water exchange based on the remobilization of elements from within the sediments and released to the overlying water through diffusive processes. Measurement of release due to resuspension and regeneration is more difficult, although tracer techniques can be used with some success. In the first instance, aliquots of sediment and overlying water (or distilled water with [Twinch and Breen, 1982; Grobler and Davies, 1979] or without [Thornton, 1979] an extractant such as EDTA) can be agitated on a shaker table in simulation of particle resuspension caused by wind or wave action. Element concentrations are then measured over time with the result reported as mg/g of sediment – measurement of suspended sediment concentrations *in situ* then allows estimation of the relative magnitude of element release due to resuspension.

Regeneration can be measured using radio-tracer techniques. Inoculation of a bacterial (Robarts, 1988), algal (Thornton, 1989), or zooplankton (Jarvis, 1986, 1987) sample with a radioactive tracer – tritium, phosphorus, nitrogen and carbon are most commonly used – 'labels' the organisms with small amounts of radioactive element which can be detected using a scintillation counter. Transfer of the organisms, using various filters or mesh screens to exclude larger or smaller organisms, to 'clean' isotope-free lake water (or distilled water, although this does create greater osmotic pressure across the cell membranes of the organisms due to the steeper concentration gradients between the organism and medium) permits estimation of biotic release – by measurement of the concentration of free isotope in the aqueous phase of the medium. Poisoning the medium causes mortality among the organisms and allows estimation of the regenerative losses due to senescence. Results of these assays are normally reported in mg/l of aqueous medium or in mg of element released. Multiplication of this mass by the volume of epilimnion or euphotic zone can produce an estimate of regenerative loading. While these loads are typically small compared to remobilization and resuspension-mediated loads, they may be significant in smaller oligotrophic lakes where such turnover rates are high and nutrient concentrations low (Lean and White, 1983; Lean and Nalewajko, 1979).

Recent advances in lake ecosystem modelling have resulted in a variety of predictive methods which can allow estimation of these in-lake processes to a degree that might be suitable for lake management purposes (although the models remain too coarse for scientific, element cycling study purposes). Janes and Hotchkiss (1990) present a compilation of such models.

DETERMINATION OF NECESSARY NONPOINT SOURCE POLLUTANT REDUCTIONS

Introduction

Thus far in our discussion of methodologies for measuring nonpoint source contaminant loadings, we have dealt with methods designed to quantify the contaminant load to a waterbody. Having derived the load, it remains to determine (1) if the load is great enough to cause impairment of the receiving water and, if so, (2) the magnitude of reduction that should be effected to reduce the contaminant load to within 'acceptable' limits – acceptable in this instance can be defined by legal requirement or waterbody–ecosystem requirements, depending on the presence or absence of any statutory requirement. These determinations will be the focus of this portion of the text, which will be devoted to providing some practical approaches to determining impairment and the degree of response necessary.

Assessing waterbody impairment

There are typically three approaches to determining waterbody impairment; namely, a regulatory approach based on the blanket application of standards or other regulatory instruments; a use-based approach; and an ecosystem-based approach. In many cases the regulatory approach parallels the use-based approach, although it can be shown in many cases that the regulatory approach is the least successful in achieving an 'acceptable' water quality. Each of these approaches is discussed in turn below.

Statutory requirements

Chapter 5 summarized some sample standards, guidelines and criteria used to assess water quality in various parts of the world. The differences in focus between drinking water standards and effluent discharge standards was pointed out. Drinking water standards are often misapplied in the assessment of ambient water quality in streams and lakes (granted, many streams are used for drinking water supply purposes, and in some instances it may be appropriate to apply drinking water guidelines to natural waters; i.e. in cases where water is used in a raw state). Thus, it is important, in the first instance, to apply the correct or applicable standard or guideline when assessing water quality. If these standards do not exist, the reader is referred to one of the other methods set forth below, or to Chapter 5 to develop a regionally relevant standard to suit their local conditions.

In the second instance, it should be recognized that the law is based on some measure of societal equity; in other words, all actors engaged in similar activities should be treated the same way. This typically results in the development of blanket standards, or legally promulgated requirements that affect the entire nation. While some countries have adopted regionally-based standards (e.g.

Alabaster, 1980; Fuggle and Rabie, 1986), with different levels of discharge permitted in sensitive areas than are permitted in other areas, most nations have approached the setting of standards on a national basis. In the United States, for example, this has resulted in application of certain water quality requirements in the semi-arid southwest that are best suited to – and, indeed, were developed in – the temperate northeast (*cf.* Thornton and Rast, 1993). While this may not seem too important, such an application of inappropriate standards can result in very large, and unnecessary, expenditure – in the situations examined by Thornton and Rast (1993), the expenditure would probably be in the form of nutrient removal, advanced wastewater treatment plants designed to reduce phosphorus concentrations in effluents to below the level generated in southwestern impoundments from undisturbed, natural lands.

Few legal requirements are based on *in situ* water quality. The State of Oregon (USA) has implemented water quality standards based on minimum permissible levels of specific water quality indices in various stretches of the eight or so river basins traversing the state. Where these waters support sensitive wildlife populations (i.e. salmon fisheries), these standards require maintenance of relatively higher levels of oxygen and lower levels of P and N than in other stream reaches. Other states and nations are experimenting with TMDLs (total maximum daily loads) or waste load allocation schemes which are basin-specific. These schemes seek to achieve a minimum level of water quality compatible with subsequent downstream usage. In some cases, these uses are 'environmental' (support of riverine life) but more commonly they are 'beneficial' (human uses). The latter schemes tend to be based on an average acceptable water quality that will permit treatment using best economically available technologies or treatment methods that can be employed at an acceptable level of cost to users. In South Africa, this level has been determined to be the level of comprehensive treatment (which includes chemical dosing and sand filtration but stops short of activated carbon filtration and reverse osmosis – see Ryding and Rast, 1989:273, for the definitions of levels of treatment; see DWA, 1986, for a discussion of waste load allocations in the Republic of South Africa). While this basis for the application of standards is related to use, it differs from the use-based assessment (see below) in that its assessment is based on use in general and not a specific use or set of uses.

To apply the method of water quality assessment to a system, samples are usually collected and analyzed in the manner specified in the legislation that created the standards and the results compared to the standards. If they exceed the standards, then the water quality is poor and mitigation is required to bring the measured values in line with the standard – occasionally standards are written in such a way as to permit some exceptions, say 2 out of 5 samples in a month can exceed the standard by 50 percent, but generally a simple comparison is needed to describe the compliance or otherwise of a water sample, and, thereby, a waterbody. This then is the second great advantage of water quality assessment based on standards – not only is it equally and equitably applied across the board, it is also right or wrong/good or bad depending on whether it is below or above the standard. This system is very simple to use. Its greatest drawback lies in this same simplicity; being based on 'average' uses or some technology/cost-driven compromise, it rarely satisfies anybody in reality and engenders a situation where compliance is all that is required of a contaminant generator (see Ryding, 1992).

Use-based requirements

The concept of beneficial use underlies use-based water quality criteria and, generally, refers to a desired use which may or may not be the existing use of a stream segment. Determination of beneficial use considers existing use, anticipated future use, various environmental factors that may place use constraints on the stream segment (or, *vice versa*, which may earmark a stream segment for a particular purpose), societal desires, economic value, and the feasibility of attaining the desired water quality. Beneficial uses can be categorized as follows (Ryding and Rast, 1989:272–3):

(1) Drinking or potable water;
(2) Industrial water;
(3) Cooling water;
(4) Irrigation or agricultural water;
(5) Recreational (partial contact) waters;
(6) Swimming or bathing (full contact) recreational waters;
(7) Fishery waters;
(8) Waterfowl or migratory wildlife waters;
(9) Boating or navigation-related waters; and
(10) Miscellaneous waters (any other purpose not included above; e.g. hydropower, etc.).

The drinking water standards given in Chapter 5 are one example of use-based water quality requirements. Use-based water quality criteria can be based either on the quality of water prior to treatment, if any, or following various levels of pre-treatment (as suggested by Ryding and Rast, 1989). Criteria are generally applied such that the quality requirements of the highest use are satisfied; for example, if water in a multi-purpose impoundment is used for agriculture, industrial process water and recreation, the recreational use standards would apply as these are likely to be the most stringent. Quality criteria are established for an entire basin or discrete stream reach. If they are established for a discrete stream reach, water quality criteria for each segment are usually structured in such a way that the highest quality reaches flow into lower quality reaches. Beneficial use criteria can be used in association with waste load allocations or TMDLs.

The benefit derived from employing use criteria is that they tend to be more site specific. They can, for this reason, exert some influence over the situation of industries and other large generators of contaminant loads by making it too expensive to treat process or storm waters to the level required for discharge into receiving waters with a high quality use designation. They also ensure an acceptable quality for all users located on the watercourse. Where this system breaks down is in terms of changes in highest beneficial use over time – if the highest use diminishes, there is no harm, but if the highest use increases it can be difficult to amend controls already in place. The application of beneficial use criteria becomes unwieldy in the face of changing conditions, and can be administratively difficult to manage even under normal conditions; determining where to draw the boundaries between use areas can appear arbitrary, even if drawn on geographical grounds, and hence the system is open to political, legal and societal abuse – especially if the legal system, as is the English legal system, is based on the concept of social equity. While variants of this scheme have been used (see, for example, Thornton *et al.*, 1995: a user survey was combined with regulatory and observer-assessed water quality criteria to derive a management plan for Zandvlei, an urban, South African coastal lake), few countries seem to have adopted the concept of beneficial use in any serious way.

Environmentally-based requirements

At first glance, environmentally-based water quality requirements appear to be a special case of use-based requirements, and, indeed, they could well be – environmentally-based use requirements are typically a cross between the use-based system and regulatory systems. These criteria have been proposed with respect to water quality, and variants have been proposed with respect to flow (Gore, 1989). Environmentally-based criteria can be applied in an across-the-board manner, regionally, or on a basin–stream reach basis. The major difference between environmentally-based criteria and other water quality assessment systems is that these systems are based on the natural environment, not some human use or arbitrary compromise value determined through the legislative process. If the former are set by technocrats and the latter by politicians, environmentally-based standards are determined by scientists, primarily by limnologists. They may even be species specific criteria established for extremely limited stream reaches. For this reason, environmentally-based criteria tend to be the most stringent of the three types of criteria described (see the compilation of critical levels in Chapter 5). While these criteria offer the best protection for aquatic and other wildlife and the best chance for achieving sustainable water-related development, they go well beyond the level at which best economically available technologies operate and may push the extremes of best available technologies. Few nations have even experimented with these types of criteria, and there are no known cases where these standards have been employed.

Determining contaminant reductions

Using whichever criteria have been established or may be appropriate for a particular situation, a determination is made of whether or not a particular contaminant is a pollutant; i.e. that its concentration or presence exceeds the levels established for a particular waterbody or watercourse. The need for such a determination can be initiated by several actions; a public complaint, an agency observation, an abnormal result from a monitoring sample (exception reporting), or other request. Once the determination of pollution has been made, a several step process should be undertaken. These are enumerated below.

(1) *Quantification* of the level of exceedance establishes the margin by which loads must be reduced to bring the basin into conformance with the requirements. This step may initiate several subsequent actions on the part of the responsible agency.

   (i) Rarely do standards specify desirable loading rates. Generally, the standards are specified in terms of a concentration. Thus, to establish a quantification of the magnitude of reduction necessary for compliance, some further manipulation of the information is required. This manipulation is usually performed with mathematical models. The range of models available is reviewed in Chapter 8; suffice it to say here that these models range from the very simple, 'black box' empirical models that can be used on a hand-held calculator (such as the OECD phosphorus budget models; OECD, 1982) to sophisticated multi-box or dynamic models requiring a mainframe computer (such as SWMM, AGNPS and CREAMS, which are also available in PC versions; US EPA, 1987). While many of the latter models are specifically designed for agricultural applications, recent urban applications have shown their utility in built situations (e.g. CREAMS: James and Niemczynowicz, 1992). The models can be used to back-calculate the maximum permissible loading rate from concentration data.

   (ii) It is also useful in many situations to calculate a natural or background loading rate based on unit area loads (UALs) for forested or natural bush lands (see Chapter 6). Contaminant loads cannot be reduced beyond the natural background level without some large-scale, whole river intervention such as that carried out at Wahnbach Reservoir in Germany (see Ryding and Rast, 1989). These two loads, the loads required to meet the standard and the natural background load, will determine the constraints within which manipulation of the loading regime can be expected.

   (iii) Where appropriate standards do not exist, or where the applicability of existing standards does not meet community or agency aspirations, it may be necessary to create or amend the basis for the assessment of pollution. This can be done by applying regionally relevant criteria (such as the regional/climatic zone-based phosphorus–trophic state indices shown in Chapter 5); by conducting user surveys to identify particular concerns and desires in the community (e.g. Thornton, 1993; Thornton *et al.*, 1989; Thornton and McMillan, 1989); or by using various types of public and/or agency-based information gathering and goal setting exercises such as the Delphi (brainstorming) technique, analytical hierarchy process (AHP), or strategic planning approach (PEST – political, economic, social and technological external environmental analysis – and SWOT – strengths, weaknesses, opportunities and threats internal environmental analysis) (see standard business management texts for descriptions of these methods; e.g. Newman and Warren, 1977). Heiskary *et al.* (1992) present an overview of these various approaches.

(2) As water quality management planning proceeds, it is often possible to *refine predictions* based on land-use inventories (these need not be too detailed if one is using unit area loads, as the UALs are relatively coarse figures – a break down into high and low density residential, industrial and commercial, agricultural, pasture, forest and natural lands is usually sufficient detail). Some degree of ground-truthing information obtained from aerial or satellite photography will be necessary; it may be advisable to walk or canoe the length of the watercourse to identify sensitive areas such as steep slopes, erodible soils, and sites of cultural or historical significance. Care should also be taken to recognize the presence of wetlands and/or channelized portions of the watershed and to introduce delivery ratios or transmission coefficients where appropriate into the models (Ryding and Rast, 1989: 138ff). If necessary, stream flows can be estimated from rainfall–runoff relationships; rainfall data being generally available from airports or government meteorological agencies. Again, these relationships can be extremely simple, such as a percentage of rainfall over the watershed area, or complex, such as the SCS TR-55 method which has been computerized by various institutions (see, for example, USDA, 1975; Schmidt and Schulze, 1987). The potential impact of various watershed treatments should also be estimated through this modelling process. Ultimately, this refining process will result in a set of management interventions that will reduce the polluting load to within the limits necessary – provided that the polluting load is not derived from natural watershed processes! (This is why the background loading should be calculated early in the management process.)

The difference between the polluting load and the desired load should take into account the fact that natural systems are variable systems. Loucks (1992) notes that, for this reason, extreme events should be modelled in preference to average conditions – in this way it will be possible to design and manage a system so as to maximize conformance to the applicable standard. While this is not always economically possible, it is a good rule of thumb to follow, if possible.

(3)   Once a thorough understanding of the watershed or waterbody in question is available, it is possible to *prepare a comprehensive list of management techniques and informational programming* that will assist the agency in achieving its goals. Such a list should be all inclusive, ranging from 'the sublime to the ridiculous' (in other words, from the 'do nothing' option to the whole river treatments mentioned above) in order to ensure a full consideration of all options available to the agency and community. The process of selection of a short list of techniques and the development of implementation plans that best fit the economic, environmental and community needs is discussed further in Chapter 10.

It should be noted that the processes described above may take many different forms based upon government structures, availability of external support systems and interest groups, cultural norms, and various other factors that affect community and governmental decision-making in the various regions of the globe. However, the basic outline of quantifying the problem, conducting field studies to refine predictions, and developing a series of alternative plans is a 'tried and true' sequential approach that has universal applicability, whether conducted in the public or private sectors, or by government agencies or concerned citizens (or a combination of both).

CHAPTER SUMMARY

This chapter has presented an overview of the methodologies for measuring nonpoint source contamination from both urban and rural settings. In addition, guidance is provided on the determination of land usage. In this way, the information presented in this chapter relates directly to the discussion of the impacts of various types of land usage set forth in Chapter 6. The utility of this information lies in its use to determine the need for, and magnitude of, nonpoint source control measures that will be required to bring a waterbody into compliance with various quality demands related to beneficial use of the water or other legal requirements. Remedial actions are discussed in Chapter 9.

While this chapter, unlike the parallel chapters in its companion volume (Chapters 6 and 7; Ryding and Rast, 1989), does not present specific advice on monitoring programmes, the general principles and techniques of measuring a variety of chemical pollutants and biological responses are referred to. Similarly, the general principles and techniques for measuring runoff and water flows are also presented. Given the range of urban and rural nonpoint pollutants, and the uniqueness of each nonpoint source control programme, it would be presumptuous to prescribe more. The exact composition and frequency of measurement of parameters included in nonpoint source management studies must be determined relative to the objectives of the management programme and not be applied by rote. (In eutrophication management programmes, the cause of water quality degradation is, by definition, enrichment with plant nutrients and the aquatic plant response – this limits the scope of concern to nitrogen, phosphorus and chlorophyll *a* in most cases; similar limitation is not possible in nonpoint source pollution management programmes where pollutants of concern can span the gamut from the nutrients to more complex forms of chemical contaminants, for example.) Nevertheless, reference to Ryding and Rast (1989) can provide useful guidance to the design and administration of sampling programmes generally.

In the subsequent chapter, the selection and use of mathematical models will be discussed. These models, with their varying complexities, can provide a useful and cost-effective means of augmenting field data. As with the field programmes discussed above, the emphasis will be on the practical aspects of obtaining and using information in order to protect and rehabilitate lakes and reservoirs.

CHAPTER 8

MODELLING OF NONPOINT SOURCE POLLUTANT LOADS

G. Jolánkai, J. Panuska and W. Rast

INTRODUCTION

The previous chapter dealt with the actual measurement of pollutant loads to receiving waterbodies from various types of land uses and sources in the drainage basin. Under utopian conditions, one could continuously measure tributary flows, and the pollutants carried in the flowing water. Under such conditions, the product of these two variables would provide an extremely accurate measurement of pollutant loads. A reality, however, is that the available field, technical and financial resources are never adequate to accurately measure the pollutant load from all potential nonpoint sources in a drainage basin. The result is that individuals and agencies often must make use of mathematical models to estimate and/or predict nonpoint source pollutant loads to waterbodies (Jorgensen *et al.*, 1996). This chapter discusses this approach.

Models can be viewed as physical or theoretical tools that attempt to provide a 'picture' of the real world, without the necessity of having to actually 'construct' a representation of the real world. That is, they attempt to simulate or mimic objects or processes that occur in the real world. Thus, speaking in the broadest sense, models can include the model toy cars or planes of children, mathematical expressions of basic physical laws (e.g. Newton's laws), physical representations of real objects (e.g. dams, gates, and hydrologic structures), simple expressions relating the dependence of two variables on each other (e.g. phosphorus and chlorophyll), and sets of differential equations describing multiple, interrelated processes occurring in the environment. In one sense, therefore, the human brain is one of the best modelling tools. For example, humans can use their brains to predict the outcome of events or actions on the basis of past experiences. In recent years, however, many technologists have come to think of models solely as mathematical expressions that are used, in concert with the almost unlimited calculation ability of computers, to simulate the real world under various scenarios. Indeed, many technical people have come to think that all models are computer models. The reality, however, is that some of the simple equations expressed in this chapter can be worked with hand-held calculators, while others can be worked solely with paper and pencil. Nevertheless, it must be admitted that computer models comprise a large portion of the available modelling tools. This trend toward computer-aided modelling will be likely to continue into the future, especially as models continue to be linked to Geographic

291

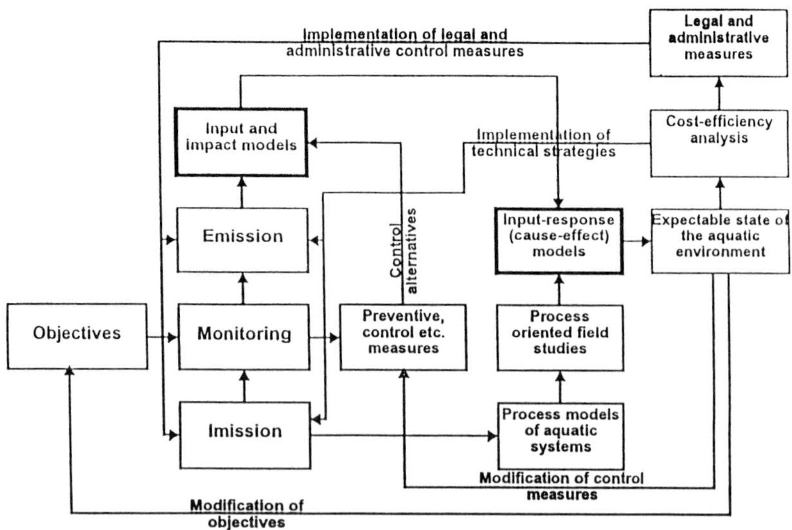

Figure 8.1 Systems approach to water quality management

Information Systems (GIS). The latter provide a means of using virtually all available data and information on a natural system.

Having characterized models in this manner, one also must admit that the development of mathematical modelling techniques (chemical, physical, or biological) has not yet reached a stage where modelling is a completely accurate and reliable basis for quantitatively describing and/or forecasting nonpoint source pollution processes. In addition, even if modelling techniques were firmly established, a lack of necessary data would still (and for a long time to come) limit the practical application of sophisticated modelling techniques.

Nevertheless, in order to design realistic nonpoint source pollutant control strategies, relevant cause and effect relationships must be quantified. Pollutant inputs from all major sources (point and nonpoint) must be related quantitatively to their measured impacts in the components of the environment (streams, lakes) receiving the pollutants. Figure 8.1 provides a schematic view of the 'systems approach' to the planning process used in developing appropriate nonpoint source control measures. As used in this chapter, modelling refers to quantification of the effects of pollutant sources and inputs on the pollutant recipients (i.e. aquatic ecosystems); the use of models has an obviously major role in this process.

THE USE OF MODELS TO ESTIMATE NONPOINT SOURCE POLLUTANT LOADS

Estimating nonpoint source pollution of aquatic systems (rivers, lakes, reservoirs) from land runoff means that the contribution of all major pollutant sources in

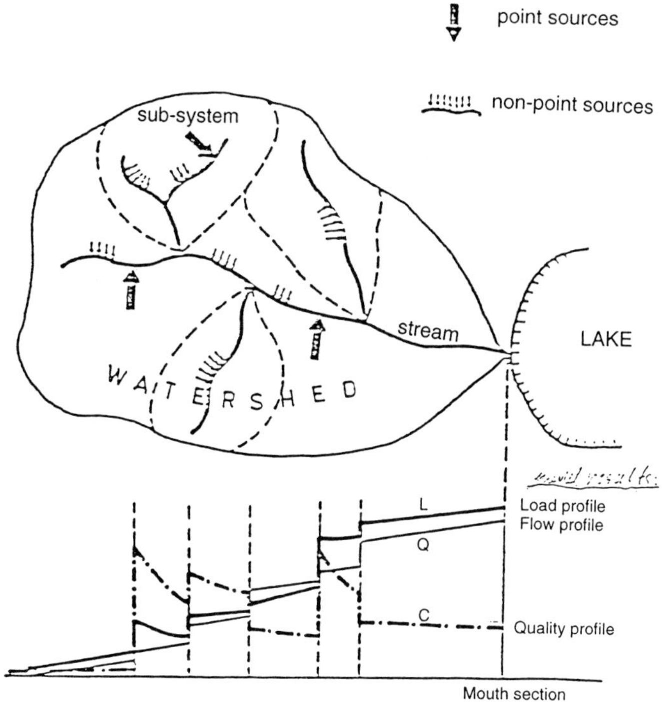

Figure 8.2 The 'missing link' problem of determining pollutant loading to a lake or reservoir

the drainage basin must be quantified to obtain an estimate of the total pollutant load at some specific geographic point, typically the river mouth (Figure 8.2). However, development of realistic strategies for controlling nonpoint source pollutants is not a simple task, largely because the relationship between a pollutant source located upstream in the drainage basin, and the resultant water quality conditions downstream in a waterbody, is affected by a variety of processes and factors. Indeed, the concept of the 'missing link problem' shown in Figure 8.2 represents the fact that the link between causes and effects of nonpoint pollution often is missing until determined quantitatively on the basis of field measurements and/or estimated by modelling efforts. This problem is sometimes also called a 'scale problem' in hydrology, or a 'delivery rate' problem when dealing with erosion and sediment (Martin *et al.*, 1996).

The main processes affecting the fate of pollutants in land runoff include: (1) surface and sub-surface runoff, (2) erosion and sediment transport, and (3) chemical, biological, and biochemical interactions within the soil–plant–water system. The hydrologic cycle (see Chapter 4) has an especially prominent role

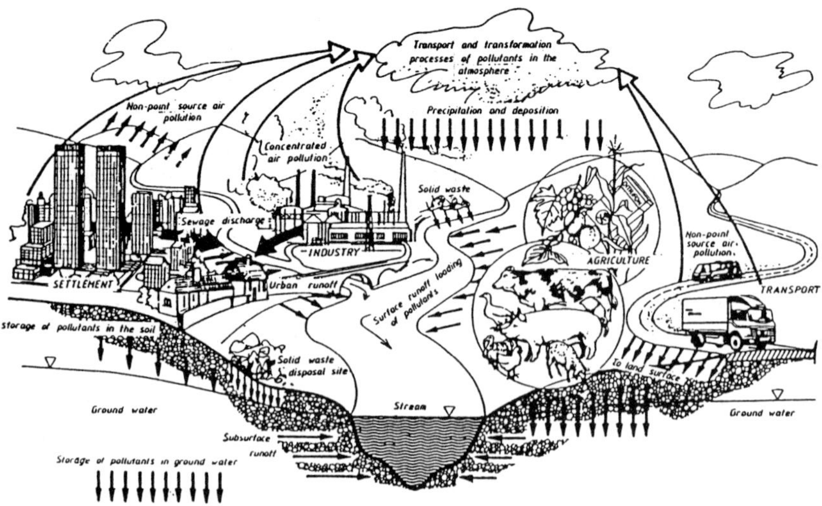

Figure 8.3 Schematic view of pollution processes

in the functioning of these processes (Figure 8.3). This means that the task of quantifying, or modelling, nonpoint source pollutant loads must include consideration of hydrology, water and soil chemistry, micro- and macro-biology, etc. In addition, the particular pollutant(s) being considered can affect the types of processes to be considered in modelling effort; for example, assessments of nutrient impacts must consider the biological cycles in which these elements are involved, while assessments of heavy metal or synthetic organic chemical impacts must consider their reactivity with the soils (Ward and Elliot, 1995; Watson and Burnett, 1995).

It is clear that the ability to accurately quantify nonpoint source pollutant loads will depend largely on the availability of relevant data. Unfortunately, due to the complexity of, and interrelationships between, the relevant physical, chemical and biological processes occurring in all major environmental components (air, land, water and biota), the experimental separation and quantification of various processes is very difficult, even in the small-scale experiments described in Chapter 7. Indeed, it is unlikely that all of the major processes affecting nonpoint source pollution of aquatic ecosystems have yet been identified, much less quantified. Accordingly, the most pragmatic advice at this stage of model development is to *Start Simple*, and proceed to more complex models as necessary to resolve the particular problem in hand, as justified by the extent of knowledge of the system being modelled.

One method for quantifying pollutant loads is to develop and/or use simple, empirical relationships and tables of pollutant loads on an areal basis for estimating the pollutant yield – the UAL method discussed in Chapter 7. At the other extreme is the use of complex, multi-parameter models that attempt to describe all the transport and transformation processes of soil–plant–water systems in a highly detailed manner. A third approach is to use a hybrid of these two extreme approaches; namely, combining empirical or experimental results with fairly sophisticated computer modelling approaches. A rationale for this hybrid approach is that it: (1) relies on readily obtainable observational data, (2) makes use of relevant information available in the scientific literature (unit area loads, for example) and (3) more closely relates to the system under investigation (when compared to the simple empirical methods) to the extent that the data and approach are system-specific. By combining the above two approaches, using models that are computationally sophisticated but theoretically simple, available data and information on a particular system can be used in an effective and understandable manner.

Based on such considerations, three basic possibilities for estimating (modelling) pollutant loads from nonpoint sources in the drainage basin can be discerned (Figure 8.4):

(1)   Estimation on the basis of unit area yields (loads);
(2)   Estimation on the basis of experimental relationships between concentrations, loads, and their affecting factors (e.g. flows); and
(3)   Estimation on the basis of simulation watershed models (including lumped parameter and distributed parameter models).

One of the most significant problems encountered in modelling nonpoint source pollution is that the relevant processes are, by definition, diffuse and spatially variable. Thus, it is difficult to accurately model the processes with 'lumped parameter' watershed models. On the other hand, spatially distributed models are very data intensive to the extent that their data needs often cannot be met

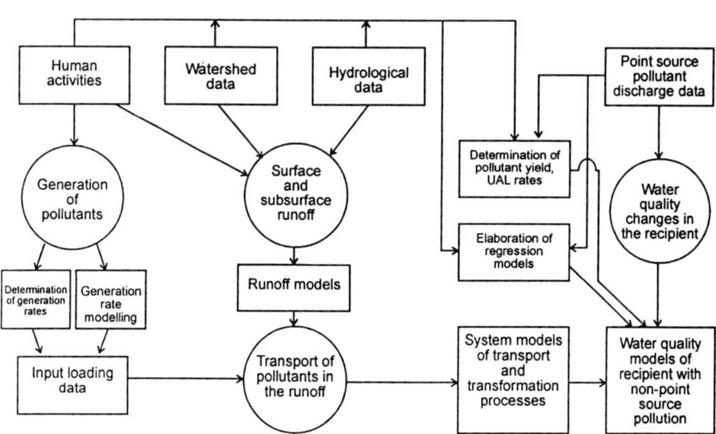

Figure 8.4   The possibilities of modelling nonpoint source land runoff pollution

within realistic time and financial constraints. To provide some practical guidance on addressing such problems, this chapter attempts to present details of modelling methods that, given the present state of knowledge, can be used most readily. Approaches that are not as readily 'usable' are discussed in less detail. More details on both approaches can be found in other reports, including those of Crawford and Linsley (1966), Free *et al.* (1975), Novotny and Chesters (1981), Jolánkai (1983), Haith (1985), Haith and Shoemaker (1987), Novotny (1988), and Novotny and Olem (1994).

ESTIMATION OF POLLUTANT LOADS USING UNIT AREA LOADS

The simplest and probably most widely used method of estimating nonpoint source pollutant loads is based on the use of unit area loads or yields. The unit area load (UAL) concept is based on the principle that, under average hydrologic conditions, land use for specific purposes (agriculture, urban, etc.) will yield a relatively constant quantity of pollutant over the annual cycle (Ryding and Rast, 1989). UALs are estimated on the basis of measurements taken from the land area of concern (see Chapter 7). The unit of a UAL is mass per area per time ($kg/ha/y$ or $g/m^2/day$). Numerous compilations of unit area load values for a large variety of land uses and water quality constituents (especially nutrients) have been reported in the literature for a variety of climatic zones. Some of these values have been reported in Chapters 5 and 6 (see also Loehr, 1974; Omernik, 1977; Novotny and Chesters, 1981; Jolánkai, 1983; Rast and Lee, 1983; Beaulac and Reckhow, 1982; Haith *et al.*, 1984; Ryding and Rast, 1989; Stahre and Urbonas, 1990; Novotny and Olem, 1994). Most of these UAL values are for nutrients; far fewer values have been reported for the more toxic and/or 'nontraditional' pollutants (e.g. pesticides, heavy metals), although some data do exist.

A common method of deriving unit area loads is with continuous, or very frequent, measurements of the flow (Q) and concentration (C) of the water quality parameter of concern in the runoff from a selected drainage area. The product of Q and C is calculated to obtain the load (flow × concentration = flux). The load is then integrated over a selected time period, and divided by the drainage basin area (A) and the time period of concern (T), as follows:

$$Y = \frac{1}{AT}\int_{T1}^{T2} Q(t)C(t)dt \qquad (8.1)$$

where: T_1 = beginning of time period T; and
  T_2 = end of time period T

Equation 8.1 can be used to determine the unit area load from nonpoint sources in the drainage basin for any given water quality parameter. When the point sources of the parameter in the drainage basin are also considered, the total yield rate can be calculated as:

$$Y = \frac{1}{AT}\left[\sum_{i=1}^{n} 86.4\, C_i Q_i \Delta T - T\sum_{j=1}^{m} L_j\right] \qquad (8.2)$$

where: Y = unit area load (kg/ha) of pollutant, over time period T;
C_i = measured or estimated pollutant concentration in time step i (g/m^3);
Q_i = measured flow rate in the ith time step (m^3/s);
A = area of the drainage basin (ha);
T = discrete time step [i.e. $n \times \Delta T = T$] (days); and
L_j = j^{th} point source load (j = 1...m) of the same contaminant (kg/day)

The pollutant yield or load value corresponding to time period T can then be converted into appropriate units (kg/ha/y, t/ha/y, etc.).

As with virtually all models for estimating pollutant loads to aquatic systems, this approach has several shortcomings. To begin with, as discussed in Chapter 7, the frequencies of measurement of Q and C usually are not identical. Typically, water quality samples (for measuring C) are taken only at certain intervals, while measurements of flow (C) may be continuous. As a result, it is necessary to use some method of estimating corresponding C_i and Q_i values. Since the concentration (C) values generally are less often measured than the flow (Q) values, the process becomes one of attempting to extrapolate C values for corresponding Q values during the period of time when C was not measured. In fact, there are several graphic and numeric methods for interpolating between measured concentration values. Because information on the actual concentration values between measurements is usually not available, no single method appears to be better than another (see Chapter 7 for a more complete discussion).

Perhaps one of the more realistic methods of attempting to relate concentration and flow values for a tributary involves the availability of reliable (measured) relationships between corresponding C_i and Q_i data pairs. The relationship can be utilized to estimate missing values for C (Figure 8.5). Unfortunately, statistically acceptable correlations often are not found, primarily because the concentration of a given water quality parameter is a function of a number of variables. That is, the C_i-Q_i relationship is not bivariate; rather, it is multivariate (see next section). This problem is especially acute in drainage basins with highly-fluctuating hydrologic regimes, in which flood waters generally can pass sampling sites unsampled. Unfortunately, from the perspective of making accurate estimates of pollutant loads, flood flows often carry the largest fraction of the total annual load of the pollutant (e.g. 60 to 80 percent of the total phosphorus and nitrogen load in small watersheds of a few hundred km^2 in Hungary and southern Africa).

Another problem is that the calculation method illustrated in Equation 8.2 is best used, in principle, only for conservative substances (materials not subject to chemical or biological transformation processes, such as decay, decomposition, settling, uptake by plants, etc.). Nevertheless, unit area loads

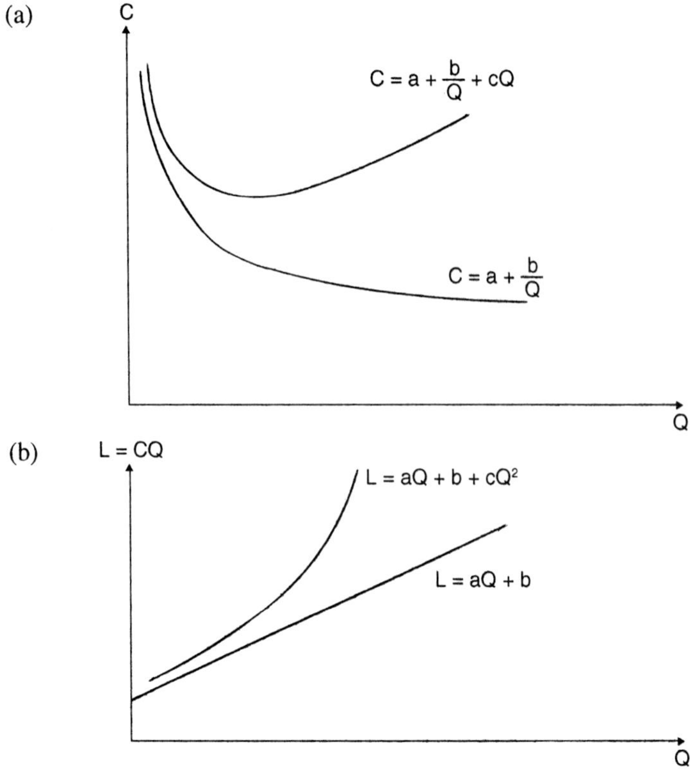

Figure 8.5 Basic relationships of flow versus concentration, and flow versus load

are commonly used for nonconservative substances, such as aquatic plant nutrients. Furthermore, although unit area loads are widely applied, strictly speaking unit area loads determined by the above-noted method correspond only to the area and time period where and when the loads were measured. They may not be applicable to other areas or time periods. Nevertheless, in many cases, such data may be the only data available for attempting to estimate nonpoint source contaminant loads.

Among the variables that affect the fate of pollutants between their source and the point of their measurement, the length of the travel pathway (the size of the drainage area and the density of the drainage system) is often the factor that most markedly determines the actual unit area load. Unfortunately, although literature unit area load values are categorized as a function of land-use types, no similar categorization usually exists in regard to drainage basin size or drainage density. Such influences must be inferred through a knowledge of local conditions, and accounted for mathematically through the use of transmission coefficients (see Ryding and Rast, 1989) or some other modifier (e.g. use of a delivery ratio).

In addition, the relationship between rainfall and runoff (involving both hydrological and hydrometeorological characteristics of a drainage basin), the main process triggering pollutant export from a drainage basin, should be considered. When applying unit area loads derived in one drainage basin to a different basin, care should be taken to ensure that such variables as the size, drainage density, land-use type(s), hydrology and climate of the two basins are similar. Because such coincidence cannot be guaranteed in all cases (and, in practice, seldom occurs), estimating nonpoint source pollutant loads on the basis of unit area loads should be recognized as providing only a rough approximation of the actual pollutant load to a waterbody. Nevertheless, unit area loads are widely used for quantifying or 'modelling' nonpoint source pollutant loads. This is due primarily to the simplicity of their use and because, in many (if not most) cases, no other information on pollutant loadings is available. The accuracy of unit area loads can be increased by:

(1)   Developing more sophisticated tables of reported values, which categorize unit area loads not only in regard to land-use types, but also as a function of other major affecting variables (soil types, drainage basin size, hydrologic characteristics, etc.);

(2)   Utilizing measured data for the site of concern in the development (or selection) of appropriate unit area load values (perhaps as mathematical functions of the affecting variables); and

(3)   Correcting or converting unit area load values to achieve loading rates corresponding to the area of concern by trying to consider the actual processes affecting the 'fate' of pollutants during their transport to receiving waterbodies in surface runoff and groundwater flows.

It is noted that item (1) above refers simply to reviewing literature data and elaborating refined tables of unit area loads. Items (2) and (3) above, however, represent refinements in the modelling of nonpoint source pollutant loads. These latter items are discussed further in the following sections.

ESTIMATION OF POLLUTANT LOADS USING EXPERIMENTAL RELATIONSHIPS

Two basic types of experimental relationships for estimating nonpoint source pollutant loads are:

(1)   Relationships between the concentration (or load) of a constituent measured in a recipient stream in a single drainage basin and the factors (or parameters) affecting or causing (e.g. flow) changes in the measured concentrations of the constituent; and

(2)   Relationships between concentrations, flows or unit area loads and the factors which can affect them, based on data from multiple drainage basins.

Characteristics of a single drainage basin

The literature contains a large number of (1) flow *versus* concentration, and (2) flow *versus* load relationships. For example, Hock (1970) distinguished two basic forms of concentration (C) *versus* flow (Q) correlations, as shown in Figure 8.5. The curves in Figure 8.5 designated as C = a + (b/Q) correspond to large streams, and to streams with heavy, but uniform, pollutant loads. The curve represented by C = a + (b/Q) + CQ corresponds to less polluted streams with a dynamic hydrologic regime (where the rising part of the curve characterizes high flow conditions). Further, the pollutant load is defined as concentration (C) times flow (Q). The upper curve in Figure 8.5b can be expressed (Hock, 1970) as:

$$L = aQ + b + cQ^2 \tag{8.3}$$

where: L = load;

        aQ = base load;

          b = point source load; and

      cQ^2 = runoff event load

Manczak (1974) presented a larger number of similar relationships (Figure 8.6). O'Brien (1972) present an even larger group of relationships (linear, logarithmic, exponential, power, and up to fifth order polynomial) describing concentrations of different ions as a function of flow.

Some researchers (Porter, 1975; Smith and Stewart, 1977) distinguished relationships for rising and falling flows, or include the rate of flow variation as a further parameter. For example, Porter (1975) presents the following equation:

$$C = a + a Q + a \, dQ/dt \tag{8.4}$$

where: dQ/dt = rate of rising flow (m^3/T)

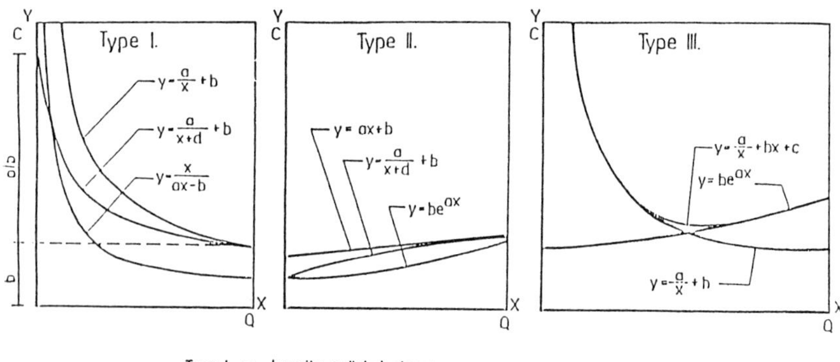

Figure 8.6   Basic types of flow *versus* concentration relationships (after Manczak, 1974)

Relationships dealing with changing flow rates (the time increment of flow, $Q_t - Q_{t-1}$) actually represent time series models. This class of models represents a sophisticated means of analyzing the records of one or more parameters. They can be used to analyze (and forecast) changes in a selected parameter(s) as a function of the previous (dynamic) condition of the system being modelled, as represented by earlier values of the parameter (autoregression analysis) or of other variables that can affect the parameter of concern (cross correlation analysis). However, these models require long periods of record on the point and nonpoint contaminant loads at a sub-basin scale, as well as in-stream information on the parameter of concern. Thus, they are useful only in special cases.

Unfortunately, regardless of the model used to analyze the relationship between concentration and flow in a given case, one is frequently confronted with highly-scattered data points, as shown in Figure 8.7. No model can easily be applied to such data. Only slightly more promising fits between the measured data and models can be obtained for flow *versus* load data points. This is especially the case for more mobile water quality constituents. This latter fit appears to be better only because the flow variable is implicitly included in both the x and y axes. Two basic reasons that flow *versus* concentration and flow *versus* load relationships frequently are difficult to identify are that:

(1) Variations in pollutant concentrations actually depend on many factors other than flow (e.g. drainage basin land usage, climatic and hydrologic conditions, pollutant inputs, soil types, etc); and

(2) Measurement data frequently do not include extreme runoff events, which often constitute a large portion of the total contaminant load to a waterbody (the only exception to this observation would be flow proportional sampling; however, this type of sampling is seldom done).

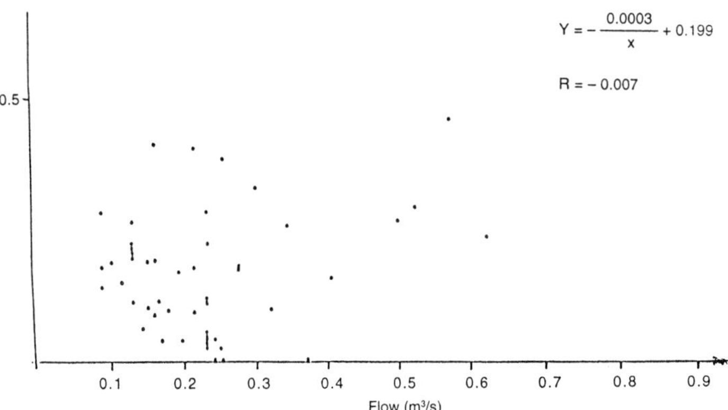

Figure 8.7 Characteristic example of the frequently encountered case of absence of correlation

Characteristics of multiple drainage basins

Experimental relationships based on data from multiple drainage basins may be more useful on a regional or larger areal basis than those based on only one drainage basin. For these models, the concentration, load or unit area load is expressed as a function of dimensionless, or normalized, hydrological and watershed parameters (e.g. the fraction of various land usages and/or soil types, the quantity of precipitation, or the quantity of runoff, etc.). Unfortunately, such relationships are not common in the literature, the main reason being the need for at least several dozen watersheds to allow accurate statistical evaluations.

One example of this type of experimental model (the Tennessee Valley Authority's Mineral Quality Model) has been presented by Betson and McMaster (1975). In this model, concentration (C) was expressed as a power function of the runoff, as follows:

$$C = a \, (Q/DA)^b \tag{8.5}$$

where: DA = drainage area;
  Q = streamflow; and
  a,b = coefficients (will vary from watershed to watershed, as a function of land use, soil and other factors)

An equation for predicting the values of these two coefficients, derived by using measures of forested land and geology as independent variables (Betson and McMaster, 1975), is given below.

$$a,b = N_1 F + N_2 C + N_3 S + N_4 I + N_5 U \tag{8.6}$$

where: $N_1...N_5$ = regression coefficients;
  F = fraction of watershed that is forested;
  C = fraction of watershed overlying carbonate rock;
  S = fraction of watershed overlying shale–sandstone rock;
  I = fraction of watershed overlying igneous rock; and
  U = fraction of watershed overlying unconsolidated rock

This model was calibrated with data from 66 drainage basins, ranging from 14.2 km² to 1980 km² in size. There were no significant point sources of pollution in the basins. Model coefficients (a) and (b) were determined for silica dioxide, iron, calcium, magnesium, potassium, bicarbonate, sulfate, chloride, total dissolved solids, calcium carbonate, specific conductance, pH and colour. Although many other factors could have been considered in this relationship (i.e. ion content of precipitation, soil types, etc.), it might be considered a 'first generation' approach to two dimensional modelling.

Another shortcoming of the above relationship is that it does not distinguish between different agricultural land uses (i.e. crops, fertilization rates, etc.); all of these factors can have significant effects on the nonpoint source pollutant load. Chesters *et al.* (1978) subsequently considered these latter variables with a multiple regression model based on data from 30 watersheds in the North

American Great Lakes Basin. As one example from this study, an equation for the unit area load for total nitrogen (TN) was derived as follows:

$$TN\ (kg/ha/y) = 0.117\ (manure\ N) + 0.0016\ (manure\ N^2 +$$
$$(fertilizer\ N + manure\ N)) + 26.0\ (\%\ corn + potatoes)$$
$$+ 3.6\ (\%\ cereals + soybeans + vegetables)$$
$$+ 0.1\ (\%\ pasture + hay) \tag{8.7}$$

For this relationship, manure and fertilizer nitrogen are expressed in kg/ha/y, while % refers to the percentage of land of the given crop in the drainage basin.

A similar equation for total phosphorus (TP) is:

$$TP\ (kg/ha/y) = -0.094 + 0.00085\ (\%\ clay) + 0.00021\ (\%\ row\ crops) \tag{8.8}$$

It should be noted that the measured and predicted annual nutrient loads from the PLUARG study were in reasonably good agreement for most cases. However, some very significant discrepancies also can be found in the tables reported by Chesters *et al.* (1978). These high deviations or low predictive capabilities of the model appear to be associated with unusual runoff conditions, ultimately being due to the fact that land runoff was not included in the regression equation.

Although Equations 8.7 and 8.8 above consider manure and inorganic fertilizer applications and major crop types, they do not distinguish between soil types and their nutrient contents. They also do not account for hydrologic conditions (runoff, precipitation). On the other hand, as a practical matter, the inclusion of too many factors and parameters in such regression models can also be questioned. This is primarily because the actual effect of some of these factors probably would be lost or masked by data collection and/or analytical errors, as well as by the inherent shortcomings in the statistical analysis. Indeed, a general observation on most regression analyses of this type is that a few dominant factors or parameters seem to explain the larger part of the total load in most cases. Accordingly, it is recommended that a maximum of no more than five or six parameters be included in such regression equations. Nevertheless, some attempts have been made to include a large number of parameters in such empirical models. For example, Dávid Telegdi (1986) included more than 50 watershed development factors and indices, relating them to the degree of eutrophication of receiving waterbodies, in a regression model. (By so doing, this model actually bypassed the need for a quantitative determination of the nutrient load.)

Perhaps the most widely used, multiple-parameter empirical runoff load model is the Universal Soil Loss Equation or USLE (Wischmeier and Smith, 1978). The basic form of this equation is:

$$A = R.K.L.S.C.P \tag{8.9}$$

where: A = quantity of soil lost from a unit area during a
unit period of time ($t/km^2/y$);
R = rainfall factor (erosion potential of rainstorms
expected for a given locality);

K = relative soil erodibility factor;
L = slope length factor;
S = slope steepness factor;
C = land cover and management factor; and
P = supporting practices (amelioration, conservation, etc.)

The variables in the USLE have been determined for various regions of the world, using climatic, meteorologic, hydrologic, geographic and soil science data, as well as the results of specific field studies.

The factors R, K, L and S are relatively fixed for a given locality. The product of these factors constitutes the basic erosion potential index (I) for a particular combination of rainfall patterns, soil properties and topographic features. This index is equivalent to the average, annual soil loss that would occur for a given region, assuming no vegetation cover or erosion reduction practices.

The rainfall factor (R) is the sum of the rainfall erosion indices (EI) for all storms during the period of prediction. EI is defined as the kinetic energy of rainfall multiplied with the maximum 30-minute rainfall intensity. Values of EI or R are generally presented in tabular or 'isoerodent' map forms for various regions of a given country. For regions with snow, an additional factor (R_s) is determined on the basis of experimental results to account for snowmelt, thaw and/or rain on frozen soil.

The soil erodibility factor (K) is determined for different soils by experimental studies on unit-erosion plots (all other factors equal the unit). This factor is generally expressed in tabular or nomogram form as a function of various soil types, organic matter content, soil structure, clay content, particle size, etc.

The dimensionless, topographic variable, which is the product of the variables L and S, is also given generally in tabular form, as a function of the length and steepness of the slope. Experimental values are restricted to slope lengths of 100 m or less, and slopes of 18 percent or less. Beyond these values, the factors must be determined by extrapolation.

The values of the cropping management (or vegetation) factor (C) range from 0.01 (for well-managed woodlands) to 1.0 (for tilled, continuously fallow fields). Values for different combinations of crop types, productivity levels, fertilizer applications, etc., are tabulated in various studies.

The supporting control factor (P), also called the amelioration factor, has a value of 1.0 where no control practices are employed. It expresses the ratio of soil loss to that of tillage (straight rows or fall-line tillage) up and down the slope. Values of P are also presented in tabular form as a function of the percent land slope and the conservation technique applied. With appropriate contour terracing, the value of P can be reduced to 0.2 or lower.

Although the factor R is used to describe hydrometeorological conditions in the USLE, it may not account for storm-generated runoff. Consequently, several modifications of the USLE were developed in an effort to include runoff. An example is (Williams, 1975):

$$A = 11.8 \, (Vq_p)^{0.56} \, K.L.S.C.P \qquad (8.10)$$

where: V = runoff volume; and,
           q_p = peak rate of runoff flow

Another modification was presented by Foster and Meyer (1975):

$$A = (aR + bcV \, q_p^{1/3}) \, K.L.S.C.P \qquad (8.11)$$

where: a,b,c = experimental coefficients and factors

It also is noted that the literature provides models and calculation methods for determining various components of the sheet erosion process, including rainfall detachment, runoff detachment, deposition, scouring, etc. However, as noted earlier, the more sophisticated the model, the less likely its ability to accurately determine the values of an ever increasing number of model parameters. Therefore, the best estimates of pollutant loads based on experimental relationships are still those provided by the USLE and its modifications. However, its applicability is limited to countries where the relevant tables, isoerodent maps and nomograms have been developed on the basis of local measurement data. (Fortunately, few countries lack this type of information.) Tables, nomograms, maps, and expressions needed for actual soil loss estimation can be found in appropriate manuals or reports, examples being those of Wischmeier and Smith (1978), Novotny and Chesters (1982), and Kiss *et al.* (1971).

Empirical nonpoint source relationships or models are not without their problems, however. For example, one problem with such relationships is that of hydrologic scale, or the 'delivery enrichment' effect. All other natural and man-made factors being equal, the runoff export (unit area load) from a given area can differ from that of other larger or smaller areas. This can occur because, proportional to the time of travel, substances carried in the runoff flows may be subjected to various losses (decay, deposition, uptake by vegetation, etc.). Furthermore, additional inputs of substances during the water transport process may result in enriched concentrations of the substances. Simply stated, usually the larger the watershed area, the smaller the unit load may be. This variation can be expressed functionally as a delivery ratio, DR (where $0 < DR < 1.0$), which is a function of the drainage basin size (see Figure 4.9, for example) or the drainage density.

Some empirical nonpoint source relationships attempt to consider the delivery ratio by taking into account the size (area) of the drainage basin. For example, Prairie and Kalff (1986) presented several relationships for total phosphorus (TP; kg/y) for various land uses (km^2), using data compiled from 38 different literature sources. Their derived relationships are as follows:

Forest:	log TP export = 0.914 + 0.986 log watershed area	(8.12)
Agriculture:	log TP export = 1.818 + 0.773 log watershed area	(8.13)
Pasture:	log TP export = 1.562 + 0.589 log watershed area	(8.14)
Row crops:	log TP export = 1.880 + 0.589 log watershed area	(8.15)
Non-row crops:	log TP export = 1.880 + 0.899 log watershed area	(8.16)
Mixed crops:	log TP export = 1.880 + 0.937 log watershed area	(8.17)

In most cases, phosphorus export from drainage basins is a linear function of the size of the drainage basins. Based on their data base, however, Prairie and Kalff (1986) concluded that previous literature values on unit area exports may have overestimated the phosphorus export for a couple of agricultural land uses. Specifically, Prairie and Kalff (1986) reported that, for row crops and pasture watersheds, the export of phosphorus declined per unit area of land as the watershed area increased. They attributed this decline to the fact that the experimental watersheds from which these values were generally derived were very small. However, they did not observe a similar decline for other types of land uses or management practices.

A similar analysis of data from 23 small watersheds in Finland was presented by Kauppi (1978). The watersheds represented typical nonpoint source areas. Kauppi (1978) found significant relationships between concentration or load *versus* flow. The percentage of cultivated land in the watershed explained most of the variance in both the phosphorus and nitrogen concentration values. The equations derived by Kauppi (1978) are as follows:

$$\text{YP conc} = 46.4 \log (\text{FP} + 1) + 7.2 \tag{8.18}$$
$$\text{YN conc} = 30 \, \text{FP} + 500 \tag{8.19}$$
$$\text{YP load} = 15.1 \log (\text{FP} + 1) + 1.9 \tag{8.20}$$
$$\text{YP load} = 9.8 \, \text{FP} + 180 \tag{8.21}$$

where: YP conc = phosphorus concentration in streamflow (μg/l);
YN conc = nitrogen concentration in streamflow (μg/l);
YP load = phosphorus unit area loads (kg/m^2/y);
YN load = nitrogen unit area loads (kg/m^2/y); and
FP = percentage of cultivated land in the watershed

Location and runoff also may be important factors determining the export of nutrients from drainage basins. For example, analyzing the annual nitrogen (Figure 8.7) and phosphorus (Figure 8.8) export data for 20 sub-watersheds of the Lake Balaton (Hungary) basin, Jolánkai (1983) derived the following relationship for the mostly agricultural southern watersheds and for the River Zala:

$$Y_p = 0.0425 \exp^{0.0233} L \tag{8.22}$$

where: Y_p = unit area load for phosphorus (kg/ha/y); and
L = annual runoff (mm)

In contrast, he found a different relationship for the mostly forested, northern sub-watersheds:

$$Y_p = 0.012 + 0.00134 \, L \tag{8.23}$$

Based on the above considerations, there are several recommendations regarding the use of empirical relationships to estimate nonpoint source pollutant loading rates from one or more watersheds of varying land uses. To begin with, in order

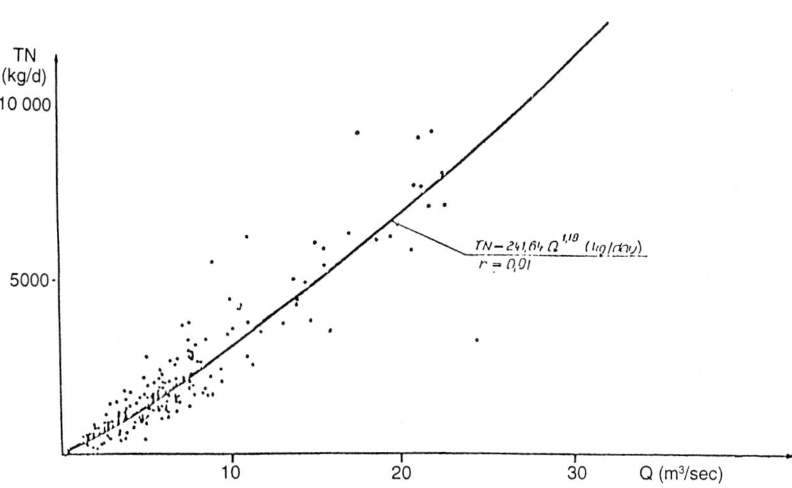

Figure 8.7 Total N load versus flow relationship for the Zala River at Fenekpuszia

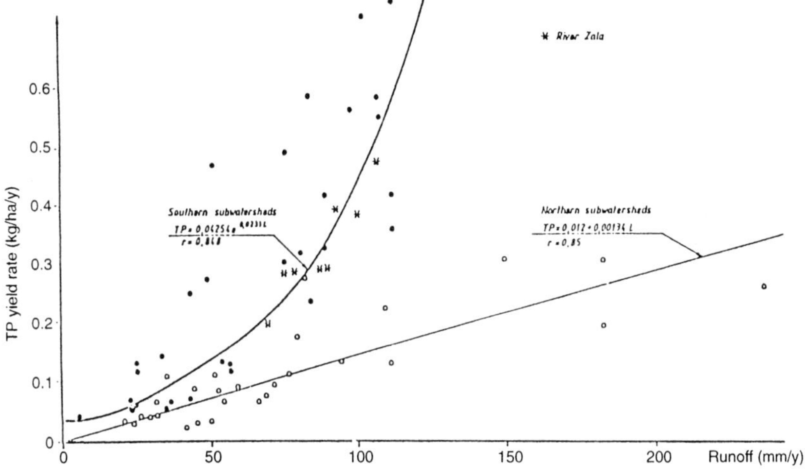

Figure 8.8 Total P area yield rate and annual runoff, Lake Balaton, Hungary

to ensure a reasonable level of reliability of the estimates based on empirical relationships, the user should have at least two to three years of flow and concentration data for at least one outlet site in the watershed of concern. In addition, the user should have topographic, land use, and soils maps of the watershed. Such data and records are required to test and verify the validity of any of the types of models or relationships identified above.

In addition, the user must consider the possible impacts of hydrologic scale and/or delivery ratios in estimating the total load of a pollutant. For example, estimating the total pollutant load at a given river mouth by calculating or modelling the loads from the sub-watersheds in a drainage basin is not simply an exercise of algebraically summing the calculated pollutant loads from the sub-watersheds. Indeed, sometimes even sophisticated computer model algorithms used to integrate spatially varying sub-watershed pollutant loads do not take into account relevant chemical and biological transformation mechanisms and/or do not consider delivery losses or enrichment gains. Such omissions of environmental realities obviously can result in misleading conclusions regarding nonpoint source management scenarios based on empirical pollutant load estimates.

ESTIMATION OF POLLUTANT LOADS USING CONCEPTUAL WATERSHED MODELS (SIMULATION MODELS)

The models discussed in this section generally provide more detailed simulations, in time and/or space, of nonpoint source pollutant transport and transformation mechanisms (Figure 8.9). As a group, these models include some or all of the compartments illustrated in Figure 8.9, coupled in a hierarchical, interactive manner (Figure 8.10). A common feature of these models is that they integrate several sub-models derived from various disciplines (hydrology, agriculture, soil science, limnology, etc.). Several categories of these models can be distinguished. For example, Donigian (1988) categorized the nonpoint source models available in the United States on the basis of factors such as: (1) land uses and load sources, (2) hydrologic processes, (3) water quality components, (4) time scale, (5) data demands, and (6) spatial scale. In contrast, dealing solely with surface and sub-surface runoff models, Novotny and Olem (1994)

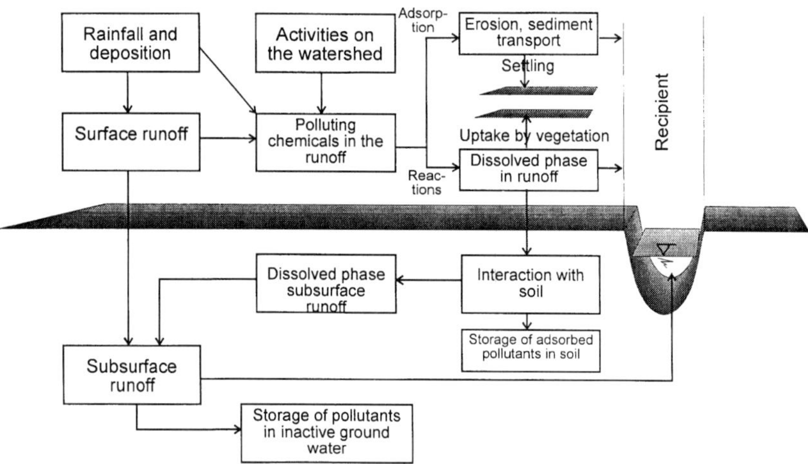

Figure 8.9 Main transport and transportation processes of pollutants in the land runoff

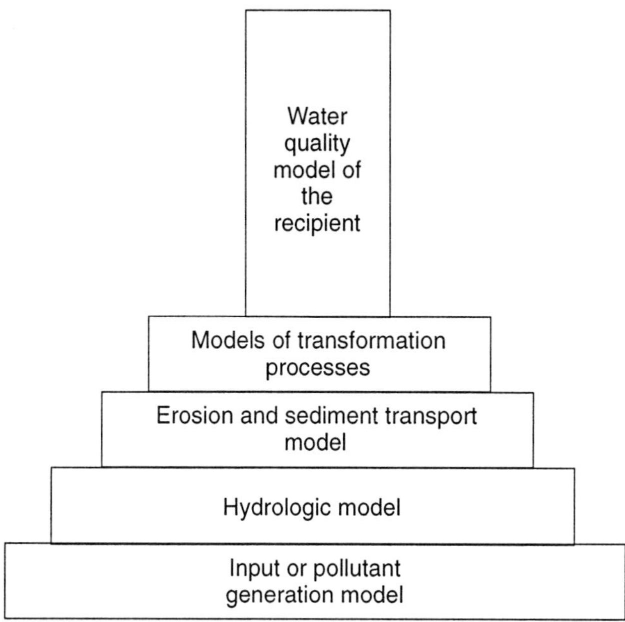

Figure 8.10   The hierarchy of sub-models playing a role in integrated watershed models

considered only two basic categories; namely (1) screening or planning models (using unit area loads), and (2) hydrologic models.

The types of models illustrated in Figure 8.10 include:

(1)   *Pollutant generation models* – which describe pollutant sources and loadings. They include atmospheric deposition models, unit area loading functions for different land uses (multiple regression models) and pollutant accumulation and release models;

(2)   *Hydrologic models* – which describe elements of the rainfall–runoff process (surface and sub-surface flow) that account for runoff generation sub-processes, including interception, evapotranspiration, infiltration, and surface detention storage. They explicitly or implicitly consider flow components (either with mass balances or hydraulic equations), such as overland flow, interflow, near-surface flow, sub-surface flow and channel flow;

(3)   *Erosion and sediment transport models* – which describe (either empirically or with semi-theoretical equations) the detachment, movement, storage and detention of soil particles;

(4)   *Material transformation models* – which describe the chemical, biological, biochemical, and physical reactions of substances that occur in, and between, the solid and liquid phases of runoff. These reactions include adsorption, desorption, decay, chemical and biological

decomposition, uptake by plants, mineralization, dissolution, volatilization, etc. In one temporal and/or spatial step of the computation, they calculate the amount (concentration) of various pollutants in the solid and/or liquid phase of the runoff; and

(5) *Water quality models* – which describe the physical, chemical and biological responses to external stimuli. Responses can include the influence of the external stimuli on light penetration and water transparency, temperature, algal or plant growth, secondary producers, and fishes, and on water quality variables such as nitrogen and phosphorus concentrations, carbon concentrations, pH, etc. Water quality models generally require detailed knowledge of, or assumptions concerning, the rates at which the element(s) upon which the model is based (known as the 'model currency') is circulated in the system, and the size and lo-cations of the various pools (sources and sinks) which affect these rates.

All these models, including their sometimes indistinguishable coupled versions, display varying levels of complexity and sophistication. Their data requirements and applicability in regard to temporal and spatial scales also differ accordingly.

In regard to the spatial scale of model simulations, one can distinguish three basic types of sub-models:

(1) *Lumped-parameter models* – which treat the drainage basin as one unit. The various characteristics of the unit are lumped together, often with an empirical equation, and the watershed is considered a uniform, homogeneous system for all parameters (Novotny and Olem, 1994). Examples of this type of model include the Unit Hydrograph Model, and the water quality applications of the same unit response and convolution concept (see Jolánkai, 1983; Jolánkai and Pesti, 1984). This latter model, termed the Unit Mass Flux Response function (UMR) was used to describe nonpoint source runoff load dynamics in several Hungarian agricultural sub-watersheds (Figure 8.11);

(2) *Distributed parameter models* – which divide a watershed into sub-components, each with uniform parameters and factors (land use, slope, crop, etc.). The flow and mass transport continuity and motion equations are solved numerically for the sub-components in two or three dimensions. Such models can make use of digital terrain models, the latter constructed on the basis of topographic, soil and land use maps. Thus, they are useful for modelling spatial changes throughout the watershed. The GIS based techniques allow very detailed description of flow and mass transport pathways as shown in Figure 8.12. (Jolánkai and Bíró, 1996). This allows the simulation of the contribution of non-point sources, such as fertilizer and sewage sludge application, to the in-stream loads in spatially detailed manner, thus forming the basis of modelling tools which can support the making of drainage basin wide NPS management decisions;

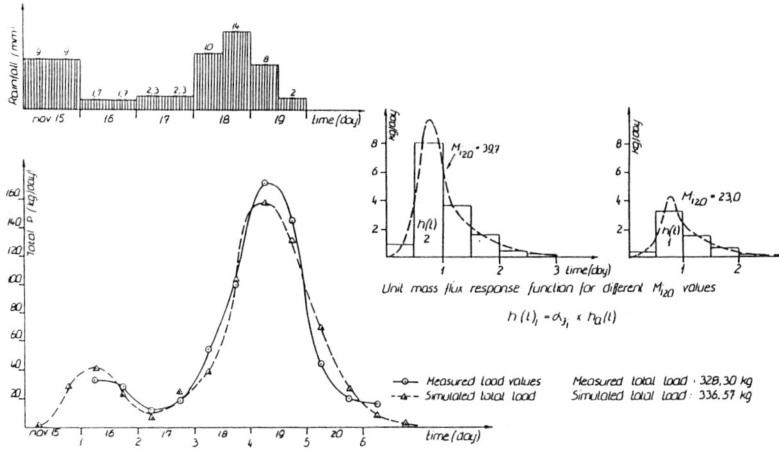

Figure 8.11 Unit mass flux response function for total P and the simulation of a combined runoff event on the Tetves Creek

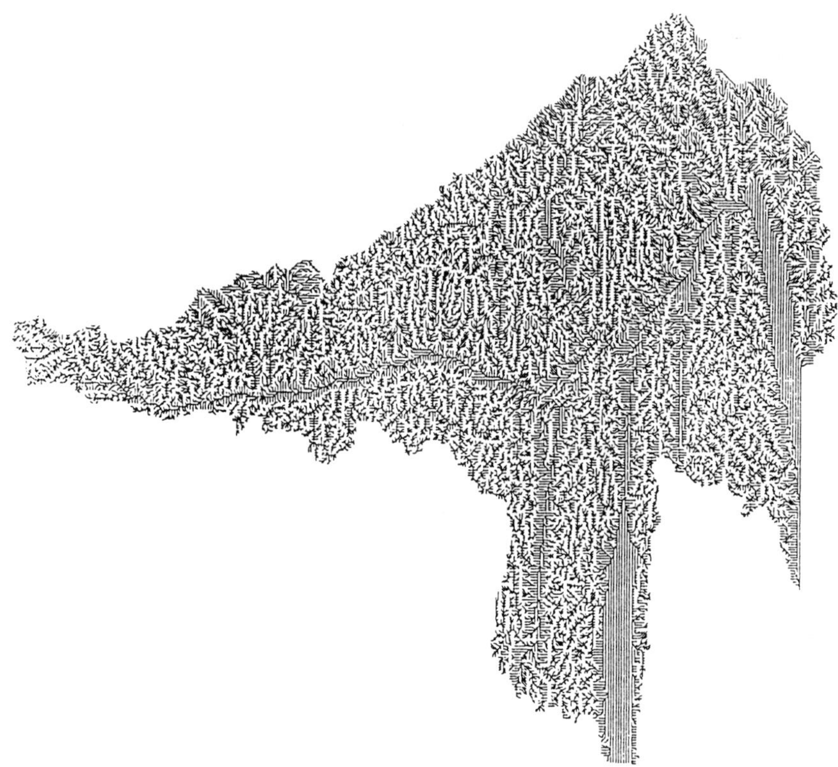

Figure 8.12 Conceptual scheme of rainfall-runoff simulation model SDWM

(3)   *Transition (between lumped and distributed parameter) models* – in which lumped parameter models are coupled, on an input–output basis, for various sub-watersheds, or else homogeneous nonpoint source regions are identified on a watershed map and their outputs (calculated mostly with empirical regression models) are routed through the watershed drainage channels with some type of stream model. A more detailed discussion of this latter type of model is presented below. It illustrates a reasonably practical solution to this type of nonpoint source modelling effort where relevant data and/or finances are limited.

In regard to the temporal scale of model simulations, one also may distinguish three basic types of sub-models:

(1)   *Long-term, average loading models (steady-state models)* – which describe the spatial variation of load and flow conditions along the drainage network of a watershed. A characteristic example of this type of model is the original form of the Universal Soil Loss Equation (Wischmeier and Smith, 1978), which can be used to calculate annual sediment loads. The model discussed in a later section of this chapter provides another example of this type of model. Although these models are not suitable for simulating runoff–load events, or for reproducing time series, they are very useful for analyzing the effects of proposed watershed management strategies, even in a spatially-varied manner;

(2)   *Event-based models* – which can simulate the response of a drainage basin to major rainfall and/or snowfall/snowmelt events. Various land and water management strategies can be compared with this type of model. Thus, nonpoint source pollution control options can be analyzed for preliminary design conditions and/or requirements, and with a relatively moderate hydrometeorological data base; and

(3)   *Continuous application watershed models* (including those used for event-based simulation) – which have long computational time requirements, especially in the case of distributed parameter models. They also require long hydrometeorological data records, often at very small time intervals (for example, hourly rainfall data).

Both (1) simple and (2) complex models can be distinguished. The simple models actually consist of computational methods or algorithms used to evaluate data and empirical information over the drainage basin scale in order to calculate the resulting nonpoint source pollutant loads, while the more complex models describe hydrologic, transport and transformation processes in more detail. Examples of both are presented in the following sections.

Simple integrated watershed models

Heidtke *et al.* (1986) presented a simple calculation method for determining the cumulative nonpoint source pollutant load from a large drainage system.

This approach simply summed up the pollutant contributions of major sub-basins:

$$TP = \sum_{i=1}^{n} TP_i = \sum_{i=1}^{n} \sum_{j=1}^{m} \sum_{k=1}^{l} UAL_{jk} A_{ijk} \qquad (8.24)$$

where:
TP = annual nonpoint pollutant load for entire drainage system;

TP_i = annual nonpoint source pollutant load from sub-basin i;

$UAL_{j,k}$ = annual nonpoint source total load per unit of land use j and soil texture k; and

$A_{i,j,k}$ = area of land use j on soil texture k within sub-basin i

A more characteristic example of this approach was presented by Haith and Dougherty (1976). The loading of an agricultural nonpoint source waste to a waterbody was calculated as the product of the runoff water volume and the concentration of the pollutant in the runoff water:

$$L = 10 \sum_{i=1}^{M} \sum_{j=1}^{N} \sum_{k=1}^{T} C_j Q_{ijk} A_{ij} \qquad (8.25)$$

where:
L = pollutant loading (kg/y);

Q_{ijk} = runoff from crop management practice j on soil type i due to precipitation event K (cm);

A_{ij} = area of land of soil type i with crop management practice j contributing runoff to a stream (km^2); and

C_j = pollutant concentration in runoff (mg/l)

With this integrating load calculation, runoff is estimated using the US Department of Agriculture Soil Conservation Service semi-empirical runoff equation (USDA SCS, 1975). The pollutant concentrations (C_j) were experimentally determined from plot-scale experiments with plant nutrients.

Both models illustrated above appear to neglect transformation processes that may occur during the time of travel between the contributing area ($A_{i,j,k}$) and the receiving waterbody. That is, the delivery or enrichment effects are neglected.

A slightly more sophisticated approach is the Simple Experimental Nonpoint Source Model (SENSMOD; Jolánkai, 1986). With this model, relationships between runoff and unit area loads corresponding to eleven land use categories were used to determine the nonpoint pollutant load from homogeneous land use sources within a drainage basin. The land runoff is determined from hydrological (measured flow) data and/or specific runoff maps. The delivery or enrichment factor is calculated as a simple exponential function of the time of travel. This latter component was based on topographic maps and hydrologic formulae. A water quality model of the main stream, including point and nonpoint pollutant inputs, subsequently was used to calculate the resultant load at the downstream-most outlet of the drainage basin. A single, lumped rate coefficient was used in this model to account for materials lost or gained within the various segments of the stream.

Another relatively simple and general model is the Generalized Watershed Loading Functions model (GWLF; Haith and Shoemaker, 1987). This model can be used to estimate monthly nitrogen and phosphorus fluxes in streamflows. Rural runoff nutrient loads were computed from daily runoff and erosion calculations, and monthly sediment yield calculations. The latter parameter was determined with a modification of the USLE model, as previously presented by Haith (1985). Urban runoff nutrient loads were computed from daily nutrient accumulation rates and exponential washoff functions. Groundwater discharge was determined by lumped parameter soil moisture balances for the saturated and unsaturated zone. Default values for chemical parameters were estimated from literature values. The model also accounted for watershed delivery effects.

A common feature of the models discussed above is that (1) they are relatively easy to use, and (2) they require relatively small quantities of measurement and watershed data. However, a disadvantage of these models is that they either ignore transformation processes (chemical and biological reactions), or replace them with lumped parameter, empirical relationships. Nevertheless, these models appear to be useful in assessing basin-scale nonpoint source pollution management problems and treatment alternatives.

Hydrologically-based, multiparameter watershed models

The basic types of hydrologically-based, multiparameter models are surface runoff models and groundwater models. Both are described below, although emphasis is placed on the surface runoff models which tend to be more advanced (and, thus, more commonly available) than their groundwater counterparts.

Surface runoff models

The purpose of surface runoff models is to simulate the movement of water and associated nonpoint source pollutants over and through the soil to a receiving stream or river channel. The most general runoff model probably is the Hydrologic Simulation Program Fortran model (HSPF), a modification of the Stanford Watershed Model IV (Figure 8.13; Crawford and Linsley, 1966). HSPF simulates the movement of dissolved oxygen, organic matter, temperature, pesticides, nutrients, salts, bacteria, sediment, pH and plankton from the land surface through a channel and reservoir system (including groundwater). A major disadvantage of this model, however, is that many of the hydrologic parameters are empirical, thereby requiring several years of record to obtain reliable values. This restricts the applicability of HSPF to those situations where a comprehensive hydrological data base is available (or can be obtained).

Several other hydrologically-based nonpoint source models with less general applicability include: (1) the Areal Nonpoint Source Watershed Environment

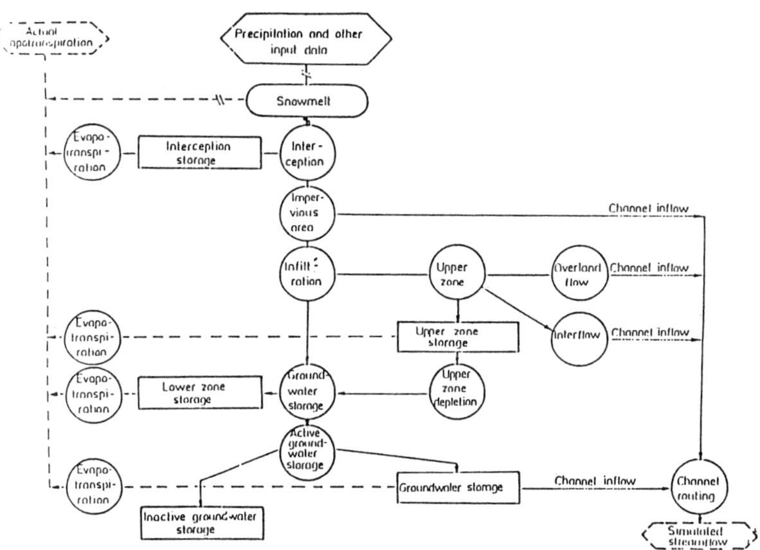

Figure 8.13 Flowchart of the Hydrocomp (Stanford) simulation model (from Crawford and Linsley, 1966)

Response Simulation model (ANSWERS; Beasley *et al.*, 1985); (2) the Precipitation-Runoff Modelling System (PRMS; Leavsley *et al.*, 1983); and the Simulator for Water Resources in Rural Basins (SWRRB; Williams and Nicks, 1994). SWRRB was developed from an earlier model, the Chemical Runoff and Erosion from Agricultural Management Systems (CREAMS; Knisel, 1980). CREAMS was designed specifically to simulate field-sized areas in order to evaluate such variables as type of tillage, crop growth, erosion, sediment yield, and nutrient and pesticide movement in runoff. In other words, the model was designed to evaluate the impact of alternative land use practices and conservation measures on the transport of nonpoint source pollutants. The modifications that created SWRRB were developed to predict the effect of management decisions on water and sediment yields within larger, more complex drainage basins, although SWRRB does not simulate chemical transport in as much detail as CREAMS.

Models which cover larger areas, thereby including landscape, management, and soil variability concerns, include the Agricultural Nonpoint Source Pollution Model (AGNPS; Young *et al.*, 1985), Small Watershed Model (SWAM; DeCoursey, 1982), and the Soil and Water Assessment Tool (SWAT; Arnold *et al.*, 1996). Two versions of AGNPS were developed; AGNPS I is used for areas from 500 to 25 000 acres, while AGNPS II is used for areas from 2.5 to 500 acres. SWAM is useful for areas up to 26 km². These latter models usually allow a more reliable representation of the field situation, and provide more useful predictions for action agency users (soil conservationists, farm advisors, and agricultural extension agents, etc.). An urban version of AGNPS is being produced.

The Erosion-Productivity Impact Calculator (EPIC; Williams *et al.*, 1984) has been developed to determine the relationship between soil erosion and soil productivity. EPIC is composed of physically-based components for simulating erosion, plant growth, and related processes and economic components for assessing erosion costs and for determining optimal management strategies. EPIC components include weather simulation, hydrology, erosion–sedimentation relationships, nutrient cycling, plant growth, tillage, soil temperature, economics, and plant environment controls. While EPIC was not developed specifically as a nonpoint source pollution model, the extensive data bases generated for use in other (nonpoint source) models can supply the necessary input parameters for EPIC and its sub-models, which can then be used to evaluate the economic impact of specific agricultural interventions. EPIC provides yet another tool for use in identifying and implementing agricultural best management practices (BMPs; see Chapter 9).

Several models describing the movement of heavy metals also have been developed. An example is the Unified Transport Model (UTM; Huff *et al.*, 1977). The UTM simulates the atmosphere–soil–plant water system, plant growth, the soil-chemical exchange of heavy metals, and metal uptake by vegetation.

Models which simulate urban nonpoint source pollution include the Storm Water Management Model (SWMM; Metcalf and Eddy, 1971), the Storage, Treatment, Overflow, Runoff Model (STORM; US Army Corps of Engineers, 1977); the Better Assessment Science Integrating Point and Nonpoint Sources Model (BASINS; Lahlou *et al.*, 1996), and the P8 Urban Catchment Mode (IEP, 1990). These urban models generally are simpler than those developed for agricultural areas, primarily because most urban runoff and pollutant loads come from impervious surfaces (Nix, 1994). Thus, the complex process of soil infiltration observed for agricultural areas is considered to be small or negligible in urban environments.

Groundwater models

As noted, groundwater modelling efforts generally have lagged behind those devoted to surface waters. A significant reason for this observation is the lack of detailed knowledge of the transformation processes that take place as nutrients, pesticides and other nonpoint source pollutants move through soils (and their parent materials) to the groundwater table. Adding to this problem is the complexity and heterogeneity of soil and geological layers through which water must pass *en route* to an aquifer or surface waterbody. Javandel *et al.* (1984) and van der Heijde *et al.* (1985) provide a review of available groundwater models and their capabilities.

Recent developments in groundwater modelling include the Leaching Evaluation of Agricultural Chemicals Handbook model (LEACH; Dean *et al.*,

1984) and the Depth to water, Recharge, Aquifer material, Soil media, Topography, Impact of vadose zone, and Conductivity of the aquifer model (DRASTIC; Aller *et al.*, 1985). The acronym, DRASTIC, represents the seven parameters that the US National Water Well Association found to be most important (and mappable) in controlling groundwater pollution potential. LEACH concentrates on the movement of pesticides through soil, whereas DRASTIC focuses on classical hydrologic parameters. Both models allow users to evaluate the probability of agricultural chemicals contaminating groundwater resources at any given location. However, the accuracy of these models has not yet been tested by comparisons with measured field data.

Another recent groundwater model, the Pesticide Root Zone Model (PRZM; Carsel *et al.*, 1985), is a simulation model developed to evaluate pesticide leaching potential under field conditions. It considers interactions of pesticides in runoff, advection in percolating water, molecular diffusion, dispersion, uptake by plants, sorption to soil, and biological and chemical degradation.

The Agricultural Chemical Transport Model (ACTMO; Free *et al.*, 1975) was developed by the US Department of Agriculture Soil Conservation Service. The hydrological component of ACTMO is a lumped-parameter sub-model (based on USDAHL-70; Holtan and Lopez, 1971), which continuously accounts for soil moisture by balancing infiltration, evapotranspiration, and seepage into the lower soil layers. A modified kinematic wave concept is used to route excess rainfall. Erosion is calculated with a modified version of the USLE. The chemical sub-model of ACTMO traces the movement of a single application of an agricultural chemical through the watershed. It uses a linear isotherm model to calculate sorption and desorption. In addition, the model calculates both the solid and dissolved phases of the waterborne chemical transport.

The Agricultural Runoff Management Model (ARM; Donigian and Crawford, 1976) simulates nutrient and pesticide runoff. The hydrologic sub-model (LANDS) is similar to the previously-noted Stanford Watershed Model IV (see Crawford and Linsley, 1966). ARM is also incorporated into the previously-referenced Hydrologic Simulation Program-Fortran model (HSPF; see Crawford and Linsley, 1966). ARM simulates sediment transport with the SED sub-model, which models pesticide adsorption, degradation and removal, and nutrient transformations and removal. Detachment and transport of soil particles is simulated as a function of the rainfall and runoff using an experimental power function, with coefficients to account for crop cover and soil conditions. It is noted that the model contains 42 parameters and coefficients, and thus requires extensive input data and calibration.

The Cornell Nutrient Simulation model (CNS; Tubbs and Haith, 1981; Haith *et al.*, 1984) considers only the transport and chemical phenomena that occur within 30 cm of the soil surface. The hydrologic transport sub-model of CNS relies on the US Department of Agriculture Soil Conservation Service curve number equation (USDA SCS, 1975). It computes daily soil moisture, percolation, evapotranspiration, and runoff (including snowmelt, where applicable). Rainfall-induced soil erosion is computed on a daily basis with the USLE, as modified by Williams (1975). The soil chemistry sub-model of CNS computes the monthly balances of the nitrogen and phosphorus forms in the soil. The dissolved and solid phases of nitrogen and phosphorus in the runoff also are modelled on a monthly basis.

A number of additional nonpoint source models, many of which are used in the United States and elsewhere (see US Environmental Protection Agency, 1987; Anderson and Woessner, 1991), are identified and briefly described in Appendix 8.1.

HOW TO SELECT AND USE NONPOINT SOURCE POLLUTION MODELS

Having described a number of commonly-used models for determining nonpoint source pollutant loads (from literally hundreds of models of lesser or greater complexity), one of the principal difficulties associated with the use of such models is the selection of the appropriate model to use in a given situation. This is important because misapplication of any model can result in erroneous, and often misleading, results and conclusions. Model selection must be based on (1) the specific objective(s) of the study to which the particular model will be applied; and, (2) the criteria to be applied to, and/or the situation to be assessed by, the model. Examples of the range of objectives and situations likely to be subject to modelling include:

(1) An agricultural extension agency attempting to help a farmer increase his crop yields might need information on the potential water pollution impacts of using increasing quantities of agricultural chemicals;

(2) An urban planning agency attempting to evaluate the impact(s) of expanding urban areas and their potential to contaminate groundwater as a result of liquid waste disposal in confined areas; or

(3) On the larger scale, an environmental agency acting on a request from policy-makers at the regional or national level for a regional or national assessment of the pollution potential of certain nonpoint source pollutants (as the basis for preparing pollution control legislation).

Clearly, satisfying each of these diverse objectives will require a different type and complexity of model. While it is not possible to unequivocally prescribe a formula for selecting the 'right' or best model for a given application, it is possible to provide at least some practical guidelines for selecting nonpoint source pollution models in specific situations based on the foregoing examples:

(1) For agricultural planning purposes, models that illustrate the relative advantages of different management practices may be more useful than models which produce absolute values (which may have little meaning for farmers and/or decision-makers);

(2) For urban development planning purposes, simple models that provide a preliminary assessment of potential nonpoint source pollution problems may be more useful in the initial stages of a development project, while more complex, simulation models may be required later to formulate specific pollution control programmes and/or treatments; and

(3) In generating an overview of nonpoint source pollution potentials, determination or prediction of long-term, annual average values may be more useful than forecasts of single, storm-specific values. (However, site- and event-specific assessments could be important in investigating the source(s) of surface or groundwater pollutants and the extent of pollution – this would be the case where the legislation is applied to specific situations, such as in examples (1) and (2) above.) Generally, the model requirements for a local *versus* national assessment of nonpoint source pollution potentials are likely to be very different; detailed models generally require too much information and/or data for application on a national scale.

Thus, model selection and application is dependent in part on:

(1) A clear definition of the specific objectives in evaluating nonpoint source pollution problems (for example, are absolute values or relative comparisons required?);

(2) The spatial (field, farm, basin or country) and temporal (single event, monthly values, or long-term, annual averages) scales of the problem being investigated; and

(3) The data needs of the model being considered *versus* the data available to the model user (a model clearly is only as reliable as the input data).

The data requirements of a model can be simple, but extensive. For example, Tables 8.1, 8.2, and 8.3 list the hydrologic, erosion, and soil parameters, respectively, required to run the CREAMS model. Although the model contains a large number of parameters, much of the necessary input data can be obtained, in most cases, relatively easily. For example, detailed weather and soil data usually can be obtained from national weather records and soil survey information. However, it is equally clear that complete records are sometimes not available (historical weather data and soil surveys may be fragmented or nonexistent). Unfortunately, this means that the application of detailed (simulation) nonpoint source pollution models may be severely limited in such situations.

Based on the above considerations, there are four basic aspects to consider when selecting a model for quantifying pollutant loads from nonpoint sources:

(1) *The objective of the modelling effort* – this can vary from (a) assessing the impacts of a single action on pollutant generation (spreading of manure, waste disposal, landfill leaching, etc.); (b) quantifying the total pollutant load from one or more drainage basins (developing input data

Table 8.1   Hydrology for CREAMS

Parameter symbol	Parameter name, units	Source
DACRE	Catchment area, ha	Field data
RC	Effective hydraulic conductivity of soil, mm/h	Calibration parameter
GA	Effective capillary tension of soil, mm	Calibration parameter
FUL	Fraction of porespace filled at field capacity	Field data
BST	Fraction of plant available water storage filled when simulation begins	Field data
CONA	Soil evaporation parameter	Field data and model manual
POROS	Soil porosity of root zone, mm/mm	Field data
BR15	Immobile soil water content at 15 bars tension, mm/mm	Field data
DS	Depth of soil surface layer, mm	Field data and model manual
DP	Depth of maximum root growth layer, mm	Field data and model manual
RMN	Manning's n for overland flow	Model manual
SLOPE	Effective hydrologic slope, m/m	Field data
XLP	Effective hydrologic slope length, m	Field data
GR	Winter cover	Snow cover condition
TEMP	Monthly mean temperature, °C	Weather records
RAD	Monthly mean radiation, tyd^{-1}	Weather records
LAI	Leaf area index for corn	Field data

required for modelling in-lake pollutant impacts); (c) selecting a nonpoint source pollution control strategy (land use and/or amelioration planning of one or more land uses); (d) forecasting and predicting future pollutant loads, including trend and/or scenario analysis;

(2)   *The spatial scale of the study* – which can vary from a small watershed, to a large, multi-basin watershed, to a regional or national reconnaissance;

(3)   *The time scale of the study* – which may require long-term, annual averages, single event comparisons, monthly balances, long-term changes, or an assessment of trends and/or tendencies; and

(4)   *The availability of data* – which is limited by the availability of financial and manpower resources for gathering (collecting and measuring) the necessary data, and time constraints relating to the collection and analysis of data (i.e. the length of the study and urgency of providing a solution to a nonpoint source pollution problem).

It is noted that the literature offers some elaborate and sophisticated methods to assist in the selection of an appropriate model for a given situation. An example is the decision flowcharts (Figure 8.14; Donigian, 1988). Such sources usually

Table 8.2 Erosion parameters for CREAMS

Parameter symbol	Parameter name, units	Source
NEARCV	Manning's n for overland flow over bare soils	Field data
WTDSOI	Weight density of soil, kg/m^3	Field data
KR	Soil erodibility by erosion by concentrated flow	Model manual
NEARCH	Manning's n for channel flow over bare soil	Model manual
YALCON	Yalin constant for sediment transport	Model manual
SOLCLY	Clay fraction in the original surface soil layer exposed to erosion	Soil data
SOLSLT	Silt fraction in the original surface soil layer exposed to erosion	Soil data
SOLSND	Sand fraction in the original surface soil layer exposed to erosion	Soil data
SOLORG	Fraction of organic matter in the original surface soil layer exposed to erosion	Soil data
SSCLY	Specific surface area of clay particles, m^2/g	Soil data
SSSLT	Specific surface area of silt particles, m^2/g	Soil data
SSSND	Specific surface area of sand particles, m^2/g	Soil data
SSORG	Specific surface area of organic matter particles, m^2/g	Soil data
DATOV	Area represented by overland flow profile, ha	Field data
SLNGTH	Slope length of representative overland flow profile, m	Field data
AVESLP	Average slope of representative overland flow profile, m/m	Field data
	Slope at the	
SB	Upper end of profile, m/m	Field data
SM	Midsection of profile, m/m	for uniform
SE	Lower end of profile, m/m	slope
	Distance from	
XIN(3)	Top of slope where mid uniform section begins, m	Field data for
XIN(4)	Top of slope where mid uniform sections begin, m	uniform slope
	Elevation above	
YIN(3)	Lower point where mid uniform section begins, m	Field data
YIN(4)	Lower point where mid uniform section begins, m	Field data
NK	Number of slope segments differentiated by changes in soil erodibility	Field data

Table 8.3 Soil parameters for CREAMS

Parameter symbol	Parameter name, units	Source
SOLPOR	Soil porosity, m³/m³	Soil data
FC	Field capacity, m³/m³	Soil data
OM	Organic matter available for denitrification (% of soil mass)	Soil data
SOLN	Soluble nitrogen, kg/ha	Model manual
SOLP	Soluble phosphorus, kg/ha	Plot data
NO3	Nitrate, kg/ha	Model manual
SOILN	Soil nitrogen, kg/kg	Model manual
SOILP	Soil phosphorus, kg/kg	Plot data
EXXN	Extraction coefficient for nitrogen	Model manual
EXXP	Extraction coefficient for phosphorus	Model manual
AN	Enrichment coefficient for nitrogen	Model manual
BN	Enrichment exponent for nitrogen	Model manual
AP	Enrichment coefficient for phosphorus	Plot data
BP	Enrichment exponent for phosphorus	Plot data
RCN	Concentration of nitrogen in rainfall, mg/l	Model manual

also provide tabulated assessments of the capabilities of alternative models. Notwithstanding, however, the selection of an appropriate model remains a cumbersome and uncertain task, due, in part, to the fact that the data and information needs of the various models vary greatly. A further consideration would be the accessibility of the models (computer packages, user's manuals, etc.) and their costs.

Additional rules of thumb and practical guidelines in regard to model selection and application are as follows:

(1) Whatever the purpose of a modelling exercise, and whatever the model selected, relevant data and information usually are required (especially for simulation models), on (a) hydrometeorological conditions (rainfall–runoff); (b) basic water quality (pollutant concentrations measured in runoff); (c) soils (physical and chemical soil parameters); (d) topographic mapping; (e) land use mapping; and (f) human activities which would directly affect nonpoint source pollution (manure and fertilizer application rates, location, number and types of landfills, waste disposal sites, etc.);

(2) If sources of data (or resources for gathering data) are very limited, and/or when no flow and concentration records are available, rough estimates (order of magnitude or less) can be attained with the use of unit area load values as reported in this document and/or the various cited references (detailed summaries of many unit area load values can be found in the reports of Loehr (1974), Reckhow *et al.* (1980), Rast and Lee (1983), Jolánkai (1983) and Ryding and Rast (1989), for example);

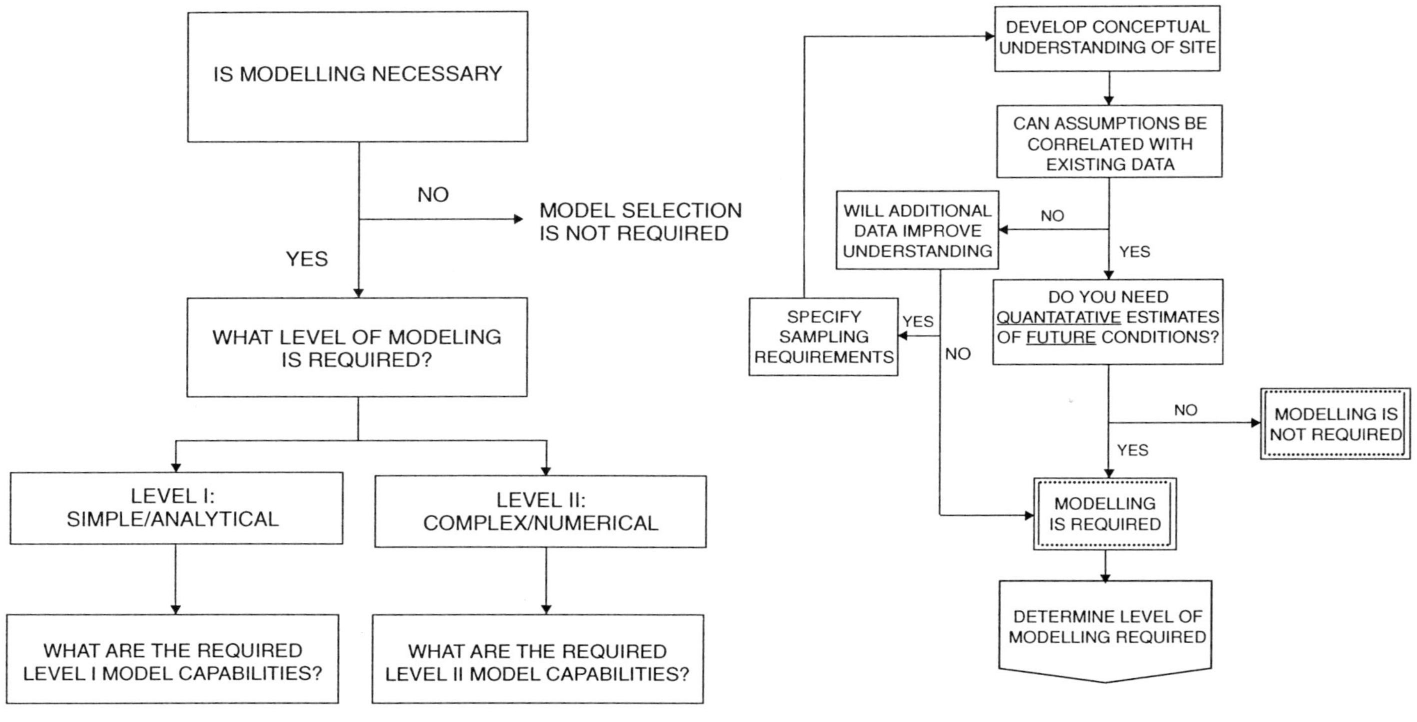

Figure 8.14 Decision flowcharts for model selection and need for modelling (cited by Donigian, 1988)

(3) If flow and concentration records are available, and the study objective is to estimate only the total pollutant load at an outflow section, then flow *versus* load, or flow *versus* concentration, regression models can be developed from the data (time series analysis also could be used if the records are of sufficient duration); and

(4) If flow and concentrations records are available, and the study objective is to estimate the nonpoint source pollutant contribution of various sub-units (sub-catchments, farms, disposal sites, etc.) of a watershed, use should be made of either (a) multivariate regression models for weighting the contributions of various sources (see Betson and McMaster, 1975; Chesters *et al.*, 1978; Kauppi, 1978), or (b) spatial 'screening' models (see Haith and Dougherty, 1976; Jolánkai, 1986), perhaps in combination with unit area loads *versus* runoff relationships or other regression models, for weighting the impacts of specific land uses or anthropogenic activities; and

(5) For analyzing the detailed behaviour of small watersheds (a few hectares to a few km^2), hydrologically-based, multiparameter, multi-component, conceptual watershed models are useful (although, because of their considerable data requirements, their large-scale application is often very cumbersome, and both labour- and time-intensive; thus, these types of models are generally used only for modelling relatively small watersheds – in fact, no really successful, large-scale application of these types of models has yet been reported in the literature).

Theoretically speaking, while the best 'all-purpose' modelling tool is likely to be the multiparameter, multicomponent, spatially distributed model, their present stage of development generally restricts their use to experimental/research applications. Indeed, the availability of space photo-interpretation and map digitization facilities, along with the existence of densely set and continuously operated multi-parameter monitoring networks, is a pre-requisite for the practical use of these very sophisticated and complex models. Thus, for most practical applications, some less complex model, requiring less extensive data, will probably be used.

CHAPTER SUMMARY

In this chapter, a range of mathematical models has been described, ranging from the very simple unit area load-based relationships that can be used in obtaining an order of magnitude estimation of the nonpoint source pollutant loads from various types of land use to the most sophisticated simulation models that account for the physical and biogeochemical processes that act upon pollutants as they are transferred from the land surface to watercourses. While the latter are largely used solely for research and experimental purposes, numerous intermediate models, offering a more elegant elaboration of the simple empirical relationships used in the unit area load-based models are in common use. These models are generally based on the TR-55 rainfall-runoff relationships derived by the USDA SCS (1975), the Universal Soil Loss Equation (USLE; Wischmeier and Smith, 1978), and unit area load determinations. Examples include the AGNPS (Young *et al.*, 1985) and SWMM (Metcalf and Eddy, 1971) models commonly used in the United States for rural and urban application, respectively. These models provide a more realistic representation of actual events than the simple empirical relationships without requiring an excessive amount of measured input data.

As in all modelling exercises, the output of these various models is simply a representation of the actual conditions, ranging from annual average conditions in the case of simple empirical models to daily (or some other period, more frequent than annual) estimations in the case of the more sophisticated models. Where more precise data are required, measured field data over a sufficiently long period to account for inter-annual variability should be obtained – generally over at least a three-year period. Field investigations of this nature are typically labour- and cost-intensive, requiring continuous sampling of stream flows and frequent sampling of the chemical parameters of concern. It is these requirements that often provide the rationale for using models in the assessment and control of nonpoint source pollutants. Nevertheless, field verification of model output is recommended in order to permit such output to be used with some degree of confidence in the design and installation of BMPs.

Models used in nonpoint source pollution investigations should be as complex as required by the problem being investigated, and as time and finance allow. Because of the wide range of conditions and circumstances surrounding nonpoint source pollution investigations, it is impossible to prescribe more than general guidelines for the selection and use of models. In this regard, it is always a useful 'rule of thumb' to start simple and add levels of sophistication as required by the problem and as permitted by available facilities, data, manpower and funds. As noted above, model outputs are only as 'good' as the input data. Attempting to use simulation models with data suited to simple empirical models (i.e. using inappropriate models and/or data) can only result in erroneous and misleading results.

Generally, the basic data requirements for any modelling exercise include meteorological data (specifically precipitation data), information on runoff conditions (i.e. to determine how much of the precipitation is likely to be reflected in the overland flow), land use (which determines the UALs to be employed), and other information on human usage of the watershed (e.g. point sources). As model sophistication increases, further data, such as data on channel length, slope, and nature and composition of channel lining, become necessary. Ultimately, in the most complex models, continuous flow gauge data, detailed land use and watershed mapping data, frequent water chemistry measurements, and knowledge of the transfer rates between the various (identified) sources and sinks of pollutants, *inter alia*, may be required.

Mathematical models can be used in both diagnostic and analytical/assessment modes. In the management of nonpoint source pollution, models can be used to identify the most significant sources of pollution in a watershed. Even simple empirical relationships, such as those employing UALs, can provide guidance on the relative magnitudes of pollution contributed from various land uses within a watershed. Such information will allow treatment of the most severe problems as a matter of priority (see Chapters 9 and 10). Once these priorities have been established, models can provide guidance on the relative effectiveness of the various BMPs (see Chapter 9) available to reduce nonpoint pollution. Modelling scenarios can be constructed wherein the relative pollutant reduction capacities of individual BMPs can be forecast. Care should be taken to ensure that such forecasts also take into account predicted changes in land usage during the period in which the BMP is being implemented (e.g.'worst case' scenarios should generally be constructed for planning purposes). This approach can narrow the field of appropriate BMPs to those that will be most effective; these measures can then be more closely examined, costed, and evaluated prior to design (and construction) of measures to be installed. In this regard, Chapter 9 presents a hierarchical approach to nonpoint source management which emphasizes both structural (engineered) and non-structural (socio-legal) techniques of pollution control.

Appendix 8.1   Examples of the nature, capabilities, data requirements and resource needs of some commonly available nonpoint source pollution models

Simple Experimental Nonpoint Source Model, SENSMOD

Description
SENSMOD is a complex river basin model system especially designed for larger drainage basins for supporting decisions on water pollution control strategies of both point and nonpoint (thus agricultural) sources of practically any contaminating substances. A special feature of the model, as contrasted to many other existing watershed models, is that it provides for full calibration on the basis of existing hydrological and water quality monitoring data, making use also of practically all kinds of available geographical, topographical, land use, etc. information. The basic feature of the model is that it searches and finds the pathways of runoff and thus contaminant transport, and utilizes this generated information in the model calculations.

Capabilities
- Utilizes GIS based information (land use, topography, application rates, etc), can be applied to very large river basins (have been tested to such basins)
- Determines flow and transport routes from digital terrain model DTM, (no need for entering stream network as lines)
- Estimates water quality model parameters on the basis of measurement data (self calibration)
- Determines the extent of diffuse pollution loads in a spatially detailed manner (when run in calibration mode for a specified condition, e.g. with annual runoff and in-stream pollution load data).
- Verification of the model relatively easy when records of several years length are available (e.g. the model as calibrated against the data of a selected year, can be verified by running the model in scenario mode with the data of another year, while preserving the parameter values, and by comparing the measured and simulated data).
- Applications included several water quality constituents, such as P and N forms, BOD-DO, COD, Cadmium, etc. and can include new ones, provided measurement data exist.
- Can be efficiently used for assessing the expected results of overall basin management strategies, point source control, diffuse source control and runoff modification strategies inclusive. Quite reassuring results in this respect have been achieved when analyzing the effects of strategies that had been already implemented.

Limitations
- Does not simulate single events. The model system works on the steady state basis, which means that hydrological and pollutant concentration profiles of a given larger period of time (that must be longer than the time of concentration) for any sub network (or for the whole of) the river system, is a function of the runoff conditions and of the data on point and nonpoint source loading.
- It can handle only such polluting constituents for which monitoring data exist.
- Does not differentiate between surface and subsurface runoff.

Input data
Classified in two categories: 1. Drainage basin data, digital maps (GIS) of the terrain, land use, soil types, application rates (e.g. fertilizers, atmospheric deposition rates, etc.). 2. Data of hydrological and water quality monitoring stations and of point source discharges. Time of travel data (nomograms) of the stream network is useful additional information.

Appendix 8.1   Continued

Output description
Two categories: 1. Hydrological (flow) and water quality profiles of the stream network in tabulated and graphical form. 2. Digital maps of diffuse loading rates for the selected constituents (those of the actual, compute-simulated contributions and not the export rates).

Availability
The application of SENSMOD needs the active contribution of the developers (e.g. Géza Jolánkai and co-workers, Institute for Water Pollution Control of the Water Resources Research Centre Plc. VITUKI, H-1095 Budapest Kvassay u. 1, Tel: 361 2155360, Fax: 361 2161514, E-mail: Jolankai@ibm.net), that is the system is not sold or disseminated as software, but is used as a tool in research/development ventures.

Resource requirements
The model was written in programme language 'C' and can be run in IBM or compatible PCs (486 or Pentium) in conjunction with a special GIS that the developers do not wish to advertise here. Graphic output needs VGA colour monitor.

Agricultural Nonpoint Source Pollution Model (AGNPS)

Description
AGNPS is a single event based model intended to simulate sediment and nutrient transport from agricultural watersheds in Minnesota. It is also being used and tested in neighbouring states (principally Nebraska and Iowa), with the intention of adding pesticide simulation. The model works on a cell basis, and the watershed (ranging from 2.5 to 23 000 acres) can be divided into 1-acre elements. The model predicts runoff volume and peak rate, eroded and delivered sediment, nutrient (nitrogen and phosphorous) concentration, and chemical oxygen demand in the runoff, and the sediment for single storm events for all the cells in the watershed.

Capabilities
- Compares the effects of various BMPs,
- Analyzes pollutant loads from feedlots,
- Estimates water quality parameters at intermediate points throughout the watershed network,
- Estimates erosion for five different particle sizes (clay, silt, small aggregates, large aggregates, and sand),
- Subdivides transport portion into soluble pollutants and sediment-attached pollutants.

Limitations
- Used only for single events,
- Has undergone only limited testing for pollutant transport,
- Not adequately tested for particle size distribution during transport,
- Does not simulate receiving waters.

Input data
Classified into two categories: watershed data that includes information that applies to the entire watershed and to the storm event to be simulated, and cell data that includes physical information and parameters based on the land conservation practice. Data required may be obtained through Minnesota Land Management Information Service (LL45 Metro Square, 7th and Robert, St. Paul, Minnesota 55101), visual analysis, maps, topographic and soils data, technical publications, or the AGNPS manual.

Appendix 8.1   Continued

Output description
The basic output from AGNPS includes hydrology; runoff; and sediment, nutrient, and chemical oxygen demand. The output can be examined for a single cell or for the entire watershed. Detailed sediment and nutrient analyses (weighted and mean) are also available.

Availability
AGNPS is a fairly new model. The manual written by Robert A. Young and others in 1986 provides a list of references for additional data and a number of individuals to contact for further information. The Guide to Model Users is in press. For further information, contact Robert A. Young, Agricultural Research Service, USDA, North Central Soil Conservation Research Lab, Morris, Minn. 56267; telephone 612/589-3411.

Resource requirements
AGNPS is written in FORTRAN IV and developed on a Hewlett-Packard 1000 computer. An IBM-PC compatible version requiring 256K memory is available for monochrome screen (version 1.0) or graphics (version 1.1).

**Areal Nonpoint Source Watershed Environment
Response Simulation (ANSWERS)**

Description
ANSWERS is an event-based surface hydrology model that estimates hydrologic and erosion response of agricultural watersheds. The area to be studied must be subdivided into a finite number of square grids, with parameter values specified for each grid. The model simulates interception, infiltration, surface storage, surface and subsurface flow, and sediment detachment, transport, and deposition. Has been extensively validated in the Midwest.

Capabilities
- Can evaluate different erosion control management practices for agricultural lands and construction sites,
- BMPs can be evaluated by modifying soil infiltration values and surface conditions,
- Modular program structure permits modification of existing program code and the addition of user-supplied algorithms,
- Grid analysis allows for consideration of spatial variation of hydrologic and sediment processes.

Limitations
- Simulates only single events,
- Pesticide fate/transport and snowmelt processes cannot be simulated,
- Watershed size is small,
- Does not process on a land management unit basis.

Input data
Input information for the ANSWERS model contains simulation requirements, rainfall information, soils data, land use and surface information, channel descriptions, and individual element information that includes BMPs.

Appendix 8.1   Continued

Output description
The output listing consists of input data, watershed characteristics, flow and sediment information at the watershed outlet, effectiveness of structural BMPs, net transported sediment yield or deposition for each element, and channel deposition.

Availability
Documentation and user support, including a User's Manual, are available free from US EPA Region V. The ANSWERS program was developed by D. B. Beasley and L. F. Huggins at Purdue University under EPA sponsorship. For further information, contact Professor Beasley, Department of Agricultural Engineering, Purdue University, West Lafayette, Ind. 47907; telephone 317/494-1198.

Resource requirements
Written in Fortran, the program will run either on a PC or a mainframe. Requires a large memory, especially to simulate large watersheds. The data files have been designed to use Soil Conservation Service soil surveys, US Geological Survey topographic maps, and crop and management surveys.

Agricultural Runoff Management Model (ARM)

Description
The ARM model simulates the hydrologic, sediment production, pesticide, and nutrient processes on the land surface and in the soil profile that determine the quantity and quality of runoff in small agricultural watersheds. The major components of the model individually simulate the hydrologic response of the watershed, sediment production, pesticide adsorption/desorption, pesticide degradation, and nutrient transformations.

Capabilities
- Can simulate surface runoff, subsurface flow, and snowmelt,
- Both event-based and continuous simulations are available options,
- Includes different management practices.

Limitations
- Application is limited to agricultural watersheds less than 5 km^2 (1.9 sq. miles),
- No channel routing procedures are included,
- Model does not link the cost associated with different BMPs to pollutant loadings.

Input data
Detailed input data are required for simulating hydrology, snowmelt, sediment, pesticides, and nutrients. Some data must be generated from physical watershed and pollutant characteristics, land surface conditions, agricultural cropping, and management practices. Calibration and verification data are also required.

Output description
The printout provides summaries of runoff, sediment, pesticides, and nutrient loss in addition to nutrients remaining in the various soil zones. Generally, the output summaries are printed daily or monthly, but can be obtained hourly.

Appendix 8.1   Continued

Availability
The model was developed by Hydrocomp, Inc., Palo Alto, Calif., and is available through Tom Barnwell at the Water Quality Modeling Center, Environmental Research Laboratory, US EPA, College Station Rd., Athens, Ga. 30613; telephone 404/546-3175. The model is adequately documented.

Resource requirements
The model has been tested on the IBM 370/168 using the FORTRAN H compiler. The program requires approximately 360 K of storage for compilation of the largest subroutine, whereas program execution requires up to 230 K of storage, depending on the model options selected. However, Version II of the ARM model has been adapted to run on a Hewlett-Packard 3000 Series II computer, which is substantially smaller than the IBM machines on which the model was developed and tested.

Chemicals, Runoff, and Erosion from Agricultural Systems (CREAMS)

Description
CREAMS and CREAMS 2 are field-scale models that simulate surface and subsurface runoff, evapotranspiration, erosion, sediment yield, and plant nutrient and pesticide delivery. One purpose of these models is to evaluate BMPs. User-defined management activities simulated by CREAMS 2 include aerial spraying or soil incorporation of pesticides, animal waste management, and alternative agricultural practices such as minimum tillage and terracing. USDA is modifying CREAMS 2 to make the model more user friendly. The model does not need to be specifically calibrated for a given watershed. Most of the required parameter values are physically measurable.

Capabilities
* Represents soil processes with reasonable accuracy,
* Simulates continuously: considers event loads,
* Can simulate up to 20 pesticides at one time,
* Includes BMPs.

Limitations
* Subsurface drainage is not simulated,
* Data management/handling capabilities are limited,
* Maximum size of simulation area is limited to field plots,
* Receiving waters are not simulated.

Input data
CREAMS and CREAMS 2 require extensive data on meteorology, hydrology, erosion, and chemistry of the pollutants.

Output description
Output can be very detailed. Erosion data are available for each element considered.

Availability
Program manuals, tapes, and floppy disks can be obtained from the USDA-ARS Southeast Watershed Research Laboratory, P.O. Box 946, Tifton, Ga. 31793; telephone 912/386-3462.

Resource requirements
The model can be run on mainframe computers (IBM) or on personal computers (IBM-PC, AT&T).

331

Appendix 8.1   Continued

Groundwater Leaching Effects on Agricultural Management Systems (GLEAMS)

Description
GLEAMS is an extension of USDA's CREAMS models. GLEAMS simulates leaching of pesticides and nutrients from agricultural watersheds. The leaching behaviour of pesticides in root zones has been tested and validated. The model is being modified to incorporate subsurface nutrient transport and the effects of variability in soil porosity at different depths. The GLEAMS model is in development/testing stage. For more information, contact Walter G. Knisel, Jr., USDA-ARS, Southeast Watershed Research Laboratory, P. O. Box 946, Tifton, Ga. 31793; telephone 912/386-3462.

Cows and Fish (COWFISH)

Description
The COWFISH model is designed to assist resource specialists analyzing the condition of the riparian environment in relation to past and current livestock grazing management and to estimate the compatibility of grazing with associated aquatic resources. It is not intended to replace presently used stream surveys or fish population analyses. Rather, it uses existing information to derive an initial indication of how livestock grazing may be affecting trout populations.

The model considers six variables in determining a stream's suitability to support trout: (1) the extent of the streambank which is undercut, (2) the extent of the stream edge with vegetational overhang, (3) the extent of the streambank showing bare soil or trampling, (4) stream embeddedness, (5) stream width, and (6) stream depth. Two additional variables, stream gradient and the drainage soil type, are used to calculate fish production and recreational and economic value.

The field value obtained for each variable is converted to a parameter suitability index (PSI) based on principles similar to those developed by the US Fish and Wildlife Service in its habitat suitability index models. The PSI values are then averaged to compare the stream's existing habitat conditions with its potential habitat suitability index.

Capabilities
- Although originally developed for the mountainous regions of central Montana, after some adjustments this model can be used throughout the western United States,
- Can be used to determine stream habitat productivity any time during the season prior to snow cover,
- Can accurately assess current habitat conditions, provided the sampling area is at least 100 feet long,
- Can be used to evaluate larger sections of uniform streams. Data from five sites per stream mile would be needed to provide a 10 percent sampling of the study area,
- Can analyze a wide variety of riparian and stream types (the variation being in dimensions, flow conditions, streambank conditions, and surrounding environment).

Limitations
- Accuracy diminishes when the estimated analysis of grazing effects on fish production does not immediately follow the modelled livestock use,
- Less accurate for use along streams with rocky streambanks that do not follow the natural development of undercut banks,
- When sample areas smaller than 100 feet are used, the results will reflect population numbers only for the immediate area.

Appendix 8.1   Continued

Input data
Requires field data that include descriptive information about the stream, allotment, and sample size being evaluated. Specifically, information is needed on sample size, vegetative type, side valley slope gradient, percentage of undercut banks and banks supporting vegetative overhang, embeddedness, streambank alteration, width/depth ratio, and stream gradient.

Output description
The printout is in tabular form and contains information on the optimum number of catchable trout per 300 m of stream for (1) optional conditions for this stream and (2) existing conditions. The losses from optional conditions are also displayed, that is, the number of trout per 300 m of stream per year, recreation loss in wildlife and fish user days, and economic loss in dollars per 300 m of stream per year.

Availability
The model was developed by the US Forest Service's Northern Region Wildlife and Fish Habitat Relationships Program. Copies are available from USDA Forest Services, Federal Bldg., P.O. Box 7669, Missoula, Mont. 59807; telephone 406/329-3101.

Resource requirements
The user may obtain the results either manually by following the Guide to the Field Form or through a computer by recording the variables in the field and using the Data General software program developed for this model. The manual procedure provides the flexibility of obtaining the results while still on the site. No programming knowledge is required.

Evaluation System To Rate Feedlot Pollution Potential (ESRFPP)

Description
The animal lot evaluation system, developed to evaluate and rate the pollution potential of feedlot operations, consists of two parts: (1) a simple screening procedure that evaluates the potential pollution hazard associated with the feedlot, and (2) a more detailed analysis that is better able to identify feedlots that are not potential pollution hazards. The Soil Conservation Service's curve number method is used to estimate the runoff. Chemical oxygen demand and phosphorus are the two parameters used as the pollutant indicators. The ESRFPP model has been tested in several states. Currently, the Minnesota Pollution Control Agency requires that all animal lots in the State be rated by employing this model.

Capabilities
* Estimates pollutant discharge using simple techniques,
* Can be used as a screening procedure,
* Considers both surface and groundwater pollution potential,
* Evaluates the effects of different land management practices.

Limitations
* Runoff calculations may not be valid for large tributary areas (more than 100 acres),
* Model does not deal with receiving water bodies,
* The discharge point defined in the model may be difficult to apply in the field,
* Potential pollution threats to ground water are treated lightly.

Appendix 8.1   Continued

Input data
Data requirement is minimal. Most of the required data are presented in the ESRFPP manual.

Output description
The model estimates the concentration of pollutant indicators at the discharge point.

Availability
The method is well documented in the ESRFPP manual. Further assistance is available from Robert A. Young, North Central Soil Conservation Research Lab, Agricultural Research Service, USDA, Morris, Minn. 56267; telephone 612/589-3411.

Resource Requirements
All the calculations can be performed using a small desktop calculator. Programs have been developed to use with Hewlett-Packard 67/97/41C. Monroe 325, and Compucorp 327 calculators.

Hydrological Simulation Program – Fortran (HSPF)

Description
The HSPF is a series of fully integrated computer codes that simulate watershed hydrology and the behaviour of conventional and organic pollutants in surface runoff and receiving waters. The processes that affect the fate and transport of pesticides and nutrients from agricultural land are derived from the Agricultural Runoff Management (ARM) model, whereas the sediment delivery algorithms are from the Nonpoint Source (NPS) model. The HSPF contains three application modules: the PERLND (pervious land) and IMPLND (impervious land) modules perform the land and soil simulation for those land surfaces; the RCHRES (reach/reservoir) module simulates the processes that occur in a single reach and the bed sediments of a receiving water body (a stream or well-mixed reservoir). Extensive and flexible data management and statistical routines are available for analyzing simulated or observed time series data.

Capabilities
- Consists of systematic modular framework that allows a variety of operating modes, including continuous hydrologic simulation,
- Nonpoint source loading (including alternative control practices) and receiving water quality simulation are integrated into a single package,
- HSPF can analyze relative contributions and impacts of both point and nonpoint source loadings; offers options to use either simplified or detailed representations of nonpoint source runoff processes,
- Models risk assessment of the exposure of aquatic organisms to toxic chemicals delivered to receiving waters,
- By adjusting parameters, can include different agricultural management practices.

Limitations
- Calibration is usually needed for site-specific applications,
- A long time period (2–3 months) may be required to learn the operational details of applying HSPF if the user has no prior experience,
- Model does not link the cost associated with different BMPs to pollutant delivery,
- Depending on the extent of model use, computer costs for model operation and data storage can be a significant fraction (10–15 percent) of total application costs.

Appendix 8.1   Continued

Input data
If all modules are selected for model implementation, the HSPF requires an extensive amount of data. The meteorological and some hydrologic data are time series inputs.

Output description
The HSPF output includes system state variables, temporal variation of pollutant concentrations at a given spatial distribution, and annual summaries describing pollutant duration and flux. A summary of time-varying contaminant concentrations is provided along with the link between simulated receiving water pollutant concentration and risk assessment.

Availability
HSPF is in the public domain and can be obtained from the Center for Water Quality Modeling, Environmental Research Laboratory, US EPA, College Station Road, Athens, Ga. 30613; telephone 404/546-3175. The EPA Water Quality Modeling Center provides user assistance on an ongoing basis, periodically scheduling free training sessions.

Resource requirements
HSPF requires a FORTRAN compiler that supports direct access I/O. Twelve external files are required. The system requires 128 K of instruction and data storage on virtual memory machines or about 250 K with extensive overlaying on overlay-type machines. It has been installed on several systems, including IBM, DEC, VAX, System 10/20, Data General, MV4000, CDC Cyber, HP3000, HP1000, and Burroughs and Harris.

Nonpoint Source Loading Model (NPS)

Description
The NPS model continuously simulates hydrologic processes, including snow accumulation and melt, and pollutant accumulation, generation, and washoff from the land surface. Sediment and other suspended material are used as basic indicators of nonpoint pollutants. Simulates erosion on both pervious and impervious areas.

Capabilities
* Can simulate urban, agricultural, and silvicultural nonpoint source pollution,
* Both event-based and continuous simulations are available options,
* Can simulate nonpoint pollution from a maximum of five different land use practices in a single simulation run,
* Different agricultural and construction management practices can be simulated.

Limitations
* Does not consider subsurface flow, groundwater pollution, or channel processes,
* Simulates only sediment and nutrient transport processes; does not estimate pesticide delivery,
* Does not simulate the relationship between cost of BMPs and runoff as pollutant loadings.

Input data
Requires extensive input data, including those related to model operation, parameter evaluation, and calibration.

Appendix 8.1   Continued

Output description
The output from the NPS model includes (1) output heading (summary of the watershed characteristics, simulation run characteristics, and input data), (2) time interval output (can be for 15-minute intervals) and storm summaries, (3) monthly and yearly summaries, and (4) output to interface with other models (optional).

Availability
This model, developed by A. S. Donigian and N. H. Crawford of Hydrocomp, Inc., Mountain View, Calif., under US EPA sponsorship, is available through EPA's Water Quality Modeling Center, Environmental Research Laboratory, College Station Rd., Athens, Ga. 30613; telephone 404/546-3175. The model is well documented and has been tested for the simulation of nutrient loadings in surface runoff.

Resource requirements
The NPS model was developed on IBM 360/67 and 370/168 computers. Later, it was adapted to UNIVAC 1108, CDC 6000, and Honeywell Series 32. The computer core requirements for compilation and execution can be as high as 194 K and 144 K, respectively. With a reasonable level of technical support, it is expected that 2 to 3 man-months will be required to use and apply the NPS model.

Stormwater Management Model (SWMM Simplified)

Description
The simplified SWMM simulates runoff and nutrient transport in an urban watershed. The five tasks performed in this model include data preparation, rainfall characterization, storage-treatment balance, overflow-quality assessment, and receiving water response. Each task generally combines small computer programs with hand calculations.

Capabilities
- Can be linked to simplified, single-event receiving water model,
- Available options include continuous and event-based simulations,
- Models effect of pollutant delivery on receiving waters.

Limitations
- Does not consider water quality change or treatment during storage,
- Does not simulate snowmelt and sediment transport,
- Overflow quantities and qualities must be measured to calibrate model.

Input data
Input data include hourly precipitation, runoff coefficient, treatment rate, storage volume, and receiving water characteristics.

Output description
Output contains time-varying overflows and runoff and summation of these data. Pollutant loadings and receiving water response also are included.

Availability
The various SWMM models are available from EPA's Nonpoint Source Branch, 401 M St. SW, Washington, DC 20460. However, incomplete documentation and inadequate user support are problems with using the model.

Appendix 8.1   Continued

Resource requirements
The computer programs for simplified SWMM have been developed on an IBM 360/67
digital computer. The storage-treatment program has been used successfully on Xerox
560 and IBM 1130 computers.

Stormwater Management Model (SWMM)

Description
SWMM is a comprehensive, mathematical model capable of representing urban
stormwater runoff and combined sewer overflow phenomena. Level surface runoff
generated by precipitation is routed through channels and pipe networks. The analysis
uses finite difference approaches. The water quality parameters simulated in SWMM
include sediment and nutrients.

SWMM contains its own receiving water model, RECEIV. It also can be linked to
STORM, QUAL-II, and other simplified receiving water models.

Capabilities
- Both combined and separate sewerage systems may be evaluated,
- Considers treatment in five different storage systems,
- Includes several physical and chemical treatment options to evaluate water quality
  changes,
- Includes capital and operating and management costs for treatment options,
- Can simulate the effects of pollutant delivery on the quality of receiving waters,
- Different storage and treatment techniques may be considered as options for controlling
  nonpoint source pollution.

Limitations
- Is a large model with complex and detailed input requirements,
- Statistical summaries are limited,
- Uses a monthly flow routing method that can be costly because it requires short time
  steps,
- It is comprehensive, but data management capabilities are not advanced.

Input data
Detailed input data sets are required. Some of the input information typically required
includes precipitation, air temperature, wind speed, channel, pipe networks, land use
patterns, and storage and treatment facilities.

Output description
Output consists of summaries of treatment options and costs, variation of water quality
with time, and hydrographs and pollutographs with daily and hourly variations.

Availability
The model is well documented and available from EPA in four versions: Final Report,
Verification and Testing, User's Manual, and Program Listing. The existence of
established user groups that meet semi-annually is considered an important factor that
promotes its use. For a copy, write to EPA's Nonpoint Source Branch, 401 M St. SW,
Washington, DC 20460.

Appendix 8.1   Continued

Resource requirements
The computer hardware system should be the equivalent of the IBM 360/65 with peripheral storage devices and a usable core capacity of no less than 360 Kbytes. Data requirements are common to engineering design and analysis, and are mainly descriptive of the real system.

Small Watershed Model (SWAM)

Description
SWAM is a continuous simulation model that estimates change in hydrologic, sediment, and chemical characteristics of a small agricultural watershed in response to different land use and management practices. The overland flow and pollutant transport are estimated by CREAMS 2. The movement and interactions of sediments, pesticides, and nutrients in the surface drainage (channel network), and surface retention basins (reservoirs) are simulated. The model also can simulate surface/groundwater interactions.

Capabilities
- Is an integrated watershed model that uses a dynamic version of CREAMS 2 together with channel, reservoir, and groundwater routing of water, sediment, nutrients, and pesticides,
- Incorporates backwater effects in channel routing,
- Provides a detailed representation of soil and watershed processes.

Limitations
- A complex model to use,
- Watershed area is less than 10 km^2,
- Not practical for long-term (20 years or more) simulations.

Input data
Input data include rainfall, soil characteristics, topography, land use, and management practices.

Output description
The model developers are still working on output structure.

Availability
Model is still being tested and has not been released yet. Intended as a research model that can later be upgraded to basin scale. The model will be available from Dr. Donald DeCoursey, USDA-ARS, P.O. Box E, Fort Collins, Colo. 80522; telephone 303/221-0578.

Resource requirements
Initially written in FORTRAN for mainframe computers.

CHAPTER 9

AVAILABLE NONPOINT SOURCE POLLUTION CONTROL MEASURES

W. B. Clapham Jr, T. Davenport, M. M. Holland, W. Rast, S.-O. Ryding and J. A. Thornton

INTRODUCTION

There are several general approaches to be considered in attempting to address nonpoint source pollution in aquatic ecosystems. These approaches essentially comprise:

(1) Reducing the generation of nonpoint pollutants at the source;
(2) Inhibiting the transport of nonpoint source pollutants from their source to receiving waters; and
(3) Treating the resultant water quality degradation of the receiving waters.

From an overall perspective, it is usually more effective to attempt to treat or correct the basic underlying and/or more readily controllable nonpoint pollutant sources, rather than simply treating the symptoms. As noted by Ryding and Rast (1989), this approach is usually the most effective strategy over the long term. As noted above, this can be achieved either by reducing the generation of the polluting substances, or reducing their discharge or drainage at the point of origin. Another approach is to inhibit the movement of the nonpoint pollutants from their point of origin to susceptible receiving waters. This latter method involves altering the water drainage patterns, by such measures as installing structures in a watershed to inhibit water movement, or using various soil covers (e.g. grasses) for essentially the same purpose.

In a given situation, one can attempt to control nonpoint source pollutants with either structural or non-structural measures. An example of the former would be construction of an animal feedlot waste treatment facility, while an example of the latter would be to change the application methods of agricultural fertilizers or chemicals. The latter approach may even result in a reduction in the costs of specific treatment activities. For example, the Pollution from Land Use Activities Reference Group (PLUARG, 1978) – a study group formed under the auspices of the United States and Canadian International Joint Commission (IJC) – found that simple measures, such as changes in the timing of fertilizer applications, not applying fertilizers in excess of the actual crop needs, and ploughing manure fertilizers into the soil, actually resulted in monetary savings

Table 9.1 Summary of potential nonpoint source pollutant control measures for various land use activities (modified from US EPA, 1988)

Forested lands	*Agricultural lands*
Maintenance of ground cover	Conservation tillage
Management of roads and trails	Contour farming/strip cropping
Management of riparian zones	Integrated pest management
Management of biocides	Management of pasture/range lands
	Crop rotation
Multiple category	Terraces
Grassed waterways	Management of animal wastes
Management of stream bank areas	Management of fertilizers
Water interception/diversion	Livestock exclusion
Stabilization of stream banks	
Detention/sedimentation basins	*Urban lands*
Vegetative stabilization	Porous pavements
	Flood water detention/storage
Construction sites	Street cleaning
Non-vegetative soil stabilization	
Disturbance limitation	*Waste disposal sites*
Land surface roughening	Impervious liners/caps
Stormwater diversion	Flood routing

to the farmers in the North American (Laurentian) Great Lakes Basin. Such measures were considered to be 'good stewardship' of the land, with attendant positive benefits.

Nevertheless, one can also simply treat the symptoms of nonpoint source pollution. This latter approach may be the most logical approach in situations where the costs of treating the basic cause(s) are excessive, or if additional measures are required in a given situation.

The focus of this chapter is to examine alternative measures for attempting to control nonpoint source pollution of aquatic ecosystems. In fact, there is a surprisingly large number of alternative control measures for application in various situations (Tables 9.1 and 9.2). Some measures are more effective in developed countries, while others are most useful for developing nations. Thus, an attempt will be made to point out the strengths and limitations of the alternative methods, and to provide an indication of the expected costs and duration of effectiveness.

It is also noted that the large number of control options precludes a detailed description of any individual method. Rather, the various options will be identified and briefly discussed, in accordance with the above objectives. References for more detailed information are also provided. The reader is referred to these references for further details regarding specific control measures of interest.

A HIERARCHICAL APPROACH

The adoption of a hierarchical approach to nonpoint source management suggests that source minimization should always be a priority in any nonpoint source management programme. Where this has been accomplished, the hierarchical approach suggests that every effort should then be made to recycle or re-use potential contaminants in some other beneficial operation. Disposal should be viewed as the option of last resort. Some nations have pursued this philosophy more aggressively than others; Western Europe has a far lower waste load per unit of production than the United States, for example (Ryding, 1992). The United States, in turn, has far more stringent disposal regulations and protocols than many other nations; such 'end-of-pipe' approaches often place the burden of enforcement on governments, giving industry little incentive to engage in any other environmental protection actions (such as waste minimization), although waste recycling and re-use may be stimulated in order to minimize 'end-of-pipe' expenses (Gaba and Stever, 1995). While few countries have successfully introduced the hierarchical approach set out in this manual, the exigencies of international trade and competition may be powerful incentives to encourage nations to do so in the very near future (Higgins, 1995; Voorhees and Woeller, 1998).

Source reduction

Source minimization is an imprecisely defined term that means different things to different people. It is most useful to restrict it to on-site efforts taken to reduce the amount of material that may enter the environment as a potential contaminant. It has a number of dimensions. It may include changes in site design to reduce runoff, or eliminate certain kinds of contamination altogether. This may be relatively simple on some sites. On others, operations can be adjusted to increase the efficiency of materials uses to reduce environmental exposure of contaminants. Finally, some potential contaminants that might have entered the environment can be recycled in place. The operative notion here is that the recycling takes place in a way that is integrated into the site operations rather than being separated totally from it. Because of this, both the commitment and the economics of source reduction are integrated into the site operations, with the result that potential contaminants do not leave the site. While most often practiced in the industrial sphere, source minimization can be as simple as composting household vegetable waste and gardening debris, or using a rain barrel to collect roof-top runoff that might otherwise carry soil and contaminants into streams and lakes.

Recycling and re-use

There are few notions in environmental science which have the aura of recycling. In this particular case, recycling refers to any process by which a potential contaminant is shipped off the site producing it to a recycler, who makes a usable product from it. The 'usable product' may or may not be equivalent (or even similar) to the original product generating the 'waste' material (Manser and Keeling, 1996). One of the most straightforward examples of recycling is motor oil reclamation. Here, the oil reclaimer takes the spent motor oil and removes the contaminants, to produce other oil-based products for re-sale. While this is often possible and extremely appropriate for many potential contaminants currently disposed of as waste, it may not always be possible, especially with hazardous materials. Some materials have been so contaminated that they cannot feasibly be separated using standard reclamation techniques. In such cases, recycling could mean the manufacture of a new product with different specifications from the original – in the case of recycled motor oil, such re-use may take the form of oils used to restore and maintain road surfaces or incorporation into boiler fuels. This class of recycling is sometimes known as remanufacture, since there is a change in the actual specifications or use of the material.

Clearing houses

Materials produced in relatively small quantities may have a problem finding or developing a market. Recycling these materials often requires an active mechanism to find a potential user. This is the function of a waste exchange. The concept began in 1972 in Norway, but dozens of waste exchanges now exist, primarily in the industrialized countries of the world. A typical waste exchange is a clearing house for information. It is often run by a government agency or a non-profit organization whose function is to assist in bringing parties together to negotiate the purchase or exchange of the materials. It gets a small commission for bringing the buyer and the seller together.

Exchanges

Some waste exchanges take a more active role. Rather than simply providing information, they attempt to create a market for waste materials. They help companies producing wastes find ways to process those wastes into saleable form, and they may actively search out buyers for those materials. The commission charged for this service is obviously higher than that charged for the information exchange, but the more active approach may be needed when it is not immediately apparent how to get a potential buyer and seller together.

Destruction/detoxification

Should source minimization and recycling not be possible, perhaps due to the nature of the potential contaminant, recourse may be had to techniques which destroy either the contaminant or its toxic or hazardous properties. Physical and chemical treatments are the typical means of accomplishing this and these treatments cover a very broad range of processes. In general, chemical treatment processes involve chemical reactions of some sort, and physical treatment processes do not. In other respects, the two are similar and typically used in tandem. Some treatment processes are oriented toward recovery of potential contaminants; others are simply involved with reducing its hazard. They range from the re-refining or re-manufacturing of materials to the separation of contaminants from the aqueous solutions or slurries in which they are found. Physical and chemical treatments in general rely on processes that are well known and well established in the fields of solid waste and wastewater management, such as the neutralization of acidic or alkaline wastes, the aeration of highly organic wastes, or the incineration of solids.

Incineration of hazardous contaminants is very different from the simple burning of waste materials that used to be common in many areas. Temperatures are maintained at very high levels, so that even the most stable organic molecules are broken down into simple substances such as carbon dioxide, water, nitrogen, chlorine, etc. Most hazardous waste incinerators operate at temperatures in excess of 1100°C. Molecules that can be destroyed, which include virtually all organic materials, are destroyed at these temperatures, producing a noncombustible residue and gaseous exhaust. These gases and some very light particles may have some hazardous characteristics of their own and must be removed either in an electrostatic precipitator or in a scrubber (depending on the nature of the parent material: for example, hydrogen chloride (HCl), which forms hydrochloric acid upon solution in water, is produced when chlorinated hydrocarbons – e.g. some solvents, pesticides, and PCBs – are burned). Other intermediate combustion products may leave the combustion chamber before they are completely destroyed. These need to be broken down still further before they can safely be discharged into the atmosphere. Thus, many hazardous waste incinerators have an afterburner adjacent to the main combustion chamber.

Assuming that further treatment or recycling is not feasible, contaminants can be removed from the environment by means of detoxification or solidification. Sludges, for example, may be suitable for ultimate disposal if they are simply dewatered and ground into small pieces of uniform size prior to disposal in a secure landfill. It is also possible to use much more powerful solidification techniques, as is done with radioactive contaminants, by incorporating them into a stable material with a very high structural integrity and very low potential for leaching.

Disposal

The technique of final resort, disposal, is in itself a type of source minimization with respect to contaminants, proper disposal measures effectively removing contaminants from direct contact with the biosphere. Two types of disposal are commonly practiced depending on the nature of the materials being disposed of. The more common type, the sanitary landfill, is used primarily for nonhazardous materials, although various potential contaminants are frequently co-disposed in this type of landfill. The preferred technique for the disposal of hazardous materials is the secure landfill, specially designed and engineered to minimize the escape of contaminants to the environment (Blackman, 1996). Both types of landfill are briefly discussed below.

Sanitary landfill

Sanitary landfills often contain both domestic garbage and industrial wastes. In their most basic form, without any liners, leachate collection and treatment systems, or cover requirements (i.e. simply a 'hole in the ground'), these facilities are synonymous with dumps or middens; areas of the land surface where solid waste disposal is practiced. In their engineered form, where an attempt is made to manage and control the generation of contaminants produced by and from the solid waste, these facilities are known as landfills. As the latter facilities are gaining currency in both developed and developing nations, our discussion focuses on engineered solid waste disposal facilities as opposed to uncontrolled dumping.

Wastes deposited in these facilities break down anaerobically (see Chapter 6). After a period of time, the remaining wastes consist of refractory organic molecules, similar to those found in soils. Leachates from sanitary landfill sites, including industrial waste disposal sites, generally have such a complex composition that only multi-step treatment can be effective in neutralizing them. These wastes can, of course, contain any type of pollutant (see Chapter 5); however, heavy metals are usually the main pollutants in most relevant cases. These can be reduced by limiting the co-disposal of industrial and municipal wastes. Nevertheless, several different studies have indicated that, if the mass of metal hydroxides in industrial sludges is lower than 1 percent of the mass of municipal solid waste, no detrimental effects are likely to occur.

Secure landfill

A secure landfill is a carefully engineered burial site designed to accept hazardous wastes and to retard their escape into the environment. Underneath the wastes typically lie several liners of materials, such as compacted clay, high density polyethylene, and other artificial impervious fabrics, which are designed to prevent groundwater from entering the landfill and to prevent polluted water from within the landfill from reaching the water table or surface watercourse. Between the liners and the wastes themselves is a leachate collection system designed to collect any liquid which does accumulate in the landfill and allow it to be pumped out quickly and efficiently for treatment. When the landfill has been filled, it is covered with a compacted clay cap designed to prevent water from entering any cell within the structure containing hazardous material.

The impermeability of the cap and of the foundation underlying the landfill are the keys to minimizing releases of hazardous materials into the environment. If water could percolate freely through the landfill, it could pick up and transport contaminants as it came in contact with them. This phenomenon is called leaching. An impermeable cap overlying the storage cells minimizes the amount of water that can enter the landfill after closure. Likewise, the impermeable foundation is designed to prevent contaminated water from flowing through the soil into the groundwater system. Instead, it can be channeled to the leachate collection system and removed. Further, wastes are deliberately emplaced in such a way as to minimize their change through time. They are solid or semi-solid to begin with, and they are typically enclosed in steel drums. Soil is placed between the drums as the storage cell is filled, so that there is a clay barrier between waste masses even if the drums break down. This acts to insure the cell's structural integrity in the event of problems. Nevertheless, it is questionable just how 'secure' a landfill can be: a corollary of the logic of the secure landfill is that the structure must be managed as long as the materials emplaced therein remain hazardous – that is, forever, a daunting prospect that few permitting authorities have taken seriously. In many ways, to paraphrase Weinberg (1972), secure landfills represent a Faustian bargain with society: a place to put dangerous contaminants, but at the cost of eternal vigilance.

Underground injection

Underground injection is the disposal of contaminants in very deep wells that extend into permeable aquifers well below the level at which potable groundwater is found. These aquifers are separated from usable groundwater supplies by thick impermeable strata. This approach has been used for many years to dispose of oil-well brines and other wastes. It is a relatively inexpensive way to get rid of hazardous contaminants: it is roughly one tenth the cost of disposal in a secure landfill. While no form of land disposal puts hazardous wastes more out of society's reach, deep well injection has both advantages and disadvantages. On the positive side, a well operated injection well can effectively remove contaminants from contact with the human environment. On the other hand, one is relying on the underground geology to insure the isolation of the contaminants from the biosphere, even though it is impossible to assess the local geology directly. One cannot, for example, place a monitoring well in the vicinity of an underground injection well, because this would breach precisely the impermeable layers that must not be breached for the injection well to operate safely. Thus it is all but impossible to estimate the actual distribution of the contaminants in three dimensional space into which they have been injected by means of the well. It is entirely possible that contaminants could migrate through the subsurface into an area of cracked or otherwise breached strata that would enable them to come into contact with the biosphere. And because the contaminants are located so far from the surface, it would be very difficult to respond to a problem, even if one is identified. It is extremely expensive to clean up a contaminated underground disposal site as experiences with the Superfund programme in the United States have demonstrated.

The technologies of ultimate disposal – secure landfills and underground injection – have the capacity to provide releases into the environment over very long periods of time. The level of the release may be very low, indeed it may be very difficult to measure, and it may be so low that it would not be noticed without a detailed epidemiological study. The issues in this case are the nature and pathway of the release, the possibility of detecting it, and the length of time over which it can occur. For modern secure landfills with adequate leachate detection systems and with adequate enforcement of leachate detection, it is highly likely that any failure, either of the cap or of the liner system, would be noticed in time to prevent a large amount of waste from migrating outside of the landfill boundary. The same clearly cannot be said for underground injection systems. Nor can it be said for a secure landfill in which the leachate detection system has ceased to be monitored because of the expiration of the post-closure monitoring interval. It is not until some specific problem is noticed that further detailed monitoring is done, and only then do we notice that a leak has, indeed, occurred over some unknown interval of time. Unfortunately, the most likely event which triggers this sort of detailed monitoring is a rash of birth defects, cancers, or some other disease which is localized in a particular area.

STRUCTURAL CONTROL MEASURES

Introduction

In contrast to the development of a nonpoint source contaminant control philosophy, such as has been described above, the remainder of this chapter will deal with a range of structural and non-structural approaches to managing nonpoint source pollution. These approaches are often referred to as 'best management practices' or BMPs. Table 9.2 provides a summary of various structural nonpoint source pollution control techniques that will be described in the following sections. The Table is arranged in terms of the generalized types of control measures described, and human use activities as set out in Chapter 6. The body of the table contains the various specific control measures – primarily abstracted from the PLUARG (1977) report – and includes an indication of the possible control objective set out in terms of the pollutants described in Chapter 5. Thus, Table 9.2 is intended to be integrative, linking the topics covered in Chapters 5 and 6 with the control measures that form the subject of this chapter – whose selection and application form the basis for the subsequent chapters of this manual.

This chapter does not purport to be a construction or design handbook, nor a 'cook book' from which a 'recipe' for nonpoint source pollution control can be extracted at random. Rather, our intention is to provide a broad range of possible options, describing the basic principles behind the various pollution control

Table 9.2 Nonpoint source contaminant remedial measures (after PLUARG, 1977; PLUARG reference number given where appropriate). Upper case lettering = significant reduction in loading. Lower case lettering = moderate reduction in loading. A = acidification; B = macrobiota; E = sediments and erosive materials; H = heavy metals and radionuclides; L = macropollutants and litter; M = microorganisms and metabolic products; N = nutrients; O = synthetic organic chemicals; P = airborne pollutants; and S = salts (see Chapter 5 for detailed descriptions of these contaminants)

Remedial technique	Land use activity							
	Atmosphere	Silviculture	Agriculture	Transport	Residential	Industrial	Waste	Stream
Soil and bank stabilization								
1 Chemical soil stabilizers	p	e	e,n	e	E		E	E
16 Organic mulch		E	E	E	e	e	e	E
17 Netting/matting				E	e	e	e	E
35 Cover crops	p	e	E					
39 Grassed outlets		E	E	e	e	E		
79 Reforestation		E	E					e
82 Rip-rap armour								E
85 Shoreland slope design								E,h,n,o
86 Revegetation						E		
92 Stream bank vegetation								E
99 Livestock exclusion		e	E,n					
107 Hydroseeding		e	e	e	e			
109 Plant stabilization		e	e		e		e	

Table 9.2 Continued

Remedial technique	Land use activity							
	Atmosphere	Silviculture	Agriculture	Transport	Residential	Industrial	Waste	Stream
Runoff inhibition								
2 Rooftop ponding	p				e,n	e,n		
14 Temporary chutes		e	e	e	e	e	e	e
15 Check dams			e	e	e	e		e
19 Open furrow contouring			E					
22 Diversion terraces		e	e	e	e	e		
29 Surface roughening			e					
30 Soil aggregation			e					
38 Contouring			E					
77 Check dams			e	e	e	e		e
78 Retaining walls		e		E		e		
83 Culvert outlets		e	e	e	e	e		e
84 Dolos (tetrapods)				E	e	e		E
87 Low slopes (3–5:1)							E	E
90 Head gradient control							E	
95 Jetties, bank armour								E
98 Runoff re-routing				e	e	e		
100 Land smoothing		e						
101 Gabion baskets		e	e	E	e	E		E
102 Geofabrics				e	E	e	E	e
Infiltration								
3 Dutch/French drains					e,h,N	e,h,N		
4 Porous paving				e,h	e,h	e,h		
5 Lattice pavers				e	e,n	e,N		e

No.	Measure								
6	Infiltration basins					e,h,N	e,h,N		
7	Detention basins			e,h,N		e,h,N	e,h,N		
8	Dry wells					e,h,N	e,h,N		
9	Gravity shafts					e,h,n			
10	Injection wells					h,n,o	h,n,o		
20	Spray irrigation			O,N		O,N	O,N	O,N	
21	Surface water diversion		e	e	e	e	e	e	
42	Wet weather ponds	p				N	N		
55	Seepage reduction							O,h,N	
56	Underdrains				h,N	h,N	h,N		
57	Evaporation ponds	(p)					h,O,S	h,O,S	
66	Overburden segregation				e,N	e,N	e,N	e,N	
81	Sediment basins		E	E	E	E	E	E	

Soil management

No.	Measure								
11	Conservation construction		e	e	E	E	E		E
12	Temporary mulching–seeding		e		e	e	e	e	
13	Steep slope conservation		e	e	e	e	e		
23	No-till cultivation			E					
31	Strip-cropping		E	E					E
32	Alternative tillage			E					
33	Conservation tillage			E					
34	Sod-based crop rotation		e	E					
36	Enhanced soil fertility			E					
75	Road design	p	e		E				

Chemical applications management

No.	Measure								
24	Pesticide application	p	O	O	O	o	o		o
25	Non-chemical pesticides		O	O	O	o	o		o
26	Pellet fertilizers	p	N,o	N,o					
27	Fertilizer placement	P	N,o	N,o					
28	Timed applications	P	N,O	N,O					

Table 9.2 Continued

		Land use activity							
Remedial technique		Atmosphere	Silviculture	Agriculture	Transport	Residential	Industrial	Waste	Stream
37	Timed field operations	p	e,n,o	E,n,o					
62	Land-sludge disposal			n,o		n,o	n,o	N,O	
63	Manure storage			N					
96	Deicing salt reduction				S,h	S	S		
Runoff enhancement/liners									
74	Reduced infiltration							h,n,o	
106	Landfill liners							a,h,o	
Source reduction/control									
58	Street cleaning				E,h,L,n	E,L,N	E,h,L,n,o	E,h,L,N,O	
59	Aquifer purging							H,N,O	
64	Sewer flushing					E,h,n	E,h,n		
89	Resource recovery						h,o	h,o	
108	Catch basin cleaning				E,h,n,o	E,h,n,o	E,h,n,o		
Water/contaminant treatment									
40	Alum dosing			N		N	N		N
46	Treatment lagoons			e,N,o		E,N,o	e,N,O		e,n
50	Dissolved air flotation					E,h,n,o	E,h,n,o		
51	Phys-chem precipitation					E,N	E,N		e,n
52	Reverse osmosis							H,O	
53	Ion–clay exchange		O	O					
60	Chemical neutralization	P					A	A	
61	*In situ* liming	p					A,n,o	A,N,o	
91	Biological treatment							E,H,N,O	

		P							
In-sewer application/on-site wastewater treatment									
18	Aerobic septic systems			n,o		n,o	n,o		n,o
41	Swirl concentrators			E,n		E,N	E,N		
43	Stationary screens					e,h,n	e,h,n		
44	Horizontal rotary screens					e,h,n	e,h,n		
45	Vertical rotary screens					e,h,n	e,h,n		
47	Rotating bio-filters			e,N		e,N	e,N		e,N
48	Trickling bio-filters			e,N,O		e,N,O	e,N,O		e,N
49	Contact stabilization					N	e,N		e,M,N
65	Stormwater CSO regulators					e,n	e,n		
88	Multi-family septic systems			e,N,O		e,N,O			e,N,O
97	Septic tank			N		N	N		N
103	Waste incineration/evaporation	P		N		N	N	N	
104	Waste composting			N		N		N	
Modified mining operations									
67	Low wall mine barriers							E,h,n	
68	Longwall strip mining							e,h,n	
69	Block cut mining							E,h,n	
70	Head-of-hollow filling							E,h,o	
71	Box cut mining							e,h,n,o	
72	Area mining							e,h,n	
73	Auger mining							h,n,o	
76	Blocking infiltration							a,o	
Diversion									
54	Surface water diversion		e,h,n,o	e,h,n,o		e,h,n,o	e,h,n,o	e,h,n,o	
59	Aquifer interception							h,n,o	
94	Runoff diversion		h,n,o	h,n,o	h,n,o	h,n,o	h,n,o		h,n,o
105	Runoff controls			N					

techniques, whose selection must always remain site-specific, reflecting the combined influences of culture, cost and control objective(s). Non-structural options, which can be used either independently or in conjunction with these structural options, are discussed later in this chapter.

Soil and bank stabilization

Control objective

This group of nonpoint source contaminant control measures includes those techniques designed to minimize the movement of soils and attached pollutants to watercourses, usually by reducing flow velocities within stream beds and waterbodies.

Control technique

These techniques make use of a variety of materials for stabilizing the land surface or stream banks including vegetation, organic matter, man-made fibres and rocks. The most common of the older materials used for this purpose are the rock or concrete rip-rap materials used to line a stream bed or channel (Figure 9.1). Generally, these materials present a rough surface that acts to dispel the

Figure 9.1   Rip-rap used to stabilize a roadside drainageway

Figure 9.2 Wooden revetment used to stabilize a road cut; similar structures can be used to stabilize lake shores

kinetic energy of the flowing water, thereby reducing the erosive potential of the fluid. Wooden or metal revetments along the shoreline of a waterbody (Figure 9.2) and rock-filled gabions can also be used in this manner, especially where erosion and pollutant transport occurs along a watercourse boundary. More recently, there have been moves toward the use of vegetative cover to not only roughen the land or watercourse surface but also to provide a measure of biological uptake of nutrients or contaminants (Figure 9.3). This group of techniques is often referred to as 'green belts'. A variant of this technique is to provide organic material to the soil in the form of mulch which also acts to inhibit particle washoff. In situations where there is high flow or rapid changes in water levels, the use of man-made fibre netting or matting is sometimes used to stabilize the soils and permit plant rooting or placement of rip-rapping materials (Figure 9.4). Occasionally natural fibres, such as jute or hemp, are used instead of the man-made materials.

Where these techniques are not readily employable, chemical soil stabilizers can be used. In addition, the simple expedient of excluding cattle and other traffic from the watercourse and its vicinity can also prove effective. In the former case, the exclusion of cattle necessitates the provision of off-site watering points.

Figure 9.3 (A) Grassed waterways stabilize drainage way soils. (B) Grassed buffer strips reduce the transfer of agro-chemicals from cropped fields to neighbouring waterways

Figure 9.4 (A) Hay and nylon mesh provides protection for disturbed ground and allows natural vegetation to take root. (B) Impervious (polyethylene) sheeting prevents the loss of soil particles from material stockpile sites by containing the eroded particles on-site

Control costs

While any indication of cost is always relative to location in the world, the cost of labour and the relative costs of materials, it would be fair to state that the natural revegetation of watercourses and erodible lands usually offers the most cost-effective solution to erosional problems. In severe cases, a combination of measures such as filling erosional gullies with gabion baskets covered with top soils and stabilized with fibre netting and planted may have to be employed. The extensive use of engineered and man-made materials, obviously, is at the higher end of the cost spectrum. Generally, placement and maintenance of adequate vegetative cover and/or soil organic content is an effective means of reducing soil erosivity. Maintenance requirements vary from the need for regular watering and (perhaps) fertilization of nascent plant materials (for up to two years after planting; Wali, 1992) until they are well established and viable, to on-going mowing of grassed swales and waterways (preferably with the new generation of mulching mowers which return the organic matter thus harvested to the soil, thereby continuing the process of organic enrichment), to remedial grouting and maintenance of concrete structures and gabion baskets. Thus, revegetation techniques probably have somewhat higher operating costs over the longer term than the engineered methodologies which may off-set their initial low capital cost. Again, the economics of any given treatment technique will vary in terms of labour costs, site accessibility and degree of mechanization.

Effectiveness

Soil and bank stabilization is generally an effective technique to control soil erosion and can contribute to the reduction in loading of those pollutants which adhere to soil particles – such as nutrients, many synthetic chemicals and biocides, and some heavy metals. These techniques, in the agricultural sphere, can also ensure the long-term yields of the land, and contribute to the sustainability of future agricultural operations. These techniques are, however, usually ancillary to other types of control measures aimed at pollutant source reduction when employed in the control of contaminants other than sediments. Good sediment/soil management practices are strongly recommended as a component in all nonpoint source contamination control programmes.

Runoff inhibition

Control objective

Runoff inhibition, as a group of management techniques, is designed to modify the quantity of water flowing off the land surface or through a watercourse.

Control technique

The techniques placed in this group range from detention basins and measures to the actual re-routing of runoff. Typically, urban stormwater detention practices and agricultural runoff management practices involve the ponding of water and its gradual release over time – instead of the rapid runoff curve that has become typical in urban and rural settings where impervious surfaces and direct drainage systems are common, these methods aim for the slow release of water (Figure 9.5). Generally, these methods also involve a measure of water conservation or infiltration (the latter are discussed as a separate control technique below), such as the traditional rain barrel which collects water for family use, and the various contouring and ploughing techniques that promote water use by crops and (landscape) plants (Figure 9.6). In more hardened environments, such as in highway engineering applications or industrial settings, these structures can take the form of check dams or energy dissipators within culverts. In these cases, the use of geofabrics and man-made fibre nets, as discussed above, augment the inhibiting structures to minimize soil erosion and material transport (Figure 9.4).

The design of runoff inhibiting structures and systems can take many forms and provide for a number of scenarios (Debo and Reese, 1995). In a practical handbook on the design of (particularly) engineered structures, Stahre and Urbonas (1990) suggest the following methodologies:

(1) *Storage*, including inlet regulation (such as roof-top ponding and the use of parking/playing fields for temporary stormwater storage); open or 'wet' ponds; concrete basin, in-sewer, in-pipe and tunnel storage; treatment plant storage; and, various submerged storage systems for retaining stormwater within a lake basin or other watercourse; and

(2) *Flow regulation*, including a variety of fixed or moveable flow regulators (which typically consist of weirs, gates or fixed diameter outlets), flow regulating devices (such as the 'Steinscrew', 'Hydrobrake' or 'Wirbeldrossel' which are designed to pass a fixed volume per unit time; some devices are also designed to reduce the kinetic energy of the stormwater flow through oblique water entry ports, for example), and basin-wide planning which allows stormwater to be diverted to portions of the watershed having a lesser volume of precipitation (e.g. flood routing).

Such methodologies can be fairly simple, in terms of providing a storage area (see below), or relatively complex, such as flood routing which requires a knowledge of the storm characteristics of the entire watershed and probably some degree of pumpage to distribute the runoff through the system. While there are examples of the latter – the most notable being Seattle, Washington (USA) and Bremen (Germany), most municipal and other stormwater management systems make use of the less complex management techniques. In other words, storage tends to be the methodology of choice for most situations – it is most easily understood, visible, and, nevertheless, effective if properly sited and sized. Stahre and Urbonas (1990) give the following guidelines for sizing storage facilities:

Figure 9.5 (A) Stormwater detention basin. (B) Close-up view of the outlet control structure

Figure 9.6 Contour strip-cropping reduces the length of drainageways

(1) *Select a 'design storm' model* from among the available techniques which include the use of intensity–duration–frequency (I–D–F) curves (calculated from Eltinge's general formula) – which maximize depth of runoff per storm event or block rainstorm curves – which assume a constant rainfall intensity per storm event, or various modifications of these approaches which seek to mimic actual rainfall characteristics (e.g. the 'Chicago' and 'Illinois State Water Survey' design storms). Generally, a 2-, 5- or 10-year return period storm is selected as the basis of the structural design, although other periods can be used and may be specified in various national legislation.

(2) *Calculate the 'design flow'* using the basic relationship, $Q = C \cdot I_T \cdot A$, where Q is the runoff volume in m^3/s, C is the runoff coefficient, I_T is the rainfall intensity in $m^3/s/km^2$ (commonly converted from l/s/ha), and A is the watershed area in km^2, all multiplied by the 'design storm' duration in s to generate a volume, V_{in}. If using the block rainstorm method, Stahre and Urbonas recommend application of the Rational Formula, a multiplier of 1.25, to the calculated volume to estimate total runoff volume for a given storm event.

(3) *Determine the 'storage volume' for detention basins* by calculating the outflow volume of the pond, V_{out}, using the relationship, $V_{out} = A_d \cdot K \cdot t$, where A_d is the area of the discharge pipe or weir crest in m², K is the desired discharge rate in m/s and t is the time required for the pond to empty in s. The 'storage volume' is the difference between V_{in} and V_{out}. This general relationship can be modified to take into account the time that it takes specific pollutants to travel through the watercourse. Typical modifications include the 'time-area' method, the 'rain point' method, and the 'cumulative curve' method. Design algorithms for detention pond systems have been computerized, including the most common computer model of the US EPA, SWMM (Storm Water Management Model) (US EPA, 1975) (see Chapter 8). However, Stahre and Urbonas caution that this particular public domain program is highly complex (although good user-support is available from the US EPA) and that potential users should develop experience and familiarity with its use before applying the model in a problem-solving situation. It is, nevertheless, 'the model of choice for analyzing the performance of complete storm sewer systems, which may include detention facilities within such systems' (Stahre and Urbonas, 1990:263).

(4) *Select an appropriate site* for the facility which will incorporate the necessary area, soil characteristics and safety features (such as degree of public accessibility, basin shape and slopes, and public health risk) to maximize the efficacy of the basin and minimize public impacts. This site selection process can be done prior to step 3 in order to allow calculation of the precise operating characteristics of the facility, or done at this point after step 3 has created some site screening guidelines (Stahre and Urbonas refer to this process as 'superficial sizing'). In any event, once a site is selected, the calculations given above will have to be repeated during the design and implementation phase of the nonpoint source control programme (the 'detailed calculation' procedure of Stahre and Urbonas). It should be noted that, in terms of selecting a suitable site, multiple use of the basin area is becoming more common, where weather conditions permit – generally, this multiple use takes the form of playing/parking fields and other infrastructure that has a low degree of wet weather usage and can accept periodic inundation. Such schemes make maximum use of limited urban or farmyard areas.

Control costs

As above, the more natural the method employed, the more cost-effective it is likely to be. Engineered structures, such as stormwater drainages, culverts and various armouring techniques, generally require a more capital-intensive input than softer structures such as swales, reduced slope angles, and soil roughening techniques, although these latter may require more frequent maintenance for best operation. It should be noted that in the engineered structures, because of their cost and mass, care should be taken in sizing and siting these structures as it may prove difficult or impossible to return to a site to correct problems or deficiencies. With the exception of roof-top ponding, these techniques tend to be land intensive, although some of these deficiencies can be overcome through multiple use facilities or locating the facilities within (or under) existing infrastructure or rights-of-way (in the case of transportation corridors, for example).

Effectiveness

Runoff inhibition techniques probably represent the most widely used methods of managing stormwater runoff currently in use. While the methods do have some drawbacks (e.g. siltation in the case of detention ponds), the general level of satisfaction experienced by communities and individuals using these techniques has been consistently high, and provision for implementation of this type of control technology is being increasingly written into stormwater and nonpoint source pollution control legislation and policy around the world (e.g. in the United States, Europe, southern Africa, and Australia). These techniques offer a high degree of flexibility of application for either public, corporate or private use; good relative cost effectiveness within the constraints imposed by the technique(s) selected, and acceptable levels of flood and pollutant control for many contaminants of concern (Dennison, 1996).

Infiltration

Control objective

Infiltration, also referred to as 'local disposal' or retention (to distinguish this option from detention, or runoff inhibition, discussed above), includes a variety of techniques designed to retain and dispose of runoff on-site, usually through some form of underground or atmospheric treatment (Ferguson, 1994).

Control technique

Infiltration is similar in many respects to runoff inhibition, discussed above. Both methods typically require the storage of large volumes of water on-site or within a sub-drainage basin. Infiltration differs from stormwater detention in that the runoff is retained on site. The most common methods of infiltration are related in this manner, and include wet ponds, evaporation pans, injection wells and septic tank drain fields – the so-called 'French' or 'Dutch' drains (Figure 9.7). While these latter are not typically thought of as nonpoint source control measures – rather they are usually considered as nonpoint sources in many developed countries! – they do represent one of the oldest technologies still in

Figure 9.7 An infiltration bed located adjacent to a wet detention pond

common use throughout the world. The principle underlying this technique is one of soil filtration (see Chapter 4). Such filtration or treatment is effective for nutrients, especially phosphorus, and biodegradable matter, but can provide a direct pathway to the groundwater for more persistent contaminants such as pesticides and nitrogen fractions. Where such contaminants are to be disposed of, deep well injection, usually into geologically-stable strata such as salt domes, is the preferred method of containment. In certain situations, primarily associated with the disposal of toxic chemicals, hydrocarbon residues, and brines, this method has been used with a high degree of success. Nevertheless, in-ground disposal has also created some of the most well known environmental catastrophes of this century as in the case of Love Canal, New York (USA) where contaminants were disposed of in a former canal at shallow depth. Because of the difficulties in monitoring subsurface disposal, it was only through surfacial seepage and epidemiological studies spurred by a high incidence of public health problems in the area that this situation was discovered.

In contrast to these below-ground treatments, atmospheric disposal *via* evaporation ponds is possible where circumstance or geological instabilities make deep disposal impossible. In the case of brines and similar particulates, evaporation ponds can concentrate and solidify wastes into a form that is more easily disposed of (or re-used in the case of salts). For industries such as the leather tanning industry, evaporation ponds can dramatically reduce the volume of saline runoff from manufactories. In other industrial applications, evaporation and drying of organic materials such a brew mash or sewage sludge permits re-use of the organic residues as cattle feedstuffs or fertilizer, respectively (provided the latter does not contain other contaminants such as metals, etc.). Of course, there is a risk of airborne contaminant disposal from such facilities, especially if residues are burnt or incinerated (Ryding, 1992).

Other modifications of on-site disposal techniques include spray irrigation, diversions, covering stockpiles and using pervious pavers. Spray irrigation employs a combination of infiltration and evaporation, although the former is the preferred route for disposal of nutrients. Generally, wastewaters are commonly treated in this manner. Underdrains and similar collection mechanisms are often used to prevent groundwater contamination by seepage waters. Surface water diversions also contribute to this technique by routing stormwaters away from areas where contamination can occur. For example, simple steps such as covering material stockpiles can significantly reduce product loss or surface water and groundwater contamination. In urban environments, the use of porous pavers can enhance infiltration in situations where previously impervious paving would have been employed.

While these techniques cover a mixed bag of methodologies, most require on-site storage of runoff waters. As previously, the volume of these waters can be calculated in much the same manner as for detention ponds:

(1) *Select a 'design storm' model* as above.

(2) *Calculate the 'design flow'* as above.

(3) *Determine the 'storage volume' for infiltration basins* by applying Darcy's Law to the "design flow" and generating the outflow volume from the basin, V_{out}, using the relationship, $V_{out} = k \cdot i \cdot A_{perc}/2 \cdot t$, where k is the hydraulic conductivity of the soil in m/s, i is the hydraulic gradient in m/m (which generally is assumed to be 1.0 if the basin is situated at least 1 m above the groundwater table level), A_{perc} is the area of the *sides* of the infiltration facility in m², and t is the period over which infiltration occurs in s. The 'storage volume' is the difference between V_{in} and V_{out}. This formula can be modified to account for varying degrees of impervious surface in the watershed and rainstorm characteristics using appropriate multipliers. Site selection is obviously an important step in completing this calculation given the variation in infiltration rates, based on soil types, across most watersheds. Stahre and Urbonas present guidelines for site selection using a point-ranking system based on the ratio of impervious to infiltration surface areas, topsoil and underlying soil types, hydraulic gradient (i), and nature and use of the infiltration surface.

(4) *Select an appropriate site* as above (Stahre and Urbonas, 1990).

Control costs

Unlike detention facilities, retention facilities such as those described above and listed in Table 9.2 require more expensive construction techniques, and while evaporation and wet ponds are constructed on the sediment surface they typically require lining or other special construction techniques. In addition, these facilities are also relatively expensive to monitor and additional measures may have to be taken or implemented to avoid seepage or contamination of the underground water supply. In 1992, for example, the US EPA cited underground water contamination from leaking underground storage tanks, landfills and other disposal facilities, and agrochemical applications as the worst remaining water quality problem in the United States, and the most expensive and difficult to clean up (US EPA, 1989). Nevertheless, the use of these techniques may often be less space-intensive than surface treatments, although many do require storage area for runoff waters prior to infiltration. Infiltration, without supplemental injection capability, often functions least efficiently under wet weather conditions when the soil is moist or saturated, and when the water table is high.

In contrast, the use of evaporation ponds can result in cost savings through the re-use of salts and product recovered through this means. Such savings can offset at least some of the extra investment in land and liners. In fact, wastewater evaporation is one of the more effective disposal techniques in hot, dry climates such as in southern Africa.

Effectiveness

Although infiltration spans a broad range of applications and techniques, its application to surface runoff water management is an effective means of removing many contaminants of concern, including nutrients, metals, oxygen-consuming substances, and some organics. The versatility of these structures often permits their placement under parking areas or within playing fields and other public areas, such as rights-of-way along transportation corridors. When these techniques are used for disposal purposes in connection with process water disposal (a point source) or for disposal of process water contaminated stormwater, there may be a significantly increased chance of groundwater contamination. While industry spokespersons in the United States claim that underground disposal of contaminated waters is safe and has a clean operating record, much of this claim is due to intensive site investigation prior to facility placement (and possibly to the lack of monitoring after site closure!). Even where monitoring requirements are in place, these typically fail to consider that pollutants injected into underground strata are expected to remain sealed in those strata forever – monitoring requirements usually lapse after a few years.

One drawback to this type of runoff management is the eventual need to reconstruct or replace the system, which in terms of life expectancy for infiltration systems is due to siltation of the interstices between sand grains or in terms of life expectancy for dry wells and deep well injection areas is limited by the capacities of such systems.

Soil management

Control objective

Like soil stabilization, soil management practices are designed to conserve soils in place within watersheds, to prevent erosion and transport of soils to watercourses. In contrast to bank stabilization, however, soil management attempts to reduce erosion while the soils are still being utilized for economic benefit, typically associated with agriculture.

Control technique

Many of the same techniques used for bank stabilization can be employed in soil management. Generally, the same principles apply; namely, maintenance of ground cover and enrichment of the soils with organic matter to enhance their erosion resistance. Thus, while it is an accepted axiom that agriculture will disturb the soil surface, the degree of disturbance can be managed by using various ploughing techniques. While the chisel plough was the basis for the agricultural revolution that paved the way for the rise of European civilizations (Claiborne, 1983), this plough disturbed the soil to well below the A horizon and opened highly erodible soils to the transporting action of wind and water – this single agricultural implement created measurable increases in sedimentation within lake basins, which, in Wisconsin, has been measured in terms of feet of accumulation in the last century (natural sedimentation rates in the period preceding the introduction of the chisel plough were measurable in inches per century). In some nations, especially in Africa and other semi-arid areas, the very implement that made mass food production possible resulted in the loss of topsoil and the destruction of the agricultural industry (Monem Balba, 1995). Thus, adoption of soil management measures by the agricultural industry should be an essential part of any rural nonpoint source contaminant control plan if for no other reason than to save the future of the agricultural industry!

Soil management practices, as mentioned, revolve around actions taken to minimize soil disturbances leading to erosion. Conservation tillage (Figure 9.8), mulching, no-till agriculture (Figure 9.9) and sod-based crop rotation (Figure 9.10) all contribute to the maintenance of vegetative cover and organic enrichment of agricultural soils. In urban settings, conservation construction techniques and construction erosion prevention measures provide a parallel to rural soil conservation measures (see also mining operations below). One of the major attractions of this type of practice is the reduced amount of work necessary to prepare fields or construction sites – one of the disadvantages is that the large investments in chisel ploughs and similar machinery must recur

Figure 9.8 Conservation tillage leaves the crop residues at the land surface to reduce the erosive energy of rainfall on disturbed soils

Figure 9.9 With no-till planting, crops (winter wheat in this case) are introduced into soils using a 'seed drill' or similar device that does not disturb the soil surface

Figure 9.10 (A) Maintaining a cover crop can protect the soil surface from the erosive energy of rainfall. (B) Changing ploughing practices to leave the soil surface undisturbed during the non-growing season can help prevent soil loss during 'spring'

as it does not seem feasible to modify these existing implements for use with the less disruptive techniques. Notwithstanding this, the actual implementation of these techniques is not significantly different from the point of view of the operator; good soil management practices in the agricultural industry, for example, can be implemented by the current operators (provided they are informed of the methodologies) with little or no re-training required.

Control costs

Aside from the major capital expenses of re-tooling, the actual implementation costs are little different from those already incurred in the agricultural, construction and extraction industries. This group of control measures should be implemented in terms of land and water management practices regardless of the existence of any nonpoint source control programme. Tax incentives or outright government manipulation of the mechanical equipment and implement markets could effect the necessary replacement of equipment when existing equipment is replaced or (substantively) repaired. Ryding (1992) outlines a number of market-related strategies designed to accomplish such a shift in production technologies. Once in place, the benefits of good soil management practices will serve future generations as well as our own.

Effectiveness

Good soil management practices are extremely effective in the control of particulates and adsorbed materials. Combined with the chemicals application practices outlined below, these practices can even result in significant cost savings as well as pollutant load reductions, the magnitudes of which were recently responsible for wide-ranging changes in the United States stormwater management provisions (Chapters 301 and 402 of the US Federal Water Pollution Control Act, as amended; Davidson and Delogu, 1992).

Chemicals applications management

Control objective

The objective of chemicals application management techniques is the minimization of pollutant availability and reduction of pollutant loads carried into water courses.

Control technique

Like the soil management practices discussed above, chemicals application management is primarily an area in which the agricultural industry is impacted, although there are urban applications as well. Generally, these techniques can be divided into groups related to application rates and groups related to application times. A smaller group of techniques within this category also deal with the substance applied. This latter group relates to the form in which fertilizers and agrochemicals are applied; to wit, pelletized chemicals can deliver more chemical over a longer period to the area of the crop where it is needed than can similar quantities of liquid or granular chemicals in many cases. While generalization is difficult due to the wide range of crop types, some of the chemical loss experienced with granular and liquid chemical forms can also be managed by placing the chemicals into the root zones, for example, rather than employing broadcast application methods which result in a less directed placement of chemicals. Similarly, the use of alternative, generally less damaging chemical forms can reduce nonpoint impacts. An example of this has been the change in composition of road de-icing salts from sodium chloride to less environmentally-damaging calcium chloride.

Studies have also shown that the timing of chemical applications is critical to ensuring that the chemicals reach their intended destination, rather than being washed into the nearest watercourse or into the groundwater aquifer. In Europe, the traditional deposition of manures onto the frozen land surface in anticipation of the spring thaw has been shown to underlie the rapid wash-off of this material – and its nutrients and oxygen-consuming substances – in the snow meltwater, causing a significant part of the spring nutrient peak recorded in many temperate zone lakes. Ploughing the material into the ground after the thaw minimizes loss and enhances the nutritive value of the manure for crops rather than waterbodies. Attention to weather forecasts can also ensure that those chemicals that are applied to fields (or lawns, in urban settings) remain on the lands.

Finally, it is human nature to exceed recommended dosages on the basis that 'if a little does good, a lot will do more good'. This is as true of agrochemicals as it is of household chemicals. Good land stewardship would suggest that chemicals be used per the label directions and in quantities dictated by crop needs rather than by some artificial schedule. Thus good agricultural extension programmes and field soil testing programmes should be developed as an essential element of any good chemicals application management plan. Curiously, this would seem to be most relevant to agri-business operations; subsistence farming operations tend to be more balanced in their use of soil additives, although some forms of 'slash and burn' agriculture can be equally destructive to soils (Wali, 1992). Thus, perhaps, the situation of agri-business operations in regions of the world best suited to the highly technical nature of this type of integrated pest and nutrient management is more than fortuitous.

Control costs

Generally, integrated nutrient and pest management (INPM) programmes have a negative cost – they yield positive savings in agrochemical costs over traditional mechanized agriculture. Studies in the temperate zone have shown similar or increased yields from a significantly reduced input of chemicals. Not only have these studies shown increased yields for reduced input costs but they have typically resulted in reduced sediment loss from the farming operations. Thus, application of integrated nutrient and pest management practices have the double benefit of good soil stewardship and lower costs. In effect, INPM is an artificial attempt to mimic natural processes in the context of unnatural (virtual) monocultures without resorting to excessive chemical usage (Reuveni, 1995).

The success of INPM in the temperate zone is also due, in part, to ready access to products such as pelletized fertilizers, alternative ploughs, and the like, which may not be readily available elsewhere where agriculture is less mechanized or less of a big business. (Despite such availability, INPM can be capital intensive during the transitional period – especially if INPM is integrated with soil management techniques – although not necessarily so if initially adopted using traditional farm implements.) One of the drawbacks to INPM is the need for close monitoring of crops for pest infestations, soil nutrient status, etc. – services which can be provided by the farmer but are more efficiently provided by co-operatives, agricultural extension departments or other agencies. The skills necessary and investment in test kits, etc., may be out of reach of many in developing nations at the present time (Thornton *et al.*, 1991).

Effectiveness

INPM is proving to be a cost-effective means of enhancing (or, at least, maintaining) crop yields without resorting to excessive and costly application of artificial fertilizers and other chemicals, many of which lead to groundwater contamination in farming areas, suspected public health problems in farming areas and their markets, and as yet unknown impacts on humans and their environment. One of the primary foci of INPM is the close and careful monitoring of crops, yet, surprisingly, such close attention to immediate details often begets much longer term benefit. In addition to reduced input costs, experience with INPM practices have typically shown that soil losses are also reduced – a fact that ensures the long-term viability of the farming operation as well as significant short-term gains. In addition, in developed countries, the product that results from farms using INPM practices is often more acceptable to the 'informed consumer' (i.e. those persons who have legitimate concerns over the degree of potential agrochemical contamination of their foodstuffs). INPM practices also dovetail well with the soil management practices discussed above.

Runoff enhancement

Control objective

While most nonpoint source pollution control techniques seek to reduce the volume of runoff, certain circumstances demand the opposite approach. In terms of runoff enhancement, the control objective is to route waters away from contaminants and prevent pollution by minimizing contact between runoff waters and potential pollutants.

Control technique

This group of nonpoint source control techniques includes the use of impervious liners, caps and surfaces designed to prevent infiltration of the storm waters into underground aquifers or to prevent surface overflows from coming into contact with contaminants. In its most simple form, this group of techniques includes such measures as placing tarpaulins or other covers over material dumps, stockpiles or other exposed storage areas. In its most complex form, this group of techniques includes the creation, containment and closure of secure landfills, radioactive waste dumps and toxic chemical disposal sites. In most of these more complex situations, the use of clays and impervious liners both to line the disposal site prior to disposal taking place and to seal the top of the site after closure to prevent groundwater seepage/leakage and rainwater infiltration, respectively. In these latter situations, the key to successful establishment and closure rests upon exhaustive pre-construction site investigations, a stable surfacial geology, and careful construction, waste placement and sealing. Generally, the cap or seal placed over the site after closure must be less permeable than the liner to prevent infiltration into the sealed basin and creation of a 'bath tub' effect within the dump. In such situations, seepage waters can contact and transport the materials confined in the waste disposal facility, and potentially enter ground or surface waters elsewhere in the watershed. In addition, therefore, the clay and geofabric seal is usually mounded to promote surface runoff rather than ponding and infiltration into the facility. Guttering or other surface channels may also be included in the final design, and under-drains may also be provided to facilitate the collection and treatment of any seepage water that may enter the system. As noted, underground water contamination remains the most serious problem facing the United States, where as many as 20 000 secure landfills remain in operation. In Europe, such sites range in number from about 500 in Scandinavia to over 4300 in the Netherlands. In addition, abandoned sites number between 3000 and 35 000 in Europe and over 75 000 in the United States – it is these latter sites, typically constructed before stringent permitting requirements were promulgated, that form the primary concerns over environmental contamination as waste inventories and detailed locational information are lacking for many of these sites.

Any comprehensive nonpoint source contaminant management plan should make provision for solid waste disposal activities, whether they be rural (where manure, agrochemical residues, and household wastes are the primary solids) or urban (where industrial, commercial and household chemicals and solids are the principal wastes). Promulgation of recycling ordinances and waste disposal site regulations can reduce the degree of threat to human and other life posed by ground and surface water contamination arising from these sources.

Control costs

As mentioned, the techniques included under this group of control options range from the very simple to the very complex. In their simplest form, these techniques can actually result in cost savings – simple techniques such as covering materials dumps are actually good housekeeping practices which minimize loss of materials while protecting water quality. The investment in roofing, coverings or other shelter materials is usually rapidly repaid through the reduced washoff of stockpiled salts and solids. At the other extreme, that of secured landfills, costs can be very high depending on the nature of the materials being disposed of, the site characteristics, and the climatic conditions. Sites having stable surfacial geology and soil profiles, and low rainfalls, are best suited to landfill development at relatively low cost. Of course, the lack of investment in such landfills often results in contaminated waters, especially groundwaters, which are extremely expensive to rehabilitate, if such rehabilitation is possible. In the case of toxic wastes, such prevention 'is worth a pound of cure'. In addition to the establishment and closure costs, costs of secure landfills are high due to the intensive siting studies that should be done prior to establishing the landfill, and to the extensive monitoring requirements that many countries impose on closed sites – and also on operating sites in many cases. Analytical costs are a principle component of these pre- and post-establishment investigations.

Effectiveness

Given the high cost of rehabilitation associated with groundwater contamination, the use of secure landfills provides an effective means of minimizing the possibility of such contamination. As with most 'good housekeeping' practices, this suite of techniques can result in middle to longer-term savings over more traditional storage and disposal methods.

Source reduction

Control objective

Similar to the runoff enhancement techniques described above, source reduction attempts to remove potential contaminants from contact with runoff and other waters. These techniques also attempt to minimize the transport of contaminants to watercourses by removing the contaminants (rather than the transport medium, as was the case with the runoff enhancement methods).

Control technique

Source reduction is a particularly effective means of reducing nonpoint source water pollution, although some techniques such as street washing may actually enhance the movement of contaminants into the stormwater system or nearby watercourses. The preferred techniques include street sweeping – particularly effective against coarse particulates and macro-pollutants – and resource recovery/recycling. These techniques provide a solid link between water pollution control and the more familiar anti-littering and recycling campaigns waged in many nations throughout the world. Such housekeeping minimizes the amount of contaminants entering the aquatic ecosystem. Other techniques, such as sewer flushing and catch basin cleaning, aim to do this same job by seasonally removing accumulated sediments and solids from those areas where they collect during low flow periods or the dry season(s). However, unless these systems form part of a combined sewerage system, flushing simply enhances the throughput of pollutants to aquatic systems. Where storm sewers are made up of channelized sections, placement of sedimentation basins and litter racks in the channels can overcome some of the disadvantages of flushing in separated sewer systems (which are the far more common type in much of the world). Notwithstanding these engineered devices, however, regular cleaning and maintenance of these sediment and litter collection devices is essential to ensure their consistent performance (and to minimize possible public health problems due to accumulated refuse).

At the high end of the scale, source reduction also provides some degree of resolution to contamination problems. In the case of groundwater contamination, aquifer purging through high volume pumping to waste or treatment can effectively rehabilitate some types of contaminated aquifers. Where large-scale pumpage is required, considerable costs can be incurred.

Control costs

As noted above, costs can range from moderate to high, with street cleaning costs being at the lower end of the scale. Many developing nations use manual street sweeping methods as a means of promoting more complete employment among the populace. While these methods are moderately efficient at best, they do minimize the high capital and operating costs associated with mechanical sweepers.

Resource recovery, contrary to popular belief, does incur a cost even though the waste haulier often resells the materials recovered to third party manufacturers. The cost element is incurred for collection and storage of the materials until there is sufficient mass for re-sale. While resource recovery usually results in a lower cost than traditional refuse disposal techniques, it does promote good stewardship, and, thereby, good water quality management. In effect, the relatively slight costs associated with this type of technique make these rather neutral from the cost point of view, especially since most urban areas provide for some form of refuse collection and removal.

On the higher end of the cost scale, aquifer rehabilitation is perhaps the most costly and time consuming of all the available techniques. As noted under the section on runoff enhancement, it is generally more cost-effective to prevent aquifer contamination than to attempt to mitigate it later.

Effectiveness

While these techniques are usually very effective against macropollutants – litter and trash – they become increasingly less effective against small particulates and dissolved constituents. Studies conducted in North America have suggested that mechanical street sweeping has little effect on the proportion of dissolved and fine particulate matter reaching a waterbody, e.g. silt-sized particles and adsorbed contaminants, such as nutrients, metals, toxics, and oxygen-consuming substances (Ryding and Rast, 1989).

Water treatment

Control objective

The objective of this group of techniques is to remove pollutants from waters after the waters have been contaminated.

Control technique

These techniques are the standard wastewater engineering techniques that have been developed over the last century. They include filtration (ranging from bar screens and trash racks to sand filters and reverse osmosis [RO] techniques), flocculation (alum dosing and liming), flotation, and chemical treatments (waste/ acid neutralization and ion exchange). While comprehensive physical/chemical treatment represents a high-tech advance on the more standard engineering practices, biological treatments, such as lagoon systems and biological treatments, represent new technologies that are presently being developed and implemented (Kadlec and Knight, 1996). With regard to these, the creation of artificial wetlands and the use of aquatic plants to treat effluent streams represents an unusually cost-effective means of treating wastes, especially in warmer climates. These latter techniques are particularly effective in removing nutrients, metals and some organics from water. Virtually all of the technologies, biological as well as engineered, are well proven (US EPA, 1993; Kent, 1994; Moshiri, 1994; Hammer, 1996; Kadlec and Knight, 1996).

Control costs

The most significant drawback to the engineered solutions is their cost. While the technologies exist to turn any waste stream into distilled water anywhere in the world, few nations can afford to subject storm waters or other waters to such intensive and expensive treatments. Hence, the engineered solutions are usually used in conjunction with combined sewer systems – in which stormwater is routed through wastewater treatment plants, specific industrial applications, or cases of last resort – as in the Wahnbach Phosphate Elimination Plant (PEP; see Ryding and Rast, 1989).

Biological treatments, on the other hand, represent a more cost effective means of treating contaminated waters on a larger scale than is commonly possible using engineered techniques. For example, the wastewater system at Walt Disney World's EPCOT Center in Orlando, Florida (USA), is totally based on biological treatment and is capable of consistently maintaining an effluent quality that is consistent with stringent state requirements. For these reasons, it is becoming more common to add a biological component to detention pond systems (see above) – in effect creating artificial wetlands to replace or augment wetlands lost through development. This addition also has the desirable effect of permitting the management agency to select against nuisance vegetation that often infests pond systems. While such plantings require somewhat more management intervention to maintain, as opposed to more sterile pond environments, the water quality benefits are usually worth this effort. While not effective or appropriate in all cases, biological treatments are adding another dimension to stormwater and runoff treatments.

Effectiveness

While the removal of contaminants from water that has already been allowed to become polluted is the least desirable of the many nonpoint source management techniques, it remains a highly effective means of restoring contaminated waters to a higher level of utility. The technologies involved, with the possible exception of the biological techniques, are all well proven and highly effective. As more experience is gained in biological treatments, these too are proving to be very effective in managing water quality problems. The latter are particularly suited to use in the inter-tropical region where most plants enjoy a year-round growing season and high rate of productivity. Under such conditions, biological treatments enjoy significant cost advantages over the engineered options.

In-sewer applications

Control objective

Literally falling somewhere between the high-tech wastewater treatment methods described above and the source-reduction techniques described earlier, in-sewer applications aim to concentrate and treat (or remove to treatment) contaminants being carried in the stormwater stream. Thus, these methods have applicability in those situations where established stormwater drainage systems exist, where such systems are planned, or where other techniques are not possible due to space or other limitations.

Control technique

The techniques included under the general heading of in-sewer/on-site treatments can be grouped into three basic categories; namely, private sewerage systems (e.g. septic tanks, multi-family systems and waste incinerators), public sewerage systems (biofilters and screens), and in-sewer treatments (concentrators and regulators). The former generally rely on soil filtration to remove nutrients and oxygen-consuming substances from domestic wastewater ('grey water'), and can be built as either below-ground (traditional systems, such as French drains) or within mounds at the ground surface ('mounded systems') where the water table precludes construction below ground level. The ability of the soils to percolate or filter the effluent is the critical factor in this treatment process; many types of soil, especially glacial tills, are unsuitable for this type of treatment. In such circumstances, the use of waterborne sewerage systems, referred to as public systems above, is usually recommended. These systems are better suited to treating 'stronger' effluents (e.g. those effluents having a high, > 10 mg P/l nutrient content as opposed to those with a more dilute nutrient concentration). Augmenting simple screening methods for the removal of coarse solids and floatables are the biofilters or trickling filters that form the hallmark of secondary treatment plants. These filters make use of microorganisms living on the surfaces of stones to remove oxygen-consuming, organic matter from effluents prior to discharge. Flow rate and the 'strength' of the effluents determine the efficacy of these systems – slow flows and high concentrations of nutrients and organic matter make these units very efficient. Unfortunately, these units are susceptible to poisoning (usually *via* some toxic discharge dumped into the sewer system) and catastrophic die-offs that can seriously affect their performance. The ultimate form of public treatment, tertiary treatment, requires some form of nutrient removal, now increasingly provided using activated sludge – e.g. a portion of the organic solids contained in the sewage is used as a substrate for microbial growth, such microbes using the available nutrients for growth. Excess sludge is siphoned off, dried and (typically) land disposed. These systems can use either aerobic, anaerobic or alternating aerobic/anaerobic zones to accomplish nutrient removal. The centralized treatment facilities that contain these units are best suited to urban environments with closely situated properties, unlike the private systems that generally require larger, more widely-spaced properties. Where public systems exist, the potential for employing additional treatment techniques to the in-pipe or in-sewer flows exists, and can prove effective especially in the case of combined sewer flows subject to extensive seasonal variation. These methods primarily store water for later treatment, although swirl concentrators also begin the process of sediment removal.

Control costs

Since these options are again primarily engineered options, they do carry a relatively substantial price tag. Indeed, it is partially the cost of private septic systems that has provided some of the impetus for communal, multi-family systems; that and the fact that in the developed nations there has been a move toward planned urban developments (PUDs) that place moderate to high density housing into larger acreages that retain some degree of function as ecological units. Under these higher densities, multi-family systems are more efficient and effective than a series of smaller, individual ones. On the other hand, in most developing nations, private septic systems consist of various forms of latrines which provide a significant increase in the level of sanitation. These systems are essential in order to break the cycle of many waterborne parasites that utilize human hosts and to limit the spread of waterborne diseases previously rampant due to contaminated water supplies (obviously, the success of these latrines depends on adequate soil filtration and separation from water wells or water supply points). Because they are not as technologically sophisticated as septic tank systems, the latrines are much less costly and have proven to be culturally acceptable to many nations. A variant on the private sewerage system is the waste incinerators pioneered in Scandinavia. These, however, are relatively expensive units.

Public systems require an extensive infrastructure demanding on-going maintenance and servicing for best effect; leaking sewers negate any benefit gained in terms of groundwater protection in many cases. In addition, the degree of technical control required to ensure optimal operating conditions increases as the degree of treatment increases. The same is true of construction costs and complexity of operation. Highly skilled technicians are required to effectively operate an activated sludge treatment plant, for example. The benefits are essentially in terms of being able to spread the financial burden across a whole community of users who typically pay for the service through a graduated tariff scale that places a premium on difficult or intractable wastes, occasionally requiring pre-treatment prior to the waste being accepted for treatment – an example of this is the requirement for neutralizing strong acids or precipitating heavy metals that would otherwise kill the microorganisms in the biofilters. These plants can also produce a more acceptable effluent, if properly run, than many septic systems.

Other waste treatment options include waste composting or recycling by biological means (e.g. aquaculture) which has been practiced extensively in Asia. Many of these systems make use of all waste materials to successively farm pigs, fish, and fowl in an almost closed-loop system. Until recently, the high nutrient content of the water leaving these facilities has been ignored, but this is now becoming increasingly subject to some degree of management to optimize nutrient utilization and conversion to useable biomass (i.e. fish flesh or plant biomass other than algae). These systems can actually generate income!

Effectiveness

Depending on locale, the various levels of technology and the various techniques mentioned above can be effective in reducing the contaminant load contained within wastewater flows. Selection of an appropriate technique is usually guided by legal requirements (effluent discharge standards, public health standards, zoning requirements, etc.) and, in part thereafter, by cost considerations. Most methods tabulated in Table 9.2 are effective in removing particulates and, to lesser degrees, nutrients, oxygen-consuming substances, and pathogens. Some in-sewer treatments are also designed to remove metals and toxicants adsorbed to particulates. Most of these methods would be considered point source control measures which affect nonpoint source contamination only peripherally since many of the contaminants would become nonpoint if not treated in this manner.

Mining operations

Control objective

Extractive industries typically generate significant quantities of materials that are of little or no import to their operations – materials which are referred to as 'spoils' or, in the case of surfacial mining operations, overburden. These materials are usually stockpiled or dumped to 'slimes dams' (in the form of a slurry) which can contribute to aquatic pollution through direct runoff of process waters, leaching or contamination of stormwaters (erosion, etc.). By modifying the extraction process, in ways analogous to the integrated nutrient and pest management practices suggested for application in the agricultural industry, the volumes of water used and the potential for contamination can be reduced. Thus, the objective in most cases is to minimize water contact with potential pollutants.

Control technique

Modified mining operations is a special case application of many of the principles discussed above, including source reduction, water treatment, and runoff enhancement. The use of water treatment techniques applies primarily to those contaminants that would be considered point sources; i.e. process waters used in mineral separation, acidic mine waste, chemically contaminated wastewaters from finishing operations. These types of waters are best treated using wastewater treatment techniques such as neutralization, flocculation and oxidation. Similarly slurry waters (return water) from slimes dams and similar disposal areas should also be treated as a point source, particularly to reduce the high suspended (and dissolved) solids loads commonly carried by such waters. A second major group of polluting activities can be best addressed by improved housekeeping practices, whether these be runoff enhancement (to remove water from contact with potential contaminants) or source reduction techniques (to remove potential pollutants from contact with water). In many cases, good housekeeping practices can reduce production costs (e.g. through less wastage) or enhance product yield. Such techniques have also been shown to have significant impacts on improved worker safety which also produces a positive effect on production. Many of the techniques listed in Table 9.2 refer to extraction methodologies which involve back-filling surface-mined areas, leaving a certain amount of potentially extractive materials *in situ* to act as barriers not only to potential pollutant movements but also to 'frame' the mining operations. This has the effect not only of reducing contamination but also of decreasing the potential for cave-ins or other dangerous situations to develop. Thus, the majority of these techniques have identifiable benefits in addition to environmental preservation.

Control costs

Some of the methods that can be employed in the modification of mining operations will require significant capital investment as they involve major changes in the way resources are typically extracted. An example would be auger mining instead of the more usual open-pit mining. Most of the methods suggested, however, are mere modifications of existing techniques that can be easily (and probably cost-effectively) implemented. The benefits of good housekeeping practices in terms of both environmental protection and worker safety have been mentioned. Other benefits can accrue from adoption of backfilling techniques; i.e. instead of having to use mechanical or other equipment to transport waste or overburden to separate disposal areas (or stockpiles) where they consume valuable land area, extraction equipment emplaces the waste materials into the open areas created to the rear of the working face. Instead of two or three separate actions, backfilling reduces spoil movements to a single operation. In this way, the potential for spills, erosion and water contamination must be reduced, the amount and types of equipment required may be reduced, the exposure of workers to hazards is lessened, and the number of temporary structures (whether these be for environmental pollution control, spoil containment or other purposes) is minimized – all of which must lead to some degree of cost savings. While not all types of mining are amenable to backfilling, other techniques may prove equally efficient. As has been the case in the agricultural industry, there has usually been considerable opposition to such 'newfangled ideas', but those companies that have adopted these techniques have generally had good experience with them.

Effectiveness

While extractive industries will always create some degree of disturbance (= contamination of waters, in this instance) simply due to the nature of their activities, adoption of modified extraction techniques can significantly moderate the intensity of those impacts. In addition to moderating the environmental impacts of their operations, such techniques commonly have both direct and indirect financial benefit to those industries. In many instances, adoption of modified techniques will not involve any substantive change in equipment (unlike many of the changes proposed above in terms of the agricultural industry) and can be implemented without further investment in ancillary studies, technologies or monitoring – mining already consuming many of these services in simply locating extractable deposits. Given their effectiveness (and their neutral or even positive cash flow potentials), adoption of appropriate alternative mining techniques should be considered even where the nonpoint source pollution impacts of such operations are negligible.

Diversion

Control objective

The final major group of nonpoint source pollution control techniques may be considered as something of an 'if all else fails . . .' group of techniques whose control objective is to manage the movement stormwaters or groundwaters.

Control technique

While bearing some resemblance to the flood routing techniques discussed earlier (see runoff inhibition), this group of techniques seeks not so much to control the quantity or rate of runoff but rather its location or direction relative to sources of possible contamination or areas of contaminant impact. Interbasin transfers (IBTs) and 'flood control' schemes that channel or pipe runoff waters around sensitive waterbodies (whether they be sensitive to flooding or environmental damage) are examples of techniques contained within this category. A recent example of such a scheme is Lake Delavan, Wisconsin (USA), where the nonpoint source contaminant load carried by flood waters was identified as the cause of water quality degradation in the lake. In response, such inflows are diverted by channelization (actually by an artificial peninsular) away from the main body of the lake to the outlet – in effect 'short-circuiting' the inflow–outflow process (UW-IES, 1986). These techniques can also be applied in reverse; i.e. to divert flushing waters through systems already impacted by excessive contaminant loads arising from either point or nonpoint sources (Cooke *et al.*, 1993). Traditionally, diversion techniques have been applied for flood control purposes to reduce flood peaks downstream and for pollution control purposes following oil or chemical spills to ensure downstream water supplies. The latter have also been used extensively in the case of groundwater contaminant containment, where polluted groundwaters are intercepted by down gradient well fields to prevent contaminant spread (Figure 9.11).

Control costs

Given that these techniques are usually based on extensive engineering interventions in natural systems, the costs associated with their construction are usually high, although more recent movements back to earthen berms instead of mass concrete structures has resulted in some cost savings – these earthen structures normally make use of geofabrics or other strengthening materials to

Figure 9.11 A diversion scheme that intercepts barnyard runoff can protect down-gradient watercourses from contamination

ensure their protection from erosion and resistance to seepage losses and undermining. While it may have been implied that these methods are those of last resort, they do have a significant place in the repertoire of nonpoint source control techniques and can prove to be cost effective 'solutions' where constraints of topography or land usage apply. In the Lake Delavan example mentioned earlier, the degree of watershed development and environmental sensitivity of the headwater marsh area precluded other management options. Thus, while expensive, diversion proved to be the best available technique in this situation.

Effectiveness

Diversion is usually the least effective of all the various control options discussed. It simply shifts the problems or potential problems elsewhere without addressing either their causes or their consequences. Thus, in this sense, diversion does represent the choice of last resort. While this option is usually available in nonpoint source pollution control plans, it should be accepted only in those situations where other controls cannot be readily implemented. The relatively high costs and lack of treatment for contaminant removal render little benefit from adopting this option.

NON-STRUCTURAL MEASURES

Introduction

Growing environmental awareness encourages all sectors of society to try to do their part in resolving environmental problems. To be successful, however, environmental efforts must become more concentrated on preventive measures based on market forces and economic realities, in contrast to the present, mainly administrative approach which has traditionally been focused on 'tidying-up afterwards'. In many countries, the mechanism for ensuring adequate management of contaminants and the avoidance of public health and safety problems has been the use of legislative mandates or laws to control the production, use, discharge and disposition of substances that may be harmful to humans (e.g. the control of poisons, biocides, industrial chemicals, etc.) or the environment (e.g. nutrient and other effluent discharge standards, etc.). These standards continue to be the principle means by which nations regulate the environmental exposure of potential contaminants (see Schlickman *et al.*, 1994; Ryding, 1992; Fuggle and Rabie, 1986; Alabaster, 1980; and others, also van der Leeden *et al.*, 1990, for a compilation of standards from Europe and North America). Nevertheless, there is a growing recognition that environmental policy has both direct and indirect effects on economic activity, both as a direct constraint (imposing limits on the emissions of pollutants, restrictions on plant design and product characteristics, and site planning and land use) and indirect influence (on production costs, prices, competitiveness, profitability, demand and employment). These effects are manifested in associated industries (both manufacturing and agricultural) as well, affecting suppliers and retailers, and, ultimately, the 'man-in-the-street', influencing buying preferences and consumption patterns.

Conventional economic development has contributed to the betterment of mankind, but generally at the expense of environmental degradation to various degrees (see James and Niemczynowicz, 1992). Prices in all types of economies have generally failed to even partially reflect the ecological consequences of human activities (Gore, 1989; Ryding, 1992). Both socialist and developing countries are aspiring to achieve levels of material consumption typical of developed countries. If they succeed, levels of consumption of natural resources could rise roughly five-fold. However, at even the current world levels of energy, mineral and water consumption, this aspiration is unsustainable. To avoid catastrophic clashes between aspirations and ecological realities, there is a need to develop a new and sustainable concept of consumption and its relation to the standard of living to be introduced in all countries of the world (*cf.* Thornton *et al.*, 1992). However, a new pattern of resource consumption cannot be achieved without introducing a new set of policy instruments. Taxes on environmentally important resources emerge in many people's views, as the front runner among those instruments. Revenues from environmental taxes can be returned to the tax payer by lowering other taxes, thus leaving the average tax burden largely unchanged. As an example, charges on all types of polluting discharges from business ought to be neutral, with regard to trade competition, and may be returned to the polluter to more quickly enable the necessary environmental investments to be made to rectify the situation. Economic theory suggests that environmental taxes replacing productivity taxes can make economies stronger, not weaker. For these reasons, our emphasis in this portion of the chapter on contaminant control measures will be on economic, rather than the more traditional legal, mechanisms used to protect the natural environment.

Economic instruments

Many people claim that economic or market instruments are appropriate tools for ensuring that economic growth will fully account for the 'expenditures' associated with our environmental capital. It would be surprising if, in a situation where great confidence is placed in market mechanisms, such tools were not fully exploited for environmental protection. Economic incentives can help shape economic development towards environmentally-clean technologies. Used in conjunction with regulatory instruments, economic incentives can provide an additional stimulus to a discharger to improve the quality of emissions, and to find new solutions for minimizing waste. To their advantage, economic instruments are more flexible, and can prove more cost-effective, than reliance on regulations alone.

In many developed countries, economic instruments have been used since the early days of environmental policy development. However, economic instruments have remained secondary to other instruments, enabling a more directed, regulatory approach to environmental protection. Many economists have advocated greater use of economic instruments, as they are expected to provide environmental policy-makers with flexible, effective and efficient options in realizing environmental objectives. They are also expected to lead to an ongoing pressure for further reduction in emissions and the development of cleaner technologies. These aspects have frequently been put forward as features supporting a preference for economic instruments over direct regulations.

Economic instruments are tools with fairly wide application in the field of environmental policy. With these instruments, financial burdens are put on polluters, in some cases these costs provide a clear incentive for the polluter to reduce pollution loads. Alternatively, polluters can be offered financial incentives if they modify the environmental impacts of their activities.

Charge-based systems

Charges can be considered as the 'price of pollution' or a fee for environmental 'services' (Nelson, 1995). Typically, charges induce the polluter to reduce emissions to the level at which the unit cost of in-house abatement equals the marginal cost of external treatment. Beyond this level it is cheaper for the polluter to pay the charge than to continue abatement measures. The higher the charge, the greater the incentive for internal remedial measures. Even if the initial effects of using a charge system as a stimulant for investment in waste treatment and alternative technical solutions might not be the same as those of regulations, there is almost certainly to be a greater benefit over time. This is because technological progress in the field of environmental control would receive continuous stimulation under this scenario. Through innovations in abatement techniques, in processing, in the forms of (new) products, in consumer sacrifices for the benefit of the environment through changes in their patterns of consumption, and in other non-environmental utilities that are gained, the effects of charges on the economy can be supposed to be reduced considerably over the long term. However, as the practical possibilities of using this approach are not generally available, and because immediate results are badly needed in some problem areas, regulatory measures are often called for as a complementary measure to economic charges. Further, and probably most important, international trade and economic co-operation is a necessity to ensure that all producers have a fair and equitable chance to market their goods and services.

A distinction can be made between effluent charges (based on the quantity and/or quality of discharged pollutants), user charges (tariffs that are uniform or differ according to the amount of effluent treated), product charges (laid upon the price of products), administrative charges (authorization fees) and tax differentials (allowing more favourable prices for 'environmentally-friendly' products):

(1) Effluent charges are common in water pollution control policy and, to some extent, in relation to aircraft noise abatement policy. In contrast, effluent charges do not play an important role in air pollution control or waste management.

(2) User charges are common with respect to collection and treatment of municipal solid wastes and of wastewaters discharged into municipal sewers. User charges are most effective where society considers collective action more desirable than (diffused) private action and where their introduction enhances their environmental effectiveness and economic efficiency.

(3) The majority of product charges are applied to cover environmental expenditures relating to potentially harmful products. Their environmental effectiveness depends on the extent to which necessary expenditures are covered: the higher the sales of such products, the higher the revenues from the charges, and the better they serve their purpose.

(4) Administrative charges are mainly intended to finance direct regulatory measures, such as licensing and control activities on the part of authorities; thus they have a redistributive effect. Administrative charges may be environmentally effective if the revenue facilitates a better performance of the control tasks by the authorities.

(5)   Tax differentiation as an instrument of environmental policy has not been widely used. In a number of countries, however, tax differentiation has been applied to fuels to encourage consumption of unleaded fuel, and on car prices to favour the sales of 'clean' cars.

Subsidies

Subsidies is a general term for various forms of financial assistance. Subsidies can either take the form of grants (a non-repayable form), soft loans (where interest rates are set below the market rate), tax allowances (accelerated depreciation), charge exceptions, or rebates. Another means of applying subsidies is through mandatory investment in environmental protection technologies. While production costs may be increased as producers invest in waste treatment, or shift to more expensive production or distribution methods, the environmental effects are more immediate and both direct – in terms of modified environmental demands – and indirect – because price increases are likely to reduce consumption and waste production that can damage the environment. However, such regulations, whether across-the-board or point-by-point, are not well-suited for reducing environmental damages at low cost, particularly since producers, who have been given a sort of a 'certificate of satisfactory performance', have almost no incentive to find new production methods or products less detrimental to the environment.

Deposit–refund systems

Deposit–refund systems were originally introduced voluntarily for purely economic reasons. These systems are widely applied in the beverage and automotive industries with respect to beverage bottles, aluminium cans, batteries and car bodies. Due to their nature, deposit–refund systems can be relatively attractive. They might be considered promising because deposit–refund systems 'reward good behaviour' (whereas charges can be regarded as 'penalties for bad behaviour'). Deposit–refund systems are typically an instrument that aims at prevention of pollution. However, they generally are not effective enough in cases of major environmental problems as a consequence of their largely voluntary character and the relatively low levels of the deposits. Full collaboration between producers and retailers is necessary for deposit–refund systems to operate satisfactorily.

Market-related schemes

One economic instrument that has recently attracted the attention of many economists is the so-called 'emissions-trading programmes' which involve transferable pollution rights within a free market setting. This instrument appears to combine the advantages of charges with the functions of a commodity exchange – including daily market price quotations – thereby stimulating cost-effective solutions to reducing pollution discharges. The rationale is that polluters who can reduce their emissions to below the established limits can make application to an authority to be granted credit for the additional amount of reduction beyond that required by law. The resultant 'pollution right' or 'pollution credit' is entered into their 'emission reduction account' and can either be used later if an additional pollution source is brought on line, or sold to another polluter who is unable or unwilling to reduce the levels of their emissions. The timing and necessity for using transferable pollution rights are regulated by:

(1)   Offset policies, where new or expanding emission sources in areas which do not comply with given quality norms (non-attainment areas) must acquire pollution rights so that improvements in the overall pollution situation can be obtained;

(2)   Waste load allocation policies, where discharges from all of the various emission sources in a given area are considered to be enclosed in a 'bubble' ('the bubble concept') such that the total sum of emissions (the total maximum daily load, or TMDL; US EPA, 1992) within the bubble cannot exceed the sum of permitted discharges within the bubble area;

(3)   Net discharge policies, where new or expanded emission sources undergo an abbreviated administrative procedure which otherwise would have had to have been completed in full if the emission increase exceed a given limit; and

(4)   Emission banking, where possibilities exist for saving pollution rights for later use or sale to newcomers.

Market intervention could also take the form of subsidies in case market prices fall below certain levels, or of price guarantees that would create or facilitate the continued existence of a market (e.g. based on potentially valuable residuals and recycled materials). Such market interventions are similar to product charges. However, the latter are applied in already existing and well-functioning markets.

Further, liability insurance coverage could be required of polluters in an amount that would cover the costs of possible environmental damages or clean-ups associated with their emissions and the storage of their wastes. Here the risk is transferred to insurance companies, whose premiums will reflect the probability of environmental damage and/or clean-up costs in a market-related manner.

Enforcement

Enforcement incentives may be considered a legal, rather than an economic, instrument (see below). There are two major types of enforcement incentive. These are non-compliance fees (imposed when polluters do not comply with certain regulations) and performance bonds (payments to authorities in expectation of compliance with imposed regulations). Both are commonly used in many free market economies for a variety of purposes.

Legal instruments

As mentioned above, legal or regulatory measures, backed up by engineered facilities, have formed the traditional approach to environmental protection – ever since primitive man first buried his/her waste away from the area of habitation. While these early attempts at waste/contaminant management were initially enshrined in folk lore and tradition, they were eventually codified into law as we now know it; uniform, written rules of conduct. Some of the earliest rules governing public health and sanitation are contained in the Hebrew Bible (*Leviticus* Chapters 11, 13, 14 and 15). Today, most nations have legislation covering issues relating to drinking water supplies, wastewater and solid waste disposal, and public health; generally, these laws forbid the contamination of natural waters as the result of the discharge of wastes. Such prohibitions can be a simply worded statement contained within a water act or similar piece of legislation (i.e. 'no person shall discharge or cause to be discharged any substance into the public water as will, or is likely to, create a nuisance or render the public water, private water or underground water, as the case may be, detrimental, harmful or injurious to the health, safety or welfare of the public or any section thereof, or to any consumer or user of the public water or to any birds, fish or aquatic life, livestock or wild animals'; paraphrased from Rowe, 1982) or more elaborate and specific statute. These latter are generally more robust than the general prohibitions, specifying, as they typically do, actual contaminants, permissible levels (if any) and required actions to be taken in the control of those contaminants released to the environment. This specificity enables enforcement of the provisions of the law. Generally, in practice, it is common to find that the law, itself, authorizes some government agency or agencies to promulgate regulations, standards and/or guidelines to achieve the legislative objectives (see Chapter 5). Often the agency or agencies promulgating these rules will also be charged with their enforcement.

As noted above, the efficacy of the legal/regulatory approach to contaminant containment is only as effective as the level of enforcement devoted to the laws governing contaminant releases. Even in cases when such laws are stringently enforced, compliance can be determined by economics – if it is less expensive to pay the penalty than to implement pollution abatement procedures, then many or even most will opt to pay a fine, if caught and convicted, rather than incur 'unnecessary' expense. (While many readers may immediately think of industrial dischargers in this context, the foregoing generalization can also be applied to individual citizens who may find it more convenient to litter rather than expend the energy to deposit waste in an appropriate receptacle!) Thus, if regulatory measures are adopted, they must be:

(1) Equitable – legally fair and applicable to all parties;
(2) Comprehensive – the enabling legislation usually contains the general prohibition language to close any possible 'loopholes' or oversights (i.e. the reservation of governmental powers to protect the health, safety and welfare of its citizens);

(3)   Appropriate – more stringent water pollution control requirements may be needed in water-poor countries, for example (although many authorities are now calling for global standards to limit the potential for transboundary pollutant exports; see Ryding, 1992);

(4)   Scientifically-based – considering both human and environmental uses of water; and

(5)   Reviewed regularly – the content of whatever form of regulation that may be enacted will change as more information becomes available on various contaminants, especially the man-made organic compounds (see Chapter 5); as information on critical levels/environmental impacts of contaminants changes; as new contaminants are introduced; as the value of money changes over time (to maintain effective penalties); and as societal values and needs/uses change over time.

A detailed consideration of the drafting and content of particular pieces of legislation is beyond the scope of this manual. Numerous texts and compilations exist which outline the content of various water and water pollution control acts, and the reader is referred to these for advice on content-related issues (see Schlickman *et al.*, 1994; Davidson and Delogu, 1992; Goldfarb, 1988; Fuggle and Rabie, 1986; etc.). In addition, the decision-making format and form of such legislation differs markedly between nations, making any universal generalizations, other than those given above, less than meaningful. Notwithstanding this, however, while many nations will probably find it useful to draft, *de novo*, enabling legislation suited to their particular needs, regulatory agencies may find it useful, and convenient, to consult existing compilations of standards, guidelines and criteria when formulating specific regulations under this enabling legislation (see Chapter 5). Additional economic benefit may be had from this latter step as nations move toward the application of internal environmental requirements to imports received from other nations (e.g. EEC biocide content limits apply equally to imports and locally grown produce), and the globalization of water and air quality standards. For convenience, some selected standards and criteria are reproduced at the end of Chapter 5.

Experiences with economic and legal instruments

When environmental pollution control began to be discussed in the late 1960s, the typical reaction often was to 'let the polluter pay'. This approach was expected to result, comparatively cheaply, in cleaner production due to competitive market forces and would, it was felt, fully integrate the use of the environment into the economic sphere. The 'polluter-pays-principle' (PPP) was so broadly accepted by the early 1970s that it has become the central dogma of modern environmental policy.

The theoretical basis of the PPP is that, ideally, the price of goods or services should fully reflect the costs of both production and the resources consumed.

This approach should include environmental (natural) resources as well as the use of the air, land and water for storage, transmission or discharge of emissions. It should further characterize environmental resources in the same manner as other, more conventional factors of production, such as labour and materials, that are part of a corporate balance sheet. This approach recognizes that, if natural assets are not considered as production costs, they are likely to be wasted, degraded, or destroyed – it is often this 'free' use of resources that leads to environmental degradation. In theory then, the polluters pay the full cost of any environmental damages caused by their operations, creating an incentive for them to reduce the degree of damage, at least to a level where the cost of pollution reduction is similar to the marginal cost of the damage caused by the pollution. However, by meeting certain standards of environmental quality for emissions generated during the production process, the polluters are usually considered to have 'paid the price'. Further, in practice, the PPP becomes a 'beneficiary pays' principle since any pollution control costs are generally passed directly to the consumers through price increases.

While the 'polluter-pays-principle' has proven to be successful in contributing to the reduction of point source pollution, several years of international experience with the PPP system have, nevertheless, revealed some shortcomings:

(1) Not all polluters comply with the standards, partly because enforcement and control is nearly impossible to achieve to the full;

(2) Not all polluters make use of the 'best available technology' (BAT), believing that environmental pollution will not occur if they comply with the standards (it will; the standards usually reflect the 'best *economically* available technology' at the time they were written – not necessarily the latest or best technology – and can be quickly outdated);

(3) Standards are usually arbitrary political compromises; poor countries typically have poor standards since their apparent wealth of natural resources is often outweighed by their perceived lack of economic development;

(4) Not all areas of measurable environmental degradation can be adequately addressed by setting standards; e.g. car use, water consumption, energy consumption, land use, etc., create environmental damages that are quite measurable but poorly controlled by standards as presently defined; and

(5) Not all environmental damages are measurable in economic terms – indeed, such costs are usually extremely difficult to determine – so that polluters have often had to cover only the cost of pollution control relative to their immediate production process, while the public (the beneficiaries) often has to cover the environmental cost.

After 15 years of environmental policy development under the PPP principle, many people believe it is still far from being fully implemented. While few policy analysts believe that the PPP will ever be fully implemented, given the current escalation in global environmental problems, most would contend that it remains part of a multi-facetted, hierarchical pollution control programme making full use of legal, economic and technical instruments to protect our natural environment.

CHAPTER SUMMARY

In this chapter, the basic philosophies of nonpoint source pollution control have been briefly reviewed. Although this chapter does not present a 'cook book' approach to solving water quality problems, the basic approaches and concepts underlying those techniques that have been used, planned or proposed for the control of nonpoint source contaminants are outlined. This information has been presented in various ways. Table 9.1 presented an abbreviated list of control techniques in terms of common land uses. Table 9.2 presented a more extensive list of control techniques in terms of both land use and control methodology. These controls can be implemented through both structural (built) and non-structural (socioeconomic) means. To summarize and synthesize this other information given in the chapter, Table 9.3 presents a partial list of control techniques in terms of their control objectives; e.g., Table 9.3 can be used to find those techniques that would be available using water quantity management as a control mechanism, etc. Table 9.3 also attempts to provide some indication of relative cost, effectiveness and principle means of implementation associated with the various types of control measures.

While no value judgements of the various techniques described in this chapter are presented, a hierarchical approach to dealing with water quality degradation arising from nonpoint source contamination has been utilized; to wit, it is recommended that the water manager first consider treating and ameliorating the causes of the water quality problems; should this prove difficult or impossible to accomplish, the manager is urged to treat the symptoms of the problem (*cf.* Ryding and Rast, 1989, for a discussion of in-lake treatment methods addressing the problem of nutrient enrichment); and, should even this be impossible to accomplish, only then should the manager consider diversion. In terms of relative costs, the more natural approaches can be more cost-effective than engineered solutions, although there are on-going maintenance costs associated with the use of vegetation or similar natural materials. However, there is also considerable visual amenity value associated with such measures (*cf.* McHarg, 1969) that have significant, if unquantified (unquantifiable?), impacts particularly in urban settings. In no case should only certain courses of action with regard to the use of the numerous control techniques mentioned in this chapter be prescribed. Rather, the selection of a particular control strategy should be based on a thorough review of all available technologies and techniques. In addition, the selection of a particular control strategy should also take cognizance of cultural considerations, the public health (i.e. wet ponds may not be ideal in areas where malaria is prevalent), and cost factors. Thornton *et al.* (1991) provide persuasive evidence of the 'costs' involved in neglecting such considerations, while Ryding

Table 9.3 Summary of approaches to nonpoint source pollution control

Objective	Means[a]	Cost	Effectiveness	Example
Reduce water velocity	Structural	Moderate–high	Good–excellent	
	– engineered			Rip-rap, chemical stabilizers
	– non-engineered			Re-vegetation
	Non-structural	Low–moderate	Fair-good	
	– legal			Cattle exclusion from streams
	– economic			Tax differentials for planning
Reduce water quantity	Structural	Moderate–high	Good	
	– engineered			Check dams, energy dissipators
	– non-engineered			Contour ploughing
	Non-structural			
	– legal			Zoning, no net increase in runoff
	– other			Rain barrels
Increase percolation	Structural	Moderate–high	Fair	
	– engineered			Septic tanks, injection wells
	Non-structural	Low–high	Fair–excellent	
	– legal			Zoning, sewer requirement
Reduce pollutant exposure	Structural	Moderate	Good	
	– engineered			Plough design, secure landfills
	– non-engineered			Conservation tillage, INPM
	Non-structural	Low–moderate	Good	
	– other			Recycling, anti-littering
Reduce pollutant quantity	Structural	Low–high	Fair–good	
	– engineered			Wastewater treatment
	– non-engineered			Greenbelts, wetlands
	Non-structural	Moderate	Fair	
	– economic			Differential fee structures
Divert water flow/ divert pollutants	Structural	High	Poor	
	– engineered			Storm sewers, flood controls

[a]Various methodologies are summarized as structural (= built) and non-structural (= socioeconomic), and further divided into engineered (= constructed of man-made materials, including concrete), non-engineered (= constructed of natural materials), economic (= having financial implications), legal (= laws and regulations, including trade and professional codes of practice), and other. See text and Table 9.2 for details

(1992) argues persuasively for consideration of the total cost of a project (human and environmental as well as economic) in selecting a control strategy. We endorse these views (see Chapter 10). Similarly, legal mechanisms are not enough to safeguard the environment in the absence of concern, commitment and enforcement, yet these are also necessary to provide consistent compliance and a standard of measurement against which success or failure can be measured.

Finally, it should be noted that structural controls represent only one facet of the nonpoint source pollution control programme. Non-structural approaches, such as providing economic incentives and conducting public education and information campaigns, can also greatly assist in ameliorating nonpoint source pollution problems. Especially in terms of public wastewater treatment and industrial pre-treatment of effluents, taxing and investment incentives and market interventions can greatly influence choices being made by producers and consumers, while concepts such as 'adopt-a-stream' can provide a hands-on alternative and learning experience for scholars and adults. Some of these latter activities can also contribute significantly to the process of community-building in both developed and developing countries. Increased awareness, together with anti-littering/recycling campaigns, can significantly reduce macropollutants especially in urban areas that impact natural waters and significantly reduce the public cost of pollution abatement projects. Some of these issues, associated with selecting an appropriate mix of structural and non-structural techniques, are also discussed in Chapter 10 in terms of identifying an appropriate strategy.

CHAPTER 10

SELECTION OF A NONPOINT SOURCE POLLUTION CONTROL PROGRAMME

M. M. Holland, W. Rast and J. A. Thornton

GENERAL CONSIDERATIONS

In order to successfully control nonpoint source pollution of waters, the planning aspects upon which the remedial measures are based ought to follow a well thought-out and structured scheme. It is an advantage if the various planning aspects are approached in a step-by-step manner. Such an approach is facilitated by a question-and-answer approach to the planning and decision-making process. Such an approach is presented below. By adopting and following this approach, a comprehensive assessment/diagnosis of the nonpoint source pollution problem may be completed, and an appropriate response/action plan can be formulated and implemented – nothing will be overlooked. Further, by adopting a standardized approach to diagnosis and correction of nonpoint source pollution problems, comparisons between various issues will be facilitated and the quality of the dialogue between the various environmental actors improved. A systematic approach to information gathering, evaluation and use, on nonpoint source management issues may facilitate priority setting and decision-making about other/future environmental management measures.

No single approach or control measure will successfully treat all cases of nonpoint source pollution. As noted earlier in the description of the nonpoint source pollution process and the various factors which can affect it (see Chapters 4 through 7), the extent of present scientific knowledge is insufficient to be able to devise a completely fool-proof, standardized nonpoint source pollution control programme (the 'cook book' solution mentioned in Chapter 9). Nevertheless, the present state of knowledge and experience in nonpoint source pollution control is sufficient to develop a generalized approach which, if used in conjunction with an adequate monitoring programme and continuing scrutiny of the measured data, will usually work in the majority of cases likely to be encountered. The use of statistically analyzed data bases and derived quantitative relationships allows for the development of a reasonable, generalized approach for attempting to assess and control nonpoint source pollution of lakes and reservoirs. Of course, one should always remain aware of the uncertainty and potential error associated with such data bases, in order to use them effectively for predictive purposes.

The most feasible control option in a given situation can vary from location to location, depending on the circumstances. As noted earlier, it is generally believed that the control of external pollutant inputs represents the most effective, long-term strategy for attempting to control nonpoint source pollution of both natural lakes and reservoirs. Nevertheless, it is important to be realistic in selecting specific control measures, in terms of how much reduction in the pollutant input can be expected, which pollutant(s) will be affected (and to what degree), and how much such control measures will be likely to cost. An unrealistic management plan can undermine popular support for pollution control efforts if it is observed that a given plan will not achieve the desired pollution control goals, or that it is inappropriate from the point of view of cost-effectiveness.

It must be recognized that, in a given situation, the observed differences in the characteristics of natural lakes and reservoirs may affect the selection of control programmes based on reduction of the external pollution load. For example, compared to a natural lake, the relatively larger size of a reservoir drainage basin may require that one concentrate initially on land significantly more distant from the reservoir. This is because the nonpoint source pollution impact of distant pollutant sources may not be lessened by the 'effective transmission' of pollutant (see Chapter 4) to the same extent as in a lake basin. Furthermore, the larger basin size of a reservoir may cross more administrative boundaries, thereby complicating implementation of necessary control measures.

It must also be recognized that it may not be possible to achieve the desired water quality and trophic conditions in all cases, even after implementation of realistic pollution control efforts. If so, additional control efforts (often considerably more expensive) will be necessary to achieve the desired in-lake conditions. Otherwise, the in-lake conditions resulting from the achievable pollution control efforts will have to be accepted as the best that can be obtained under the circumstances. This decision should be made by those who are most familiar with the specific circumstances.

WATER QUALITY AS RELATED TO DESIRED WATER USE

As noted previously, 'good' or 'bad' water quality is often defined on the basis of the desired uses to which the water is to be placed (see Appendices 5.1 to 5.3). The suitability of waters for various uses is determined by several in-lake parameters that can be related to the trophic status of a lake or reservoir. Trophic status, in turn, is defined by specific boundary conditions for several of these parameters (see Ryding and Rast, 1989). Because of such relationships, it is also possible to relate desired water uses to the optimal (or minimally-acceptable) water quality for such uses. Therefore, a logical approach for establishing an effective nonpoint source pollution control programme is to determine the necessary water quality and/or trophic conditions for a desired water use (or uses), and design the programme to achieve these necessary conditions.

It is important to remember that one cannot always define the trophic status or intended water use of a lake or reservoir in an unequivocal manner (see Thornton and Rast, 1993). Further, the same waterbody can exhibit conditions indicative of one trophic state based on one water quality parameter, and another trophic state based on a second parameter. Notwithstanding, an ideal or acceptable water quality state can usually be identified for a given desired water use by using a combination of several parameters. For example, a lake or reservoir used as a drinking water supply ideally should have water of such good quality that it can be treated easily, using standard inexpensive methods, to yield water suitable for human consumption. The content of phytoplankton and their metabolic products in waterbodies used for such purposes should be as low as possible to facilitate this goal. Similarly, water used for swimming and other recreational pursuits requiring full body contact with the water should be free from nuisance blooms of planktonic organisms which can cause such physical symptoms as allergic skin reactions and conjunctivitis.

If a waterbody has one primary use, the control measures for achieving the desired water quality can be based on this single use. In many cases, however, there may be multiple competing uses for the same waterbody. In these cases, determination of the desired water quality should be based on the highest use, or the use requiring the least impaired quality. In areas of water scarcity, this 'rule of thumb' may be modified to provide water of a quality suited to the highest priority use. This use may require less stringent water quality in some cases. However, in such cases, the use of water quality standards less stringent than those required for the most sensitive water use may produce water quality conditions unsuitable for the most sensitive use over the longer term. Thus, decisions on the desired or primary water use of a waterbody used for multiple purposes are best made on the basis of specific knowledge of the lake or reservoir in question.

The foregoing descriptions point out only a few types of water use impairments related to nonpoint source pollution. To reiterate, the 'usability' of a waterbody is dependent on its water quality, which, in turn, is directly influenced by its trophic state and, ultimately, determined by both external and internal contaminant loadings. Both biotic (e.g. phytoplankton) and abiotic (e.g. suspended sediments) factors influence these loads which are intimately related to the hydrology of the watershed under investigation (see Chapter 4).

Based on practical experience with temperate zone lakes and reservoirs, a summary of intended water uses and their optimal ('required') and minimally acceptable ('still tolerable') trophic states is provided in Table 10.1. In all cases, the waters should be free of toxic substances and pathogens.

Table 10.1 Intended lake and reservoir water uses as related to trophic conditions (adapted from Bernhardt, 1981)

Desired utilization	*Trophic state*	
	Required	*Still tolerable*
Drinking water production	oligotrophic	mesotrophic
Bathing purposes	mesotrophic	slightly eutrophic
Low-water improvement		
– with long distance supply line	–	mesotrophic
– without long distance supply line	–	slightly eutrophic
Fish culture		
– salmonid waterbodies	oligotrophic	mesotrophic
– cyprinid waterbodies	–	eutrophic
Providing process water	mesotrophic	slightly eutrophic
Cooling water production	–	eutrophic
Water sports		
– without bathing	mesotrophic	eutrophic
– with bathing	oligotrophic	mesotrophic
Landscaping in recreation areas	–	slightly eutrophic[1]
Irrigation (by means of channels)	–	strongly eutrophic
Energy production	–	strongly eutrophic[2,3]

[1]within the scope of landscaping, a eutrophic state caused by the natural aging process, can even be desirable; [2]without consideration of the eventual water quality requirements for the receiving canal; [3]not valid for river power plants, which may be impaired by macrophyte and algal growths

A SIMPLE APPROACH FOR DEFINING AND SELECTING AN EFFECTIVE NONPOINT SOURCE CONTROL PROGRAMME

A logical sequence of decisions to be made by a water manager is outlined in Figure 10.1. The items considered in Figure 10.1 are similar to the steps outlined in Chapter 3 in regard to formulation of nonpoint source pollution control policies and programmes. The final decision on an appropriate control strategy should be (but rarely is) a consensus judgement, based on an holistic and dispassionate assessment of the relevant social, technical, economical and ecological aspects of the perceived problem – the decision to establish such a programme generally lies with political decision-makers. Nevertheless, it is incumbent upon technical staff to provide the best possible advice to these decision-makers. This advice should be based on factual information if a protection and rehabilitation programme is to be successful. Thus, if one is not already in place, a responsive monitoring programme should be established to both define the pre-treatment condition of the waterbody (the problem) and to properly evaluate the final outcome of the remedial measures enacted (the solution).

One point previously made in Chapter 3 (and again in Chapter 8) is reiterated here; namely, start with a simple approach, and then add more detail and

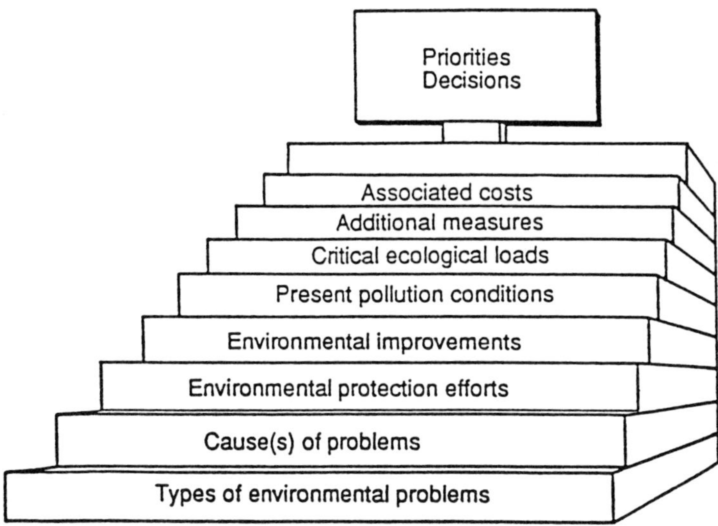

Figure 10.1 A systematic approach to information about environmental issues may facilitate priority setting and decision making about future environmental conditions

complexity as further knowledge and experience is gained. In this way, a comprehensive pollution control programme can be built on success, which generally reinforces societal (and political) goals as well as quality of life.

A simplified and practical approach for selecting appropriate nonpoint source pollution control measures is outlined in Figure 10.2. A 'decision-tree' approach is taken, with the answers to key questions dictating the direction to be taken. While the non-technical decision-maker may gain some insight into the complexity of the diagnosis and remedial planning process by reviewing this simple approach, the individual components of Figure 10.2, discussed below, are directed primarily toward technical staff. The fundamental basis of this approach relies on the control of nonpoint source pollutant inputs to a lake or reservoir. The rationale for such an approach was discussed in Chapter 9.

Step 1: Assess the problem

One must first determine the existence of a nonpoint source pollution problem. Identification of the existence of a problem is generally based on a combination of factors related to a perception of impaired use or reduced quality of the waters (see Chapter 5 for discussion of the nature of nonpoint source pollution and its associated water quality deterioration). For example, a pollution problem may be identified, *inter alia*, by the public, based on aesthetic criteria such as excessive growths of algae and/or macrophytes, fish kills, decreased water transparency, or hypolimnetic oxygen depletion (identified as odour problems related to the

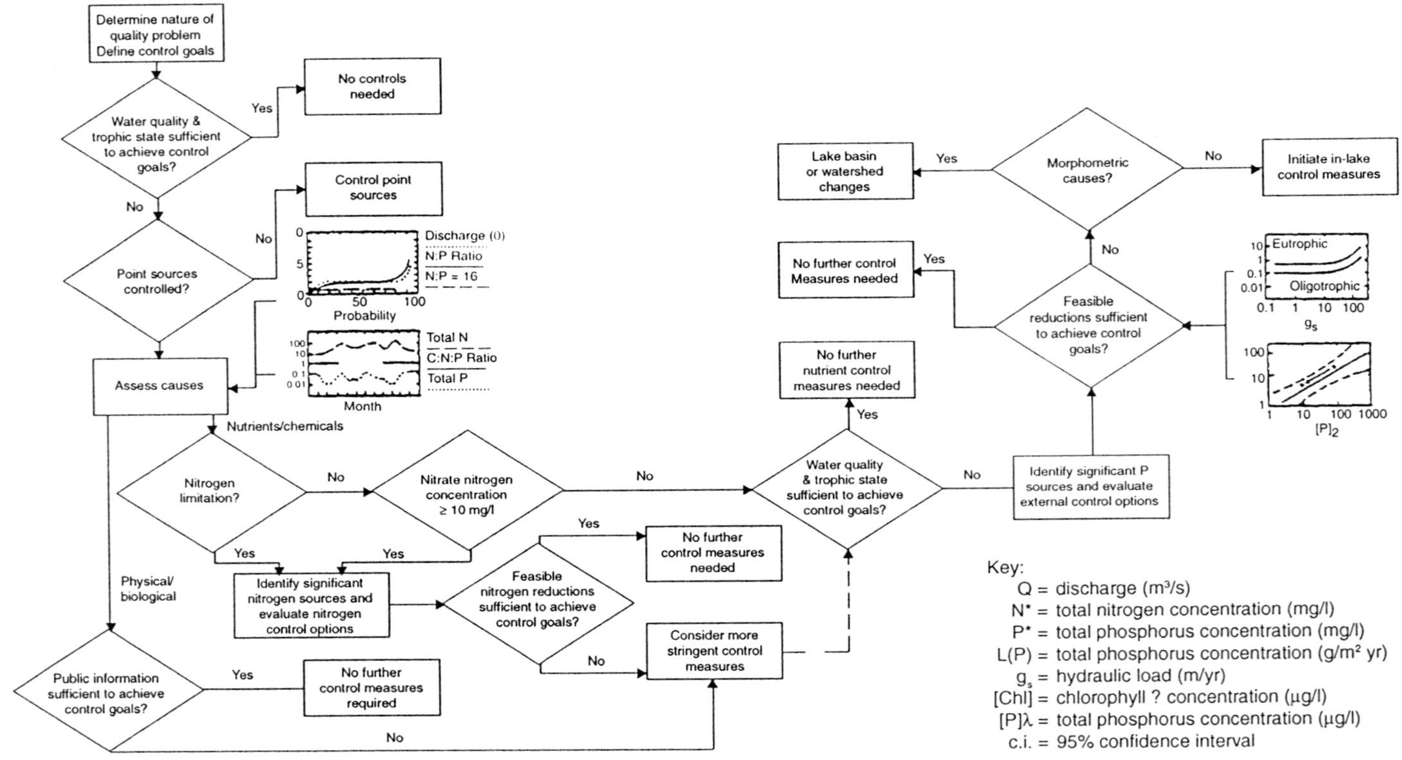

Figure 10.2 The nonpoint source control decision tree process

presence of hydrogen sulfide; see Thornton and McMillan, 1989; Quick and Johansson, 1992); by potable water providers, based on taste and odour problems in drinking water supply reservoirs or treatment difficulties such as excessive chemical usage or reduced back-wash times; or by environmental agency staff, based on the results of routine environmental monitoring programmes which have detected levels of contaminants in specific water courses that exceed established standards or criteria.

When a problem is reported, it should be confirmed without delay. A timely response to a reported problem may permit identification of the contaminant of concern (Step 2) and determination of its source (Step 3). For this reason, it is important that an effective 'exceptions' reporting and response mechanism be established by the responsible agency both as part of any existing monitoring programmes (so that problems can be identified, and responded to, in a timely fashion), and as a prerequisite for managing any environmental emergency. This procedure should be established whether or not obvious cases of environmental pollution exist.

Step 2: Determine the contaminant of concern

As noted, once the existence of a problem has been determined and confirmed, it is important to define the nature and origin of the problem – what are the symptoms and likely causes of the problem? In making such a determination, a knowledge of the watershed is important. Initially, an assessment must be made of the likely persistence of the problem – is it a one-time spill or an on-going problem? Obviously the response to a one-time spill will differ from the longer term approaches necessary to resolve more persistent problems. Generally, a one-time spill can be identified with more certainty and contained more quickly than more persistent problems. Similarly, point sources of contamination can be identified and controlled more readily than nonpoint sources of pollution. Given the variety of possible contaminants that can reduce water quality, identification of the specific cause or causes is critical to a determination of the nature of a response, if any.

If it is decided that a response is necessary to achieve a desired water quality in a lake or reservoir, an assessment of the magnitude of the response must be made. To this end, a control objective or specific goals should be established for the control of the pollutant of concern.

Step 3: Define the goals

Generally, the goals of a remedial action programme will be determined by the major water use (or uses) of the lake or reservoir. These uses define the necessary water quality to support such uses (see Table 10.1). Obviously, if the existing trophic state and quality of a waterbody is compatible with the water use, no further action is necessary in regard to pollution loading conditions. If not, both point and nonpoint pollution control measures may be necessary. As 'rules of thumb' it is most beneficial to treat point sources before treating nonpoint sources of contaminants, and to treat sources of contamination in closest proximity to the waterbody rather than treating sources far removed from the area of concern. These sources can be determined by a routine sampling programme that gathers chemical and biological data on an aquatic system, or estimated using various watershed-based modelling techniques (see Chapter 8).

Usually, a combination of monitoring data and modelling results will be necessary to define quantitative pollution reduction goals for a particular pollutant of concern. Establishment of quantitative goals is essential if a meaningful control programme is to be implemented and if its efficacy in controlling the contaminant of concern is to be assessed (see below). Since an effective, long-term control measure is usually to control the external pollutant load (see Chapter 9), the next step is to determine the likely extent to which a pollutant can be controlled.

Step 4: Identify alternative pollution control options

An important yardstick for assessing the efficiency of external pollution control measures is the determination of whether or not an economically feasible means of reducing the pollutant content of an effluent or stream exists. This assessment can only be made once the problem is clearly understood and defined. Only at this stage can a determination be made as to whether or not the problem is severe enough (or will become severe enough) to warrant implementation of a control programme.

At this point in the diagnostic study of a pollution episode, the existence of a problem and its cause(s) has been confirmed. Data on the watershed, its population, land use, point sources and jurisdictional attributes have been tabulated, and, where appropriate, trends have been identified. While the principal focus has been on the particular contaminant of concern identified in Step 2, the watershed analysis may have highlighted other less obvious problems and further contaminants of concern. These, too, should be taken into consideration in the formulation of a response strategy. This is especially true in situations where the identified trends suggest rapidly changing land uses or increasing population pressures in specific portions of the watershed. These incipient problems and invidious pollutants should also be quantified to the extent possible and their control included in the remedial action programme being developed. The control of nonpoint source pollution problems, unlike the control of point source problems (where structural controls generally can be added as situations change and develop), requires a more far-reaching vision of the entire watershed and its dynamics since imposition of nonpoint source controls generally entails acquisition of lands for structural control measures or the conduct of public information campaigns to effect changes in societal behaviour. Such actions typically take a longer time to come to fruition than the construction or modification of the wastewater treatment facilities used to treat many point sources.

In identifying the available control options, two issues merit special mention; namely, the inclusion of the 'do nothing' option, and the need for as comprehensive an assessment of options as possible. Although often overlooked, the 'do nothing' option is always the default position or consequence of not acting. In terms of this option, conditions can only stay the same or grow increasingly worse. However, this option requires no additional investment of resources to control or correct a problem and only public tolerance to live with the consequences. Quantitative statement of these consequences is often beneficial in establishing the need for implementation of corrective actions. Similarly, consideration of the complete range of applicable options is often precluded by a premature determination that specific actions would be too costly or not acceptable. Experience has shown that many of these options may indeed be inappropriate, but their inclusion and quantitative evaluation provides decision-makers with the necessary evidence to support their selection of another course of action. Not only does this bolster the confidence of the decision-makers in the advice that they receive from their technical staff, but it also underlines the fact that a comprehensive assessment was indeed made and allows an informed choice of options to be made. For example, the use of complicated wastewater treatment methods to remove pollutants from river water may appear to be too costly and/or too ambitious an undertaking for use in the control of nonpoint source pollutants, but such an option has been employed at the Wahnbach Reservoir (Germany) as the most cost effective means of removing phosphorus from the inflowing waters to the reservoir (Bernhardt, 1983). This option was selected only after careful consideration of other watershed-based options, but has rarely been included in other lake and reservoir nonpoint source control plans.

Before adequate control programmes can be implemented, however, it is often necessary to use generally applicable predictive tools to evaluate both the problems and expected results of remedial measures. Through the use of quantitative techniques such as those described in Chapter 8, it is possible to calculate: (1) the natural background load and expected response in the waterbody, (2) the present load and existing response – an important cross-check to ensure the accuracy of the model predictions, and (3) the expected load and in-lake pollutant concentration resulting from a given control programme, based on a knowledge of the watershed characteristics, pollutant loading, water retention time and physical attributes of a lake. For example, by employing known relationships between phosphorus and common eutrophication response variables such as chlorophyll and Secchi disc transparency found in multi-lake studies, algal biomass (in terms of chlorophyll) and water transparency can be predicted (e.g. OECD, 1982). The transparency of the water is an easily understood parameter, and it is informative when explaining lake water quality data to laymen (see Thornton *et al.*, 1989). A more detailed approach, such as computer-based scenario analysis, may also be used to assess the available nonpoint source pollution control options, if the data warrant.

Simple models often provide sufficient information for planning and policy-making purposes in many situations; nevertheless, the use of a more temporally or spatially detailed dynamic model is also useful in some situations, as discussed in Chapter 8. More complex models may permit assessment of water quality or responses to pollution control actions that have greater relevance to users than the outputs of simple empirical models. For example, with regard to standing waters, the public interest is usually focused on water quality during the summer period, the period of maximum usage, while, in flowing waters, recreational use may be prolonged throughout the entire year. In contrast, other water users may have a year round interest in water quality (e.g. water supply operators, hydropower generators). This interest may also be site-specific; for example, the utility operators may be interested in the water quality at the point of abstraction, while recreational users may be more concerned with water quality throughout an aquatic system. In the more specific situations, more complex modelling approaches would be more appropriate and useful.

As an aid in the management policy-making process, there is often a need to use these relationships and model-based approaches to forecast future trends in water quality for a given waterbody under a changing pattern of contaminant inputs. Such an assessment typically is not only in terms of the average values for a specific time period, but also for extreme situations (e.g. worst possible conditions) which tend to be of special interest to both lake/reservoir managers and the public. Extreme situations can be hard to identify through routine monitoring programmes; consequently, relationships between the average and maximum values of chlorophyll (OECD, 1982) and other nonpoint source pollution parameters (Ryding, 1981) can be very useful for making predictions for water management purposes.

Step 5: Prepare an action plan

Once the available options have been compiled and evaluated, an agreed course of action should be decided upon, and a management plan completed. Publication of the agreed nonpoint source management plan is an essential step in the process of implementing remedial measures. Too often the step of writing and publishing a management plan is overlooked or dispensed with as an unnecessary delay in proceeding with a project. In cases where this lapse has occurred, there has been significant confusion, particularly in the later stages of project implementation, when the original proponents of the project have moved on or are no longer available for consultation. Nonpoint source pollution control plans are an important means of communicating the desired course of action and of evaluating the success of the project in later years. (These plans are of great import to nonpoint source management projects given the typically long periods – ten or more years – over which nonpoint source controls are implemented.)

Also, the publication of a management plan is an important consensus building technique whereby interested parties are informed of the outcome of the inventory and analysis process completed during Steps 1 to 4, above. By compiling the information gathered during these phases of the project into a management plan, agency staff can ensure that complete consideration has been given to the diagnosis of the problem and evaluation of alternatives.

Step 6: Consider the need for further (in-lake) control measures

If the expected improvement in water quality and/or trophic conditions from external pollution control measures will not be sufficient (based on model predictions or post-treatment monitoring) to achieve the nonpoint source pollution control goals established in Step 3, in-lake control methods can be considered as supplemental measures. The expected water quality improvement, for example, following a pollution load reduction of 75 to 90 percent may still fall short of that required to meet the desired water-use objectives. Such a situation is especially relevant to shallow waterbodies because their water mass is more susceptible to mixing by wind action, etc., which can re-introduce accumulated contaminants to the water column, or keep accumulated contaminants circulating in the water column, long after controls have been introduced in the watersheds. In such cases, options such as alteration of the lake basin morphometry (e.g. dredging) or initiation of in-lake pollution control measures would be appropriate. As pointed out in Chapter 6, the latter measures are usually only temporarily effective since they generally treat the symptoms of contamination rather than its sources. However, such measures can be very useful when the primary method of external nutrient control alone is either inadequate to achieve the goals, or is too expensive to be implemented in a given situation. In-lake controls include such measures as pollutant inactivation, hypolimnetic aeration, harvesting of macrophytes, application of algicides, etc. Biological controls (e.g. enhancement of certain food chain pathways by introduction or replacement of specific food chain organisms) may also be considered. Nevertheless, it should be borne in mind that the long-term, ecological effects of in-lake management approaches, and especially of biological manipulation approaches, are largely unknown at present.

Step 7: Evaluate the effectiveness of control programme

In most of the cases studied so far, economic optimization with respect to water quality is primarily concerned with control measures in three major areas: (1) source control of pollutants in the watershed (external control); (2) temporal detention in the waterbody (internal control); and (3) treatment plants in the case of waters used for water supply (off-line control). The ultimate benefit that can be realized will usually be substantially higher if optimization is related to all three control categories as a whole. This integrated approach is useful for the control of most nonpoint source contaminants. As noted earlier, it is preferable over the long term to reduce or eliminate the sources of the substances (e.g. phosphorus) causing nonpoint source pollution, rather than temporarily ameliorating their symptoms. Finally, if neither the external or internal control measures are sufficient to achieve the nonpoint source pollution control goals, the achievable water quality may have to be accepted as the best that can be attained under the circumstances. Nevertheless, practical experience suggests that, over the long term, the basic condition of the aquatic ecosystem will usually be improved as a result of implementing nonpoint source pollution controls, even though all of the nonpoint source pollution control goals have not been achieved.

This type of post-implementation evaluation is an important part of the management process. Like plan writing, post-implementation assessment is often overlooked or not budgeted for during the diagnostic and planning phases of the project. However, post-implementation monitoring and evaluation provides important guidance for future remedial management operations by identifying both strengths and weaknesses in the diagnostic and treatment technique selection processes. Post-implementation evaluation also provides important feedback to decision-makers and the public regarding the use of (public) resources in environmental management projects. Not only does this encourage subsequent actions elsewhere, but it encourages continued vigilance over, and maintenance of, the remedial measures adopted in the target system as a result of the expenditure of those resources. Indeed, maintenance of the installed practices can be considered the final step in the nonpoint source pollution control process.

WHAT IF THE WATERBODY DOES NOT RESPOND AS EXPECTED?

Differences in the internal structure and/or morphometric/hydrologic environment of a waterbody may result in unexpected variations in water quality response between seemingly similar lakes and reservoirs. This concern applies to both the differences between the observed and predicted responses to nonpoint source pollution models, and the observed and predicted responses of waterbodies to specific control measures. Other factors which potentially can cause complications in predicting the responses of a lake or reservoir to nonpoint source pollution control measures include the effects of toxic substances on aquatic life, changes in the level of zooplankton predation by fish, alterations in algal growth limitation by elements other than phosphorus, and modifications to the light attenuation capacity of the water column.

Consequently, as a general rule, one should attempt to identify as many as possible of the factors that may be responsible for uncertainties in the predictive capabilities of existing nonpoint source pollution models. Some of the most important elements of uncertainty to be considered in the management of polluted lakes and reservoirs are identified and discussed below. Not all of these elements can be quantified given the present state of knowledge of waterbody responses to external stimuli, but most can, at least, be qualitatively assessed on the basis of current limnological theory. This situation reinforces the need for comprehensive, post-implementation assessments such as those recommended above.

Assess the response time ('lag period') of the lake or reservoir

If a contaminant behaves as a chemically conservative substance (i.e. a substance that does not undergo any biological or chemical transformations within a watercourse), the time necessary for a lake or reservoir to respond to a control programme can be calculated solely on the basis of the flushing rate of the waterbody. This 'response time' constitutes the so-called 'lag period' between the time that an input of a contaminant to a waterbody is altered, such as through the initiation of a nutrient control programme, and the time that the waterbody reaches a new steady state condition (i.e. the time that the waterbody 'responds' to the control programme). This lag period is dependent on the hydraulic residence time, and, in the case of a conservative substance, is equal to approximately three times the hydraulic residence time (Vollenweider, 1969; Sohzogni *et al.*, 1982).

However, some contaminants, such as the aquatic plant nutrients, are non-conservative substances in aquatic systems. They undergo various biogeochemical transformations in the water column; i.e. they are assimilated by phytoplankton, interact with the bottom sediments, etc. In such cases, the residence time of the substance is used to calculate the response time (or lag time) of the lake or reservoir to altered inputs (increases or decreases). This is in contrast to the use of the hydraulic residence time in the case of conservative substances. Sohzogni *et al.* (1976) provide a detailed derivation of the non-conservative pollutant residence time calculation, which Rast and Lee (1978) calculate as follows:

$$R(P) = [P]_l/[P]_{in} \tag{10.1}$$

where: $R(P)$ = the pollutant (phosphorus) residence time;
$[P]_l$ = the annual mean total pollutant mass in the waterbody
(e.g. kg P); and
$[P]_{in}$ = the annual total pollutant input to the waterbody
(e.g. kg P/y)

The basis for this equation is that the annual average mass of total pollutant in the waterbody will differ from the annual input as a direct function of the combined impact of the various in-lake transformations and losses, and the flushing rate of the nutrient from the waterbody. A similar expression can be developed for the nitrogen residence time. However, in the latter case, the relationship between the nitrogen levels and the nitrogen residence time would necessarily be more complex and uncertain, since the gaseous phase in the aqueous chemistry of nitrogen must also be considered. Rast and Lee (1978) have shown that, in most cases, the pollutant residence time is shorter than the hydraulic residence time (usually by several-fold) because of the environmental aqueous chemistry of pollutants.) Nevertheless, the same three-times the residence time (of the pollutant) 'rule of thumb' applies to the determination of the expected 'lag period'.

Exceptions to this 'rule' appear to be relatively unproductive, oligotrophic lakes and reservoirs, whose response times approach their hydraulic residence times; eutrophic waterbodies appear to exhibit the shortest pollutant residence times and often exhibit a rapid 'response' to the initiation of control programmes. However, the magnitude of the response (especially the publicly perceived response) must be compared to the initial degraded condition of the waterbody when assessing the extent of water quality improvement due to a control programme.

Reassess the contaminant of concern and models used

If symptoms of contamination persist, a re-assessment of the contaminant(s) of concern identified during the diagnostic study should be made. In particular, an assessment of co-pollutants generally associated with the identified contaminant of concern should be conducted. Guidance in identifying co-pollutants is given in Chapter 6, which sets forth the various pollutant groups associated with varying human activities. The identification of co-pollutants is also important to the selection of appropriate models, upon which forecasts of future conditions are typically based. As noted above, the presence of both conservative and non-conservative substances in the pollutant load can result in differing response times; the overall response time of a waterbody to a mixed pollutant load may be related to either the hydraulic residence time or to the pollutant residence time, whichever is the greater.

Reassess pollutant loading rates

In addition to analyzing the water for its contaminant content, measurement of the discharge will always give the most accurate nutrient loading values for a given waterbody. In the absence of directly measured loading values, a traditional approach is to estimate the annual contaminant load from a drainage basin using unit area loads for specific land-use patterns in the drainage basin (see Chapter 4). These estimates, however, are valid only for 'normal' conditions during an average hydrologic year and can vary significantly in a given multi-year period (this variation is given in the coefficient of hydrological variation, which, in arid regions, can range up to 35 percent or more). Likewise, the effects of the estimated pollutant loads can be modified in lakes with short water residence times (i.e. a fast water flushing rate). A certain portion of the contaminants entering the lake may be flushed through the lake before they have any discernable effect. Therefore, consideration of the actual hydrological conditions is most important in regard to assessing the ecologically relevant contaminant inputs. This topic was previously discussed in Chapter 8, in regard to estimation of the nutrient load.

A comparatively high contaminant concentration in the water column, compared to the observed external load, may be indicative of the existence of internal loading; e.g. loading from groundwater seepage, nitrogen fixation or phosphorus release from the sediments. Comparing a given data set to the generally observed relationship between the loading and the in-lake concentrations of nitrogen and phosphorus, for example, can indicate the occurrence of such conditions. Predictive lake models should not be applied to lakes affected by internal sources of nutrients, unless the internal load can be estimated and added to the external load value, so as to obtain a realistic total load estimate.

As pointed out in Chapter 4, various physical, chemical and biological processes can alter contaminant concentrations during the period of downstream travel from the source areas in the upper part of a drainage basin to the downstream receiving lake or reservoir. As a result, water quality can exhibit considerable variation along a river stretch due to such processes, as well as to successive inputs of additional contaminants from other sources along the tributary. Consequently, contaminant sources in the upper parts of a drainage basin can have a lesser impact on the water quality of a downstream lake or reservoir than sources closer to the waterbody. This necessitates a selective approach to the implementation of nonpoint source pollution controls that incorporates an assessment of the relative importance of the effects of the external controls in various portions of the drainage basin on the observed water quality of the receiving water as the basic component of an effective water management programme, augmented by additional measures as needed.

Reassess the sampling programme

Due to potentially large and intermittent variations in water quality often observed in many polluted waterbodies, an efficient and economically feasible sampling programme is not always a routine matter (see Ryding and Rast, 1989). It is logical, therefore, to suggest that samples be taken as frequently as possible to properly describe the real conditions. Average annual values of any relevant parameter cannot be obtained if the sampling effort is distributed irregularly over the year in response to hydrological events. However, difficulties in forecasting hydrological and meteorological conditions from one year to another emphasize the need for evenly distributed sampling intervals over the annual cycle as a baseline sampling/monitoring programme. To overcome these limitations, a regular sampling programme, supplemented by sampling of spring flood and major rainfall events, is recommended. Furthermore, because large year-to-year variations can make predictions based on only one sampling year rather suspect, a sampling period of at least three years normally should be regarded as the minimum effort for the proper assessment of a waterbody. By employing such a programme, limitations of the monitoring data may be overcome, and predictive capabilities enhanced.

Another potential problem with a sampling programme is the 'patchiness' phenomenon, often noted in planktonic organisms. Because plankton may not be uniformly distributed throughout a waterbody, several samples distributed over the lake surface area, especially for large lakes, are necessary to overcome this problem. Similarly, if planktonic organisms are irregularly distributed vertically, sampling at many depths, or use of depth integrated samples, may be necessary.

Other considerations

Various other factors can impinge on the accuracy of pre-implementation forecasts, and assessments of response times, and can affect the response of aquatic systems to nonpoint source pollution controls. Several of these are briefly discussed below:

(1) *Geographic factors* Different physical factors may have a significant impact on the ultimate bioproductivity of a waterbody (e.g. mean air temperature, rainfall, snow cover, over-land flow, soil erosion, etc.). Stratified waterbodies in tropical regions, for example, may have hypolimnetic temperatures of greater than 20°C, which can promote microbial conversion of contaminants (e.g. denitrification).

(2) *Morphometric factors* The shape of a lake basin can substantially influence its response to changes in contaminant loading. Generally it is difficult to define the 'average conditions' for waterbodies with substantial longitudinal gradients in water quality. In reservoirs where the main station used for comparison is normally near the water outlet, in lakes consisting of several sub-basins, and in hypertrophic lakes exhibiting extreme 'patchiness', it may be difficult to obtain a characterization of water quality in the lake as a whole. The morphometric conditions of lake basins are also of vital importance in terms of stratification, wind induced mixing, and internal loading. Reservoirs, in particular, that are subject to periodic drawdown can appear to have several, widely differing morphometries during an annual cycle.

(3) *Hydrodynamic factors* The mixing regime of a waterbody is affected by such factors as seasonal differences in wind velocities and directions, and the ratios of hypolimnetic volume to epilimnetic volume, of mixing depth to mean depth, and of littoral area to epilimnetic area. Resuspension of sediments by wind action not only increases the mobilization of pollutants, but also enhances denitrification, causing elements other than

phosphorus (e.g. nitrogen) to act as growth limiting nutrients, for example. In addition, uncertainties about the potential effects of control measures often increase with water residence times, but decrease with depth. Thus far, few critical pollutant loading levels have been identified for oligomictic or for polymictic tropical lakes; pollutant loading capacities in these two cases may be quite different from those of temperate waterbodies (see Thornton and Rast, 1993).

(4) *Geochemical and chemical factors* The relationship between contaminant loads and trophic reactions is different for lakes with hard waters, and for drainage basins with igneous substrata. Because this is of practical importance in water quality management, the hardness of the water should be included in assessments of the water quality impacts of reactive contaminants; in hard water lakes, for example, the biota are not as sensitive to the presence of toxic elements as in soft water lakes.

(5) *Biological factors* The significant role of the biological community structure in the pollution response of a waterbody must not be overlooked. In this regard, the selective feeding of fish on invertebrates (zooplankton and bottom fauna), or, in the absence of fish, 'overgrazing' of algae by zooplankton can influence the response of an aquatic system to the implementation of nonpoint source control measures.

DETAILED SCENARIO ANALYSIS AS AN AID IN DECISION-MAKING REGARDING NONPOINT SOURCE POLLUTION CONTROL MEASURES

The strengths and limitations of dynamic mathematical models as assessment and predictive tools in the management of nonpoint source pollution were discussed previously in Chapter 8. Prominent limitations were the increased data requirements (compared to the simpler empirical models) and the difficulties in applying such models to different waterbodies without considerable revision of the types of necessary model variables and/or values of these variables. One of the strengths of these dynamic models, however, is their usually greater ability to provide insight into the internal processes controlling nonpoint source pollution (e.g. why or how a relationship or process occurs). They also allow multiple changes to be considered simultaneously. Manipulation of the types and/or values of the state variables in the model can provide potentially valuable insight into what factors exert primary control in the nonpoint source pollution process. Dynamic models can also be useful tools in those situations where the selection of a nonpoint source pollution control strategy depends on controlling internal ecological processes. Through the use of dynamic models, the values and/or relationships describing the ecological processes of interest can be altered, and the model used to simulate the effects of such changes on water quality and other relevant variables. Alternatively, the simulated effects of external pollutant control programmes can be compared with those involving control of internal

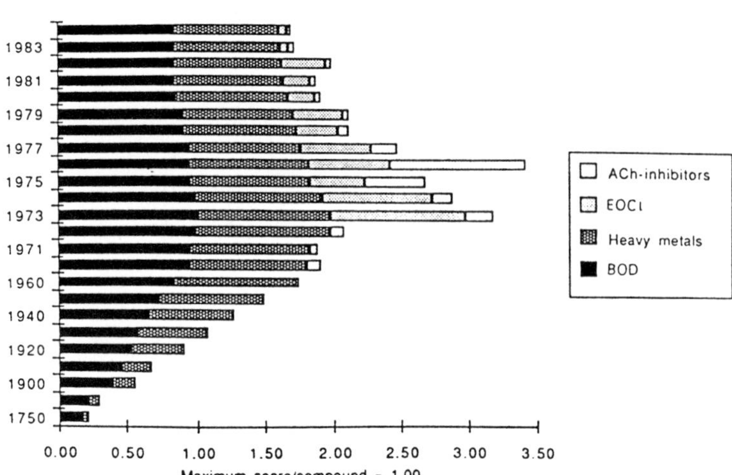

Figure 10.3   The pollution history of the Rhine

ecological processes. Based on the results of such comparisons, as well as the measured limnological conditions, one can attempt to select optimal control strategies in a given situation.

One example of the use of a dynamic model to aid in the decision-making process regarding nonpoint source pollution control alternatives is provided here. This example uses an analysis of the simulated effects of different contaminant control strategies to aid in the selection of an optimal pollution control strategy for the Rhine River (Jolánkai, 1996). The study was carried out between 1990 and 1992 as a collaborative venture between the International Institute for Applied Systems Analysis (IIASA), the Hungarian Water Resources Research Centre (VITUKI), the German Bundesanstalt fur Gewasserkunde, the University of Utrecht (Netherlands), and The Netherlands National Institute for Public Health and Environmental Hygiene (RIVM), and resulted in the publication of the Simple Experimental Nonpoint Source Model (SENSMOD) that forms the basis of this presentation. The discussion is set out using the step-wise approach to nonpoint source management planning recommended above.

Step 1: Assess the problem

The Rhine River is a major water supply source, transportation corridor and recreational area draining a large portion of western Europe. The river originates in Switzerland and flows through France, Germany, Belgium and The Netherlands. Several international river basin agreements have been concluded

410

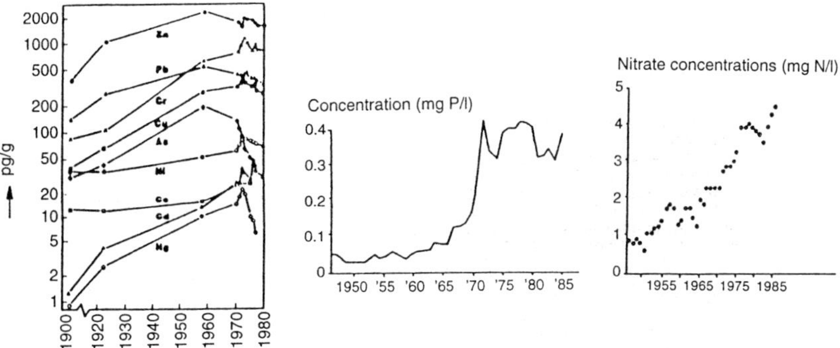

Figure 10.4   Trends in (A) cadmium, (B) phosphorus and (C) nitrogen concentrations observed in the Rhine River at Bimmen-Lobith on the Dutch-German border

(e.g. the 1963 Berne Agreement on the International Commission for the Protection of the Rhine Against Pollution; the 1976 Bonn Convention for the Protection of the Rhine River Against Chemical Pollution; and the 1992 Helsinki Convention on the Protection and Use of Transboundary Watercourses and International Lakes) which both recognize the existence of water quality problems in the river and express a commitment on the part of the riparian nations to rectify the problems. The Rhine has a long history of human use – and abuse – which has been chronicled by Petts *et al.* (1989); much of this abuse occurred during the 1900s, with the pollution impacts being of both domestic (oxygen demanding substances) and industrial (synthetic organic chemicals and metals) origin (Figure 10.3). It was also surmised that some degree of contamination was the result of natural, geological processes which introduced metals and nutrients of igneous origin into the watercourse. Because of the uncertainty surrounding this natural background contribution, and its impact on setting and achieving water quality standards (particularly in the Lower Rhine River), the present investigation was commissioned.

Step 2: Determine the contaminant of concern

Using data provided in the scientific literature, the contaminants of concern were identified as cadmium, phosphorus and nitrogen. The former was selected on the basis of its known use by industries located within the drainage basin, its presence in agricultural fertilizers, and its significant urban generation. The latter elements, the plant nutrients, were selected on the basis of their contribution to the eutrophication of the waters of both the Rhine River and North Sea. The concentrations of all three elements had been monitored at various sampling stations within the river system for some time, and intensively since 1971 (Figure 10.4).

Step 3: Define the goals

The objectives of the study were three-fold; namely: (1) the development of emissions inventories for the three elements based on a quantification of their extraction, production, consumption, and disposal within the drainage basin; (2) the development of an applications inventory based on a quantification of the mass of these elements deposited on the land surface (by any means, including atmospheric deposition, and urban and agricultural applications) within the drainage basin; and (3) the development of a nonpoint source pollution inventory based on a quantification of the mass of material transported from the land surface to the river basin and through the watershed. The latter objective is used in this example.

Step 4: Identify alternative pollution control options

In order to identify appropriate control measures, it was necessary to first estimate the total loads of cadmium, nitrogen and phosphorus in the system, and to determine the probable origin(s) of these loads. Given the magnitude of the drainage basin, it was decided that the use of a regional-scale simulation model would be an appropriate means to develop such a quantification, and to assess the movement of the contaminants of concern through the waterway. SENSMOD (Figure 10.5) was selected as the tool of choice. This model was regional in scale (i.e. the model could accommodate the multiple drainage basins within the Rhine River Basin), able to incorporate measured values of river flow, adjust UAL-based contaminant generation data for measured contaminant concentrations, and account for in-stream transformation, deposition and resuspension processes. The model could also incorporate mapping information contained in various national and local Geographical Information Systems (GIS), and make use of tabulated point source effluent discharge data. The model was calibrated using data from 1977 and verified against data obtained during 1988, the last year for which complete records were available at the time the study was initiated. The model was then used to forecast conditions in the river system during 1995. In this way, both present and future scenarios were generated; the present 'predictions', as noted previously, would allow an assessment to be made of the degree of confidence that could be placed in the model output.

The model output is given in Tables 10.2, 10.3 and 10.4 for cadmium, phosphorus and nitrogen, respectively. Based on these data, SENSMOD can be seen to have performed acceptably, with good agreement being seen between measured and predicted values. Thus, some confidence can be placed in the quantitative output which, for cadmium (Table 10.2), clearly shows the increasing dominance of nonpoint source contributions to the total load. The data also show the results of increasing public awareness of the problems caused by heavy metals, as the total load declines over the study period (Table 10.2). In the case of phosphorus (Table 10.3), the data would suggest that the greater portion of

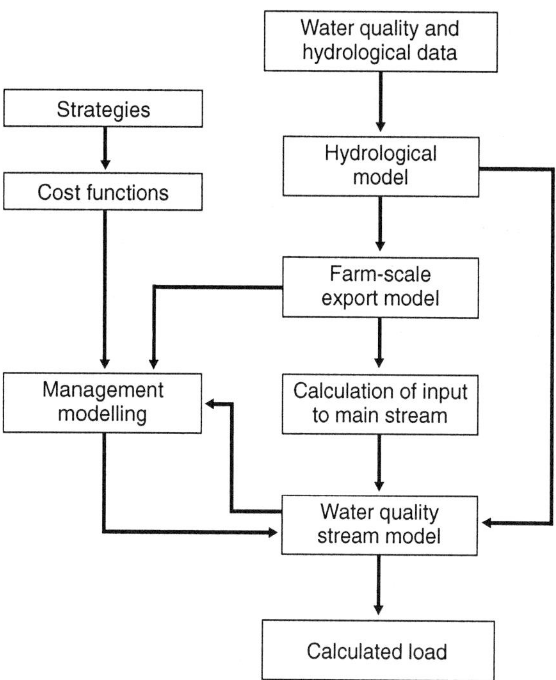

Figure 10.5 SENSMOD flow chart, showing the interactions of the various sub-models employed

the load is from point sources, although it is predicted that point and nonpoint source contributions will approach parity by the mid- to late-1990s. In contrast, the model was not able to accurately account for nitrogen transformations, either forecasting accurately on the basis of nitrate nitrogen or on the basis of ammonium nitrogen, but not on the basis of both (Table 10.4). Both nitrate and ammonium have the potential to cause significant environmental impacts in terms of the beneficial uses of the Rhine River waters; ammonium may be toxic to fishes, thus impacting the recreational industry and commercial fisheries, while nitrate, when present in high concentrations, can impact human health (see Chapter 5). Nevertheless, as might have been anticipated based on the output for phosphorus, the model suggests that nitrogen loads are dominated by point source discharges, although it is also predicted that point and nonpoint source contributions will approach parity by the mid- to late-1990s. These data, too, show an overall reduction in the mass of phosphorus and nitrogen being discharged from point sources or lost to the river from land applications.

While the data presented in this discussion do not break down the nonpoint source contributions of metals and nutrients into their various land-use categories,

Table 10.2 Summary table of calibration and scenario runs using SENSMOD for cadmium in the Rhine River Basin

		Year ('time slice')		
		1977 calibration year	*1988 verification year*	*1995 prediction year*
Measured or reported conditions	(mg/m³) (t/y)	2.6 163.3	> 0.3 16.0[+]	assumed overall reduction of both point and diffuse sources as of 1988 = 50%
Simulated/predicted conditions	(mg/m³) (t/y)	2.6 163.57	0.355 22.32	0.17 11.5
Point source input*	(t/y)	112.78*	4.44**	2.22***
Diffuse calculated input	(t/yr) (g/ha/y)	109.08 6.78	37.15 2.32	18.5 1.16
Diffuse source calculated contribution to total load	(t/y) (%)	65.43 40	20.03 89.7	

[+]order of magnitude estimate only; *point source inputs are either given in the data base (*), or calculated in scenario runs (**), or assumed in future predictions (***)

Table 10.3 Summary table of calibration and scenario runs using SENSMOD for phosphorus in the Rhine River Basin

		Year ('time slice')		
		1977 calibration year	*1985 verification year*	*1995 prediction year*
Measured or reported conditions	(mg/m³) (t/y)	0.53 31 851.4	0.73 46 988.6	assumed overall point source load reduction as of 1985 = 63%
Simulated/predicted conditions	(mg/m³) (t/y)	0.53 33 343.53	0.728 45 854.5	0.23 18 461.1
Point source input*	(t/y)	32 672.0*	51 325.9**	11 983.9***
Diffuse calculated input	(t/y) (g/ha/y)	14 249.2 0.89	14 527.5 0.90	14 249.2 0.89
Diffuse source calculated contribution to total load	(t/y) (%)	9910 29.7		calculated load reduction efficiency = 44.63%

*Point source inputs are either given in the data base (*), or calculated in scenario runs (**), or assumed in future predictions (***)

Table 10.4 Summary table of calibration and scenario runs using SENSMOD for nitrogen in the Rhine River Basin

Year (time slice)	Measured or reported conditions		Simulated/predicted conditions		Point source input*	Diffuse calculated input		Diffuse source calculated contribution to total load	
	(g/m^3)	(kt/y)	(g/m^3)	(kt/y)	(kt/y)	(kt/y)	(kg/ha/y)	(kt/y)	%
1985 NH$_4$-N	0.63	37.212	0.63	43.505	172.865*	53.102	3.3	2.607	5.9
(calibration year) NO$_3$-N	4.43	267.740	4.43	278.716	0.0*	368.2	22.9	177.477	63.7
Total = NH$_4$-N + NO$_3$-N	5.06	304.96	5.06	322.22	172.865	421.3	26.2	180.08	55.8
1977[+] NH$_4$-N	1.04	64.65	1.04	65.425	296.759**	48.09	2.99		
NO$_3$-N	3.61	242.2	5.11	315.33	0.0	333.176	20.72		
Total = NH$_4$-N + NO$_3$-N	4.65	306.85	6.14	380.75	296.759	381.26	23.71		
1977[++] NH$_4$-N	1.04	64.65	0.391	24.63	103.718**	48.09			
NO$_3$-N	3.61	242.2	3.61	227.398	0.0	333.176			
Total = NH$_4$-N + NO$_3$-N	4.65	306.85	4.00	252.03	103.718	381.26			
1995 NH$_4$-N	overall reduction, of point		0.201	12.654	58.46***	42.48***	2.64	estimated overall reduction	
NO$_3$-N	sources by 66%, diffuse		2.92	183.908	0.0	294.56***	18.31	of total in-stream load	
Total = NH$_4$-N + NO$_3$-N	sources by 20%, as of 1985		3.12	196.562	58.461	337.04	20.95	39.0%	

[+]scenario no. 1 fitting to measured ammonium; [++]scenario no. 2 fitting to measured nitrate; *point source inputs are either given in the data base (*), or calculated in scenario runs (**), or assumed in future predictions (***)

these data were generated and provide a quantitative basis on which to determine the relative effectiveness of more stringent point source controls *versus* the greater control of nonpoint sources, or the consequences of doing nothing (i.e. maintaining the present restrictions but taking no further, future action). To date, however, the determination of a probable course of action has not been completed, and further, more refined, model-based simulations have been recommended in order to better assess the interplay between point and nonpoint sources of the contaminants of concern.

Step 5: Prepare an action plan

The utility of the model in this case-study was to synthesize a great deal of data and to use the data in a predictive manner. In so doing, the relative magnitudes of point and nonpoint sources could be determined and forecast for a future condition based on the continued application of existing control strategies. This synthesis and the related forecasts can form the basis for evaluating the effectiveness of existing control programmes and the need for additional remedial action. The model output will also permit an assessment of the potential effectiveness of additional actions given the natural background levels of the contaminants of concern in the system. This can be done by forecasting the in-stream concentrations of the contaminants of concern with a 'zero discharge' point source contribution (i.e. the point source contribution would be set at zero). Additional, similar scenarios could be developed for varying levels of agricultural controls, etc. Pending the outcome of this additional modelling effort, it is likely that a refined action plan or control strategy will be completed. At this future time, consideration will doubtless be given to the need for further (in-stream/lake) control measures and to an evaluation of the effectiveness of existing and potential additional control programmes (Steps 6 and 7, respectively).

POST-TREATMENT MONITORING

In order to obtain sufficient information for a judicious selection of nonpoint source pollution control measures, extensive studies of the chemical and biological conditions of the waterbody of concern and its tributaries are usually required. However, upon completion of the studies, and after control measures have been planned and carried out, it may be concluded that further studies are not necessary. Such a conclusion is false. Even after nonpoint source pollution control programmes have been initiated (e.g. reducing the contaminant influx), post-treatment studies should be continued for several years, ideally for a period extending for three years beyond the forecast lag period during which the lake or reservoir can be expected to reach a new equilibrium state. This should be done to compare the condition of the waterbody before and after the implementation of nonpoint source pollution control measures, and to ascertain whether or not the results expected from the model calculations have actually been achieved. Only then can it be ascertained whether or not (or to what degree) the corrective action taken was correct, and whether or not the monetary investment was a financially responsible one.

The period of time necessary for carrying out such post-treatment measurements is dependent on the individual case. The longer the lake is expected to take to recover, the longer the period of time such measurements will have to be taken. Even after a prompt recovery of a given lake, it may be necessary to continue to monitor the waterbody for several years in order to be sure that its condition has been correctly assessed. This is necessary especially in those cases which have a history of large annual variations in water quality prior to the imposition of nonpoint source pollution control measures. Examples would include lakes with relatively short water retention times, and lakes that are situated in areas with very changeable weather conditions.

Should it occur that, in spite of very careful planning and use of all available knowledge, the results obtained fall short of those expected, post-treatment measurements can be used to reformulate additional control measures that might be required (see above) as well as to improve the model predictions in question. Post-treatment monitoring and evaluation provides valuable information to others concerned with similar nonpoint source pollution management problems, and can help to guide both future planning efforts (e.g. building the information and experience base for improved lake and reservoir management technology) and decrease the uncertainty of future model predictions.

CHAPTER SUMMARY

This chapter has presented a brief over view of the general approach recommended as the basis for formulating a nonpoint source pollution control programme, a seven step procedure involving: (1) definition of the problem, (2) identification of the contaminant of concern, (3) goal setting, (4) assessment of alternatives and modelling, (5) plan writing, (6) plan refinement and (7) programme evaluation. As noted earlier, it is not possible to give more precise instructions due to the very great variety of nonpoint source contaminants, contamination problems, and geographical settings. Nevertheless, a multi-phased approach of inventory, diagnosis and action planning and implementation has been shown to be generally successful when employed in such diverse settings as semi-arid southern Africa, arid Central Asia, humid South America, and temperate Europe and North America (see Thornton and Rast, 1995).

CHAPTER 11

OPPORTUNITIES FOR THE FUTURE

M.M. Holland, S.-O. Ryding and J.A. Thornton

INTRODUCTION

The assessment and control of nonpoint source pollution of aquatic environments is a rapidly changing and growing field of environmental management. As point sources of pollution have been controlled worldwide, there has been growing awareness that simply controlling these sources and ignoring the adjacent land use, and its potential for contaminating waters, will not preserve and protect global water resources and the manifold beneficial uses to which they are put. Experience has shown, and continues to show, that far more wide-ranging actions are necessary to effectively manage water resources for such purposes.

As a result of this dynamic state of affairs, it is likely that the source material which has been cited in this document may become out-dated rather more quickly than is usually the case in works of this nature. Nevertheless, the emphasis which has been placed on the practical application of the nonpoint source pollution controls that are cited in these references is likely to counter-act, to some extent, the passage of time. In addition, the definitions, management philosophy, and technical, site-specific approach to the management of nonpoint source pollution are also likely to remain current. Thus, while concern over, and treatment of, nonpoint source pollution is changing rapidly, especially in the developed countries of North America and Europe, and the techniques for controlling such contamination have become better quantified and understood, few changes in the *practical* implications of these concerns and *practical* applications of these approaches have been, and are likely to be, discerned. For this reason, some confidence may be placed in the fundamental approaches propounded in this work – and their continued currency.

Local solutions to local problems, determined on a site-specific basis using proven and reliable assessment, diagnostic and remediation techniques, must remain the best approach for the control of environmental problems – despite the similarities of the natures of these problems on a global basis. However, in order to effect such solutions, local decision-makers and their support staffs require the basic tools (information) necessary to approach the task in an efficient and effective manner. (Re-inventing the proverbial wheel certainly is not efficient.) Thus, this work has been conceived as a means of transferring information to those who can best use and implement it *on a local basis* throughout the world.

419

RESEARCH NEEDS

Notwithstanding the foregoing, however, there are, and will continue to be, research needs that must be addressed in order to ensure the continued relevance of the environmental management solutions proposed. A number of these needs have become obvious in the course of compiling this book. These are presented briefly below as a means of bringing some measure of closure to this discussion of the assessment and control of nonpoint source pollution. Many of the research topics proposed here are offered because answers to these questions will assist both the scientific and management communities to understand and control nonpoint source pollution. Thus, their presentation here, and their consideration by environmental management practitioners, is not out of place. Consistent with the philosophies expounded herein, therefore, these research needs are set forth in six specific areas of endeavour; namely: (1) needs relating to the better understanding of the processes creating and controlling nonpoint source pollution, (2) needs relating to the measurement of nonpoint source pollution, (3) needs relating to the modelling and assessment of nonpoint source pollution, (4) needs relating to the management of nonpoint source pollution, (5) needs relating to the improved public understanding and awareness of nonpoint source pollution, and (6) needs relating to the development of appropriate institutional responses to nonpoint source pollution.

Understanding processes

(1)   Understand processes through the assembly of interdisciplinary research teams (Lubchenco *et al.*, 1991) – To adequately address nonpoint source pollution, researchers from *inter alia* fisheries, agriculture, meteorology, forestry, ecology, limnology, economics, and social science, must work together;

(2)   Understand processes within a framework that considers the entire drainage basin as well as the nearshore environment as the basic unit of study – There is a need for managers, in particular, and society, in general, to look at aquatic habitats (i.e. drainage basins, wetlands, watercourses) in the context of the regional ecosystem of which they are a part (Holland, 1993). In other words, there is a need to look at any nonpoint source pollution problem as part of a larger, complex, functioning system;

(3)   Understand and describe transport processes within a drainage system with due consideration of the spatial and temporal scales at which these processes work – Along definable hydrologic gradients, determine if the aquatic landscape boundaries that are topographically higher in the landscape are more important at trapping and/or converting nutrients than boundaries that are lower in the landscape (Holland *et al.*, 1990).

(4)   Understand and describe transformation processes within a drainage system with due consideration of the spatial and temporal scales at which

these processes work – Elaborate the concept of 'effective transmission' (the quantity of a contaminant that is transmitted from the point of entry in the watershed to the point of its impact at the river mouth; Loehr *et al.*, 1979) and identify the 'transmission coefficient' (the quantity of a contaminant transmitted between any two points of entry within the watershed). Every stretch of river between any two points of entry has a transmission coefficient associated with it. By contrast, the effective transmission is the product of all transmission coefficients at a given entry point. Monteith *et al.* (1981) provide further discussion of the transmission coefficient and effective transmission; and

(5) Understand the storage capabilities of different types of ecosystems – Such key factors as species and habitat diversity, the patterns of distribution of ecological assemblages, and the differences in the productivity and storage capabilities of different types of ecosystems all influence biosphere functions on both the watershed and global scales (Lubchenco *et al.*, 1991).

Measuring processes

(1) Develop more sensitive remote sensing techniques (e.g. the use of GIS and/or the use of aerial photos) for detecting nonpoint source pollution – Such techniques form useful tools for environmental managers in the assessment and diagnosis of, *inter alia*, nonpoint source pollution;

(2) Develop long-term monitoring records for rainfall, climate and streamflow within the major river basins in each region – Existing monitoring and information gathering programmes should be continued and extended, taking care to match the technologies used in these programmes with the capacities of the countries involved (i.e. develop 'appropriate' technologies that can be expanded and extended as resources become available and capabilities expand);

(3) Develop regionally applicable, critical level standards for contaminants of concern that impact desired uses in the different climatic regions (*cf.* Thornton and Rast, 1993; Salas and Martino, 1990) – Measure environmental effects as a consequence of multi-component pollution, including gaining an understanding of the creation and existence of naturally occurring organic compounds and of the background levels of compounds naturally occurring in aquatic systems; and

(4) Develop and disseminate the results of a series of basin studies (of varying spatial scale, land use, geology, soil type, topographic condition, and climate) using standard methodologies that practically demonstrate a range of nonpoint source pollution problems and solutions – Create a reference base covering most eventualities on a global scale.

Developing models and calculations

(1) Model the capacity for nonpoint source pollutant retention in various habitats (i.e. terrestrial, wetland, and aquatic) – Vegetation, geology, soil type, physiography, etc., play important roles in the retention of contaminants; however, little is actually known about how variable this characteristic is among the different types of aquatic habitats and how important habitat boundary conditions are to determining the magnitude of this effect;

(2) Model catchment characteristics and develop predictive capabilities to forecast unit area loads as a function of natural and anthropogenic catchment basin parameters (i.e. develop simple but reliable empirical relationships for substances other than nutrients) – Calculate emission rates based on different land uses, and their consequent load–response relationships, for all types of aquatic ecosystems; and

(3) Develop hydrologically-based models (distributed parameter models) and GIS-based models.

Developing management strategies

(1) Develop the equipment necessary for, and techniques essential to the cleanup of, current hazardous wastes sites to minimize future runoff; and develop feasible criteria for establishment of non-leaching disposal sites;

(2) Develop an economic analysis of the cost efficiency, and technical efficiency, of various nonpoint source control measures;

(3) Develop a more accurate understanding of the assimilative capacity of aquatic habitats (such as wetlands) – are their boundaries enhanced or maintained through management? – This question is obviously very important for understanding contaminant runoff, yet there is very little information that addresses it directly (Holland *et al.*, 1990);

(4) Identify or develop traditional, low intensity management techniques that have successfully maintained or enhanced the functions of aquatic ecosystems in the past; and

(5) Develop national monitoring efforts that will make available basic information for different environmental management needs – Participation in the Global Environmental Monitoring System (GEMS) programme of the United Nations Environment Programme is recommended as a minimum level of effort necessary to create local solutions to local problems.

Increasing public awareness

(1) Increase the dissemination of informational and educational materials to the public on all aspects of the nonpoint source pollution problem to

promote a 'grassroots', community level response to nonpoint source pollution problems – Develop easily understood, step-by-step approaches for transmitting information to the public regarding their role in pollution prevention and minimization, and to decision-makers regarding a timely and appropriate response to pollution episodes; and

(2)   Increase public understanding of the links between nonpoint source and other pollutants so that the public (and decision-makers) can identify their responsibilities in the area of nonpoint source pollution control and environmental management.

Creating institutional mechanisms

(1)   Create appropriate interdisciplinary teams to tackle ecological, economic, and social considerations in a holistic fashion (Lubchenco *et al.*, 1991);

(2)   Create appropriate institutional mechanisms to effectively handle multi-jurisdictional concerns (Holland, 1996);

(3)   Create innovative programmes that will involve more people in the implementation of nonpoint source pollution control strategies;

(4)   Create market incentives to encourage recycling, litter prevention, and good public health – Current research efforts are inadequate for dealing with sustainable systems that involve multiple resources, multiple eco-systems, and large spatial scales; moreover, much of the current research focuses on commodity-based systems, with little attention paid to the sustainability of natural ecosystems whose goods and services currently lack a market value (Lubchenco *et al.*, 1991; Ryding, 1993); and

(5)   Create a more effective means of disseminating information and data on nonpoint source pollution control projects in developing countries – End literary imperialism, publish case studies from less developed countries.

CONCLUDING REMARKS

As the foregoing research needs illustrate, the identification and resolution of nonpoint source contamination problems is a 'team' effort. Not only is this team composed of professionals from many disciplines, but it is very much more of a public process than the related management of point source contamination. Likewise, nonpoint source contamination is a much more individual type of problem, being far more site-specific than the measures used to control point source pollution. Each case of nonpoint source pollution must be examined individually, and there must be a definition of the priorities for each pollution problem, in accordance with the needs of, and socioeconomic conditions in, each country – and in each watershed. Thus, new mechanisms and new approaches to managing nonpoint source pollution problems must be developed (or modified) for each situation.

Many of the research needs identified during this survey relate to the creation and utilization of human resources, both professional and public. Indeed, the whole *raison d'etre* of nonpoint source assessment and control revolves around human use of both terrestrial and water resources for economic purposes, even when these purposes are passive or aesthetic. Thus, while there can be public health considerations and water-use problems associated with nonpoint source pollution, as with point source pollution, these form a component of the issues to be addressed rather than the issues themselves. In short, nonpoint source pollution is more wide ranging both in scope and in planning/implementation needs. Because of these complexities, the assessment and control of nonpoint source contamination is less easy to prescribe – the solutions needed are much more specific and specialized. Nevertheless, by adopting a consistent approach such as that identified in Chapter 10, these complexities can be reduced to more manageable proportions.

In this publication, much attention has been focussed on the two principal elements of nonpoint source pollution; namely, the pollutants themselves (Chapters 5 and 6), and their transport mechanism, the hydrological cycle (Chapter 4). These two components form the basis not only for problem identification and quantification, but also for the management practices that can be employed to reduce the impacts of nonpoint source contamination of waters. By developing an understanding of the types of pollutants and their origins, water resource managers can contribute to land-use and development planning by proposing environmentally friendly methods of water resource development – siting homes and industries in such a manner as to provide for streambank buffers, providing stormwater treatment, and encouraging good housekeeping practices through alliances with health care and sanitation professionals. By developing an understanding of the transport mechanism, water resource managers can contribute to development planning by proposing flood control and allied measures to minimize the deleterious effects of urbanization – related to the expansion of impervious surfaces and modification of the hydrology of urban waterways. In rural areas, the objectives of water resource managers in controlling nonpoint source pollution parallel those of agricultural extension specialists and agencies in preserving top soils, delivering water to crops and pastures, and increasing agricultural productivity. Such goals are also consistent with national development plans and global environmental goals (UNCED, 1993).

This similarity in objectives permits water resource managers involved in the control of nonpoint source pollution to build natural alliances with other resource management professionals in achieving control of nonpoint source pollutants. These alliances have been highlighted, together with other 'nonstructural' approaches as nonpoint source pollution control measures in Chapter 9. Chapter 9 also details, in a general sense, the structural measures for effecting pollution control that are more familiar to water resource managers. Both approaches are based on the need to interdict either the contaminants (at source or during transport) or the transport mechanism. This principal forms the basis of every control measure and best management practice so far identified.

Obviously, the identification of the contaminant of concern and its pathway from source to receiving water may be a complex process. Both field observations and model simulations may be necessary to determine the sources and/or means of transport of various contaminants. The means of constructing appropriate monitoring and modelling programmes are described in Chapters 7 and 8. Because of the close relationship between terrestrial land usage and nonpoint source pollutant generation, compilation of land use inventories forms a useful starting point for most nonpoint source pollution assessment projects; these inventories permit the use of unit area loading (UAL) relationships that can provide order of magnitude estimations of the relative contributions of various land-use activities to the pollutant load of a receiving water. Such estimations provide a basis for more detailed investigations and the ultimate identification of applicable remedial measures. Such an approach can be used to assess both the occurrence and potential for nonpoint source contamination of waters, and thus has both practical and predictive value. Use of nonpoint source pollution models in a predictive sense can often result in development activities that have a minimal impact on the environment, and thus save public funds in the longer term. Treat problems at their source, and control sources before they become problems, are two further axioms of nonpoint source pollution control – both contribute greatly to the successful implementation of their parent axiom, develop local solutions for local problems.

Finally, it would be well to return to the discussion, begun in Chapter 1, of the relevance of nonpoint source pollution control to developing countries. As stated in Chapter 1, nonpoint source contamination of waters is not just a problem of the developed world. It is a global problem. While the contaminants of concern may change with location, and while the pollution control approaches adopted may vary in extent and complexity depending on degree of socioeconomic well being, the basic principles of nonpoint source pollution control, and the good housekeeping practices that underlie them, form sound planning practice in every country – not only contributing to improved public health but also to quality of life. Maintenance of buffer strips and natural corridors provide an enhanced ambiance conducive to peaceful urban life and contribute to sustainable agriculture. Granted some techniques discussed in Chapter 9 may be most applicable where agriculture is mechanized and practiced on a large scale, but the basic philosophies behind such practices are often transferrable (and, in many cases, drawn from) traditional approaches common to many cultures: practices such as no-till planting reflect an older farming tradition that dates back to before the development of the mold board plough, for example. Thus, it will be for each nation to determine the extent and nature of its nonpoint source management programme, and the means of its implementation. In the meantime, however, inclusion of an assessment of point and nonpoint source pollution of waters at a river basin scale in the national development plan would be an appropriate first step.

CHAPTER 12

REFERENCES

J.A. Thornton

Aalderink, R.H., Lijklema, L., Breukelman, J., van Raaphorst, W. and Brinkman, A.G. (1984). Wind induced resuspension in a shallow lake. *Water Science and Technology* **17**, 903–14

Ackermann, W.C., White, G.F. and Worthington, E.B. (1973). Man-made lakes: Their problems and environmental effects. *Geophysical Monogr.* **17**, 847 pp.

Alabaster, J.S. (1980). *Draft review of the state of aquatic pollution of East African inland waters.* Report No. CIFA/80/8, Food and Agriculture Organization of the United Nations (FAO), pp 37

Ali, W., O'Melia, C.R. and Edzwald, J.K. (1984). Colloidal stability of particles in lakes. *Water Science and Technology* **17**, 701–12

Aller, L., Bennett, T., Lehr, J.H. and Petty, R.J. (1985). *DRASTIC: A standardized system for evaluating ground water pollution potential using hydrogeologic settings.* EPA/600/2-85/108. 163 pp.

American Public Works Association (APWA) (1969). *Water pollution aspects of urban runoff.* Fed. Water Pollut. Control Adm. Rep. WP-20-15

Ammon, D.C. (1979). *Urban stormwater pollutant buildup and washoff relationships.* Master of Engineering Thesis, University of Florida, Dept. of Environmental Engineering Sciences, Gainesville, Florida

Anderson, J.R., Hardy, E.E., Roach, J.T. and Witmer, R.E. (1976). *A land use and land cover classification system for use with remote sensor data.* Professional Paper No. 964, US Geological Survey, Denver, Colorado, pp. 28

Anderson, M.P. and Woessner, W.W. (1991). *Applied groundwater modeling: Simulation of flow and advective transport.* Academic Press, New York. 381 pp

Anonymous (1977). *Control of Reentrained dust from paved streets.* EPA 907/9-77-007. US EPA Kansas City, MO

APHA (American Public Health Association, American Water Works Association, and Water Pollution Control Federation) (1985). *Standard methods for the examination of water and wastewater, 16th edition,* American Public Health Association, Washington DC

APHA (American Public Health Association, American Water Works Association, and Water Pollution Control Federation) (1992). *Standard methods for the examination of water and wastewater, 18th edition,* Water Environment Federation, Alexandria, Virginia, pp. 1042

Arnold, J.G., Williams, J.R., Srinivasan, R. and King, K.W. (1996). *SWAT: Soil and Water Assessment Tool.* US Department of Agriculture, Agricultural Research Service, Temple, Texas

Assman, W.H., Drukker, B. and Janssen, A.J. (1988). Modelled historical concentrations and depositions of ammonia and ammonium in Europe. *Atmospheric Environment* **22**, 177–87

Aubertin, G.M. (1977). *Proceedings of the '208' symposium on nonpoint sources of pollution from forested land.* University of Illinois, Carbondale, Illinois, pp. 129

Bailey, D.B. and Waddell, T.E. (1979). Best management practices for agriculture and silviculture: an integrated review. In: *Best Management Practices for Agriculture and Silviculture.* Ann Arbor Science, Ann Arbor, MI. 37 pp.

Baker, J., Eisenreich, S., Johnson, T. and Halfman, B. (1985). Chlorinated hydrocarbon cycling in the benthic nepheloid layer of Lake Superior. *Environmental Science and Technology* **19**, 854–61

Balci, A.N., Ozynvaci, N. and Ozhan, S. (1981). *Sediment and nutrient discharge through streamflow from the Ortadere experimental watersheds in mature oak-beech forest ecosystems near Istanbul, Turkey.* University of Istanbul, Istanbul, pp. 15

Balon, E.K. and Coche, A.G. (1974). *Lake Kariba, a man-made tropical ecosystem in Central Africa.* Vol. 24, Monographiae Biologicae, Junk, The Hague

Banks, R.B. (1975). Some features of wind action on shallow lakes. *Journal of the Environmental Engineering Division of the American Society of Civil Engineers* **101**, 813–27

Barcelona, M.J., Helfrich, J.A., Garske, E.E. and Gibb, J.P. (1984). *Practical guide for ground-water sampling.* State Water Survey Contract Report No. 374, US Environmental Protection Agency, Ada, Oklahoma

Barcelona, M.J., Helfrich, J.A., Garske, E.E. and Gibb, J.P. (1985). A laboratory evaluation of ground water sampling mechanisms. *Ground Water Monitoring Review* **4**, 32–41

Barica, J. and Mur, L.R. (1980). *Hypertrophic ecosystems.* Developments in Hydrobiology Vol. 2, Junk, The Hague

Barrie, L. and Vet, R. (1984). The concentration and deposition of acidity, major ions and trace metals in the snow pack of the eastern Canadian Shield during the winter of 1980–81. *Atmospheric Environment* **18**, 1459–69

Beasley, D.B., Monke, E.J., Miller, E.R. and Huggins, L.F. (1985). Using simulation to assess the impacts of conservation tillage on movement of sediment and phosphorus into Lake Erie. *Journal of Soil and Water Conservation* **40**, 233–7

Beaulac, M.N. and Reckhow, K.H. (1982). An examination of land use–nutrient export relationships. *Water Resources Bulletin* **18**, 1013–24

Bernhardt, H. (1981). Reducing nutrient inflows. In: Rast, W. and Kerekes, J.J. *Proceedings, International Workshop on the Control of Eutrophication.* International Institute for Applied Systems Analysis (IIASA), A-2361 Laxenburg, Austria, October 12–15, 1981. pp 43–51

Bernhardt, H. (1983). Input control of nutrients by chemical and biological methods. *Water Supply* 1:187–206

Betson, R.P. and McMaster, M. (1975). Non-point source mineral water quality model. *Journal of the Water Pollution Control Federation* **47**

Bezuidenhout, L.M., Schoonbee, H.J. and de Wet, L.P.D. (1990). Heavy metal content in organs of the African sharptooth catfish, *Clarias gariepinus* (Burchell), from a Transvaal lake affected by mine and industrial effluents. Part 1. Zince [sic] and copper. *Water SA* **17**, 125–30

Binnie and Partners (1986). *Water and waste-water management in the malt brewing industry.* Natsurv 1. Water Research Commission, Pretoria. ISBN 0-908356-63-3

Binnie and Partners (1987a). *Water and waste-water management in the soft drink industry.* Natsurv 3. Water Research Commission, Pretoria. ISBN 0-908356-90-0

Binnie and Partners (1987b). *Water and waste-water management in the metal finishing industry*. Natsurv 2. Water Research Commission, Pretoria. ISBN 0-908356-81-1

Biro, P. (1977). Effects of exploitation, introductions and eutrophication on percids in Lake Balaton. *J. Fish. Res. Bd. Can.* **34**, 1678–83

Blackie, J.R, Ford, E.D. and Horne, J.E.M. (1980). *Environmental effects of deforestation: an annotated bibliography*. Occasional Publication No.10, Freshwater Biological Association, Ambleside, pp. 173

Blackman, W.C., Jr. (1996). *Basic hazardous waste management*, 2nd edition. Lewis Publishers, Boca Raton. 416 pp.

Bormann, F.H., Likens, G.E. and Melillo, J.M. (1977). Nitrogen budget for an aggrading northern hardwood forest ecosystem. *Science* **196**, 981–3

Bosman, H.H. and Kempster, P.L. (1985). Precipitation chemistry of Roodeplaat Dam catchment. *Water SA* **11**, 157–64

Bostrom, B., Jansson, M. and Forsberg, C. (1982). Phosphorus release from lake sediments. *Archiv fur Hydrobiologie, Beih. Ergebn. Limnol.* **18**, 5–59

Boulding, J.R. (1995). *Practical handbook of soil, vadose zone, and groundwater contamination: Assessment, prevention, and remediation*. Lewis Publishers, Boca Raton. 960 pp.

Bowonder, B. (1987). Management of environment in developing countries. *The Environmentalist* **7**, 111–22

Boyle, E., Collier, R., Dengler, A.T., Edmond, J.M., Ng, A.C. and Stallard, R.F. (1974). On the chemical mass-balance in estuaries. *Geochimica Cosmochimica Acta* **38**, 1719–28

Brinkman, A.G. and van Raaphorst, W. (1986). *De fosfaathuishouding in het Veluwemeer*. Thesis, Twente University, Enschede

Britton, G. and Marshall, K.C. (1980). *Adsorption of microorganisms to surfaces*. Wiley and Sons, New York, pp. 439

Brown, L.R., Chandler, W.U., Flavin, C., Jacobson, J., Pollock, C., Postel, S., Starke, L. and Wolf, E.C. (1987). *State of the world, 1987: A Worldwatch Institute report on progress toward a sustainable society*. W.W. Norton and Co., New York, pp. 268

Brunner, P.G. (1975). *Die Verschmutzung des Regenwasserabflusses im Trennverfahren*. Dissertation, Munich

Bruton, M.N. (1985). The effects of suspensoids on fish. *Hydrobiologia* **125**, 221–41

Bryan, R.B. (1976). Consideration on soil erodibility and sheet wash. *Catena* **3**:99–112

Bryan, R.B., Yair, A. and Hodges, W.K. (1978). Factors controlling the initiation of runoff and piping in Dinosaur Provincial Park Badlands, Alberta, Canada. *Seitch. für Geom. Suppl. Bd.*, **29**:151–68

Bubenzer, D.G. and Jones, B.A., Jr. (1971) Drop size and impact effects on the detachment of soils under simulated rainfall. *Trans. ASAE*, **14**:625–28

Buecking, W. (1974). Die Beeinflussung von chemischen Wassereigenschaften durch forstliche Düngungsmassnahmen. *Allg. Forstzeitschrift* **29**, 1077–9

Buringh, P. and van Heemst, H.D.J. (1977). *An estimation of world food production based on labour-oriented agriculture*. Center for World Food Market Research, Wageningen

Burk, C.J. and Holland, M.M. (1979). *Stone walls and sugar maples: an ecology for northeasterners*. Appalachian Mountain Club, Boston, Massachusetts

Busch, W.-D.N. and Sly, P.G. (1992). *The development of an aquatic habitat classification system for lakes*. CRC Press, Boca Raton, 240 pp.

Cadle, S.H., Countess, R.J. and Kelly, N.A. (1982). Nitric acid and ammonia in urban and rural locations. *Atmospheric Environment* **18**, 2501–6

Callender, E. (1982). Benthic phosphorus regeneration in the Potomac River estuary. *Hydrobiologia* **92**, 431–46

Canter, L.W. (1996). *Nitrates in groundwater*. Lewis Publishers, Boca Raton, 320 pp.

Cappenberg, Th.E., Hordijk, K.A., Jonkheer, G.J. and Lauwen, J.P.M. (1979). Carbon flow across the sediment water interface in Lake Vechten. *Hydrobiologia* **91**, 161–8

Carmichael, W.W. (1981). *The water environment, algal toxins and health*. Plenum, New York

Caro, J.H., Freeman, H.P. and Turner, B.C. (1974). Persistence in soil and losses in runoff of soil incorporated carbaryl in a small watershed. *J. Agric. Food Chem.* **22**, 860–63

Carol, D. and Rubin, K. (1985). *Hazardous waste management: recent changes and policy alternatives*. Congressional Budget Office, Washington, DC

Carsel, R.F., Mulkey, L.E., Lorber, M.N. and Baskin, L.B. (1985). The pesticide root zone model (PRZM): a procedure for evaluating pesticide leaching threats to groundwater. *Ecological Modelling* **30**, 49–69

Chan, E. Bursztynsky, T.A., Hantzche, N. and Litwin, Y.J. (1987). *The use of wetlands for water pollution control*. Municipal Environmental Research Laboratory, US EPA, Cincinnati, Ohio. 214 pp.

Chan, W., Vet, R., Chul-Un Ro, Tang, A. and Lusis, M. (1984). Long-term precipitation quality and wet deposition fields in the Sudbury basin. *Atmospheric Environment* **18**, 1175–88

Chandler, J. (1988). Personal communication. International Joint Commission, Washington, DC

Chapman, D. (1992). *Water quality assessments: a guide to the use of biota, sediments and water in environmental monitoring*. Chapman and Hall, London, pp. 585. ISBN 0-412-44840-8

Chesters, G., Stiefel, R., Bahr, T., Robinson, J., Ostry, R., Coote, D.R. and Whitt, D.M. (1978). *Pilot watershed studies summary report*. Pollution from Land Use Activities Reference Group, International Joint Commission, Great Lakes Regional Office, Windsor, Ontario

Chow, V.T., Maidment, D.R. and Mays, L.W. (1988). *Handbook of applied hydrology*. McGraw-Hill, New York, pp. 572

Church, T., Taramontano, J., Schalark, J., Jickells, T., Tokos, J., Knap, A. and Galloway, J. (1984). The wet deposition of trace metals to the western Atlantic Ocean at the mid-Atlantic coast and on Bermuda. *Atmospheric Environment* **18**, 2657–64

Claiborne, R. (1983). *Our marvelous native tongue: the life and times of the English language*. Times Books, New York, pp. 339

Coke, M.M. (1988). Freshwater fish conservation in South Africa: a rising tide. *Journal of the Limnological Society of Southern Africa* **14**, 29–34

Cooke, G.D., Welch, E.B, Peterson, S.A. and Newroth, P.R. (1993). *Restoration and management of lakes and reservoirs*. Lewis, Boca Raton, Florida, pp. 548. ISBN 0-87371-397-4

Cordery, I. (1977). Quality characterization of urban storm water in Sydney, Australia. *Water Res.* **13**, 197–202

Cowen, W.F. and Lee, G.F. (1973). Leaves as a source of phosphorus. *Environ. Sci. Tech.* **7**(9), 853–54

Crawford, H.H. and Linsley, R.K. (1966). *Digital simulation in hydrology: Stanford watershed model IV*. Technical Report No. 39, Stanford University, Stanford, California, pp. 210

Daniel, T.C., Wendt, R.C. and Konrad, J.G. (1978). *Nonpoint pollution: runoff in urban areas*. University of Wisconsin Extension, Madison, Wisconsin, 8 pp.

Davalos, L., Lind, O.T. and Doyle, R.D. (1989). Evaluation of phytoplankton limiting factor in Lake Chapala, Mexico. *Lake and Reservoir Management* **5**, 99–104

Davenport, T.E. (1983). *Soil erosion and sediment transport dynamics in the Blue Creek Watershed, Pike County, Illinois*. IEPA/WPC/83-004. Springfield, Illinois, 224 pp.

Dávid, L. and Telegdi, L. (1986). The influence of watershed development on the long-term eutrophication of Lake Balaton. In: Somlyódy, L. and van Straten, G. (eds) *Modelling and Managing Shallow Lake Eutrophication*. Heidelberg, Springer Verlag, 386 pp.

Davidson, J.H. and Delogu, O.E. (1989). *Federal environmental regulation. Volume 1 and Volume 2*. Butterworth, Salem, New Hampshire

Davidson, J.H. and Delogu, O.E. (1992). *Federal environmental regulation. Volume 1: Part 2*. Butterworth, Salem, New Hampshire, USA, pp. 228

Davies, B.R. (1979). Stream regulation in Africa: a review. In: Ward, J.V. and Stanford, J.A. (eds), *The ecology of regulated streams*, Plenum Press, New York, pp. 112–42

Davies, B.R. and Walmsley, R.D. (1985). *Perspectives in southern hemisphere limnology*. Developments in Hydrobiology Vol. 28, Junk, The Hague, pp. 263

Davis, F.E. (1989). *Water: Laws and Management*. Special Pub. No. 89–94, American Water Resources Association, Bethesda, Maryland

Davis, W.S. and Simon, T.P. (1995). *Biological assessment and criteria: Tools for water resource planning and decision making*. Lewis Publishers, Boca Raton, 432 pp.

Davison, W. (1985). Conceptual models for transport at a redox boundary. In: Stumm, W. (ed), *Chemical processes in lakes*, Wiley, New York, Chapter 2

Dawson, G. (1984). Tropospheric ammonia. In: Aneja, V. (ed), *Environmental impact of natural emissions*, Transactions of the APCA

de Wet, L.P.D., Schoonbee, H.J., Pretorius, J. and Bezuidenhout, L.M. (1990). Bioaccumulation of selected heavy metals by the water fern *Azolla filiculoides* Lam. in a wetland ecosystem affected by sewage, mine and industrial pollution. *Water SA* **16**, 281–6

Dean, J.D., Jowist, P.O. and Donigian, A.S. Jr (1984). *Leaching evaluation of agricultural chemicals (LEACH) handbook*. Report No. EPA-600/3-84-068, US Environmental Protection Agency, Athens, Georgia

Debo, T.N. and Reese, A.J. (1995). *Municipal storm water management*. Lewis Publishers, Boca Raton, 768 pp.

DeCoursey, D.G. (1982). Mathematical models for non-point source pollution control. *Journal of Soil and Water Conservation* **40**, 408–13

Dejoux, C. (1988). *La pollution des eaux continentales Africaines: Experience acquise situation actuelle et perspectives*. Editions de l'ORSTOM, Collection Travaux et Documents No. 213, Institut Francais de Recherche Scientifique pour le Developpement en Cooperation, Paris, pp. 513

Delumyea, R.G. and Petel, R.L. (1978). Wet and dry deposition of phosphorus into Lake Huron. *Water, Air and Soil Pollution* **10**, 187–98

Dennison, M.S. (1996). *Storm water discharges: Regulatory compliance and best management practices*. Lewis Publishers, Boca Raton, 464 pp.

Dick, T.M.M. and Marsalek, J. (1979). Importance des charges unitaires de pollutants dans le ruissellement urbain d'eaux pluviales. *Eau du Quebec*, **12**(4), 262–67

Dillon, P.J. and Reid, R.A. (1981). Input of biologically available phosphorus by precipitation to Precambrian lakes. In: Eisenreich, S. (ed), *Atmospheric pollutants in natural waters*, Ann Arbor Science Publishers, Ann Arbor, Michigan

Dolske, D.A. and Sievering, H. (1979). Trace element loading of southern Lake Michigan by dry deposition of atmospheric aerosol. *Water, Air and Soil Pollution* **12**, 485–502

Donigian, A.S. (1988). Selection, application and validation of environmental models. In: *Proceedings of the International Symposium on Water Quality Modeling of Agricultural Non-point Sources*, Utah State University Press, Logan, Utah

Donigian, A.S. and Crawford, N.H. (1976). *Modelling pesticides and nutrients on agricultural lands*. Publication PB-250-566, Hydrocomp Inc.

DWA (Department of Water Affairs) (1986). *Management of the water resources of the Republic of South Africa*. Government Printer, Pretoria ISBN 0-621-11004-3

Dyck, S. (1980). Angewandte Hidrologie, Teil 2: *Der Wasserhaushalt der Flussgebiete*. Berlin, Verlag für Bauwesen

Edmondson, W.T. and Winberg, G.G. (1971). *A manual on methods for the assessment of secondary productivity in fresh waters*. Handbook No. 17, International Biological Programme. Blackwell, Edinburgh, pp. 358. ISBN 0-632-06430-7

Edzwald, J.K. and Toensing, D.C. (1976). Phosphate adsoprtion reactions with clay minerals. *Environmental Science and Technology* **10**, 485–90

Einsele, W. (1936). Uber die Beziehungen des Eisenkreislaufs zum Phosphatkreislauf im eutrophen See. *Archiv fur Hydrobiology* **29**, 664–86

Einsele, G. (1986). Das landschattsdkologische Forschungspr-projekt Naturpark Schönbuch- Wasser- und Stotthaushalt, Bio-, Geo- und Forstwirtschaftliche Studien in Südwestdautschland. *Forschungsbericht DFG*, VCH Wainheim, pp. 636

Eisenreich, S.J., Looney, B.B. and Thornton, J.D. (1981). *Airborne organic contaminants in the Great Lakes Ecosystem*. American Chemical Society, Environmental Science and Technology, **15**, 30–8

Ellis, J.B. (1976). Sediments and water quality of urban storm water. *Water Serv.* **80**, 730–34

Ellis, J.B. (1986). Pollutional aspects of urban runoff. In Torno, H.C., Marsalek, J. and Desbordes, M. (eds) *Urban Runoff Pollution*. NATO ASI Series G, vol 10, 1–38 Springer Verlag, Heidelberg, Germany

Ellis, J.B., Hamilton, R.S. and Roberts, A.H. (1982). Sedimentary characteristics of suspensions in London stormwater. *Sediment Geol.* **33**, 147–54

Enell, M. and Löfgren, S. (1987). Phosphorus in interstitial waters. In: Persson, G. and Jansson, M. (eds), *Phosphorus in freshwater ecosystems*, Volume 48, Developments in Hydrobiology, Kluwer, Dordrecht, pp. 103–32

Evans, D. and Rigler, F. (1985). Long-distance transport of anthropogenic lead as measured by lake sediments. *Water, Air and Soil Pollution* **24**, 141–51

Evans, R. and Dillon, P.J. (1982). Historical changes in the anthropogenic lead fallout in southern Ontario, Canada. *Hydrobiologia* **91**, 131–7

Falconer, I.R., Beresford, A.M. and Runnegar, M.T.C. (1983). Evidence of liver damage by toxin from a bloom of the blue-green alga, *Microcystis aeruginosa. Medical Journal of Australia* **1**, 511–14

FAO (Food and Agriculture Organization of the United Nations) (1965). *Soil erosion by water, some measures for its control on cultivated lands*. FAO Agricultural Development Paper 81, Rome

Ferguson, B.K. (1994). *Stormwater infiltration.* Lewis Publishers, Boca Raton, 288 pp

Ferm, M., Samuelsson, K., Sjödin, A. and Grennfelt, P.-I. (1984). Long-range transport of gaseous and particulate oxidized nitrogen compounds. *Atmospheric Environment* **18**, 1731–5

Fogg, G.E. (1975). *Algal Cultures and Phytoplankton Ecology.* University of Wisconsin Press, Madison, Wisconsin

Foree, E.G. and McCarty, P.L. (1970). Anaerobic decomposition of algae. *Environmental Science and Technology* **4**, 842–9

Forster, D.L., Bardos, C.P. and Southgate, D.D. (1987). Soil erosion and water treatment costs. *J. Soil and Water Cons.* **42**, 349–52

Forster, D.L. and Becker, G.S. (1979). Erosion control strategies: Honeycreek watershed. *North Central Journal of Agricultural Economics* **1**, 53–60

Forster, D.L. and Southgate, D.D. (1988). Nonpoint source pollution: economic perspectives. Dept. of Agricultural Economics, Ohio State University, Columbus, OH

Foster, G.R. and Meyer, L.D. (1975). Mathematical simulation of upland erosion by fundamental erosion mechanics. In: *Present and prospective technology for predicting sediment yields and sources.* Report No. ARS-S-40, US Department of Agriculture, Agricultural Research Service, Washington, DC

Förstner, U., Ahlf, W., Calmano, W., Kersten, M. and Salomons, W. (1984). Mobility of heavy metals in dredged harbor sediments. In: Sly, P.G. (ed), *Sediment and water interactions*, Springer Verlag, New York, Chapter 31

Fournier, F. (1960). *Climat et erosion: la relation entre l'erosion du sol par l'eau les precipitations atmospheriques.* UNESCO, Paris

Free, M.H., Onstand, C.A. and Holtan, H.N. (1975). *ACTMO – An Agricultural Chemical Transport Model.* Report ARS-H-3, US Department of Agriculture, Agricultural Research Service, Washington, DC

Fuehrer, H.W. (1983). *Bioalementbilanzen von vier Wassereinzugsgebieten des Krofdorfer Buchenwaldes in Hessen.* Diplomarbeit, Ludwig-Maximilian Universitat, Munchen, pp. 92

Fuggle, R.F. and Rabie, M.A. (1986). *Environmental concerns in South Africa: technical and legal perspectives.* Juta, Cape Town, South Africa, pp. 587. ISBN 0-7021-1404-9

Fukuda, M.K. and Lick, W. (1980). The entrainment of cohesive sediments. *Journal of Geophysical Research* **85**, 2813–24

Gaba, J.M., and Stever, D.W. (1995). *Law of solid waste, pollution prevention and recycling.* Clark, Boardman, Callaghan, Chicago

Galloway, J.N. (1985). The deposition of sulfur and nitrogen from the remote atmosphere – Background paper. In Galloway, J.N. *et al.* (eds) *The Biogeochemical Cycling of Sulfur and Nitrogen in the Remote Atmosphere.* Riedel Publishing, pp. 143–75

Galloway, J.N., Thornton, S.P., Norton, S.A., Volchok, H.L. and McLean, R.A. (1982). Trace metals in atmospheric deposition: a review. *Atmospheric Environment* **16**, 1677–700

Galloway, J.N., Eisenreich, S.J. and Scott, B.C. (1979). *Report of a workshop on toxic substances in atmospheric deposition.* National Atmospheric Deposition Program, pp. 45

Galloway, J. and Likens, G. (1978). The collection of precipitation for chemical analyses. *Tellus* **30**, 71–82

Gatz, D.F., van Bowersox, C., Su, J. and Stensland, G.J. (1988). *Great Lakes Atmospheric Deposition (GLAD) network, 1982 and 1983: Data analyses and interpretation.* Report No. EPA-905/4-88-002 GLNPO Report No. 2, US Environmental Protection Agency

Geldreich, E.E. (1996). Pathogenic agents in freshwater resources. *Hydrological Processes* **10**(2), 315–33

Glibert, P.M. (1976). *Nutrient flux studies in the Great Bay Estuary, New Hampshire.* MSc Thesis, University of New Hampshire, Durham, New Hampshire

Goettle, A. (1978). Atmospheric contaminants, fallout and their effects on stormwater quality. *Prog. Water Tech.* **10**, 455–67

Goldfarb, W. (1988). *Water law.* 2nd Ed. Lewis, Chelsea, Michigan, pp. 284

Golterman, H.L. and Clymo, R.S. (1971). *Methods for chemical analysis of fresh waters.* Handbook No. 8, International Biological Programme. Blackwell, Edinburgh, pp. 166. ISBN 0-632-05540-5

Golubev, G.N. (1980). *Agricultural and Water Erosion of Soils: A Global Outlook.* IIASA WP-80-129, Laxenburg, Austria

Golubev, G.N. (1983). Economic activity, water resources and the environment. A challenge for hydrology. *Hydrological Sciences J.* **28**:57–75

Gore, J.A. (1989). Case histories of instream flow analysis for permitting and environmental impact assessments in the United States. *Southern African Journal of Aquatic Sciences* **16**, 194–208

Grabow, W.O.K. (1979). Disinfection of water: pros and cons. *Water SA* **5**, 98–108

Grabow, W.O.K., Coubrough, P., Nupen, E.M. and Bateman, B.W. (1984). Evaluation of coliphages as indicators of the virological quality of sewage-polluted water. *Water SA* **10**, 7–14

Granat, L. (1987). Sulfate in precipitation as observed by the European atmospheric chemistry network. *Atmospheric Environment* **12**, 424

Granat, L. (1988). Concentration gradients in atmospheric precipitation in areas of high annual precipitation. In: *Proceedings of the workshop on acid deposition processes at high elevation sites*

Gregory, K.J. and Walling, D.E. (1985). *Drainage basin form and process: a geomorphological approach.* Arnold, London, pp. 458. ISBN 0-7131-5923-5.

Greichus, Y.A. (1982). Insecticides in Lake McIlwaine. In Thornton, J.A. (ed), *Lake McIlwaine, the eutrophication and recovery of a tropical African man-made lake.* Vol. 49, Monographiae Biologicae, Junk, The Hague, pp. 94–100

Grip, H. (1982). *Water chemistry and runoff in forest streams at Kloten.* UNGI Rapport No. 58, Uppsala University, Stockholm, pp. 144

Grobler, D. and Davies, E. (1979). The availability of sediment phosphate to algae. *Water SA* **5**, 114–22

Grobler, D.C. and Davies, E. (1981). Sediments as a source of phosphate: a study of 38 impoundments. *Water SA* **7**, 54–60

Grobler, D.C.S.D. (1985). *Management-orientated eutrophication models for South African reservoirs.* PhD Dissertation, University of the Orange Free State, Bloemfontein

Grobler, D.C., Toerien, D.F. and Rossouw, J.N. (1987). A review of sediment/water quality interaction with particular reference to the Vaal River system. *Water SA* **13**, 15–22

Groen, P. and Dorresteyn, R. (1976). *Zeegolven opstellen op oceanografisch en maritiem gebied, dl II.* KNMI State Publication

Gunnison, D. and Alexander, M. (1975). Resistance and susceptibility of algae to decomposition by natural microbiological communities. *Limnology and Oceanography* **20**, 64–70

Guralnik, D.B. (1970). *Webster's New World Dictionary of the American Language*. 2nd Ed. World Publishing, New York, pp. 1692

Guy, H.P. (1965). Residential construction and sedimentation at Kensington, Maryland. In *Proceedings of the Federal Interagency Sedimentation Conference*, 1963. pp 30–6

Haith, D.A. (1985). An event-based procedure for estimating monthly sediment loads. *Transactions of the American Society of Civil Engineers* **28**, 1916–20

Haith, D.A. and Dougherty, J.V. (1976). Non-point source pollution from agricultural runoff. *Journal of the Environmental Engineering Division of the American Society of Civil Engineers*

Haith, D.A. and Shoemaker, L.L. (1987). Generalized watershed loading functions for stream flow nutrients. *Water Resources Bulletin* **23**, 471–8

Haith, D. A., Tubbs, L. J. and Pickering, N.B. (1984). Simulation of soil erosion and soil nutrient loss. *Simulation Monographs*, Purdoc, Wageningen, p. 62

Hammer, D.A. (1996). *Creating freshwater wetlands*. 2nd edition. Lewis Publishers, Boca Raton, 480 pp.

Hansen, D.O., Erbaugh, J.M. and Napier, T.L. (1987). Factors related to adoption of soil conservation practices in the Dominican Republic. *Journal of Soil and Water Conservation* **41**, 367–9

Harper, A.A. (1985). Fate of heavy metals from highway runoff. In *Stormwater Management Systems*. PhD Thesis, University of Central Florida, Orlando

Harrington, H.D. and Durrell, L.W. (1957). *How to identify plants*. Swallow Press, Chicago, Illinois, pp. 203, ISBN 0-8040-0149-9

Hasler, A.D. and Einsele, W.G. (1948). Fertilization for increasing productivity of natural inland waters. *Transactions of the 13th North American Wildlife Conference, March 1948*, pp. 527–54

Hatherly, R.S. and Viewing, K.A. (1982). The solid phase: a study of pollution benchmarks on a granitic terrain. In: Thornton, J.A. (ed), *Lake McIlwaine, the eutrophication and recovery of a tropical African man-made lake*. Vol. 49, Monographiae Biologicae, Junk, The Hague, The Netherlands, pp. 77–93

Hattingh, W.J.H. and Nupen, E.M. (1976). Health aspects of potable water supplies. *Water SA* **2**, 33–46

Hauer, F.R. and Lamberti, G.A. (1996). *Methods in stream ecology*. Academic Press, San Diego, California, pp. 674, ISBN 0-12-332905-1

Hayuma, A.M. (1983). The management and implementation of physical infrastructures in Dar es Salaam City, Tanzania. *Journal of Environmental Management* **16**, 321–34

Heaney, J.P. *et al.* (1975). *Urban stormwater management, modeling and decision making*. EPA 670/2-75-022. US EPA

Heaney, J.P., Huber, W.C. and Nix, S.J. (1976). *Stormwater management model: Level 1 Preliminary screening procedures*. EPA-600/2-76-275. US EPA, Cincinnati, Ohio

Heidtke, T.M., DePinto, J.V. and Young, T.C. (1986). *Assessment of annual total phosphorus tributary loading estimates: Application to the Saginaw River*. US EPA Rep., US EPA, Chicago, pp. 88

Heidtke, T.M., Auer, M.T., Canale, R.P. and Slawecki, T.A. (1986). Coupling non-point pollution and water quality models: An example for the Green Bay-Fox River watershed. In: *Proceedings of the non-point pollution abatement symposium.* pp. T-I-C 1-11

Heiskary, S., Bryant, N., Butkus, S., Dennis, J., Duda, A., Larsen, D.P., Raschke, R., Ratcliffe, S. and Smeltzer, E. (1992). *Developing eutrophication standards for lakes and reservoirs*, North American Lake Management Society, Alachua, Florida, pp. 51

Heit, M., Clusec, C. and Baron, J. (1984). Evidence of deposition of anthropogenic pollutants in remote Rocky Mountain lakes. *Water, Air and Soil Pollution* **22**, 403–16

Henderson, D.R. and Barrows, R.L. (1984). *Food, resources, and the future.* Publication No. 222, North Central Region Extension, Iowa State University

Hendry, C.D., Brezonite, P.L. and Edgerton, E.S. (1981). Atmospheric deposition of nitrogen and phosphorus in Florida. In Eisenreich, S. (ed), *Atmospheric pollutants in natural waters*, Ann Arbor Science Publishers, Ann Arbor, Michigan

Herricks, E.E. and Jenkins, J.R. (1995). *Stormwater runoff and receiving systems: Impact monitoring and assessment.* Lewis Publishers, Boca Raton, 480 pp.

Hess, K. (1993). *Environmental site assessment, Phase 1: A basic guide.* CRC Press, Boca Raton, pp. 304

Hicks, B. and Liss, P. (1976). Transfer of SO_2 and other reactive gases across the air/sea interphase. *Tellus* **28**, 348–54

Hicks, B., Wesley, M. and Durham, J. (1982). Critique of methods to measure dry deposition. In: Keith, L. (ed), *Energy and environmental chemistry, acid rain 2.* pp. 205–23

Higgins, T.E. (1995). *Pollution prevention handbook.* Lewis Publishers, Boca Raton, 592 pp.

Hites, R. (1981). *Sources and fates of atmospheric polycyclic aromatic hydrocarbons.* ACS Symposium Series 167, pp. 187–96

Ho, J.S.-Y. (1983). Effect of sampling variables on recovery of volatile organics in water. *Journal of the American Water Works Association* **12**, 583–6

Hock, B. (1970). *Water quality balance.* Report No. 27, Technical and Economic Water Management Series VMGT, VIZDOK Press, Budapest, pp. 124 [In Hungarian]

Hogland, W. (1979). *Snö och snöhantering I Lund vintern 1978–79.* Institute för teknisk vatterversurslära, University of Lund Report 3025

Holland, M.M. (1993). Management of land/inland water ecotones: needs for regional approaches to achieve sustainable ecological systems. *Hydrobiology* **251**, 331–40

Holland, M.M., Whigham, D.F. and Gopal, B. (1990). The characteristics of wetland ecotones. In: Naiman, R.J. and Decamps, H. (eds), *The Ecology and Management of Aquatic-Terrestrial Ecotones.* UNESCO MAB Vol. 4. Parthenon, Carnforth, United Kingdom pp. 171–98

Holland, M.M. (1996). Ensuring sustainability of natural resources: focus on institutional arrangements. *Can. J. Fish. Aquat. Sci.* **53** (suppl. 1), 432–39

Hollis, G.E., Holland, M.M., Maltby, E. and Larson, J. (1988). Wise use of wetlands. *Nature & Resources* **24**(1), 2–13

Holme, N.A. and McIntyre, A.D. (1971). *Methods for the study of marine benthos.* Handbook No. 16, International Biological Programme. Blackwell, Edinburgh, pp. 334, ISBN 0-632-06420-X

Holtan, H.N. (1971). *A formulation for quantifying the influence of soil porosity and vegetation on infiltration.* Third International Seminar for Hydrology Professors. Lafayette, Indiana, Purdue University

Holtan, H.N. and Lopez, N.C. (1971). *USDAHL-70 model of watershed hydrology.*

Technical Bulletin No. 1453, US Department of Agriculture, Agricultural Research Service, Washington, DC

Horton, R.E. (1945). Erosional development of streams and their drainage basins. Hydrologic approach to quantitative methodology. *Geol. Soc. A. Bull.* **56**, 275–370

Huber, W.C., Heanye, J.P., Nix, S.J., Dickinson, R.E. and Polmann, D.J. (1981). Storm water management model user's manual, version lll. WPA-600/2-81-109a. US EPA, Cincinnati, Ohio

Huber, W. (1986). Deterministic modeling of urban runoff quality. In Torno, A.C., Marsalek, J. and Desbordes, R. *Urban Runoff Pollution.* NATO ASI Series G, Vol 10, 167–242. Springer Verlag, Heidelberg

Hueser, (1977). Wald und Wasserqualitat. *Schriftenreihe DVWK* **41**, 89–107

Huff, D.D., Luxmore, R.V., Mankin, V.B. and Begovich, C.L. (1977). *TEHM: A Terrestrial Ecosystem Hydrology Model.* Report ORNL/NSF/EATC-17, Oak Ridge National Laboratory,Oak Ridge, Tennessee

Hunt, D.T.E. and Wilson, A.L. (1986). *The chemical analysis of water. General principles and techniques. 2nd edition.* The Royal Society of Chemistry, Burlington House, London, pp. 683

Hunter, J.M., Rey, L. and Scott, D. (1983). Man-made lakes – man-made diseases. *World Health Forum* **4**, 177–82

Hutchinson, G.E. (1957). *A Treatise on Limnology, Volume I: Geography, physics, and chemistry.* Wiley, New York, pp. 1015

Hutchinson, G.E. (1967). *A Treatise on Limnology, Volume II: Introduction to lake biology and limnoplankton.* Wiley, New York, pp. 1115

Hutchinson, G.E. (1975). *A Treatise on Limnology, Volume III: Limnological botany.* Wiley, New York, pp. 660, ISBN 0-471-42574-5

Hutchinson, G.E. and Edmonson, Y.H. (1993). *A Treatise on Limnology, Volume IV: The zoobenthos.* Wiley, New York, pp. 976, ISBN 0-471-54294-6

Hynes, H.B.N. (1972). *The Ecology of Running Waters.* University of Toronto Press, Toronto, Ontario, pp. 555, ISBN 0-8020-1689-8

IEP Inc. (1990). *P8 Urban Catchment Model.* IEP Inc., Northborough, Massachusetts

IIASA (1990). The price of pollution. *Options* (September), 4–8

IJC (1983). *Report on Great Lakes water quality. Great Lakes Water Quality Board, Report to the International Joint Commission.* Windsor, Ontario

IJC (International Joint Commission and Great Lakes Fishery Commission) (1990). *Exotic species and the shipping industry: The Great Lakes-St. Lawrence ecosystem at risk.* International Joint Commission, Washington, DC, pp. 74

James, W. and Niemczynowicz, J. (1992). *Water, development and the environment.* Lewis, Boca Raton, Florida, pp. 381

Janes, E.B. and Hotchkiss, W.R. (1990). *Transferring Models to Users.* American Water Resources Association, Bethesda, Maryland, pp. 404

Jarvis, A.C. (1986). Zooplankton community grazing in a hypertrophic lake (Hartbeespoort Dam, South Africa). *Journal of Plankton Research* **8**, 1065–78

Jarvis, A.C. (1987). *Studies on zooplankton feeding ecology and resource utilization in a sub-tropical hypertrophic impoundment (Hartbeespoort Dam, South Africa).* PhD Dissertation, Rhodes University, Grahamstown

Javandel, I., Doughty, C. and Tsang, C.F. (1984). *Groundwater transport: Handbook of mathematical models.* Volume No. 10, Water Resources Monographs, American Geophysical Union, Washington, DC, pp. 228

Jenne, E.A. (1968). Controls on Mn, Fe, Co, Ni, Cu and Zn concentrations in soils and water; the significant role of hydrous Mn and Fe oxides. In Gould, R.G. (ed) *Trace Inorganics in Water*. Advances in Chemistry Series, 73. American Chemical Society, Washington. pp. 337–87

Jewell, W.J. and McCarthy, P.L. (1971). Aerobic decomposition of algae. *Environmental Science and Technology* **5**, 1023–31

Johannessen, M. and Henriksen, A. (1977). *Chemistry of snowmelt water: Changes in concentration during melting*. Research Report 11/77, Project: Sur nederborsvirkning pa skog og fisk (SNSF). SNSF Project NISK 1432 As, NLH

Johnson, J.B. (1983). *Costs and returns for alternative tillage methods/cropping systems and crop mixes: In the triangle area of Montana*. Montana State University, Cooperative Extension Service, Bulletin 1294

Jolánkai, G. (1983). Modeling of nonpoint source pollution. In: Jörgensen, S.E. (ed) *Application of Ecological Modeling in Environmental Management*. Amsterdam, Elsevier, pp 283–379

Jolánkai, G. (coordinator) (1990). *Computer Analysis of the Balance Point and Diffuse Source Loading of Heavy Metals in the Rhine River Basin – Phase 1*. Interim Report. VITUKI Research Report (in association with IIASA, Laxenburg)

Jolankai, G. (1986). SENSMOD: A Simple Experimental Non-point Source Model System. In: *Proceedings of the international conference on water quality modelling in the inland natural environment, 10–13 June 1986*, Bournemouth, pp. 77–91

Jolánkai, G. and Bíro, I. (1996). *Modelling Point and Nonpoint Source Water Pollution in Large Drainage Basins*. Proceedings of the International Conference on Environmental Pollution, Budapest, April 15–19, pp. 467–76

Jolánkai, G. and Pesti, G. (1984). A rainfall *vs.* phosphorus export model for Tetves Creek, Lake Balaton. In *Proceedings of the Unesco/MAB5 workshop on land use impacts on aquatic systems, 10–14 October 1983*, Budapest, pp. 357–69

Jolánkai, G. and Pintér, G. (1982). *Case study on nonpoint source plant nutrient load calculations*. Proceedings of the International Symposium on the Effects of Waste Disposal on Groundwater and Surfacewater. Exeter, IAHS Publication no. 139, pp. 207–14

Jolánkai, G. and Roberts, G. (1987). *Summary and Conclusion of the Proceedings of the UNESCO-MAB-5 Workshop on the Contamination of Subsurface Water Resources by Nitrate and other Pollutants*. Budapest, 20–24 October 1985

Jolánkai, G., Balint, G., Fekete, B., Almassy, A. and Novaky, B. (1987). *A combined spatially distributed runoff and mass transport model system based on a digital terrain model*. VITUKI Progress Report, Research Centre for Water Resources Development, Budapest [In Hungarian]

Jorgensen, S.E., Halling-Sorensen, B. and Nielsen, S.N. (1996). *Handbook of environmental and ecological modeling*. Lewis Publishers, Boca Raton, 688 pp.

Junge, C.E. (1960). Sulfur in the atmosphere. *J. Geophys. Res.* **65**, 227

Kadlec, R.H. and Knight, R.L. (1996). *Treatment wetlands*. CRC Press, Boca Raton, 928 pp.

Kahl, J. (1982). Present and inferred historical atmospheric deposition of heavy metals at two eastern main sites. *North-east Environment Series* **1**, 3–4, 170–5

Karr, J.R. and Schlosser, I.J. (1978). Water resources and the land–water interface. *Science* **201**, 229–34

Kauppi, L. (1978). *Effect of drainage basin characteristics on the diffuse load of phosphorus and nitrogen*. Report No. 30, Publications of the Water Research Institute, National Board of Waters, Helsinki, pp. 21–41

Keller, H.M. and Strobel, T. (1982). Water balance and nutrient budgets in subalpine basins of different forest cover. In: Proceedings of the Symposium on Hydrological Research Basins. *Sonderheft Landeshydrologie Bern*, pp. 683–94

Kelly, T.J., Tanner, R.L. and Newman, L. (1984). Trace gas and aerosol measurements at a remote site in north-east US. *Atmospheric Environment* **18**, 2565–76

Kent, D.M. (1994). *Applied wetlands science and technology*. Lewis Publishers, Boca Raton, 448 pp.

Kiss, A., Primas, A. and Regös, F. (1971). *Guide to conservation planning on sloping land*. Orszagos Mezogazdasagi Minosegvizsgalo Intezet, Budapest [In Hungarian]

Klapwijk, A. and Snodgrass, W.J. (1984). Lake oxygen model. In: Sly, P.G. (ed), *Sediment and water interactions*. Springer Verlag, New York, Chapters 21–3

Kleiss, B.A. (1996). Sediment retention in a bottomland hardwood wetland in eastern Alaska. *Wetlands* **16**(3), 321–33

Knisel, W.G. Jr (1980). *CREAMS: A field scale model for chemicals, runoff and erosion from agricultural management systems*. Conservation Research Report No. 26, US Department of Agriculture, Washington, DC, pp. 640

KNMI/RIVM, (1983). *Chemical composition of precipitation over The Netherlands: Annual report 1983*. KNMI/RIVM Project, De Bilt

Knox, J. (1977). Sediment storage and remobilization characteristics of watersheds. In *Workshop on the fluvial transport of sediment – associated nutrients and contaminants*. Int. Joint Commission. Great Lakes Regional Office, Windsor, Ontario, pp 137–47

Korth, R.M. (1990). *A lake in the balance: Environmental issues in the Lake Superior basin*. MSc thesis, University of Wisconsin-Stevens Point, Stevens Point, Wisconsin, pp. 104

Kowalczewski, A. and Rybak, J.I. (1981). Atmospheric fallout as a source of phosphorus for Lake Warniak. *Ekologia Polska* **29**, 63–71

Krom, M.D. and Berner, R.A. (1980). The diffusion coefficient of sulfate, ammonium and phosphate ions in anoxic marine sediments. *Limnology and Oceanography* **25**, 327–37

Krumbein, W.C. and Sloss, L.L. (1963). *Stratigraphy and sedimentation*. Freeman and Co., San Francisco, California, pp. 660

Kusuda, T., Umita, T., Koga, K., Futawatari, T. and Awaya, Y. (1984). Erosional process of cohesive sediments. *Water Science and Technology* **17**, 891–902

Laenen, A. and Dunnette, D.A. (1996), *River quality: Dynamics and restoration*. Lewis Publishers, Boca Raton, pp. 480

Lahlou, M., Schoemaker, L., Paquette, M., Bo, J., Choudhury, S., Elmer, R. and Xia, F. (1996). *Better Assessment Science Integrating Point and Nonpoint Sources: BASINS*. US EPA, Washington, DC.

Lal, R. and Stewart, B.A. (1994). *Soil processes and water quality. Advances in Soil Science*. Lewis Publishers, Boca Raton, pp. 416

Landreth, R.E. and Rebers, P.A. (1996). *Municipal solid wastes: Problems and solutions*. Lewis Publishers, Boca Raton, pp. 288

Lannefors, H.O., Hansen, H.-C. and Granat, L. (1983). Background aerosol composition in southern Sweden. *Atmospheric Environment* **17**, 87–101

Lantzy, R.J. and Mackenzie, F.T. (1979). Atmospheric trace metals: Global cycles and assessment of mass input. *Geochimical et Cosmochimica Acta* **43**, 511–25

Larson, E.W., Pierce, F.J. and Dowdy, R.H. (1983). The threat of soil erosion to long-term crop production. *Science* **219**, 458–65

Lazrus, A.L., Lorange, E. and Lodge, J.P. Jr (1970). Lead and other metal ions in United States precipitation. *Environmental Science and Technology* **4**, 55–8

Lean, D.R.S. and White, E. (1983). Chemical and radiotracer measurements of phosphorus uptake by lake plankton. *Canadian Journal of Fisheries and Aquatic Sciences* **40**, 147–55

Lean, D.R.S. and Nalewajko, C. (1979). Phosphorus turnover time and phosphorus demand in large and small lakes. *Archiv fur Hydrobiologie, Ergebnisse Limnologie* **13**, 120–32

Leavsley, G.H., Lichty, R.W., Troutman, B.M. and Saindon, L.G. (1983). *Precipitation–Runoff Modeling System: User's Manual*. Water Resources Investigations Report 83-4283, US Geological Survey, Washington, DC, pp. 207

Lee, Y.N. and Schwartz, S. (1981). Evaluation of the role of uptake of nitrogen dioxide by atmospheric and surface liquid water. *Journal of Geophysical Research* **86**, 11971–83

Lee, G.F., Sohzogni, W.C. and Spear, R.D. (1977). Significance of oxic vs anoxic conditions for Lake Mendota sediment phosphorus release. In: Golterman, H.L. (ed), *Interactions Between Sediments and Fresh Water*, Junk, The Hague, pp. 294–306

Lee, D.R. (1977). A device for measuring seepage flux in lakes and estuaries. *Limnology and Oceanography* **22**, 140–7

Lehnardt, F., Brechtel, H.M. and Boness, M. (1983). Chomische Beschaffenheit und Nährstofftransport von Bachwässern aus kleinen Einzugsgebieten unterschiedlicher Landnutzung im Nordhossischan alintsandstaingabiet. *DVWK Schriften* **57**

Leng, M.L., Leovey, E.M.K. and Zubkoff, P.L. (1995). *Agrochemical environmental fate studies: State of the art*. Lewis Publishers, Boca Raton, pp. 432

Lerner, D. (1986). *Monitoring to detect changes in water quality series*. Publication No. 157, International Association of Hydrological Sciences, pp. 336

Lick, W. (1982). Entrainment, deposition and transport of fine grained sediments in lakes. *Hydrobiologia* **91**, 31–40

Lijklema, L. (1980). Interaction of orthophosphate with iron (III) and aluminium hydroxides. *Environmental Science and Technology* **14**, 537–41

Lijklema, L. (1977). The role of iron in the exchange of phosphate between water and sediments. In: Golterman, H.L. (ed), *Interactions between sediments and fresh water*, Junk, The Hague, pp. 313–17

Likens, G.E. (1985). *An ecosystem approach to aquatic ecology: Mirror Lake and its environment*. Springer–Verlag, New York, pp. 516

Lind, O.T. (1974). *Handbook of common methods in limnology*. Mosby, St. Louis, Missouri, pp. 154, ISBN 0-8016-3017-7

Lindberg, S. and Turner, R. (1984). Trace metals in rain of forested sites in eastern United States. In: *Proceedings of the international conference on heavy metals in the environment,* Heidelberg, September 1983, CEP Consultants, Edinburgh, UK

Lintz, J.L. and Simonett, D.S. (1976). *Remote sensing of environment*. Addison-Wesley, Reading, Massachusetts, pp. 694

Liss, P. and Slinn, G. (1983). *Air–sea exchange of gases and particles*. NATO ASI series, Reidel, Dordrecht

Liss, P. (1983). Gas transfer: experiments and geochemical implications. In: Liss, P. and Slinn, G. (eds), *Air-sea exchange of gases and particles*. NATO ASI Series, Reidel, Dordrecht, pp. 241–98

Livingstone, D.A. and Melack, J.M. (1984). Some lakes of subsaharan Africa. In: Taub, F.B. (ed), *Ecosystems of the World, Volume 23: Lakes and reservoirs*. Elsevier, Amsterdam pp. 467–97

Loehr, R.C. (1974). Characteristics and comparative magnitude of non-point sources. *Journal of the Water Pollution Control Federation* **46**

Loehr, R.C., Jewell, W.J., Novak, J.D., Clarkson, W.W. and Friedman, G.S. (1979). *Land Application of Wastes*, Vol. 1. Van Nostrand Reinhold Co., New York, pp. 308

Loelkes, G.L., Gordan, H.E., Schwertz, E.L., Lampert, P.D. and Miller, S.W. (1976). US Geological Survey Bulletin 1600, National Mapping Division, Geographic Investigations Office, Rolla, Missouri

Löfgren, S. (1987). *Phosphorus retention in sediments – Implications for aerobic phosphorus release in shallow lakes*. PhD Thesis, Institute of Limnology, University of Uppsala, Stockholm

Logan, J. (1983). Nitrogen oxides in the troposphere: global and regional budgets. *Journal of Geophysical Research* **88**, 10785–807

Loucks, O.L. (1992). Predictive tools for rehabilitating linkages between land and wetland ecosystems. In Wali, M.K. (ed), *Ecosystem Rehabilitation. Volume 2: Ecosystem analysis and synthesis*. SPB Academic Publishing, The Hague: pp. 297–308. ISBN 90-5103-071-1

Lubchenco, J., Olson, A.M., Brubaker, L.B., Carpenter, S.R., Holland, M.M., Hubbell, S.P., Levin, S.A., MacMahon, J.A., Matson, P.A., Melillo, J.M., Mooney, H.A., Peterson, C.H., Pulliam, H.R., Real, L.A., Regal, P.R. and Risser, P.G. (1991). The sustainable biosphere initiative: an ecological research agenda. *Ecology* **72**, 371–412

Luettich, R.A. (1987). *Sediment resuspension in a shallow lake*. Thesis, Massachusetts Institute of Technology, Cambridge, Massachusetts

Luizao, F., Matson, P.A., Livingston, G., Luizao, R. and Vitousek, P.M. (1989). Nitrous oxide flux following tropical land clearing. *Global Biogeochemical Cycles* **3**, 281–5

Maas, R.P., Dressing, S.A., Spooner, J., Smolen, M.D. and Humenik, F.J. (1984). *Best Management Practices for Agricultural Nonpoint Sources: IV. Pesticides*. Biological and Agricultural Engineering Department, North Carolina State University, Raleigh, pp. 83

Maas, R.P., Smolen, M.D., Jamieson, C.A. and Weinberg, A.C. (1987). *Setting priorities: the key to nonpoint source control*. US Environmental Protection Agency, Office of Water Regulations and Standards, Washington, DC, pp. 51

Macdonald, I.A.W., Kruger, F.J. and Ferrar, A.A. (1986). *The ecology and management of biological invasions in Southern Africa*. Oxford, Cape Town, pp. 324

Macdonald, I.A.W. and Crawford, R.J.M. (1988). *Long-term data series relating to southern Africa's renewable natural resources*. South African National Scientific Programmes Report No. 157, Foundation for Research Development, CSIR, Pretoria, pp. 497

Mackereth, F.J.H. (1963). *Some Methods of Water Analysis for Limnologists*. Scientific Publication No. 21, Freshwater Biological Association, Ambleside, Cumbria, pp. 70

Malmqvist, P.A. (1978). Atmospheric fallout and street cleaning effects on urban stormwater and snow. *Prag Wat Tech.* **10**(5–6), 490–505. Pergamon, London

Mancy, K.H. and Allen, H.E. (1982). Design of measurement systems. In Seuss, M.J. (ed), *Examination of Water for Pollution Control. A reference handbook, Vol. 1*. Pergamon, Oxford, UK, pp. 1–22

Manczak, H. (1974). *Statistical evaluation of water quality on the hydrochemical longitudinal profiles of rivers.* Project Report HUM PIP001 WHO/OVH, Pilot Zones for Water Quality Management, Budapest

Mandelker, D.R. (1992). *NEPA law and litigation.* Clark, Boardman, Callaghan, Chicago, Illinois, ISBN 0-87632-904-0

Manser, A.G.R. and Keeling, A.A. (1996). *Practical handbook of processing and recycling municipal waste.* Lewis Publishers, Boca Raton. pp. 560

Martin, J.L., Schottman, R.W. and McCutcheon, S.C. (1996). *Hydrodynamics and transport for water quality modeling.* Lewis Publishers, Boca Raton, pp. 768

McComb, A.J. and Lukatelich, R.J. (1990). Inter-relationships between biological and physicochemical factors in a database for a shallow estuarine system. *Environmental Monitoring and Assessment* **14**, 223–38

McHarg, I. (1969). *Design with Nature.* Natural History Press, New York

McKendrick, J. (1982). Water supply and sewage treatment in relation to water quality in Lake McIlwaine. In: Thornton, J.A. (ed), *Lake McIlwaine, the Eutrophication and Recovery of a Tropical African Man-made Lake.* Volume 49, Monographiae Biologicae, Junk, The Hague, The Netherlands p. 202–17. ISBN 90-6193-102-9

Meade, J. (1973). *The Theory of Environmental Externalities.* Institut Universitaire de Hautes Etudes, Geneva

Melack, J.M. (1985). Interactions of detrital particulates and plankton. *Hydrobiologia* **125**, 209–20

Menanen, M. (1981). *Quality of runoff water in urban areas.* Helsinki Water Research Institute, National Board of Waters. Publication 42, 123–90

Meriam, R.A. (1960). A note on the interception loss equation. *J. Geophys. Res.* **65**(11), 3850–51

Messner, J. and Brezonik, P. (1981). Importance of atmospheric fluxes to the nitrogen balance of lakes in the Florida peninsula. In: Eisenreich, S. (ed), *Atmospheric pollutants in natural waters*, Ann Arbor Science Publishers, Ann Arbor, Michigan

Metcalf and Eddy Inc., University of Florida, and Water Resources Engineers Inc. (1971). *Storm water management model*, Volumes 1 to 4, US Environmental Protection Agency, Washington, DC

Meyer, L.D., Foster, G.R. and Romkens, M.J.M. (1975). Source of soil eroded by water from upland slopes. In: USDA, *Present and prospective technology for predicting sediment yields and sources.* Publication No. ARS-S-40, US Department of Agriculture, Washington, DC, pp. 177–207

Meyers, P., Rice, C. and Owen, R. (1983). Input and removal of natural and pollutant materials in the surface microlayer on Lake Michigan: Environmental biogeochemistry. *Ecological Bulletin* (Stockholm) **35**, 519–32

Moeyersons, J. and DePloey, J. (1976). Quantitative data on splash erosion, simulation on unvegetated slopes. *Zeitschrift für Geomorphologie*, Suppl. Bd. **25**:120–31

Monem Balba, A. (1995). *Management of problem soils in arid ecosystems.* Lewis Publishers, Boca Raton, pp. 336

Monke, E.J., Marelli, H.J., Meyer, L.D. and DeJong, J.F. (1977). Runoff, erosion, and nutrient movement from interrill areas. *Trans. ASAE* **20**, 58–61

Monteith, T.J., Sullivan, R.A.C., Heidtke, T.M. and Sonzogni, W.C. (1981). *Watershed: a management technique for choosing among point and nonpoint control strategies.* US Environmental Protection Agency, Great Lakes National Program Office, Chicago, Illinois, USA

Morgan, R.P. (1978). Field studies of rainsplash erosion. *Earth Surface Processes* **3**, 295–99

Morrison, I.R. (1986). *The Pasteurisation of Sludge.* Report No. 86/1/86. Water Research Commission, Pretoria, pp. 55, ISBN 0-908356-59-5

Morrison, G.M.P., Revitt, D.M., Ellis, J.B., Balmer, P. and Svensson, G. (1984). The transport mechanisms and phase interactions of bioavailable heavy metals in runoff. In Sly, P. (ed) *Interactions Between Sediments and Water.* CEP Ltd., Edinburgh

Mortimer, C.H. (1941). The exchange of dissolved substances between mud and water in lakes. I. *Journal of Ecology* **29**, 280–329

Mortimer, C.H. (1942). The exchange of dissolved substances between mud and water in lakes. II. *Journal of Ecology* **30**, 147–201

Moshiri, G.A. (1994). *Constructed wetlands for water quality improvement.* Lewis Publishers, Boca Raton, pp. 656

Murphy, T., Formanski, L., Brownawell, B. and Meyer, J. (1985). Polychlorinated biphenyl emissions to the atmosphere in the Great Lakes region: Municipal landfills and incinerators. *Environmental Science and Technology* **189**, 942–6

Nelissen, C. (1982). *De verontreiniging van der Neerslagafvoer in stedelijke gebieden.* RIPJ report 1982-16

Nelson, A.C. (1995). *System development charges for water, wastewater, and stormwater facilities.* Lewis Publishers, Boca Raton. pp. 192

Nelson, W.C. *et al.* (1983). *Farming for profit and conservation.* North Dakota State University, Dept. of Agricultural Economics, AE84002

Nelson, D.W. and Logan, T.J. (1983). Chemical processes and transport of phosphorus. In Schaller, F.W. and Bailey, G.W. (eds) *Agricultural Management and Water Quality.* Iowa State University Press, Ames, IA, pp 65–91

Newman, W.H. and Warren, E.K. (1977). *The Process of Management: Concepts, behaviour and practice.* Prentice-Hall, London, pp. 670, ISBN 0-13-723429-5

Nicholls, K. and Cox, C. (1978). Atmospheric nitrogen and phosphorus loading to Harp Lake, Ontario, Canada. *Water Resources Research* **14**, 589–92

Niemann, B. (1984). Simulation of heavy metal deposition patterns over eastern North America using a regional climatological dispersion model. In *Proceedings of the international conference on heavy metals in the environment,* Heidelberg, September 1983, CEP Consultants, Edinburgh

NIWR (National Institute for Water Research) (1985). *The limnology of Hartbeespoort Dam.* South African National Scientific Programmes Report No. 110, Foundation for Research Development, CSIR, Pretoria, pp. 269

Nix, S.J. (1994). *Urban stormwater modeling and simulation.* Lewis Publishers, Boca Raton, 224 pp.

Novotny, V. (1988). *Nonpoint pollution: 1988 – Policy, economy, management, and appropriate technology.* Technical Publication No. TPS-88-4, American Water Resources Association, Bethesda, Maryland, pp. 314

Novotny, V. (1995). *Nonpoint pollution and urban stormwater management.* Water Quality Management Library Vol. 9, Technomic Publishing, Lancaster, 434 pp.

Novotny, V. and Chesters, G. (1981). *Handbook of nonpoint pollution: Sources and management.* Van Nostrand Reinhold, New York, pp. 555

Novotny, V. and Chesters, G. (1989). Delivery of sediment and pollutants from nonpoint sources: A water Quality perspective. *J. of Soil and Water Conservation* **89**, 568–76

Novotny, V. and Kincaid, G.W. (1982). Acidity of urban precipitation and its buffering during overload flow. In Yen, B.C. (ed) *Urban Stormwater Quality. Management and Planning.* Water Resources Publications, Littleton, Colorado

Novotny, V. and Olem, H. (1994). *Water quality: Prevention, identification, and management of diffuse pollution*. Van Nostrand Reinhold, New York pp. 1054 ISBN 0-442-00559-8

Novotny, V., Sung, H.M., Bannerman, R. and Baum, K. (1985). Estimating nonpoint pollution from small urban watersheds. *Journal WPCF* **57**(4), 339–48

Nurnberg, H., Valenta, P. and Nguyen, V. (1984). The wet deposition of heavy metals from the atmosphere in the Federal Republic of Germany. In: *Proceedings of the international conference on heavy metals in the environment,* Heidelberg, September 1983, CEP Consultants, Edinburgh

O'Brien, W.J. (1972). *Modelling discharge and conservative water quality in the Lower Kansas River Basin*. Bulletin 204, Part 3, State University of Kansas, Manhattan, Kansas

OECD (Organization for Economic Cooperation and Development) (1982). *Eutrophication of waters: Monitoring, assessment and control*. OECD, Paris, pp. 154

Ogutu-Ohwayo, R. (1992). The purpose, costs and benefits of fish introductions: with specific reference to the Great Lakes of Africa. *Mitteilungen Internationale Vereinigung fur Theoretische und Angewandte Limnologie* **23**, 37–44

Olem, H. and Flock, G. (1990). *The lake and reservoir restoration guidance manual. Second edition*. Report No. EPA-440/4-90-006, US Environmental Protection Agency, Washington, DC, pp. 326

Omernick, J.M. (1977). *Nonpoint source – stream nutrient level relationships: A nationwide study*. USEPA-600/3-77-105, Corvallis, Oregon

Onstad, C.A. and Forster, G.R. (1975). Erosion Modeling on a Watershed. *Trans. ASAE* **18**, 288–92

Pacyna, J.M. (1984). Estimation of atmospheric emissions of trace elements from anthropogenic sources in Europe. *Atmospheric Environment* **18**, 41–50

Palmer, R.S. (1963). The influence of a thin water layer on water impact forces. *International Association of Scientific Hydrology Bulletin* **65**, 141–48

Partheniades, E. (1965). Erosion and deposition of cohesive soils. *Journal of the Hydraulics Division of the American Society of Civil Engineers* **91**, 105–39

Partheniades, E. (1972). Results of recent investigations on erosion and deposition of cohesive sediments. In: Shen, H.W. (ed), *Sedimentation*, Colorado State University, Fort Collins, Colorado, pp. 20–1 & 20–39

Patterson, C. and Settle, D. (1976). The reduction of orders of magnitude errors in lead analysis in biological materials and natural waters by evaluating and controlling the extent and sources of industrial lead contamination induced during sample collecting, handling and analysis. In: *Proceedings of the 7th IMR symposium on accuracy in trace analysis*, Special Publication No. 422, US National Bureau of Standards, Washington, DC, pp. 321–51

Pedros-Alio, C. and Brock, T.D. (1983). The importance of attachment to particles for planktonic bacteria. *Archiv fur Hydrobiologie* **98**, 354–79

Penman, H.L. (1948). Natural evaporation from open water, bare soil and grass. *Proc. Royal Soc. London* A193, 120–45

Persson, G. and Jansson, M. (1988). *Phosphorus in Freshwater Ecosystems*. Developments in Hydrobiology Vol. 48, Kluwer, Dordrecht, pp. 340

Peterjohn, W.T. and Correll, D.L. (1984). The nutrient dynamics in an agricultural watershed: Observations on the role of a riparian forest. *Ecology* **65**, 1466–75

Petitjean, M.O.G. and Davies, B.R. (1988). *A review of the ecological and environmental impacts of inter-basin transfer schemes in Southern Africa: Synthesis (Part I) and international bibliography (Part II)*. Foundation for Research Development Occasional Report No. 38, CSIR, Pretoria, pp. 106, ISBN 0-7988-4509-0

Petterson, K. (1984). The fractional composition of phosphorus in lake sediments of different characteristics. In: Sly, P.G. (ed), *Sediments and water interactions*, Springer Verlag, New York

Petts, G.E. with Moller, H. and Roux, A.L. (1989). *Historical Change of Large Alluvial Rivers: Western Europe*. Wiley, Chichester, pp. 355, ISBN 0-471-92163-7

Pierce, R.S., Hornbeck, J.W., Likens, G.E. and Bormann, F.H. (1970). Effects of elimination of vegetation of stream water quantity and quality. In *International Symposium on Results of Research on Representative and Experimental Basins*. Wellinton, New Zealand, International Association of Scientific Hydrology, pp. 311–28

Pierce, R.S. and Keller, H.M. (1980). Forest land use impacts on upstream water sources. In: Duncan, A. and Rzoska, J. (eds), *Land Use Impacts in Lake and Reservoir Ecosystems*. Proceedings of the Unesco MAB 5 Project Workshop 1978, Warsaw. Wien Facultas, Vienna

Pitt, R.E. and Shawley, G. (1981). *San Francisco Bay Area NURP Report*. EPA: Washington, DC

PLUARG (Pollution from Land Use Activities Reference Group) (1977). *Evaluation of remedial measures to control non-point sources of water pollution in the Great Lakes basin*. International Joint Commission, Washington, DC, pp. 51

PLUARG (Pollution Land Use Activities Reference Group) (1978). *Environmental management strategy for the Great Lakes ecosystem*. Final Report, Pollution from Land Use Activities Reference Group to the International Joint Commission, Great Lakes Regional Office, Windsor, Ontario, pp. 115

Porcella, D.B. and Sorensen, D.L. (1980). *Characteristics of nonpoint source runoff and its effect on stream ecosystems*. EPA-600/3-80-82. Corvallis, Oregon, 110 pp.

Porter, S.K. (1975). *Nitrogen and Phosphorus, Food Production, Waste and the Environment*. Ann Arbor Science Publishers, Ann Arbor, Michigan

Porter, C.L. (1967). *Taxonomy of Flowering Plants*. Freeman and Co., San Francisco, California

Prairie, Y.T. and Kalff, J. (1986). Effect of catchment size on phosphorus export. *Water Resources Bulletin* **22**, 465–70

Proctor and Redfern Ltd. (1981). *Stormwater Management Guidelines*. Ministry of Natural Resources, Ontario

Quick, A.J.R. and Johansson, A.R. (1992). User assessment survey of a shallow freshwater lake, Zeekoevlei, Cape Town, with particular emphasis on water quality. *Water SA* **18**, 247–54

Randall, C.W., Helsel, D.R., Grizzard, T.J. and Hoehn, R.C. (1978). The impact of atmospheric contaminants on stormwater quality in an urban area. *Prog. Water Tech.* **10**, 417–31

Rast, W. and Lee, G.F. (1978). *Summary analysis of the North American (US Portion) OECD eutrophication project: Nutrient loading-lake response relationships and trophic state indices*. Report No. EPA-600/3-78-008, US Environmental Protection Agency, Environmental Research Laboratory, Corvallis, Oregon, pp. 454

Rast, W. and Lee, G.F. (1983). Nutrient loading estimates for lakes. *Journal of the*

Environmental Engineering Division of the American Society of Civil Engineers **109**, 502–5

Rast, W. and Thornton, J.A. (1996). Trends in eutrophication research and control. *Hydrological Processes* **10**(2), 295–313

Rast, W. and Holland, M.M. (1988). Eutrophication of lakes and reservoirs: a framework for making management decisions. *Ambio* **17**, 2–12

Rast, W., Holland, M.M. and Ryding, S.-O. (1989). Eutrophication management framework for the policy-maker. *MAB Digest 1*. UNESCO, Paris

Raudkivi, A.J. (1979). *Hydrology, An Advanced Introduction to Hydrological Processes and Modeling*. London, Pergamon, pp. 479

Reckhow, K.H., Beaulac, M.N. and Simpson, J.T. (1980). *Modeling phosphorus loading and lake response under uncertainty: A manual and compilation of export coefficients.* Report No. EPA-440/5-80-011, US Environmental Protection Agency, Office of Water Regulations and Standards, Criteria and Standards Division, Washington, DC, pp. 214

Reuveni, R. (1995). *Novel approaches to integrated pest management.* Lewis Publishers, Boca Raton, pp. 384

Reynolds, C.S. (1975). Interrelations of photosynthetic behaviour and buoyancy regulation in a natural population of a blue-green alga. *Freshwater Biology* **5**, 323–38

Rhyner, C.R., Schwartz, L.J., Wegner, R.B. and Kohrell, M.G. (1995). *Waste management and resource recovery.* Lewis Publishers, Boca Raton, pp. 544

Ribaudo, M.O. (1986). *Reducing soil erosion: offsite benefits.* Report No. 561, Economic Research Service, US Department of Agriculture

Ricker, W.E. (1971). *Methods for assessment of fish production in fresh waters.* Handbook No. 3, International Biological Programme. Blackwell, Edinburgh, pp. 348, ISBN 0-632-08490-1

Robarts, R.D. (1988). Heterotrophic bacterial activity and primary production in a hypertrophic African lake. *Hydrobiologia* **162**, 97–107

Robarts, R.D. and Zohary, T. (1985). *Microcystis aeruginosa* and underwater light attentuation in a hypertrophic lake (Hartbeespoort Dam, South Africa). *Journal of Ecology* **72**, 1001–17

Robarts, R.D., Zohary, T., Ojeda, F., Lovengreen, C.C.E. and Walmsley, R.D. (1985). Observations of inorganic turbidity, chlorophyll $\underline{a}$ concentration and the underwater light field of two hypertrophic, South African impoundments. *Journal of the Limnological Society of Southern Africa* **11**, 78–81

Robinson, W.D. (1986). *The solid waste handbook: a practical guide.* Wiley, New York, pp. 811, ISBN 0-471-87711-5

Rodgers, P.W. and DePinto, J. (1983). Estimation of phytoplankton decomposition rates using two stage continuous flow studies. *Water Research* **17**, 761–9

Rodhe, H. (1978). Budgets and turnover times of atmospheric sulfur compounds. *Atmospheric Environment* **12**, 671–80

Rodhe, H., Sönderlund, R. and Ekstedt, J. (1980). Deposition of airborne pollutants on the Baltic. *Ambio* **9** (3–4), 168–73

Ross, H. (1985). The importance of reducing sample contamination in routine monitoring of trace metals in atmospheric precipitation. *Atmospheric Environment* **19**

Ross, H. and Granat, L. (1985). Deposition of atmospheric trace metals in northern Sweden as measured in the snowpack. *Tellus*

Rowe, D.B. (1982). Water pollution: Perspectives and control. In: Thornton, J.A. (ed), *Lake McIlwaine, the eutrophication and recovery of a tropical African man-made lake*, Vol. 49, Monographiae Biologicae, Junk, The Hague, pp. 195–201

Ryding, S.O. (1981). Reversibility of man-induced eutrophication. Experiences of a lake recovery study in Sweden. *Int. Rev. Ges. Hydrobiol.* **66**, 449–502

Ryding, S.-O. (1982). Lake Trehorningen restoration project: changes in water quality after sediment dredging. *Hydrobiologia* **92**, 549–58

Ryding, S.-O. (1985). Chemical and microbial processes as regulators of the exchange of substances between sediments and water in shallow, eutrophic lakes. *Internationale Revue Gesamten Hydrobiologie* **70**, 657–702

Ryding, S.-O. (1992). *Environmental management handbook: The holistic approach – from problems to strategies.* Lewis, Boca Raton, Florida, pp. 777 ISBN 90-5199-062-6

Ryding, S.-O. and Forsberg, C. (1977). Sediments as a nutrient source in shallow polluted lakes. In: Golterman, H.L. (ed), *Interactions between sediments and fresh water*, Junk, The Hague, pp. 227–34

Ryding, S.-O. and Rast, W. (1989). *The control of eutrophication of lakes and reservoirs.* Man and the Biosphere Series Vol. 1, United Nations Educational, Scientific and Cultural Organization (UNESCO). Parthenon, Carnforth, Lancashire, pp. 314, ISBN 0-929858-13-1

Salas, H.J. and Martino, P. (1990). *Metodologias simplificadas para la evaluacion de eutroficacion en lagos calidos tropicales: Programma regional CEPIS/OPS/HPE 1981-1990.* Pan American Health Organization Report, Caracas, pp. 51

Sartor, J.D. and Boyd, G.B. (1972). *Water pollution aspects of street surface contaminants.* EPA-R2-72-081. EPA, Washington, DC

Schalf, M.R., McNabb, J.F., Dunlap, W.J., Cosby, R.G. and Fryberger, J. (1981). *Manual of Ground-Water Sampling Procedures,* US Environmental Protection Agency, Ada, Oklahoma, pp. 93

Schindler, D.W. (1988). Effects of acid rain on freshwater ecosystems. *Science* **239**, 149–58

Schindler, D.W. and Stumm, W. (1987). The surface chemistry of oxides, hydroxides and oxide minerals. In: Stumm, W. (ed), *Aquatic Surface Chemistry*, Wiley, New York, Chapter 4

Schlickman, J.A., McMahon, T.M. and van Riel, N. (1994). *International Environmental Law and Regulation.* Butterworth, Salem, New Hampshire, ISBN 0-88063-346-8

Schmidt, E.J. and Schulze, R.E. (1987). *SCS-based design runoff: Flood volume and peak discharge from small catchments in southern Africa, based on the SCS technique.* Report No. WRC-155/ TT 31/87, Water Research Commission, Pretoria, pp. 164, ISBN 0-908356-78-1

Semonin, R.G. and van Bowersox, C. (1983). Characterization of the inorganic chemistry of the precipitation of North America. In Pruppacher, H.R., Semonin, R.G. and Slinn, W.G.N. (eds) *Precipitation Scavenging, Dry Deposition and Resuspension.* Elsevier, New York, Amsterdam and Oxford, pp. 191–201

SEWRPC (Southeastern Wisconsin Regional Planning Commission) (1979). *A regional water quality management plan for southeastern Wisconsin – 2000: Volume 2, Alternative plans.* SEWRPC Planning Report No. 30, Waukesha, Wisconsin, pp. 617

Shaheen, (1975). *Contribution of urban roadway usage to water pollution.* EPA-600/2-75-004. EPA, Washington, DC

Shanahan, P., Harleman, D.R.F. and Somlyody, L. (1986). Wind induced water motion. In: Somlyody, L. and van Staaten, G. (eds), *Modeling and Managing Shallow Lake Eutrophication*, Springer Verlag, New York, Chapter 9, pp. 204–55

Sheng, Y.P. and Lick, W. (1979). The transport and resuspension of sediments in a shallow lake. *Journal of Geophysical Research* **84**, 1809–26

Sigg, L. (1987). Surface chemical aspects of the distribution and fate of metal ions in lakes. In: Stumm, W. (ed), *Aquatic Surface Chemistry*, Wiley, New York

Slanina, J. *et al.* (1987). The stability of precipitation samples under field conditions. *International Journal of Environmental Analytical Chemistry* **28**, 247–61

Slinn, S.A. and Slinn, G. (1980). Predictions for particle deposition on natural waters. *Atmospheric Environment* **14**, 1013–16

Smith, G.M. (1950). *The fresh-water algae of the United States.* McGraw-Hill, New York, pp. 719, ISBN 07-058809-0

Smith, G.M. *et al.* (1985). *Growing corn, grain sorghum and soybeans with conservation tillage.* Clemson University, Cooperative Extension Service, Circular 539

Smith, R.V. and Stewart, D.A. (1977). Statistical models of river loadings of nitrogen and phosphorus in the Lough Neagh system. *Water Research* **11**

Smith, S.J., Menzel, R.G., Rhoades, E.D., Williams, J.R. and Eck, H.V. (1983). Nutrient and sediment discharge from Southern Plains watersheds. *J. Range Management*, **36**:435–39

Smol, J.P., Battarbee, R.W., Davis, R.B. and Merilainen, J. (1986). *Diatoms and lake acidity: reconstructing pH from siliceous algal remains in lake sediments.* Developments in Hydrobiology, Vol. 29, Junk, The Hague, pp. 307

Söderlund, R. (1982). *On the difference in chemical composition of precipitation collected in bulk and wet-only collectors.* Report CM-57, Department of Meteorology, University of Stockholm, Stockholm, pp. 18

Somlyody, L. (1980). Preliminary study on wind induced interaction between water and sediment for Lake Balaton. In: *Proceedings of the 2nd joint MTA/IIASA task force meeting on Lake Balaton modeling*, MTA-VEAB, Veszprem, pp. 26–49

Sohzongni, W.C., Chapra, S.C., Armstron, D.E. and Logan, T.J. (1982). Bioavailability of phosphorus inputs to lakes. *J. Environ. Qual.* **11**:555–63

Sopper, W.E. and Kardos, L.T. (1975). *Recycling treated municipal wastewater and sludge through forest and cropland.* Pennsylvania State University Press, University Park and London, pp. 479

Sopper, W.E. and Kerr, S.N. (eds) (1979). *Utilization of municipal sewage sludge on forest and disturbed land.* Pennsylvania State University Press, University Park and London, pp. 537

Sorokin, Y.I. and Kadota, H. (1972). *Techniques for the assessment of microbial production and decomposition in fresh waters.* Handbook No. 23, International Biological Programme. Blackwell, Edinburgh, pp. 112, ISBN 0-632-08850-8

Southgate, D.D. and Pearce, D. (1987). *Natural resource degradation under agricultural colonization: a causal analysis.* Department of Economics, University College, London

Stabel, H. (1985). Mechanisms controlling the sedimentation sequence of various elements in prealpine lakes. In: Stumm, W. (ed), *Aquatic Surface Chemistry*, Wiley, New York

Stahre, P. and Urbonas, B. (1990). *Storm-water detention for drainage, water quality, and CSO management.* Prentice Hall, Englewood Cliffs, New Jersey, pp. 338, ISBN 0-13-849837-7

Starosolsky, Ö. (1970). *Vítépítési Hidraulika*. Budapest, Müszaki Könyvkiadó, pp. 703

Starosolszky, Ö., Muszkalay, L. and Börzsönyi, A. (1971). *Vízhozammérés*. VITUKI, Budapest

Steffen, W., Robertson and Kirsten (1989a). *Water and waste-water management in the sorghum malt and beer industries*. Natsurv 5, Water Research Commission, Pretoria, ISBN 0-947447-36-9

Steffen, W., Robertson and Kirsten (1989b). *Water and waste-water management in the edible oil industry*. Natsurv 6, Water Research Commission, Pretoria, ISBN 0-947447-37-7

Steffen, W., Robertson and Kirsten (1989c). *Water and waste-water management in the dairy industry*. Natsurv 4, Water Research Commission, Pretoria, ISBN 0-947447-35-0

Steffen, W., Robertson and Kirsten (1989d). *Water and waste-water management in the red meat industry*. Natsurv 7, Water Research Commission, Pretoria, ISBN 0-947447-38-5

Steffen, W., Robertson and Kirsten (1989e). *Water and waste-water management in the poultry industry*. Natsurv 9, Water Research Commission, Pretoria, ISBN 0-947447-41-5

Steffen, W., Robertson and Kirsten (1989f). *Water and waste-water management in the tanning and leather finishing industry*. Natsurv 10, Water Research Commission, Pretoria, ISBN 0-947447-60-1

Steffen, W., Robertson and Kirsten (1989g). *Water and waste-water management in the laundry industry*. Natsurv 8, Water Research Commission, Pretoria, ISBN 0-947447-39-3

Steffen, W., Robertson and Kirsten (1990a). *Water and waste-water management in the sugar industry*. Natsurv 11, Water Research Commission, Pretoria, ISBN 0-947447-75-X

Steffen, W., Robertson and Kirsten (1990b). *Water and waste-water management in the paper and pulp industry*. Natsurv 12, Water Research Commission, Pretoria, ISBN 0-947447-83-0

Stegmann, P. (1982). *Some limnological aspects of a shallow, turbid man-made lake*. PhD Dissertation, University of the Orange Free State, Bloemfontein

Stolzenburg, T.R. and Nichols, D.G. (1985). *Preliminary results on chemical changes in ground water samples due to sampling devices,* Report No. EA-4118, Electric Power Research Institute (EPRI), Palo Alto, California

Strachan, W.M.J. and Huneault, H. (1979). Polychlorinated biphenyls and organochlorine pesticides in Great Lakes precipitation. Great Lakes Res., Intern. Assoc. *Great Lakes Res.* **5**(1), 61–8

Strachan, W.N.J., Huneault, H., Schertzer, W.M. and Elder, F.C. (1980). Organochlorines in precipitation in the Great Lakes Region. In Afghan, B.K. and Mackay, D. (eds) *Hydrocarbons and halogenated hydrocarbons in the aquatic environment*. Plenum Press, New York, pp.387–96

Strickland, J.D.H. and Parsons, T.R. (1972). *A practical handbook of seawater analysis*. Bulletin No. 167, Fisheries Research Board of Canada, Ottawa, Ontario, pp. 310

Stumm, W. and Morgan, J.J. (1970). *Aquatic chemistry: An introduction emphasizing chemical equilibria in natural waters*. Wiley, New York, pp. 583, ISBN 0-471-83496-3

Stumm, W. and Leckie, J.O. (1971). Phosphate exchange with sediments: its role in the productivity of surface waters. *Advances in Water Pollution Research* **2**, III-27/1–16

Symoens, J.J., Burgis, M.J. and Gaudet, J.J. (1981). *The ecology and utilization of African inland waters.* UNEP Reports and Proceedings Series 1, UNEP, Nairobi, pp. 191

Tarlock, A.D. (1990). *Law of water rights and resources.* Clark Boardman, New York

Terwindt, J.H.J. (1977). Deposition, transportation and erosion of mud. In: Golterman, H.L. (ed), *Interactions Between Sediments and Fresh Water*, Junk, The Hague, pp. 19–24

Thomas, W. (1984). The atmospheric emission of heavy metals compared to selected trace organics at 14 stations in Bavaria, FRG. In: *Proceedings of the international conference on heavy metals in the environment,* Heidelberg, September 1983, CEP Consultants, Edinburgh

Thompson, J.G. (1980). Acid mine waters in South Africa and their amelioration. *Water SA* **6**, 130–4

Thornton, J.A. (1979a). Some aspects of the distribution of reactive phosphorus in Lake McIlwaine, Rhodesia: Phosphorus loading and abiotic responses. *Journal of the Limnological Society of Southern Africa* **5**, 65–72

Thornton, J.A. (1979). *Factors influencing the distribution of reactive phosphorus in Lake McIlwaine, Zimbabwe.* DPhil Dissertation, University of Zimbabwe, Harare

Thornton, J.A. (1982). *Lake McIlwaine, the eutrophication and recovery of a tropical African man-made lake.* Vol. 49, Monographiae Biologicae, Junk, The Hague, pp. 251

Thornton, J.A. (1986). Nutrients in African lake ecosystems: do we know all? *Journal of the Limnological Society of Southern Africa* **12**, 6–21

Thornton, J.A. (1987). Aspects of eutrophication management in tropical/sub-tropical regions. *Journal of the Limnological Society of Southern Africa* **13**, 25–43

Thornton, J.A. (1989). Aspects of the phosphorus cycle in Hartbeespoort Dam (South Africa): Phosphorus kinetics. *Hydrobiologia* **183**, 87–95

Thornton, J.A. (1993). Perceptions of Public Waters: Water Quality and Water Usage in Wisconsin. In: van Valey, T. and Crull, S. (eds), *The Small City and Regional Community,* Volume 10. Stevens Point Foundation Press, Stevens Point, Wisconsin, pp. 469–78

Thornton, J.A. and Boddington, G. (1989). A new look at the old problem of eutrophication in southern Africa. *The Environmentalist* **9**, 121–9

Thornton, J.A. and Day, J.A. (1989). Water laws and management: directions for the nineties. *Southern African Journal of Aquatic Sciences* **15**, 314–21

Thornton, J.A. and McMillan, P.H. (1989). Reconciling public opinion and water quality criteria in South Africa. *Water SA* **15**, 221–6

Thornton, J.A. and Rast, W. (1989). Preliminary observations on nutrient enrichment of semi-arid, man-made lakes in the northern and southern hemispheres. *Lake and Reservoir Management* **5**, 59–66

Thornton, J.A. and Rast, W. (1993). A test of hypotheses relating to the comparative limnology and assessment of eutrophication in semi-arid man-made lakes. In Straskraba, M., Tundisi, J.G. and Duncan, A. *Comparative Reservoir Limnology and Water Quality Management*, Developments in Hydrobiology Vol. 77, Kluwer, Dordrecht, pp. 1–24, ISBN 0-7923-1919-2

Thornton, J.A. and Rast, W. (1997). The management of international river basins. *Brasilian Journal of Hydrology*

Thornton, J.A., McComb, A.J. and Ryding, S.-O. (1995). The role of the sediments. In: A.J. McComb (ed), *Eutrophic Shallow Estuaries and Lagoons*, CRC Press, Boca Raton, Florida, pp. 205–23, ISBN 0-8493-6839-1

Thornton, J.A., McMillan, P.H. and Romanovsky, P. (1989). Perceptions of water pollution in South Africa: case studies from two water bodies (Hartbeespoort Dam and Zandvlei). *South African Journal of Psychology* **19**, 199–204

Thornton, J.A., Rast, W. and Ryding, S.-O. (1991). The role of socio-economic determinants in lake management in developing countries. *Lake and Reservoir Management* **7**, 115–20

Thornton, J.A., Williams, W.D. and Ryding, S.-O. (1992). Emigration, Economics and Environmental Pollution in Southern Africa. In: Herrmann, R. (ed), *Managing Water Resources During Global Change*. Technical Publication Series, TPS-92-4, American Water Resources Association, Bethesda, Maryland, pp. 609–17

Thornton, J.A., Beekman, H., Boddington, G., Dick, R., Harding, W.R., Lief, M., Morrison, I.R. and Quick, A.J.R. (1995). The ecology and management of Zandvlei (Cape Province, South Africa), an enriched shallow African estuary. In: McComb, A.J. (ed), *Eutrophic Shallow Estuaries and Lagoons*. CRC Press, Boca Raton, Florida pp. 109–28, ISBN 0-8493-6839-1

Thornton, J.D. and Eisenreich, S.J. (1982). Impact of land-use on the acid and trace metal composition of precipitation in the north central US. *Atmospheric Environment* **16**, 1945–55

Todd, D.K. (1961). Groundwater. In: Chow, T.E. (ed), *Handbook of Applied Hydrology*, McGraw-Hill, New York, pp. 13–55

Toerien, D.F., Scott, W.E. and Pitout, M.J. (1976). *Microcystis* toxins: isolation, identification, implications. *Water SA* **2**, 160–2

Tubbs, L.J. and Haith, D.A. (1981). Simulation model for agricultural non-point source pollution. *Journal of the Water Pollution Control Federation* **53**, 1426–33

Twinch, A.J. (1987). Phosphate exchange characteristics of wet and dried sediment samples from a hypertrophic reservoir: Implications for the measurement of sediment phosphorus status. *Water Research* **21**, 1225–30

Twinch, A.J. and Breen, C.M. (1982). Vertical stratification in sediments from a young oligotrophic South African impoundment: Implications in phosphorus cycling. *Hydrobiologia* **92**, 395–404

Umita, T., Kusuda, T., Awaya, Y., Onuma, M. and Futuwatari, T. (1984). The behaviour of suspended sediments and muds in an estuary. *Water Science and Technology* **17**, 915–27

UN (United Nations) (1983). *The Law of the Sea: United Nations Convention on the Law of the Sea with index and final act of the third United Nations conference on the Law of the Sea*, United Nations Publication No. E.83.V.5, United Nations, New York, pp. 224

UN (United Nations) (1993). *Report of the United Nations Conference on Environment and Development, Rio de Janeiro, 3-14 June 1992: Volume I: Resolutions Adopted by the Conference*. United Nations Document A/CONF.151/26/Rev.1(Vol. 1), United Nations, New York

UNESCO (1987a). Manual on Drainage in Urbanized Areas. Vol 1, *Planning and Design of Drainage Systems*. Studies in Hydrology no. 43. UNESCO Press, Paris

UNESCO (1987b). Manual on Drainage in Urbanized Areas. Vol 2, *Data Collection and Analysis for Drainage Design*. Studies in Hydrology no. 43. UNESCO Press, Paris

United States Army Corps of Engineers (1977). *Storage, Treatment, Overflow, Runoff Model. STORM: User's manual*. United States Army Corps of Engineers, Hydrological Research Center, Davis, California, pp. 170

United States Comptroller General (1983). *Agriculture's soil conservation programs miss full potential in the fight against soil erosion.* Report No. GAO/RCED-84-48, US General Accounting Office, Washington, DC

USDA (United States Department of Agriculture) (1975). *Urban hydrology for small watersheds.* Technical Release No. 55, United States Department of Agriculture, Soil Conservation Service, Washington, DC

USDA (United States Department of Agriculture) (1979). *Field manual for research in agricultural hydrology.* Agriculture Handbook No. 224, United States Government Printing Office, Washington, DC, pp. 550

USDA (United States Department of Agriculture) (1991). It's blowin' in the wind. *Hydata* **10**, 1

US EPA (United States Environmental Protection Agency) (1972). *Biological field and laboratory methods for measuring the quality of surface waters and effluents.* Report No. EPA-670/4-73-001, US EPA, Washington, DC

US EPA (United States Environmental Protection Agency) (1975). *Storm water management model user's manual. Version II.* Report No. EPA-670/2-75-017, US EPA, Athens, Georgia [Supplements were issued in 1976 and 1977 and Version III was released in 1981]

US EPA (1979). *Water related environmental data of 129 priority pollutants.* EPA 440/4 a79–029b. US EPA, Washington DC

US EPA (1983). *Final Report of the Nationwide Urban Runoff Program.* US EPA, Washington, DC

US EPA (United States Environmental Protection Agency) (1983). *Methods for chemical analysis of water and wastes.* Report No. EPA-600/4-79-020, US EPA, Cincinnati, Ohio

US EPA (United States Environmental Protection Agency) (1984). *Report to Congress: Nonpoint source pollution in the US,* pp. 10

US EPA (United States Environmental Protection Agency) (1985). *Final report on the federal/state/local nonpoint source task force and recommended national nonpoint source policy.* pp. 17

US EPA (United States Environmental Protection Agency) (1985a). *Methods for measuring acute toxicity of effluents to freshwater and marine organisms.* Report No. EPA-600/4-85-013, US EPA, Washington, DC

US EPA (United States Environmental Protection Agency) (1987). *Handbook on ground water.* Report No. EPA-625/6-87-016, US EPA, Ada, Oklahoma

US EPA (United States Environmental Protection Agency) (1987a). *Criteria for identifying critical aquifer protection areas.* 40 CFR Part 149, Federal Register, Vol. 52, No. 123, Friday, June 26, 1987

US EPA (United States Environmental Protection Agency) (1987b). *Guidelines for the preparation of the 1988 state water quality assessment (305(b) report),* pp. 19

US EPA (United States Environmental Protection Agency) (1987c). *Guide to nonpoint source pollution control.* US EPA, Office of Water, Washington, DC

US EPA (United States Environmental Protection Agency) (1989). *Report to Congress: Water quality of the nation's lakes.* Report No. EPA-440/5-89-003, US EPA, Washington, DC, pp. 23

US EPA (United States Environmental Protection Agency) (1990). *Monitoring lake and reservoir restoration.* Report No. EPA-440/4-90-007, Office of Water, Washington, DC, Chapter 4

US EPA (United States Environmental Protection Agency) (1991). *Methods for the determination of metals in environmental samples.* Report No. EPA-600/4-91-010, US EPA, Cincinnati, Ohio

US EPA (1992). *Compendium of watershed-scale models for TMDL development.* Report No. EPA-841/R-92-002, US EPA, Washington, DC

US EPA (1993). *Created and natural wetlands for controlling nonpoint source pollution.* Lewis Publishers, Boca Raton, pp. 224

US GS (1984). *National Water Summary 1983 – Hydrologic events and issues.* Water Supply Paper 2250. US Government Printing Office, Washington DC, pp. 243

US GS (1985). *National Water Summary 1984 – Hydrologic events, selected water quality trends, and groundwater resources.* Water Supply Paper 2275. US Government Printing Office, Washington, DC, pp. 467

Uunk, E.J.B. (1983). Pollution in urban runoff. In Jolánkai, G. and Roberts, G. (eds) *Land-use Impacts on Aquatic Systems.* Proceedings of Project 5 Workshop. MAB, pp 51–90

UW-IES (University of Wisconsin (Madison)-Institute for Environmental Studies) (1986). *Delavan Lake: A Recovery and Management Study.* University of Wisconsin Press, Madison, Wisconsin, pp. 332

van der Leeden, F., Troise, F.L. and Todd, D.K. (1990). *The Water Encyclopedia, 2nd edition.* Lewis, Chelsea, Michigan, pp. 808, ISBN 0-87371-120-3

van der Heijde, P., Bachmat, Y., Bredehoeft, J., Andrews, B., Holtz, D. and Sebastian, S. (1985). *Groundwater Management: The use of numerical models, 2nd edition.* Monograph No. 5, American Geophysical Union, Washington, DC, pp. 180

van Steenderen, R.A., Scott, W.E. and Welch, D.I. (1988). *Microcystis aeruginosa* as an organohalogen precursor. *Water SA* **14**, 59–64

van Steenderen, R.A., Theron, S.J. and Hassett, A.J. (1987). The occurrence of organic micro-pollutants in the Vaal River between Grootdraai Dam and Parys. *Water SA* **13**, 209–14

van Vaeck, L. and van Cauwenberghe, K. (1985). Characteristic parameters of particle size distributions of primary organic constituents of ambient aerosols. *Environmental Science and Technology* **19**, 707–16

Ven Te Chow (1964). *Handbook of Applied Hydrology.* McGraw Hill, New York

Vermes, L. (1987). Benefits and limitations of the agricultural use of non-rural wastes. Introductory paper of the Working Group 3 in the 4th *International Symposium of CIEC and FAL on Agricultural Waste Management and Environmental Protection.* Braunschweig-Voelkenrode, Germany, 11-14 May 1987, pp. 1–13

Vignon, B.W. (1985). *Nonpoint source pollution.* Monograph No. 3, American Water Resources Association (AWRA), Bethesda, Maryland, pp. 56

Vitosek, P. and Matson, P.A. (1984). Mechanisms of nitrogen retention in forest ecosystems: a field experiment. *Science* **225**, 51–2

Vollenweider, R.A. (1969). *Scientific fundamentals of the eutrophication of lakes and flowing waters, with special reference to nitrogen and phosphorus as factors in eutrophication.* OECD Report No. DAS/CSI/68.27, Paris

Vollenweider, R.A. (1974). *A manual on methods for measuring primary production in aquatic environments.* Handbook No. 12, International Biological Programme. Blackwell, Edinburgh, pp. 225, ISBN 0-632-00531-9

Voorhees, J. and Woellner, R.A. (1998). *International environmental risk management: ISO 14000 and the systems approach.* Lewis Publishers, Boca Raton, pp. 304.

Wali, M.K. (1992). *Ecosystem rehabilitation. Volume 2: Ecosystem analysis and synthesis.* SPB Academic Publishing, The Hague, pp. 388

Walker, W.W. (1987). *Empirical methods of predicting eutrophication in impoundments. Applications manual.* Technical Report E-81-9, Waterways Experiment Station, United States Army Corps of Engineers, Vicksburg, Mississippi

Walling, D.E. (1977). Natural Sheet and Channel Erosion of Unconsolidated Source Material. In Shear, H. and Watson, A.E.P. (eds) *Proceedings of the International Workshop on the Fluvial Transport of Sediment Associated Nutrients and Contaminants*

Walling, D.E. and Moorehead, P.W. (1989). The particle size characteristics of fluvial suspended sediment: an overview. *Hydrobiologia* **176/177**, 125–49

Wanielista, M.P. (1978). *Stormwater management quantity and quality.* Ann Arbor Science Publishers Inc., Ann Arbor, Michigan

Wanielista, M.P. (1987). *Quality considerations in the design of holding pons. Stormwater Retention/detention basins seminar.* University of Central Florida, Orlando

Ward, A.D. and Elliot, W.J. (1995). *Environmental hydrology.* Lewis Publishers, Boca Raton, pp. 496

Watson, I. and Burnett, A. (1995). *Hydrology: An environmental approach.* Lewis Publishers, Boca Raton, pp. 720

Ward, H.B. and Whipple, G.C. (Edmondson, W.T., ed) (1959). *Fresh-water Biology, 2nd edition.* Wiley, New York, pp. 1248

Webb, B.W. (1983). *Dissolved loads of rivers and surface water quantity/quality relationships.* Publication No. 141, International Association of Hydrological Sciences. pp. 441

Weinberg, A.M. (1972). Social institutions and nuclear energy. *Science* **177**, 27–34

Wells, F.C., Gibbons, W.J. and Dorsey, M.E. (1990). *Guidelines for collection and field analysis of water-quality samples from streams in Texas.* Open-File Report 90-127, US Geological Survey, pp. 79

Wetzel, R.G. (1975). *Limnology.* Saunders, Philadelphia, Pennsylvania, pp. 743, ISBN 0-7216-9240-0

Wetzel, R.G. and Likens, G. (1979). *Limnological Analysis.* Saunders, Philadelphia, Pennsylvania, pp. 357

Whelpdale, D. and Berry, R. (1980). The Canadian precipitation chemistry network. *AlCh E Symposium Series 196*, Vol. 76

Whitby, K. (1978). The physical characteristics of sulfur aerosols. *Atmospheric Environment* **12**, 135–59

Whyte, A.V. and Burton, I. (1980). *Environmental risk assessment.* SCOPE 15. Wiley, Chichester pp. 157

Williams, J.R. (1975). Sediment yield prediction with universal equation using runoff energy factors. In: *Present and prospective technology for predicting sediment yields and sources.* Report No. ARS-S-40, Department of Agriculture, Agricultural Research Service, Washington, DC

Williams, J.R., Jones, C.A. and Dyke, P.T. (1984). A modeling approach to determine the relationship between erosion and soil productivity. *Transactions of the American Society of Agricultural Engineers* **27**, 129–44

Williams, R. (1982). A model for the dry deposition of particles to natural water surfaces. *Atmospheric Environment* **16**, 1933–8

Williams, W.D. (1988). Limnological imbalances: an antipodean viewpoint. *Freshwater Biology* **20**, 407–20

Williams, W.D. and Noble, R.G. (1984). Salinization. In: Hart, R.C. and Allanson, B.R. (eds), *Limnological Criteria for Management of Water Quality in the Southern Hemisphere,* South African National Scientific Programmes Report No. 93, Foundation for Research Development, CSIR, Pretoria, pp. 86–107

Wilson, A.L. (1982). Design of sampling programmes. In: Seuss, M.J. (ed), *Examination of Water for Pollution Control. A reference handbook, Vol. 1.* Pergamon Press, Oxford, pp. 23–77

Wischmeier, W.H. and Smith, D.D. (1978). *Predicting rainfall erosion losses: A guide to conservation planning.* Handbook No. 537, US Department of Agriculture, Washington, DC, pp. 58

Wolman, M.G. (1985). Soil erosion and crop productivity. In Follett and Stewart, (eds), *Soil erosion and crop productivity,* Soil Conservation Society of America, Madison, Wisconsin

Wood, T., Bormann, F.H. and Voigt, G.K. (1984). Phosphorus cycling in a northern hardwood forest: Biological and chemical control. *Science* **223**, 391–3

Young, R.A., Onstad, C.A., Bosch, D.D. and Anderson, W.P. (1985). *Agricultural non-point source pollution models (AGNPS) I and II: Model documentation.* Minnesota Pollution Control Agency and US Department of Agriculture, Agricultural Research Service, Washington, DC, pp. 53

Zhao, D. and Sun, B. (1986). Air pollution and acid rain in China. *Ambio,* **15**(1), 2–5

Zilberg, R. (1966). Gastro-enteritis in Salisbury European children. *Central African Journal of Medicine* **12**, 164–8

INDEX